HANDBOOK OF
ELEMENTAL ABUNDANCES IN METEORITES

Series on Extraterrestial Chemistry

Sponsored by

International Association of Geochemistry and Cosmochemistry

VOLUME 1 Handbook of Elemental Abundances in Meteorites
Brian Mason

HANDBOOK OF ELEMENTAL ABUNDANCES IN METEORITES

Edited by

Brian Mason
National Museum of Natural History
Smithsonian Institution
Washington D.C.

GORDON AND BREACH SCIENCE PUBLISHERS
New York Paris London

Gordon and Breach, Science Publishers, Inc.
44 Park Avenue South
New York, N.Y. 10016

Editorial office for the United Kingdom

Gordon and Breach, Science Publishers Ltd.
12 Bloomsbury Way
London W. C. 1

Editorial office for France

Gordon & Breach
7–9 rue Emile Dubois
Paris 14e

Library of Congress catalog card number 71-148927. SBN 0 677 14950 6 *(cloth)*; 0 677 14955 7 *(paper)*. Printed in east Germany

FOREWORD

This book is a product of the activities of the Working Group on Extraterrestrial Chemistry of the International Association of Geochemistry and Cosmochemistry. The IAGC is one of the associations which form the International Union of Geological Sciences. Its Working Group on Extraterrestrial Chemistry was established to promote interdisciplinary contacts between geochemists, astronomers, physicists, and other workers in the earth and space sciences who deal in any way with the abundances and distribution of the elements in the universe.

This work on elemental abundances in meteorites has been organized by Dr. Brian Mason as part of this interdisciplinary effort. It is intended to bring together in one convenient volume the extensive information about the abundances of the elements in meteoritic samples and to give some indication of the relative quality of this data. Dr. Mason has organized the volume so that there is one chapter for each element which can be found in meteorites, and he has obtained the cooperation of an extensive list of contributors to compile these chapters. It is hoped that the volume will satisfy the need for a standard reference work on this subject for a number of years.

A. G. W. Cameron
Chairman, Working Group on
Extraterrestrial Chemistry
IAGC

CONTENTS

INTRODUCTION

Brian Mason

Smithsonian Institution, Washington, D.C.

Introduction

THE SYSTEMATIC INVESTIGATION of elemental abundances in meteorites can be said to date from 1923. In that year V. M. Goldschmidt published the initial part of his great work "Geochemische Verteilungsgesetze der Elemente", in which he pointed out the significance of meteorites for elucidating the geochemistry of the elements. He proposed a classification of the elements into four groups: siderophile, those enriched in nickel-iron; chalcophile, those enriched in a sulfide melt; lithophile, those enriched in a silicate melt; and atmophile, those existing as free gases. Meteorites, containing silicates, troilite (FeS), and free nickel-iron, provided a readily available "fossilized" experiment in the distribution of the elements among these phases. During the following years Goldschmidt and his co-workers made many determinations of specific elements in meteorites. These results, and those of other investigators such as I. and W. Noddack and G. von Hevesy, were summarized in the final part of "Geochemische Verteilungsgesetze der Elemente" (1937), and were used by Goldschmidt to prepare the first comprehensive table of elemental abundances in meteoritic matter. In this table he introduced the convention of referring atomic abundances to silicon as the reference element, primarily in order to relate terrestrial and meteoritic abundances to solar abundances, a convention which has since become standard practice.

Goldschmidt's table has formed the basis of and the stimulus for numerous more recent compilations, among which may be mentioned those of Brown (1949), Urey (1952), Levin et al. (1956), Suess and Urey (1956), Green (1959), Mason (1962), Vinogradov (1962), Urey (1964), Yavnel (1964), Ringwood (1966), Urey (1967), Cameron (1968) and Goles (1969). Goldschmidt's table is now largely of historical interest, since new and improved data are available for all the elements; vastly improved analytical procedures of emission and X-ray spectrography, colorimetry, and completely new techniques such as neutron activation and isotope dilution have been extensively applied to meteorites during the years since World War II, with an enormous accumulation of data, for which this monograph aims to be a critical review. Nevertheless, it is worth noting that Goldschmidt's estimates for the major elements are close to presently accepted values, and for most of the minor and trace elements are within an order of magnitude or considerably closer.

Phase composition of meteorites

About sixty minerals are known from meteorites (Mason, 1967), but many of these are rare accessories. The common and abundant minerals are listed in Table 1. Some contrasts to terrestrial mineralogy may be pointed out: nickel-iron is practically absent from terrestrial rocks; the common minerals

TABLE 1 The common minerals of meteorites

Mineral	Formula	Notes
Kamacite	α-(Fe, Ni)	(4–7% Ni)
Taenite	γ-(Fe, Ni)	(30–60% Ni)
Troilite	FeS	
Olivine	$(Mg, Fe)_2SiO_4$	
Orthopyroxene*	$(Mg, Fe)SiO_3$	
Pigeonite	$(Ca, Mg, Fe)SiO_3$	(About 10 mode per cent $CaSiO_3$)
Diopside	$Ca(Mg, Fe)Si_2O_6$	
Plagioclase	$(Na, Ca) (Al, Si)_4O_8$	

*) Divided into enstatite, with 0–10 mole% $FeSiO_3$, bronzite, 10–20%, and hypersthene, > 20%; these minerals are orthorhombic, and have monoclinic polymorphs known as clinoenstatite, clinobronzite, and clinohypersthene.

in meteorites are largely magnesium-iron silicates, whereas in the Earth's crust the commonest minerals are quartz and aluminosilicates; the common meteorite minerals are anhydrous, whereas hydrated minerals are common and abundant on Earth. These features indicate that most meteorites formed in a highly reducing environment, in which nickel and iron were largely in the metallic state. The carbonaceous chondrites, a small but remarkable class of meteorites, differ fundamentally; they consist largely of serpentine,

$(Mg, Fe)_6Si_4O_{10}(OH)_8$, the nickel is present mainly in silicates and sulfides, and they contain considerable amounts of organic compounds of extraterrestrial origin. A notable feature of the overall mineralogy of meteorites is the absence of phases, such as pyrope garnet and jadeitic pyroxenes, indicative of high pressures (i.e., large parent bodies); the origin of the diamond in the Canyon Diablo iron has been plausibly ascribed to the shock of impact with the Earth, which formed the Arizona Meteor Crater, and the presence of diamond in the small group of ureilites appears to be due to extraterrestrial shock effects.

The classification of meteorites

Current classifications of meteorites are based on mineralogy and structure. The major groups and classes are listed in Table 2. It is obvious from the figures for observed falls that the populations of the different classes vary

TABLE 2 The classification of meteorites (figures in parentheses are the numbers of observed falls in each class)

Group	Class	Principal minerals
Chondrites	Enstatite (11)	Enstatite, nickel-iron
	Bronzite (227)	Olivine, bronzite, nickel-iron
	Hypersthene (303)	Olivine,hypersthene, nickel-iron
	Carbonaceous (31)	Serpentine, olivine
Achondrites*	Aubrites (8)	Enstatite
	Diogenites (8)	Hypersthene
	Chassignite (1)	Olivine
	Ureilites (3)	Olivine, clinobronzite, nickel-iron
	Angrite (1)	Augite
	Nakhlite (1)	Diopside, olivine
	Howardites (14)	Hypersthene, plagioclase
	Eucrites (26)	Pigeonite, plagioclase
Stony-irons	Pallasites (2)	Olivine, nickel-iron
	Siderophyre (1) (Find)	Orthopyroxene, nickel-iron
	Lodranite (1)	Orthopyroxene, olivine, nickel-iron
	Mesosiderites (6)	Pyroxene, plagioclase, nickel-iron
Irons	Hexahedrites (7)	Kamacite
	Octahedrites (32)	Kamacite, taenite
	Ni-rich ataxites (1)	Taenite

* Sometimes subdivided into calcium-poor achondrites (aubrites, diogenites, chassignite, ureilites) and calcium-rich achondrites (angrite, nakhlite, howardites, eucrites).

widely (the figures for observed falls are used as being the best approach to actual extraterrestrial abundances; irons dominate meteorite finds, since they are resistant to weathering and are readily recognized as meteorites or at least as very unusual objects). Over 80% of meteorite falls are chondrites, and over 90% of these belong to two classes, frequently referred to jointly as the ordinary or common chondrites. Of the other classes of meteorites, some are represented by a single fall, which suggests that there may well be additional classes as yet unknown.

The classification of chondrites

Chondrites are characterized by the presence of chondrules, which are small (~1 mm diameter) spheroidal aggregates, usually of olivine and/or pyroxene. Chondrules are unique to chondritic meteorites, being unknown in terrestrial rocks, which suggests that they were formed by some exotic process. There is general agreement that they originated as molten silicate droplets. Where and under what circumstances they formed is still a controversial subject. Current ideas include volcanism on the meteorite parent bodies, splash droplets formed in collisions between asteroids, condensation of liquid droplets from a hot gas of solar composition, and fusion of dust in the primordial solar nebula.

Not only are the chondrites the most abundant meteorites, but many features indicate a primary origin for them and a derivate origin for the other meteorite groups. As a consequence, compositional data are far more extensive for the chondrites than for any other meteorite group. However, although the chondrites may have orginated from comparatively undifferentiated parent material, they can be subdivided into several classes and subclasses, marked off by distinct mineralogical and chemical hiatuses. This is illustrated in Fig. 1, which plots chemical analyses of individual chondrites in the form of weight per cent iron as metal and sulfide (i.e., reduced iron) against weight per cent oxidized iron (essentially iron combined in silicates). The trend is clear, from meteorites in which all the iron is in the reduced form (the enstatite chondrites) to meteorites in which all or nearly all is in the oxidized form (the carbonaceous chondrites). But the sequence is not a continuous one, the four classes of chondrites forming discrete clusters in this diagram. The classes are also distinguished by their total iron content. Urey and Craig (1953), in the original version of Fig. 1, noted a bimodal clustering of points corresponding to average total iron

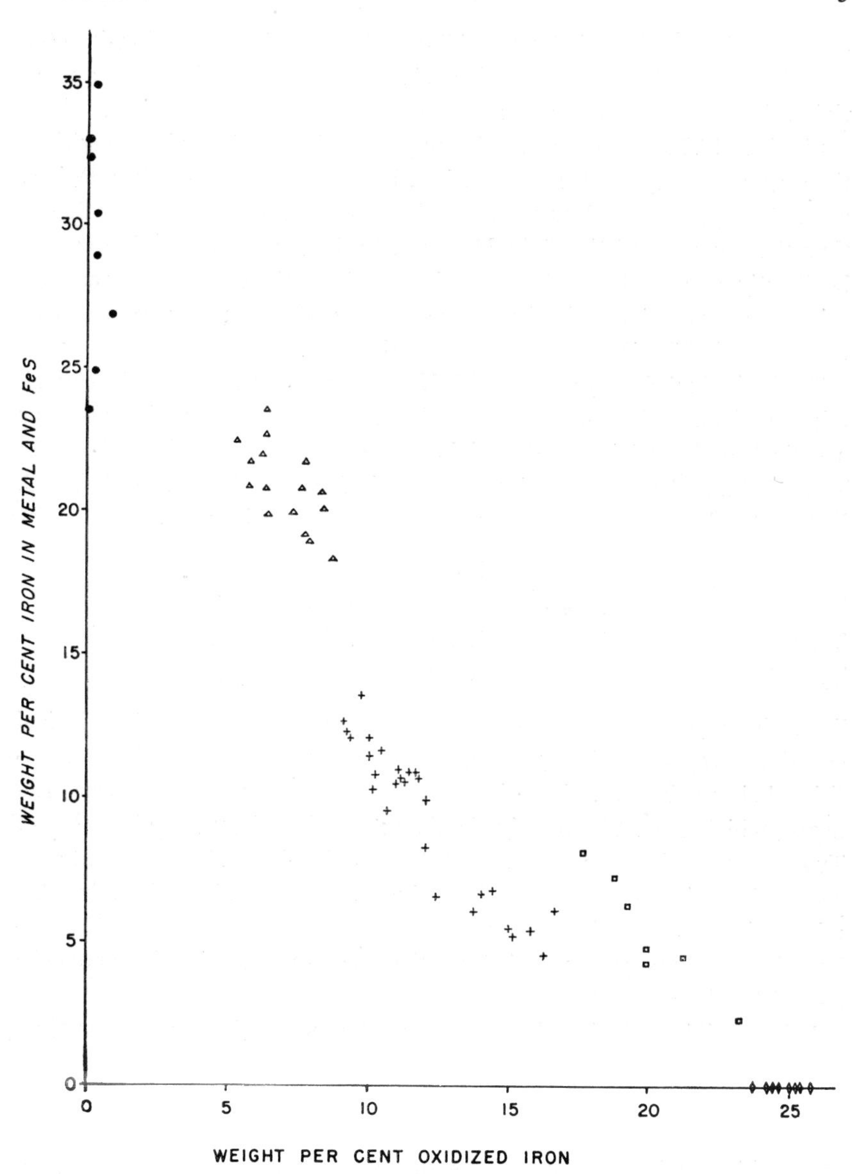

FIGURE 1 Relationship between oxidized iron and iron as metal and sulfide in analyses of chondrites, illustrating the separation into distinct classes and the variation within the classes (● = enstatite chondrites; ▲ = bronzite chondrites; + = hypersthene chondrites; □ = carbonaceous chondrites, Type III; ◆ = carbonaceous chondrites, Types I and II)

contents of approximately 22% and 28%, and named these the low-iron (L) and high-iron (H) groups respectively. The present Fig. 1, based on a more rigid selection of analyses and a considerable number of superior analyses made since 1953, shows that the H group comprises the bronzite chondrites, the L group the hypersthene chondrites. In the enstatite chondrites the iron content ranges from values corresponding to the L group to higher figures than those characteristic of the H group. Carbonaceous chondrites cannot be directly compared with the other classes of chondrites, since they contain a large amount of combined water and other volatiles; on a volatile-free basis (used in Fig. 1) they belong to the H group.

The individual classes can be divided into subclasses by the use of chemical, mineralogical, and structural distinctions. On the basis of his numerous chemical analyses Wiik (1956) divided the carbonaceous chondrites into Types I, II, and III, some of the principal distinguishing factors being C, H_2O, total S, and specific gravity, the mean values being:

	C	H_2O	S	S.G.
Type I	3.54	20.08	6.04	2.2
Type II	2.46	13.35	3.16	2.7
Type III	0.46	0.99	2.21	3.4

These three types are clearly demarcated both by chemical and mineralogical criteria, and appear to be discrete groups, meteorites of intermediate composition being unknown. There is one unique carbonaceous chondrite, Renazzo, which cannot be readily classified; its mineralogy resembles that of Type II, except for the presence of 12% free nickel-iron. Type III carbonaceous chondrites have been called olivine-pigeonite chondrites, but the mineral identified as pigeonite is now known to be a pyroxene of the clinobronzite-clinohypersthene series, so the term olivine-pigeonite chondrite should be abandoned.

It is possible to divide the hypersthene chondrites into two groups, on criteria of chemical and mineralogical composition, and to some degree on structure. Inspection of Fig. 1 shows a cluster of hypersthene chondrite analyses around 10% oxidized iron, and a smaller cluster around 15% oxidized iron, with possibly a hiatus between them. This was first perceived by Prior (1916), and he called them after two analysed meteorites the Baroti type and the Soko-Banja type respectively. In the 1920 paper in which he established the current classification, he placed these two types in the single class of hypersthene chondrites; the overall chemical composition of the two types is very similar, except for the degree of oxidation of the iron (as Fig. 1

shows, the gerater amount of oxidized iron in the Soko-Banja type is compensated by a concomitant decrease in the amount of iron in the metal phase). Mason and Wiik (1964) studied a number of meteorites of the Soko-Banja type and found that chemically and mineralogically they corresponded to the amphoterites, then considered a class of achondrites, evidently because they contain few and poorly defined chondrules; Mason and Wiik therefore considered the amphoterites as a subclass of the hypersthene chondrites. Independently, Keil and Ferdriksson (1964) pointed out some distinctive features of the Soko-Banja type chondrites, in particular ... "The total iron content of the Soko-Banja group is almost the same as in the L-group chondrites, whereas the metallic nickel-iron content is considerably lower. For this reason the group constituted of Soko-Banja chondrites should properly be designated the low-iron-low metal (or LL group) of chondrites." Thus the terms Soko-Banja type (or group), LL group, and amphoterite refer to a single group of meteorites. Whether this group should be considered a subclass of the hypersthene chondrites or an independent class is still disputed; a compositional hiatus probably exists between this group and the remaining hypersthene chondrites (Fredriksson et al., 1968), but is a narrow one, much narrower than those between the other chondrite classes, and between the Type I, II, III carbonaceous chondrites.

The bronzite chondrites form a very coherent group and are not readily subdivided on chemical or mineralogical criteria. The enstatite chondrites, however, show a wide spread in chemical composition, their total iron content ranging from 20–35%. They can be divided into two subclasses, sometimes known as Type I and Type II respectively (Anders, 1964). Type I enstatite chondrites contain more than 30% Fe and more than 5% S, the principal mineral is clinoenstatite, and chondritic structure is well developed. Type II enstatite chondrites contain less than 30% Fe and 5% S, the principal mineral is enstatite, and chondritic structure is poorly developed. Type I and Type II enstatite chondrites show characteristic differences in minor and trace element contents (Larimer and Anders, 1967).

The similarities and differences in overall chemical composition between chondrites of the different classes and subclasses is illustrated in Table 3. The variation in iron content, and the sympathetic variation in nickel and cobalt, is clearly seen. The carbonaceous chondrites, the common chondrites and the enstatite chondrites have distinctive Si/Mg ratios, as pointed out originally by Urey (1961). Fraction of the major lithophile elements between different classes of chondrites is the subject of a recent paper by Ahrens et al. (1969).

TABLE 3 Chemical analyses of chondritic meteorites

	1	2	3	4	5	6	7
Fe	0.00	0.00	0.00	2.28	8.37	16.30	22.05
Ni	0.00	0.00	0.00	0.97	1.21	1.74	1.68
Co	0.00	0.00	0.00	0.05	0.06	0.09	0.08
FeS	–	–	6.74	5.84	6.42	5.48	9.02
SiO_2	22.56	27.81	33.40	40.81	40.32	36.74	39.83
TiO_2	0.07	0.08	0.10	0.17	0.12	0.12	–
Al_2O_3	1.65	2.15	2.51	2.12	2.19	2.04	2.17
Cr_2O_3	0.36	0.36	0.52	0.51	0.52	0.55	0.21
FeO	23.70	27.34	25.43	18.51	12.43	10.24	–
MnO	0.19	0.21	0.19	0.40	0.34	0.32	<0.02
MgO	15.81	19.46	23.98	25.32	24.94	23.44	20.94
CaO	1.22	1.66	2.56	1.85	1.82	1.60	0.62
Na_2O	0.74	0.63	0.51	0.97	1.00	0.90	0.80
K_2O	0.07	0.05	0.04	0.11	0.11	0.09	0.09
P_2O_5	0.28	0.30	0.38	0.18	0.18	0.27	–
H_2O	19.89	12.86	2.07	0.15	0.05	0.15	0.12
C	3.10	2.48	0.47	0.03	0.08	0.02	0.18
NiO	1.23	1.53	1.64	–	–	–	–
CoO	0.06	0.07	0.08	–	–	–	–
S	5.49	3.66	–	–	–	–	–
Sum	96.42	100.65	100.62	100.27	100.16	100.09	99.65

Analyses in atoms per 10,000 atoms Si

	1	2	3	4	5	6	7
Si	10,000	10,000	10,000	10,000	10,000	10,000	10,000
Mg	10,400	10,400	10,700	9240	9240	9510	7760
Fe	8730	8220	7750	5380	5920	8140	7540
Al	860	910	890	610	640	660	630
Ca	580	640	820	490	480	470	350
Na	630	440	300	460	480	480	390
Ni	440	440	400	240	310	480	430
Cr	130	104	122	100	100	120	85
Mn	71	60	50	82	71	74	79
P	106	91	97	38	38	62	44
K	38	22	16	36	36	33	30
Ti	29	28	28	31	22	25	30
Co	22	19	19	12	14	25	22

Key to analyses

1) Carbonaceous chondrite, Type I: Orgueil (Wiik, 1956). Deficiency in summation can be ascribed to reporting all Fe as FeO and all S as S, although both ferric iron and sulfate are present in considerable amounts.
2) Carbonaceous chondrite, Type II: Mighei (Wiik, 1956)
3) Carbonaceous chondrite, Type III: Mokoia (Wiik, 1956)
4) Hypersthene chondrite, amphoterite: Cherokee Springs (Jarosewich and Mason, 1969)
5) Hypersthene chondrite: Leedey (Jarosewich, 1967)
6) Bronzite chondrite: Guarena (Jarosewich and Mason, 1969)
7) Enstatite chondrite: Pillistfer (Jarosewich and Mason, 1969). Includes Si 0.18, TiS 0.14, Cr_2S_3 0.29, MnS 0.26, CaS 0.90, P 0.09.

The most elaborate classification of the chondrites is that of Van Schmus and Wood (1967). They distinguish six petrologic types on the basis of mineralogical and structural criteria (Table 4). They then construct a two-dimensional classification grid (Table 5), using these six petrologic types and five chemical groupings (enstatite chondrites (E), carbonaceous chondrites (C), bronzite chondrites (H), hypersthene chondrites (L), and amphoterites (LL)). It will be noted that no carbonaceous chondrites of types 5 and 6 are known, and that there are no representatives of types 1 and 2 in the remaining chemical groups. Their C1, C2, and C3, classes correspond closely to Wiik's Type I, II, III. Their E3 and E4 classes correspond to the Type I enstatite chondrites, E5 and E6 to the Type II. The Van Schmus–Wood classification imples that each chemical group is essentially an isochemical sequence, and that the classes within each group are genetically related; they suggest that (except for the carbonaceous chondrites) the sequence may represent progressive recrystallization. This interpretation is not universally accepted. However, the classification stands independently of its genetic implications; it provides a workable scheme for subdividing the larger chondrite classes, and has shown its utility in the interpretation of minor and trace element data.

Keil (1969) has recently given an extensive discussion of meteorite composition and classification, and proposed a set of symbols for the different chondrite classes. These classes are based on similar criteria to those used by Van Schmus and Wood, but the symbols are somewhat different, as follows:

Van Schmus and Wood	C1	C2	C3,4	LL	L	H	E3,4	E5,6
Keil	Cc_1	Cc_2	CHL	CLL	CL	CH	Ce_1	Ce_2

TABLE 4 Summary of petrologic types of chondrites (Van Schmus and Wood, 1967)

	Petrologic types					
	1	2	3	4	5	6
(i) Homogeneity of olivine and pyroxene compositions	–	Greater than 5% mean deviations		Less than 5% mean deviations to uniform	Uniform	
(ii) Structural state of low-Ca pyroxene	–	Predominately monoclinic		Abundant monoclinic crystals	Orthorhombic	
(iii) Degree of development of secondary feldspar	–	Absent		Predominately as microcrystalline aggregates		Clear, interstitial grains
(iv) Igneous glass	–	Clear and isotropic primary glass; variable abundance		Turbid if present	Absent	
(v) Metallic minerals (maximum Ni content)	–	(<20%) Taenite absent or very minor	kamacite and taenite present (>20%)			
(vi) Sulfide minerals (average Ni content)	–	>0.5%	<0.5%			
(vii) Overall texture	No chondrules	Very sharply defined chondrules		Well-defined chondrules	Chondrules readily delineated	Poorly defined chondrules
(viii) Texture of matrix	All fine-grained, opaque	Much opaque matrix	Opaque matrix	Transparent microcrystalline matrix	Recrystallized matrix	
(ix) Bulk carbon content	~2.8%	0.6–2.8%	0.2–1.0%	<0.2%		
(x) Bulk water content	~20%	4–18%	<2%			

TABLE 5 Classification of the chondrites (Van Schmus and Wood, 1967)

Chemical group	Petrologic type 1	2	3	4	5	6
E	E 1 –	E 2 –	E 3 1*	E 4 4	E 5 2	E 6 6
C	C 1 4	C 2 16	C 3 8	C 4 2	C 5 –	C 6 –
H	H 1 –	H 2 –	H 3 7	H 4 35	H 5 74	H 6 44
L	L 1 –	L 2 –	L 3 9	L 4 18	L 5 43	L 6 152
LL	LL 1 –	LL 2 –	LL 3 4	LL 4 3	LL 5 7	LL 6 21

* Number of examples of each meteorite type now known is given in its box.

The classification of iron meteorites. The traditional basis for classifying iron meteorites is their structure. Given in Table 6 is the structural classification proposed by Buchwald and Munck (1965), which has been used in a number of recent publications on the iron meteorites.

A more detailed classification of the irons by Wasson and coworkers (Wasson, 1967, 1969; Wasson and Kimberlin, 1967; Wasson and Wetherill, 1969) is based on their chemical compositions. This work is a refinement of the earlier "Ga–Ge classification" proposed by Brown and coworkers (Goldberg, Uchiyama and Brown, 1951; Lovering et al., 1957). This new chemical

TABLE 6 Structural classification of iron meteorites based on that of Buchwald and Munck (1965)

Class	Symbol	Kamacite band width	Remarks
Hexahedrites	H	–	No octahedral structure, kamacite crystals generally as large as specimen
Coarsest octahedrites	Ogg	b.w. > 3.3 mm	–
Coarse octahedrites	Og	1.3 mm ≤ b.w. < 3.3 mm	–
Medium octahedrites	Om	0.5 mm ≤ b.w. < 1.3 mm	–
Fine octahedrites	Of	0.2 mm ≤ b.w. < 0.5 mm	–
Finest octahedrites	Off	b.w. < 0.2 mm	Continuous kamacite bands
Plessitic octahedrites	Opl	b.w. < 0.2 mm	Isolated spindles or sparks of kamacite
Ataxites	D	–	Fine plessite, kamacite spindles very rare

TABLE 7 Characteristics of chemically defined groups of iron meteorites

Group	Freq. (%)	Struct.	Ni conc. (%)	Ga conc. (ppm)	Ge conc. (ppm)	Ir conc. (ppm)	Ge–Ni corr.
I	15	Off–Ogg	6.4–14.4	33–97	77–519	0.6–5	neg.
IIA	11	H	5.3–5.7	57–64	169–190	2–60	pos.
IIB	5	Ogg	5.8–6.4	46–59	107–183	0.01–0.5	neg.
IIC	3	Opl	9.2–11.5	37–39	88–114	4–11	pos.
IID	3	Of–Om	9.6–11.6	69–84	83–102	3–18	pos.
IIIA	17	Om	7.4–9.0	18–22	33–46	0.1–16	pos.
IIIB	8	Om	9.0–10.7	16–20	28–38	0.02–0.1	neg.
IVA	11	Of	7.5–9.5	1.7–2.4	0.09–0.13	0.1–4	pos.
IVB	4	D	16–18	0.2–0.3	0.03–0.08	4–30	pos.

classification is based on structural data and accurate analyses of Ni, Ga, Ge and Ir. Nine groups of irons have been defined which show limited ranges of these and other elements, and very similar structures. These are listed in Table 7 along with the compositional and structural ranges characteristic of each group. Within each group correlations are found between the concentrations of individual elements with each other, and with small variations in structure. The relationship of Ge to Ga and Ni is shown in Fig. 2 of the chapter on Ga in this volume. The sign of the Ge–Ni correlation is listed in

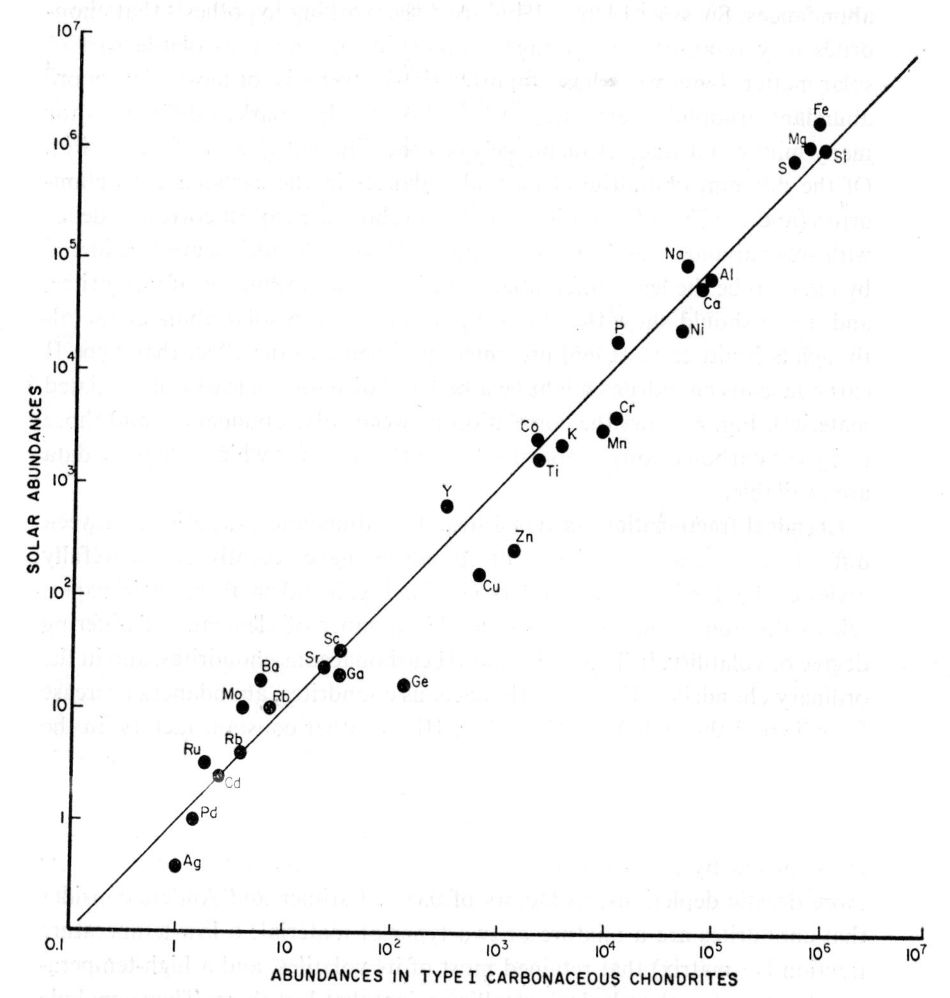

FIGURE 2 Comparison of elemental abundances (normalized to Si = 10^6 atoms) in Type I carbonaceous chondrites with those in the Sun

Table 7. Current analyses are accurate enough to resolve individual members of a group from 90% of the other members of the same group. Table 7 also lists the frequency of occurrence of irons in the various groups; about 77% of the irons are found to be members of the 9 groups. The remaining irons are currently considered to be anomalous, but the number will decrease as additional minor groups are discovered and the limits of existing groups are expanded.

Chondritic and solar abundances. In compiling their table of elemental abundances, Suess and Urey (1956) used the working hypothesis that chondrites may represent the average composition of the non-volatile part of solar matter. Later work has supported this hypothesis for most of the more abundant lithophile elements, but has also revealed marked differences for many minor and trace elements between the different classes of chondrites. Of the different chondrite classes, abundances in the carbonaceous chondrites (and the Type I enstatite chondrites) show the closest correspondence with solar abundances. The Type I carbonaceous chondrites are considered by many to be the least differentiated and most homogeneous of meteorites, and hence should show the closest approximation to solar abundances (although Schmitt et al. (1966) presented arguments to the effect that Type II carbonaceous chondrites might be a better choice for the least-differentiated material). Fig. 2 shows the correlation between solar abundances and those in Type I carbonaceous chondrites for 29 elements for which adequate data are available.

Chemical fractionations in chondrites. The abundance variations between different classes and subclasses of chondrites have recently been carefully reviewed by Larimer and Anders (1967). Fig. 3, taken from their paper, relates the abundances of a considerable number of elements, of differing degree of volatility, in Type I, II, and III carbonaceous chondrites, and in the ordinary chondrites. For the carbonaceous chondrites, abundances decrease from Type I through Type II to Type III by rather constant factors, in the ratio of 1.0:0.6:0.3; Type I enstatite chondrites, with a factor 0.7, resemble Type II carbonaceous chondrites. In ordinary chondrites and Type II enstatite chondrites, nine elements (Au, Cu, F, Ga, Ge, S, Sb, Se, and Sn) are depleted by factors of 0.2–0.5, whereas the remaining elements show more drastic depletions, to factors of 0.002. Larimer and Anders consider that chondrites are a mixture of two types of material: a low-temperature fraction (= matrix) that retained most of its volatiles, and a high-temperature fraction (= chondrules, metallic grains) that lost them, They conclude that these fractionations occurred in the solar nebula as it cooled from high

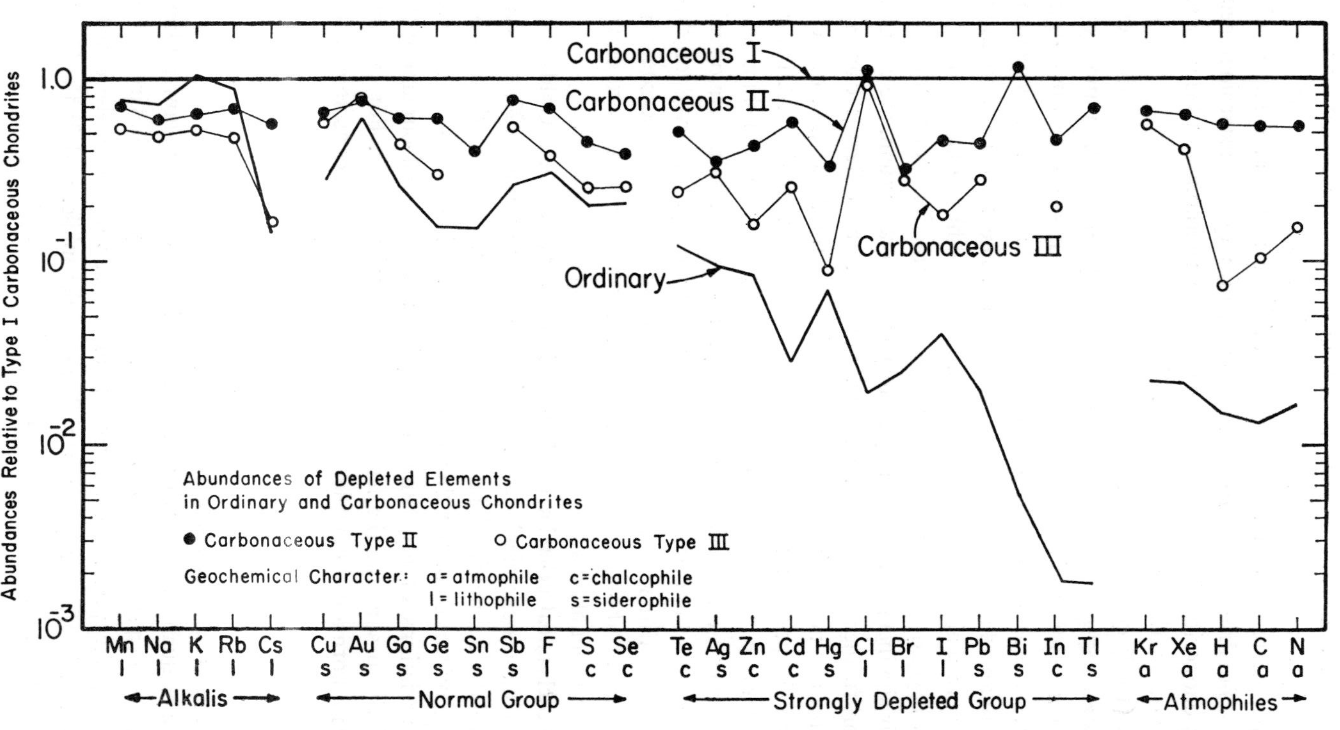

FIGURE 3 Relative abundances of selected elements in Type II and III carbonaceous chondrites and in ordinary chondrites, normalized to Type I carbonaceous chondrite abundances = 1.0 (Larimer and Anders, 1967)

temperatures, and cannot have been produced in the meteorite parent bodies. They correlate the different compositions of the chondrite classes with different regions of aggregation within the ancestral solar nebula: enstatite chondrites come from the inner fringe of the asteroidal belt; ordinary chondrites from the center and inner half; carbonaceous chondrites from the outer fringe or from comets.

Goles (1969) comments ... "While the model of Larimer and Anders (1967) is an engagingly simple approach to the systematization and, hopefully, understanding of geochemical fractionations among meteorites, much further work must be done before it can be accepted as valid." This seems to be a reasonable statement of the present situation. The existence of fractionation, both of major and minor elements, between different classes of chondrites is beyond dispute; how and where these fractionations took place, and their significance for the early history of the solar system, are certainly open for further elucidation. Clearly, however, the different classes of chondrites are not a simple genetic sequence, in the sense that one class represents the parent material from which the others were derived. Instead, they appear to be samples from different regions of an ancestral nebula which was somewhat differentiated chemically, and possibly mineralogically. Carbonaceous chondrites, usually considered more "primitive" or "undifferentiated" with respect to other chondrites, are so only to the extent that their elemental abundances approximate solar abundances more closely; they are presumably coeval, not ancestral, to the other classes.

The geochemical behavior of elements in meteorites. When Goldschmidt proposed his geochemical classification of the elements in 1923, few data were available for many of the minor and trace elements. This situation has been largely remedied, but thorough investigation has revealed some surprising variations of geochemical behavior under special circumstances, particularly in some of the less common meteorite classes. For example, potassium, normally a completely lithophile element, occurs as an essential component of the sulfide djerfisherite, $K_3CuFe_{12}S_{14}$, in some enstatite chondrites. For many elements, therefore, it is necessary to qualify their geochemical classification according to the specific environment. Table 8 gives the geochemical behavior as seen in the ordinary chondrites, with appropriate qualifications where called for. Some elements show multiple affinity, even in a single class of meteorite. In ordinary chondrites iron shows lithophile, chalcophile, and siderophile affinities, whereas in the enstatite chondrites it is essentially chalcophile and siderophile, and in the Type I carbonaceous chondrites it is essentially lithophile, these meteorites con-

TABLE 8 The geochemical behavior of the elements in chondritic meteorites

Atmophile: $\underline{\text{N}}$
Lithophile: Na
Chalcophile: $\underline{\underline{\text{In}}}$
Siderophile: $\underline{\underline{\underline{\text{Fe}}}}$

$\underline{\text{H}}$																	$\underline{\text{He}}$
Li[1]	Be											B	$\underline{\text{C}}$	$\underline{\text{N}}$	O	F	$\underline{\text{Ne}}$
Na[1]	Mg[1]											Al	Si	P	$\underline{\underline{\text{S}}}$	Cl	$\underline{\text{A}}$
K[1]	Ca[1]	Sc	Ti[1]	V	Cr[1]	Mn[1]	$\underline{\underline{\underline{\text{Fe}}}}$[3]	$\underline{\underline{\underline{\text{Co}}}}$	$\underline{\underline{\underline{\text{Ni}}}}$[4]	$\underline{\underline{\underline{\text{Cu}}}}$	Zn[1]	$\underline{\underline{\underline{\text{Ga}}}}$[2]	$\underline{\underline{\underline{\text{Ge}}}}$	$\underline{\underline{\underline{\text{As}}}}$	$\underline{\underline{\text{Se}}}$	Br	$\underline{\text{Kr}}$
Rb	Sr[1]	Y	Y	Nb	$\underline{\underline{\underline{\text{Mo}}}}$[2]		$\underline{\underline{\underline{\text{Ru}}}}$	$\underline{\underline{\underline{\text{Rh}}}}$	$\underline{\underline{\underline{\text{Pd}}}}$	$\underline{\underline{\text{Ag}}}$	$\underline{\underline{\text{Cd}}}$	$\underline{\underline{\text{In}}}$	$\underline{\underline{\underline{\text{Sn}}}}$	$\underline{\underline{\underline{\text{Sb}}}}$	$\underline{\underline{\text{Te}}}$	I[6]	$\underline{\text{Xe}}$
Cs	Ba	La-Lu	Hf	Ta	$\underline{\underline{\underline{\text{W}}}}$	$\underline{\underline{\underline{\text{Re}}}}$	$\underline{\underline{\underline{\text{Os}}}}$	$\underline{\underline{\underline{\text{Ir}}}}$	$\underline{\underline{\underline{\text{Pt}}}}$	$\underline{\underline{\underline{\text{Au}}}}$	$\underline{\underline{\underline{\text{Hg}}}}$	$\underline{\underline{\text{Tl}}}$[5]	$\underline{\underline{\text{Pb}}}$	$\underline{\underline{\text{Bi}}}$			
			Th		U												

[1] Partly chalcophile in enstatite chondrites
[2] Also moderately chalcophile
[3] Also chalcophile and lithophile
[4] Chalcophile and lithophile in carbonaceous chondrites
[5] Also moderately siderophile
[6] Possibly chalcophile

taining no free metal and little or no sulfide. Variations in behavior occur within a single class of meteorites; for example, titanium is progressively more chalcophile and less lithophile in going from the more chondritic to the more recrystallized members of the enstatite chondrites (Easton and Hey, 1968).

Location of minor and trace elements in meteorites. The location of a specific element in a meteorite is of course conditioned by its geochemical character. Siderophile elements are present in the nickel-iron, chalcophile elements in the troilite, lithophile elements in the silicates (and accessory minerals such as phosphates and oxides). However, the quantitative expression of this is only partly explored, and is a promising field for further investigation. Some of the possible ways a minor or trace element may be incorporated in a meteorite are:

a) as a minor constituent of a major phase; e.g., in the common chondrites the manganese is present in solid solution in the olivine and pyroxene

b) as a major constituent of a minor phase; e.g., zirconium has been found to occur as rare grains of zircon ($ZrSiO_4$)

c) as a minor constituent of a minor phase; e.g., most of the chlorine in stony meteorites is present in chlorapatite, $Ca_5(PO_4)_3Cl$

d) possibly as a constituent of an intergranular film; e.g., water-soluble bromine and iodine.

Other mechanisms can be postulated. For example, the rare gases have been found to be concentrated in the surface layers of meteorite minerals, and emplacement by the solar wind has been advanced as an explanation.

References

Ahrens L.H., von Michaelis H., Erlank A.J., and Willis J.P. (1969) "Fractionation of some abundant lithophile element ratios in chondrites" in *Meteorite Research* (Ed. by P.M Millman). Reidel.

Anders E. (1964) *Space Sci. Rev.* **3**, 583.

Brown H. (1949) *Rev. Mod. Phys.* **21**, 625.

Buchwald V.F. and Munck S. (1965) *Analecta Geol.* No. **1**, 81 pp.

Cameron A.G.W. (1968) "A new table of abundances of the elements in the solar system" in *Origin and distribution of the elements* (Ed. by L.H. Ahrens). Pergamon.

Easton A.J. and Hey M.H. (1968) *Mineral. Mag.* **36**, 740.

Fredriksson K., Nelen J., and Fredriksson B.J. (1968) "The LL-group chondrites" in *Origin and distribution of the elements* (Ed. by L.H. Ahrens). Pergamon.

Goldberg E., Uchiyama A., and Brown H. (1951) *Geochim. Cosmochim. Acta* **2**, 1.

Goldschmidt V.M. (1923) *Norsk Videnskapsselskap. Skrifter, Mat.-Naturv. Klasse*, No. **3**.
Goldschmidt V.M. (1937) *Norsk Videnskaps-Akad. Skrifter, Mat.-Naturv. Klasse*, No. **4**.
Goles G.G. (1969) "Cosmic abundances, nucleosynthesis and cosmic chronology". Chapter 5 of *The Handbook of Geochemistry, Part 1*. Springer.
Green J. (1959) *Bull. Geol. Soc. Am.* **70**, 1127.
Jarosewich E. (1967) *Geochim. Cosmochim. Acta* **31**, 1104.
Jarosewich E. and Mason B. (1969) *Geochim. Cosmochim. Acta* **33**, 411.
Keil K. (1969) "Meteorite composition". Chapter 4 of *The Handbook of Geochemistry, Part 1*. Springer.
Keil K. and Fredriksson K. (1964) *J. Geophys. Res.* **69**, 3487.
Larimer J.W. and Anders E. (1967) *Geochim. Cosmochim. Acta* **31**, 1239.
Levin B.Y., Kozlovskaia S.V. and Starkova A.G. (1956) *Meteoritika* **14**, 38.
Lovering J.F., Nichiporuk W., Chodos A. and Brown H. (1957) *Geochim. Cosmochim. Acta* **11**, 263.
Mason B. (1962) *Meteorites*. Wiley.
Mason B. (1967) *Am. Mineral.* **52**, 307.
Mason B. and Wiik H.B. (1964) *Geochim. Cosmochim. Acta* **28**, 533.
Prior G.T. (1916) *Mineral. Mag.* **18**, 26.
Prior G.T. (1920) *Mineral. Mag.* **19**, 51.
Ringwood A.E. (1966) *Rev. Geophys.* **4**, 113.
Schmitt R.A., Smith R.H. and Goles G.G. (1966) *Science* **153**, 644.
Suess H.E. and Urey H.C. (1956) *Rev. Mod. Phys.* **28**, 53.
Urey H.C. (1961) *J. Geophys. Res.* **66**, 1988.
Urey H.C. (1964) *Rev. Geophys.* **2**, 1.
Urey H.C. (1967) *Quart. J. Roy. Astron. Soc.* **8**, 23.
Urey H.C. and Craig H. (1953) *Geochim. Cosmochim. Acta* **4**, 36.
Van Schmus W.R. and Wood J.A. (1967) *Geochim. Cosmochim. Acta* **31**, 747.
Vinogradov A.P. (1962) *Geochemistry*, No. **4**, 329.
Wasson J.T. (1967) *Geochim. Cosmochim. Acta* **31**, 161.
Wasson J.T. (1969) *Geochim. Cosmochim. Acta* **33**.
Wasson J.T. and Kimberlin J. (1967) *Geochim. Cosmochim. Acta* **31**, 2065.
Wasson J.T. and Wetherill G.W. (1970) *Geochim. Cosmochim. Acta* **34**.
Wiik H.B. (1956) *Geochim. Cosmochim. Acta* **9**, 279.
Yavnel A.A. (1964) *Meteoritika* **25**, 75.

HYDROGEN (1)

I. R. Kaplan

Department of Geology and Institute
of Geophysics and Planetary Physics
University of California, Los Angeles

HYDROGEN IS RECOGNIZED as being the most abundant element in the cosmos and in our Sun, its concentration being $>10^4$ atoms/atom Si. In the crust of the earth, this ratio drops to 0.14, the hydrogen residing almost entirely in the ocean and to lesser extents in ice, fresh water bodies, pore water of rocks and to an even smaller extent hydrated minerals and organic matter. In meteorites, the abundance of hydrogen relative to silicon ranges from 2 to <0.01, and is present almost entirely either as water of hydration or as organic molecules.

The great problem in arriving at reliable values for the concentration of hydrogen in meteorites is the ability to differentiate (i) terrestrial from indigenous hydrogen and (ii) organic from aqueous or elemental hydrogen. There are apparently no reliable data on the presence of hydrogen gas in meteorites. Such gas would be expected to be most abundant in chemically reduced chondrites. In many of these, however, metallic iron is present, which would react with traces of atmospheric water to liberate hydrogen gas.

The problem of differentiating terrestrial hydrogen from that indigenous to the meteorite has not been adequately accomplished. Presently, we can say that hydrogen is located in water (adsorbed, hydrated in salts and in the lattice of silicates) and in organic molecules. Its greatest abundance is in Type I and Type II carbonaceous chondrites, but it is also present in almost all stony meteorites in trace amounts.

The two studies where the most extensive work on water content is published are those of Boato (1954) and Wiik (1956). The technique for water measurement was different in each investigation. In the first, it was obtained by evaporation under vacuum at various temperatures from ambient room temperature to 800 °C. The evolved water was then reduced to hydrogen gas over zinc and the pressure of gas measured. Wiik (1956) reported that "... the water is distilled out in the presence of an oxidizing agent (PbO_2). The water reported is, therefore, the total oxidized hydrogen involving the water bound as $(OH)^-$ in minerals and the water formed by combustion of organic matter."

TABLE 1 Hydrogen released as water (% by weight) from carbonaceous chondrites in three separate studies

<table>
<tr><th rowspan="2">Meteorite[4]</th><th colspan="2">Wiik (1956)[1]</th><th colspan="2">Boato (1954)</th><th>Kaplan (unpublished)[3]</th></tr>
<tr><th>% H_2O-</th><th>% H_2O+</th><th>% released up to 180 °C</th><th>% combined[2] H_2O</th><th>% H_2O</th></tr>
<tr><td>1. Tonk</td><td>10.92</td><td>10.74</td><td>—</td><td>—</td><td>—</td></tr>
<tr><td>2. Orgueil</td><td colspan="2">19.89</td><td>4.0</td><td>7.3</td><td>5.9</td></tr>
<tr><td>3. Ivuna</td><td colspan="2">18.68</td><td>4.7</td><td>7.0</td><td>4.5</td></tr>
<tr><td>4. Nogoya</td><td>3.43</td><td>10.85</td><td>—</td><td>—</td><td>—</td></tr>
<tr><td>5. Cold Bokkeveld</td><td colspan="2">10.5</td><td>2.1</td><td>7.8</td><td>2.5</td></tr>
<tr><td></td><td colspan="2">15.17</td><td>2.6</td><td>8.0</td><td></td></tr>
<tr><td>6. Mighei</td><td colspan="2">12.86</td><td>1.7</td><td>8.6</td><td>2.3</td></tr>
<tr><td>7. Nawapali</td><td colspan="2">16.41</td><td>3.0</td><td>9.0</td><td>—</td></tr>
<tr><td>8. Haripura</td><td colspan="2">13.70</td><td>2.4</td><td>7.3</td><td>—</td></tr>
<tr><td>9. Boriskino</td><td>2.96</td><td>8.72</td><td>—</td><td>—</td><td>—</td></tr>
<tr><td>10. Santa Cruz</td><td>1.10</td><td>9.23</td><td>1.8</td><td>8.4</td><td>—</td></tr>
<tr><td>11. Murray</td><td>2.44</td><td>9.98</td><td>2.3</td><td>6.8</td><td>2.2</td></tr>
<tr><td>12. Ornans</td><td>0.18</td><td>0.25</td><td>—</td><td>—</td><td>—</td></tr>
<tr><td>13. Warrenton</td><td>0.00</td><td>0.10</td><td>—</td><td>—</td><td>—</td></tr>
<tr><td>14. Lance</td><td colspan="2">1.40</td><td>0.6</td><td>0.9</td><td>—</td></tr>
<tr><td>15. Felix</td><td colspan="2">0.16</td><td>(1.0)</td><td>—</td><td>—</td></tr>
<tr><td>16. Mokoia</td><td colspan="2">2.07</td><td>0.3</td><td>0.8</td><td>0.8</td></tr>
<tr><td>17. Pueblito de Allende</td><td colspan="2">—</td><td>—</td><td>—</td><td><0.1</td></tr>
<tr><td>18. Non-carbonaceous chondrites (range)</td><td colspan="2">0.6–0</td><td>(0.1)</td><td>—</td><td>1.0–0</td></tr>
</table>

[1] Represents total hydrogen in meteorite liberated as water (including adsorbed and organic).

[2] Represents all water liberated by heating between 180 °C and 800 °C. Includes water of hydration, silicate lattice water and organic matter.

[3] Released by heating to 80 °C under partial vacuum in presence of anhydrous P_2O_5.

[4] Type I, nos.. 1–3; Type II, nos. 4–11: Type III, nos, 12–17.

The discrepancy between the two sets of results is very large, especially as the meteorites used for analysis came from the same source (Wiik, 1956). It can be seen by comparing the two sets of results in Table 1, that Wiik's values are consistently higher (in the case of type I carbonaceous chondrites by almost a factor of 2) than those given by Boato.

Boato (1954) concluded from his studies that all water liberated by heating from ambient room temperature to 180 °C under vacuum represented terrestrial water adsorbed by different minerals in the meteorite. This conclusion was reached by measuring the D/H ratio of this liberated water at a series of temperatures from 25 °C to 800 °C and comparing the δD values (see Table 2)

TABLE 2 Water fractions extracted in vacuum at different temperatures from carbonaceous chondrites, in weight per cent of meteorite sample, from Boato (1954)

Temperatures	Ivuna		Cold Bokkeveld (Paris)	
	Per cent water	δD (%)	Per cent water	δD (%)[1]
25°–180 °C	6.0	+ 1.8	2.6	−2.8
180°–260 °C	1.1	+11.4	1.1	−5.6
260°–325 °C	0.7	+19.5	1.2	−9.8
325°–400 °C	0.8	+31.5	1.9	−8.6
400°–500 °C	1.2	+37.9	2.4	−3.8
500°–800 °C	3.1	+42.0	1.6	−2.1
Total combined water (180°–800 °C)	6.9	+30.0	8.2	−5.7

1) Standard used was Lake Michigan water. By comparison the following data was obtained on terrestrial samples:
Average ocean: +4.7; Rain and snow: −14.4 to +4.8; atmospheric water vapour: down to −13.5; rivers and lakes: −10.3 to +4.1; hot spring waters: −13.0 to +3.2; Hekla lava: −15.0.

with those of terrestrial water. For Ivuna, Orgueil, and Mokoia meteorites, water released above 180°C gave values which fell outside the range of atmospheric water. According to Boato, the water removed by heating (pyrolysis) under vacuum to 800 °C is quantitatively equal to the water obtained by combustion (the "combined" water of Table 1). He concluded, "We do not yet know with certainty the chemical form in which the water exists in the carbonaceous chondrites, but the present data indicate that it is mainly combined as hydrated minerals." He apparently reached this conclusion from the fact that water was still released at temperatures above

500 °C (Table 2), thus excluding organic compounds as a source. This, however, is probably incorrect since C—H—O molecules were detected by Hayes and Biemann (1968) when Murray was heated to 285 °C, and Holbrook chondrite to 525 °C. Water could be liberated by cleavage of the C—H—O bonds or by oxidation of hydrocarbons by metal oxides at elevated temperatures.

In a separate study, the present author dried crushed meteorite fragments under partial vacuum (1–2 mm Hg) at 80 °C for 24 hours in the presence of P_2O_5 desiccant. The per cent loss of water is given in Table 1, for comparison with the data of Boato and Wiik. These results are very close to those obtained by Boato during his heating experiment up to 180 °C, and confirms his conclusion that such water has formed by adsorption onto minerals or by hydration of salts.

Absorption of water on to silicate minerals occurs very readily, as has been noted by Epstein and Taylor (1970) from analysis of Apollo 11 lunar rocks. These investigators concluded that most of the water (~400 ppm) in the fines and breccia originated in Houston in the brief time that the samples were handled in re-packaging for shipment. Probably the largest amount of terrestrial water is adsorbed on to soluble salts in Type I carbonaceous chondrites. Of these salts, magnesium sulfate appears to be the most significant (Kaplan and Hulston, 1966) and is known to exist as epsomite ($MgSO_4 \cdot 7\,H_2O$), often concentrated in veins or layers. This selective concentration may account for the heterogeneous distribution of water, and for variation in the analytical results obtained.

Hydrated silicate minerals are known to be present in carbonaceous chondrites. These have been described as serpentines or chlorites but their structural resemblance is not great (DuFresne and Anders, 1962). Differential thermal analysis (Nagy et al., 1963) yielded major inflection points 900–950 °C for Orgueil and Ivuna. According to Larimer and Anders (1967), these data coincide most closely to the properties of talc [$Mg_3Si_4O_4(OH)_2$]. The hydrated silicates are interpreted by DuFresne and Anders (1962) to have arisen from aqueous alteration of olivine within the planetary body from which the meteorite formed. An accurate estimate of the amount of such minerals is difficult, and hence it is difficult to determine the precise water content (or more correctly, OH^- in the lattice). If, however, we assume that all the adsorbed water measured by Boato was removed $<$180 °C and that between 180–500 °C, organic compounds and hydrated salts yield the major contribution of water, we are left with 3.1% "silicate lattice water" for Ivuna and 1.6% for Cold Bokkeveld (see Table 2).

Based on the fact that talc would yield about 4.8% water and chlorite and serpentine about 12.5% each, one would conclude that a substantial fraction of the carbonaceous chondrites are composed of such minerals. The data presently available are too incomplete for a more rigorous evaluation.

The most abundant water-soluble salt in the Type I and Type II carbonaceous chondrites is epsomite. It is probable that in the meteorite, prior to Earth arrival, it was present as magnesium sulfate monohydrate (DuFresne and Anders, 1962). If one assumes that the water-soluble sulfate extracted from the meteorite was originally present as $MgSO_4 \cdot H_2O$, some values for this water of hydration can be calculated from the results of Kaplan and Hulston (1966). Values for this water range from 1.4% in Orgueil to 0.07% in Mokoia (Table 3). In summary, therefore, it may be concluded that water of hydration of salts and hydroxyl radicals in silicate minerals together may contribute approximately 4.5%, 2% and <0.5% H_2O in Types I, II and III carbonaceous chondrites, respectively.

TABLE 3 Hydrogen present as water in $MgSO_4 \cdot H_2O$ and as C—H radicals in organic compounds

	A. Water of Hydration		B. Carbon Compounds		
	% S[1]	% H_2O	% C[2]	% C[2]	C/H
	(as H_2O-soluble SO_4)	(as $MgSO_4 \cdot H_2O$)	(total)	(as carbonate)	(insoluble residue)
METEORITE					
Ivuna	–	–	4.03	0.20	1.0
Orgueil	2.47	1.39	3.75	0.13	1.3
Mighei	0.46	0.26	2.85	0.21	1.1
Cold Bokkeveld	1.05	0.59	2.35	0.07	1.4
Erakot	–	–	2.30	0.05	1.3
Murray	0.39	0.22	2.24	0.13	1.3
Mokoia	0.14	0.08	0.74	0.00	3.5

[1]) Data from Kaplan and Hulston (1966)
[2]) Data from Smith and Kaplan (1970)

The second source of hydrogen is present in organic matter. A large variety of carbon compounds has been identified in carbonaceous chondrites (Hayes, 1967), ranging from oxidized carbonates (dolomite and breunnerite) to reduced hydrocarbons. A satisfactory carbon balance has not yet been accomplished on meteorites due to the presence of volatile, non-identified compounds (Smith and Kaplan, 1970). The most abundant carbon form resides in an insoluble fraction left after the meteorites have been extracted with organic solvents, water, hydrochloric and hydrofluoric acids. This residue is enriched in carbon, and its C/H ratio on an atom equivalent basis (obtained by combustion) is generally 1 or slightly higher for Type I and II meteorites. Assuming that 5–10% of carbon is in the form of hydrogen-enriched hydrocarbons or lipid-soluble compounds and that the non-identified volatile compounds, constituting 20–30% of the total carbon, are also hydrogen-rich, one may come to a maximum composition in the C/H ratio of 1/2 for the entire meteorite organic matter. On this basis, the water derived from combustion of the organic compounds, calculated from the data of Smith and Kaplan (1970) after correcting for inorganic carbon, would be a maximum of 5.7% for Ivuna to 3.2% for Murray (Table 3).

The maximum indigenous water content obtainable from combustion of Type I and II meteorites would vary between the arithmetic sums of the various forms of hydrogen and would lie in the range of approximately 10–5% respectively. On this basis, hydrogen content of these meteorites would lie in the range of 1–0.5%.

In Type III carbonaceous chondrites, no hydrated silicates have been reported, and although small amounts of water-soluble sulfate was extracted from Mokoia (Kaplan and Hulston, 1966), it is uncertain if this was indigenous to the meteorite. On the basis of the organic carbon content, and from the studies of Smith and Kaplan (1970), it may be assumed that most of the organic matter in these meteorites is in the form of a carbon-rich polynuclear complex. On the basis of a C/H ratio of 1/3, the water derived from the organic matter would be $<0.4\%$ and the hydrogen content $<0.04\%$.

The data for other meteorites are very unreliable. Most of the water measured is probably adsorbed from the atmosphere. Soluble salts are present in very low abundance, and since it is uncertain whether sulfate may have arisen from terrestrial oxidation of sulfide, these data cannot be used in estimating water of hydration. In enstatite chondrites, oldhamite (CaS) may have been hydrolysed by atmospheric water to produce hydroxides, leading to anomalously high water contents in analyses (Wiik, 1956). The amount of organic matter in non-carbonaceous chondrites is small ($<0.3\%$ C) and much of this material may be terrestrial contamination (Smith and Kaplan, 1970). No reliability can presently be ascribed to the hydrogen content of non-carbonaceous meteorites.

References

Boato, G. (1954) *Geochim. et Cosmochim. Acta*, **6**, 209–220.
DuFresne, E.R. and E.Anders (1962) *Geochim. et Cosmochim. Acta* **26**, 1085–1114.
Epstein, S. and H.P.Taylor (1970) *Science*, **167**, 533–535.
Hayes, J.M. (1967) *Geochim. et Cosmochim. Acta*, **31**, 1395–1440.
Hayes, J.M. and K.Bieman (1968) *Geochim. et Cosmochim. Acta* **32**, 239–268.
Kaplan, I.R. and J.R.Hulston (1966) *Geochim. et Cosmochim. Acta*, **30**, 479–496.
Larimer, J.W. and E.Anders (1967) *Geochim. et Cosmochim. Acta*, **31**, 1239–1270.
Nagy, B., W.G.Meinschein and D.J.Hennessy (1963) *Ann. N.Y. Acad. Sci.*, **108**, 534–552.
Smith, J.W. and I.R.Kaplan (1970) *Science*, **167**, 1367–1370.
Wiik, H.B. (1956) *Geochim. et Cosmochim. Acta*, **9**, 279–289.

(Received 23 June 1970)

THE INERT GASES:
$He^{(2)}$, $Ne^{(10)}$, $Ar^{(18)}$, $Kr^{(36)}$ and $Xe^{(54)}$ (2)

D. Heymann

Rice University, Houston, Texas

Introduction

THE INERT GASES are present in all types of meteorites in concentrations ranging from less than 10^{-10} to about 10^{-2} ccSTP/g. These elements were acquired by the meteorites by a number of processes such as trapping (e.g. absorption, occlusion, ion-implantation), the decay of natural radioactivities, and the production by particle bombardment ("early irradiation", cosmic rays). In the present review I shall restrict the discussion to the trapped gases; the reader interested in the other aspects of the inert gases is referred to the following publications: Kirsten et al., 1963; Anders, 1964; Hintenberger, König et al., 1965; Eberhardt et al., 1965a; Eberhardt et al., 1966, Wänke, 1966a; Heymann and Mazor, 1968; Zähringer, 1968.

The elements are normally determined mass-spectrometrically either by the isotope dilution technique or by peak-height comparison (i.e. calibration of instrument sensitivity before or after the sample measurement). The reported accuracies are in most instances $\pm 10\%$ or better. Isotope ratios for a given element are commonly reported to $\pm 2\%$ or better. The detailed experimental procedures are described in several of the publications listed in the preceding paragraph.

In Table 1 I have listed the inert gas contents of five representative meteorites to show the great variability in gas contents and isotopic composition: the He^4 content of the dark portion of Fayetteville is almost 3000 times greater than that of Bruderheim, the He^4/He^3 ratio varies by almost three orders of magnitude between Fayetteville-dark and Grant. With only few exceptions such isotopic variations can be explained in terms of mixtures of two or more components present in variable proportions. Since each component has its own characteristic composition, results such as listed in Table 1

TABLE 1 Inert gases in five meteorites Units 10^{-8} ccSTP/g

Meteorite	Class	He^3	He^4	Ne^{20}	Ne^{21}	Ne^{22}	Ar^{36}	Ar^{38}	Ar^{40}	Kr^{84}	Xe^{132}	Reference
Grant	Of	490	1,880	5.89	5.95	6.25	18.8	30.0	27.6			Signer and Nier, 1960
Bruderheim	L	52.4	561	8.8	9.90	10.9	1.56	1.56	1,155	0.015	0.013	Eberhardt et al., 1966
Chainpur	LL	48	2,080	13.8	10.6	11.8	51	10.5	4,940	0.31	0.24	Heymann and Mazor, 1967
Orgueil	Cl	6.1	10,800	31.2	0.80	4.3	77	15.9	518	0.58	0.18	Mazor et al., 1969
Fayetteville (light)	H	54	1,880	12	8.6	8.4	1.50	1.23	2,450	0.055	0.022	Müller and Zähringer, 1966
Fayetteville (dark)	H	474	1,410,000	5,450	21.4	450	286	58	6,100	0.17	0.10	Müller and Zähringer, 1966

Subscripts: In this review X^n (e.g. He^3) represents the measured total content of a given isotope; X^n_C represents the cosmogenic component; X^n_R the radiogenic component; X^n_T the trapped component, and X^n_F a fissionogenic component.

Summary: Grant contains only cosmogenic gas: $He^4/He^3 \sim 2$–3; $Ne^{20}/Ne^{21}/Ne^{22} \sim 1/1/1$, and $Ar^{36}/Ar^{38} \sim 0.6$. Bruderheim contains principally cosmogenic gas and He^4_R, Ar^{40}_R. However, small amounts of Ar^{36}_T, Kr^{84}_T, and Xe^{132}_T are present. Chainpur differs from Bruderheim in that it contains substantial amounts of Ar_T, Kr_T, and Xe_T (e.g. $Ar^{36}/Ar^{38} = 4.85$, close to the ratio for Ar_T, 5.20. Orgueil too contains substantial amounts of Ar_T, Kr_T and Xe_T. In addition, this meteorite contains Ne_T and He_T; the amount of He^4 is much too large for He^4_C plus He^4_R only.

Fayetteville light-dark: the light portion has gas contents similar to Bruderheim, but the dark portion contains vast amounts of He_T, Ne_T, and Ar_T.

can normally be resolved into constituent components. Details will be discussed in the following sections.

A brief historical note must begin with the pioneering work of Paneth and his collaborators, who, as early as 1928 attempted to measure U, Th—He^4 ages of meteorites (cf. Arrol et al., 1942). A number of these ages were embarrassingly high: greater than any reasonable estimate of the age of the solar system. Bauer (1947) pointed out that meteorites are exposed to cosmic radiation and that part, if not all He^4 in the iron meteorites could have arisen from cosmic-ray induced nuclear reactions. A few years later the evidence for *cosmogenic* (i.e. cosmic-ray produced) He was reported by Paneth et al. (1953). In quick succession cosmogenic neon and argon isotopes were discovered (Reasbeck and Mayne, 1955; Gentner and Zähringer, 1957). Cosmogenic krypton and xenon, normally masked by the trapped components, were detected and characterized more recently (cf. Marti et al., 1966). Meanwhile the first age determinations by the K—Ar^{40} method were reported by Gerling and Pavlova (1951), Gerling and Rik (1955), Wasserburg and Hayden (1955), Geiss and Hess (1958). These and other studies have established the *radiogenic* gases (He^4, Ar^{40}) in meteorite chronology. In 1956, Gerling and Levskii (1956) found vast quantities of He and Ne in the achondrite Pesyanoe; the abundances of these gases are so high and their isotopic compositions so different from atmospheric, cosmogenic or radiogenic, that the authors concluded that Pesyanoe contains gas of still another origin. In the early publications, this component was called *primordial*, but more recently the term *trapped* gas has become preferred because it avoids the implication that the elemental and isotopic abundance ratios of the gas are identical to those in the early solar nebula. In 1960, Zähringer and Gentner (1960) reported similar results for the howardite Kapoeta. In the same year, Reynolds (1960a, b) reported the discovery of "anomalous" xenon in two chondrites, Richardton and Murray. This discovery has stimulated intensive investigations into the isotopic composition of the heavy inert gases and gave birth to a new field of studies, "xenology". The work of Reynolds and others has firmly established that certain meteorites contain *fissionogenic* Xe; i.e. gas that has arisen from the spontaneous fission of trans-uranium nuclides. A most interesting discovery was made by König et al. (1961) who found that the dark portion of the bronzite chondrite Pantar (this meteorite consists of two portions which are quite similar in composition, structure, and mineralogy; one portion is darker than the other, whence light-dark structure) contains vast quantities of trapped He and Ne, whereas the light portion, only a few mm away, contains none or only very little. Subsequently, it was found that the dark portions also contain substantial amounts of trapped argon, krypton, and xenon (Signer and Suess, 1963), and that *gas-rich* meteorites, as they are called, occur among the amphoteric chondrites (Heymann and Mazor, 1966) and carbonaceous chondrites (Heymann and Mazor, 1967). An excellent review of this topic has been given by Pepin and Signer (1965).

The Carbonaceous Chondrites

By way of introduction I shall discuss the carbonaceous chondrites; for all but a few of these meteorites inert gas data are now available (cf. Mazor et al., 1969). Fig. 1a shows $He^4_T \equiv He^4 - He^4_C$ in all types, C1–C4. The data are somewhat arbitrarily divided into two groups, with the division at 5000×10^{-8} ccSTP/g, for the following reason. After correction for cosmogenic $He^4_C = 5.2\, He^3_C$ one must correct for He^4_R. With normal U and Th contents and an assumed age of 4.5 b.y. the maximum He^4_R content is $\sim 2000 \times 10^{-8}$ ccSTP/g. Many carbonaceous chondrites, especially the

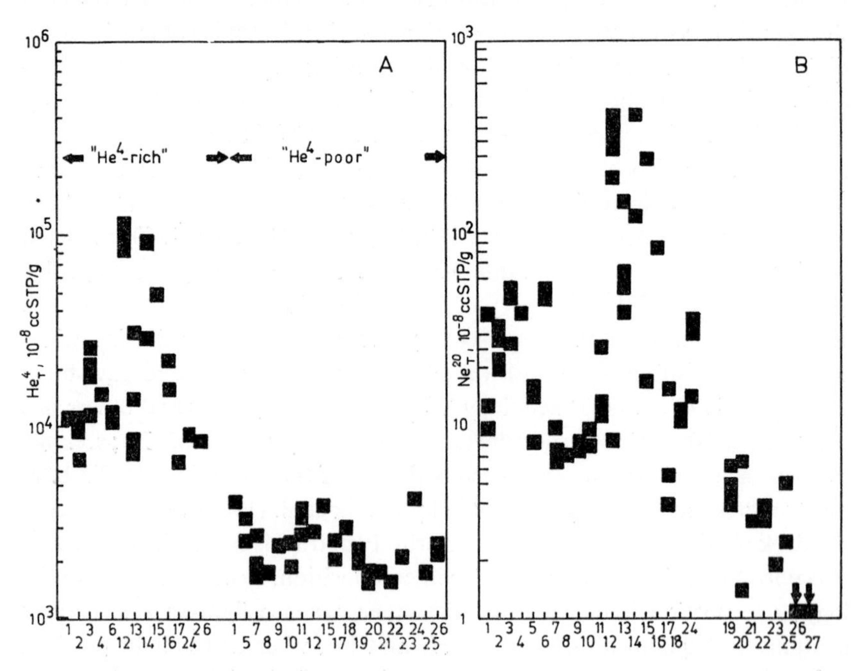

FIGURE 1

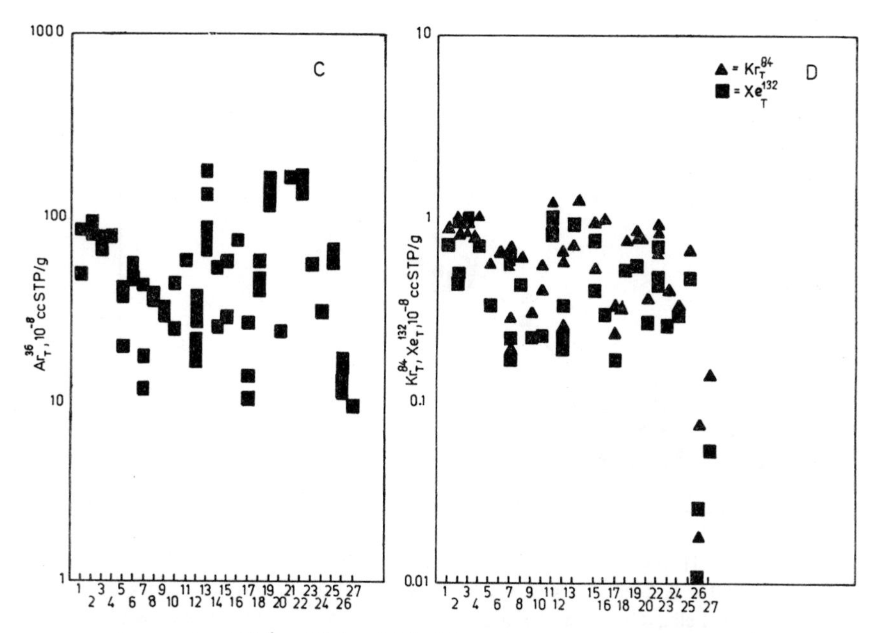

FIGURE 1 a) $He_T^4 \equiv He^4 - 5.2\ He_C^3$ in carbonaceous chondrites. "He^4-poor < 5000×10^{-8} ccSTP/g < "He^4-rich". b) Ne_T^{20} from three-component plot, Fig. 4. c) Ar_T^{36}. d) Kr_T^{84} and Xe_T^{132}. Numbers: 1 Alais, 2 Ivuna, 3 Orgueil, 4 Tonk, 5 Al Rais, 6 Boriskino, 7 Cold Bokkeveld, 8 Erakot, 9 Haripura, 10 Kaba, 11 Mighei, 12 Mokoia and Psuedo St. Caprais, 13 Murray, 14 Nawapali, 15 Nogoya, 16 Pollen, 17 Renazzo, 18 Santa Cruz, 19 Felix, 20 Grosnaja, 21 Kainsaz, 22 Lancé, 23 Ornans, 24 Vigarano, 25 Warrenton, 26 Karoonda, 27 Coolidge.
Note: In this and the following figures a number of data points "overlap" and are represented by a single symbol. The true number of measurements used in preparing the figures is therefore somewhat greater than apparent.

C1's and C2's, have suffered substantial Ar_R^{40} diffusion losses, hence it is safe to assume that the true He_R^4 content of these meteorites is much less than 2000×10^{-8} ccSTP/g. Thus, most, if not all He_T^4 in the "He^4-rich" group (Fig. 1) is genuine trapped He. For the "He^4-poor" group, the conclusion cannot be firm. However, it is interesting that on a plot of He_T^4 vs. Ne_T^{20} (Fig. 2) most points lie between the 45° lines corresponding to He_T^4/Ne_T^{20} of 200 and 400. This seems to imply that even in a meteorite such as Cold Bokkeveld with $He^4 \approx 2000$, most if not all He^4 could be of the trapped variety. The meteorites that do not conform to the trend in Fig. 2: Renazzo, Felix, Grosnaja, Kainsaz, Lancé, Ornans, and Coolidge are, with one exception, C3's and C4's.

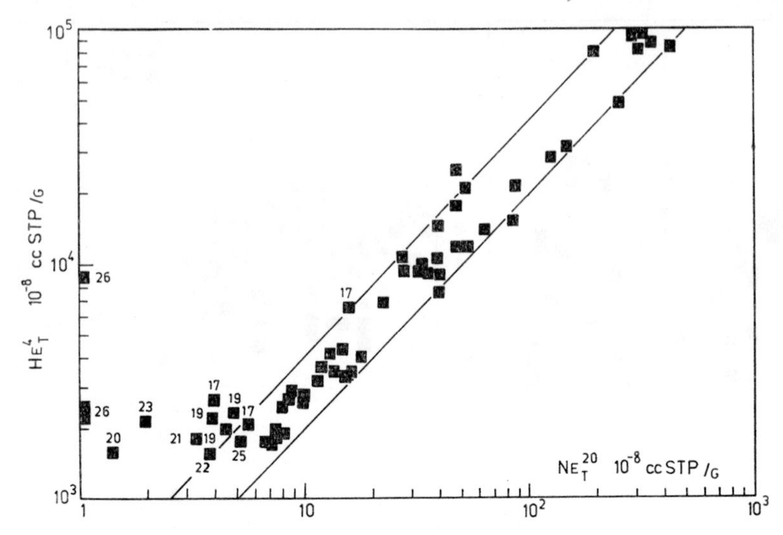

FIGURE 2 He^4_T vs. Ne^{20}_T in carbonaceous chondrites. Most points lie between $200 < He^4_T/Ne^{20}_T < 400$, including several of the "He^4-poor" variety. Renazzo, Felix, Grosnaja, Kainsaz, Lancé, Ornans, Warrenton and Karoonda do not comform.

Fig. 1b shows Ne^{20}_T; the calculation of this component is straight-forward from a three-component plot (see below). As in the case of He^4_T one observes that the content of the trapped component can vary substantially in a single chondrite (e.g. Mokoia, Nogoya). One observes further that the C3's and C4's are on the whole relatively Ne^{20}_T-poor, especially nos. 19–27 (with the exception of Vigarano). This fact strengthens the earlier conclusion that most, if not all He^4 in the latter meteorites is radiogenic rather than trapped.

Fig. 1c shows Ar^{36}_T. It is immediately apparent that Ar^{36}_T does not follow the Ne^{20}_T and the He^4_T trends; this component does not decrease as markedly in the C3's and C4's as the lighter trapped gases. In fact certain of the C3's (Felix, Kainsaz, and Lancé) contain very large quantities of Ar^{36}_T.

Figure 1d shows Kr^{84}_T and Xe^{132}_T; again the variation is considerably smaller than that of Ne^{20}_T or He^4_T: more than 80% of the points lie between 0.2 and 1.0×10^{-8} ccSTP/g.

With Figs. 1 and 2 before us, we reach the following broad conclusions:

a) All but a few of the carbonaceous chondrites contain substantial amounts of trapped gases. In terms of abundances (atoms/10^6 Si) the gas contents correspond roughly to: $He^4_T = 0 \simeq 10$; $Ne^{20}_T = 0 \simeq 4 \times 10^{-2}$; $Ar^{36}_T = 10^{-3} \simeq 2 \times 10^{-2}$; Kr^{84}_T and $Xe^{132}_T = 10^{-5} \simeq 10^{-4}$.

b) He^4_T and Ne^{20}_T vary more strongly than the heavier trapped gases; generally speaking the light gases are more abundant in the C1's and C2's, but this is not the case for the heavier gases.

c) All the trapped gases may vary appreciably within a single chondrite, up to almost two orders of magnitude. In one case, Nogoya, the variation has been shown to occur between physically distinct portions of the meteorite.

d) The abundances of all the inert gases are vastly below their expected "cosmic abundances". Using Cameron's (1963, 1967) values ($He^4 = 5 \times 10^9$; $Ne^{20} = 1.54 \times 10^7$; $Ar^{36} = 1.40 \times 10^6$; $Kr^{84} = 11.38$; $Xe^{132} = 0.846$) the "deficiency factors" for the highest gas contents are approximately: $He^4_T = 5 \times 10^8$; $Ne^{20}_T = 4 \times 10^8$; $Ar^{36}_T = 7 \times 10^7$; $Kr^{84}_T = 10^5$; $Xe^{132}_T = 10^4$. These large "deficiency factors" are obviously related to the volatile geochemical character of the elements.

The question of the origin of the trapped gases has been a matter of lively debate in the past decade, especially after more detailed information had become available on the isotopic compositions and elemental ratios. It was Stauffer (1961) who first detected a significant correlation between the abundances of the light primordial gases and their isotopic compositions. Fig. 3 (which is similar to the one in Pepin and Signer's review, but updated) illustrates Stauffer's hypothesis. In this figure, preferential Ne_T loss (with respect to Ar^{36}_T) would have displaced the points downward. Preferential Ne^{20}_T loss (with repect to Ne^{22}_T) would have displaced the points to the left. Both trends could have been produced by diffusive loss of Ne_T from the minerals with little if any Ar_T loss (which would explain the fact that Ar^{36}_T/Ar^{38}_T varies at most 10% over the entire range of Ne^{20}_T/Ne^{22}_T); the light Ne^{20}_T isotope would have diffused faster, leaving behind an isotopic mixture which is depleted in this isotope. Fig. 3 is grossly consistent with this hypothesis: the meteorites with highest Ne^{20}_T/Ne^{22}_T ratios are in the upper right-hand corner; those with the lowest ratios are generally in the lower left-hand corner. However, closer inspection of Fig. 3 reveals a very disturbing aspect: below $Ne^{20}_T/Ar^{36}_T = 1$ the neon isotopic ratio displacys a very large range for small variations of the element ratio. For example, at $Ne^{20}_T/Ar^{36}_T = 0.5$ the neon isotopic ratio varies from $\sim 6 \simeq 12$. It seems impossible that such a vast range of isotopic ratios could have been produced by simple Ne diffusion from a meteorite in the "Mokoia-cluster" ($Ne^{20}_T/Ar^{36}_T \sim 20$; $Ne^{20}_T/Ne^{22}_T \sim 12 \simeq 14$).

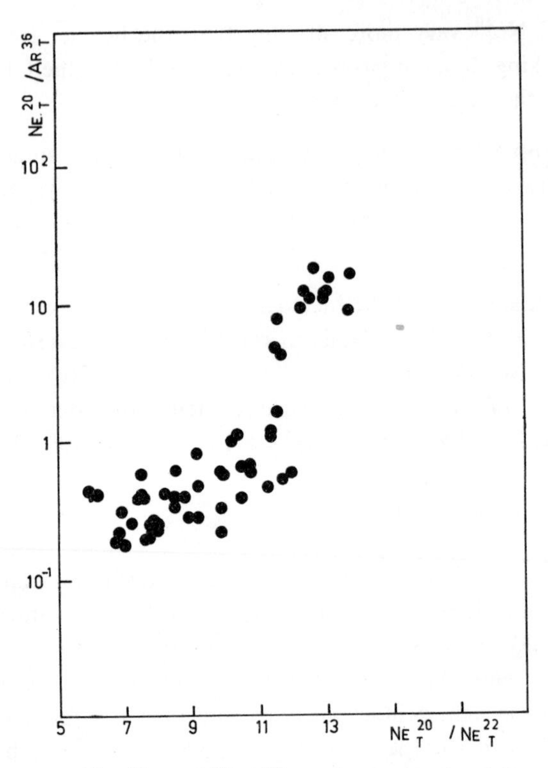

FIGURE 3 Ne^{20}_T/Ar^{36}_T vs. Ne^{20}_T/Ne^{22}_T in carbonaceous chondrites. According to the diffusion theory the original ratios were much larger than the ones now seen in the meteorites. Preferential Ne_T loss lowered Ne^{20}_T/Ar^{36}_T the attending isotope fractionation of Ne lowered Ne^{20}_T/Ne^{22}_T. The most serious objection to this hypothesis is that the range of Ne^{20}_T/Ne^{22}_T is much too great for small variations of Ne^{20}_T/Ar^{22}_T.

Signer and Suess (1963) took an entirely different approach to the problem. They suggested that meteorites in general contain a mixture of two distinct kinds of the light trapped gases, which they called "solar" and "planetary". Subsequently, Pepin (1967) has worked out the details for Ne^{20}_T. I quote Pepin: "There is definite evidence in the isotopic composition of neon that the general framework of the two-component model proposed by Signer and Suess (1963) is correct. The Murray carbonaceous chondrite contains varying amounts of 'solar' neon with isotopic composition $Ne^{20}_T/Ne^{22}_T = 14.3 \pm 0.5$, $Ne^{21}_T/Ne^{22}_T = 0.040 \pm 0.003$. Other carbonaceous chondrites contain solar-type neon with somewhat lower Ne^{20}_T/Ne^{22}_T ratios. All carbonaceous chondrites appear to contain a second trapped component with

$Ne_T^{20}/Ne_T^{22} = 8.2 \pm 0.4$, $Ne_T^{21}/Ne_T^{22} = 0.025 \pm 0.003$, here designated as 'Neon-A'. There is no evidence for extensive diffusion loss of carbonaceous chondrite neon, although relatively small fractionations of the solar component are possible."

Fig. 4 shows all the available data on a plot similar to Pepin's (1967). The composition of "solar" neon (here designated as Ne_B) was changed to

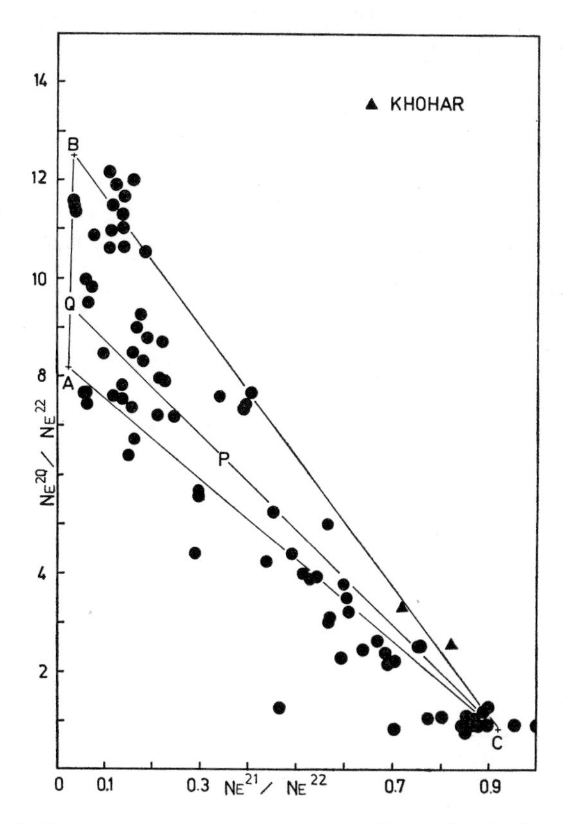

FIGURE 4 Three-component neon-plot according to Pepin. C represents cosmogenic composition, A represents Ne_A ("planetary", "fractionated" component), and B represents Ne_B ("solar", "unfractionated" component) The line CPQ is drawn through a point P within the triangle.

$$\frac{Ne_C}{Ne_T} = \frac{PQ}{PC}, \quad \frac{Ne_A}{Ne_B} = \frac{QB}{QA}.$$

The trapped component can thus be resolved into Ne_A and Ne_B. Compositions: Ne^{20}/Ne^{22} for $A = 8.2 \pm 0.4$, $B = 12.5 \pm 0.4$, $C = 0.853$; Ne^{21}/Ne^{22} for $A = 0.025 \pm 0.003$, $B = 0.036 \pm 0.003$, $C = 0{,}918$.

12.5 ± 0.4; 0.036 ± 0.003 (Pepin, private communication). Point C represents cosmogenic neon with isotopic ratios 0.853; 0.918. A line drawn through point P within the triangle and point C intersects AB in Q; this intersect represents Ne_T; its isotopic composition as well as its proportion to cosmogenic Ne: Ne_T/Ne_C = CP/PQ. Likewise Ne_A/Ne_B = BQ/AQ. Mazor et al. (1969) who have resolved the Ne_T-data into Ne_A and Ne_B find that Ne_A is, on the whole, less variable than Ne_B and that Ne_A correlates remarkably well with Ar_T^{36} in the C1's and C2's.

Mazor et al. (1969) have also addressed themselves to the question whether He_T in the carbonaceous chondrites is a two-component mixture. They argue that since the Ne_T^{20}/Ne_T^{22} ratio is appreciably different between the two components, one might expect to find a substantial variation in He_T^4/He_T^3. Unfortunately He_T^3 is masked by He_C^3 in most of the carbonaceous chondrites. Fig. 5 shows a minimum He_T^4/He_T^3 ratio ($He_T^3 \equiv He^3$, assuming that

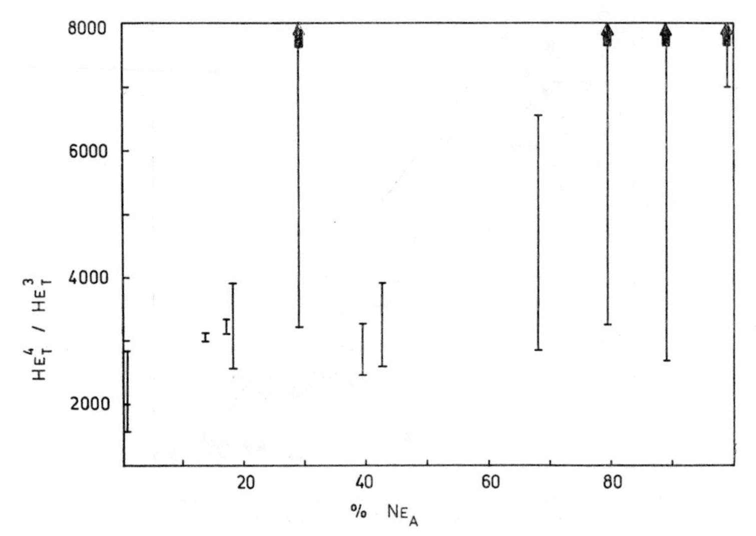

FIGURE 5 He_T^4/He_T^3 vs. % Ne_A in carbonaceous chondrites. Because He_C^3 often masks He_T^3 the errors are large. Minimum value: all He_C^3 assumed lost by diffusion; maximum value: He_C^3 calculated from Ne_C^{21}, quantitative retention assumed. For 100% He_C^4: $He_C^3/He_C^3 \approx 2600 \pm 100$; for 100% Ne_B: $He_T^4/He_T^3 \approx 6000 \pm 600$.

all He_C^3 was lost by diffusion) and a maximum ratio (He_C^3 calculated from Ne_C^{21}, quantitatively retained) plotted against % Ne_A. Because of the uncertainty in He_T^3 the data scatter. Nevertheless there appears to be a trend of increasing He_T^4/He_T^3 ratio toward 100% Ne_A. The authors have given the following values: "solar" (0% A) $He_T^4/He_T^3 = 2600 \pm 100$; "planetary" (100% A) $He_T^4/He_T^3 = 6000 \pm 600$.

If He_T, like Ne_T is a two-component system, then Fig. 2 contains two populations: He_A^4 vs. Ne_A^{20} and He_B^4 vs. Ne_B^{20}. Mazor et al. (1969) have attempted to resolve these populations by plotting He_T^4/Ne_T^{20} vs. Ne_T^{20}/Ne_T^{22} (the latter ratio reflects the percentages of Ne_A and Ne_B). Unfortunately, most of the chondrites which contain a high proportion of Ne_A belong to the "He^4-poor" group for which, as we have seen, the true He_T^4 content is rather uncertain. The only clear-cut trend is that the C1's have considerably higher He_T^4/Ne_T^{20} ratios (300–400) than the C2's (250–325) at comparable Ne_T^{20}/Ne_T^{22} ratios (8–9). With increasing Ne_B content the He_T^4/Ne_T^{20} ratio decreases among the C2's from ~300 at 0% Ne_B to 200–250 at ~90% Ne_B; however, Mokoia is unusual in this respect: its He_T^4/Ne_T^{20} ratios are considerably higher (~250 ≃ 400) than those of the other gas-rich C2's (Murray, Nawapali, Nogoya, and Pollen).

Fig. 6 shows Ne_A^{20} vs. Ar_T^{36}. The unnumbered points scatter about an average of $Ar_T^{36}/Ne_A^{20} = 3.7$ within a factor of two. Considering the errors in

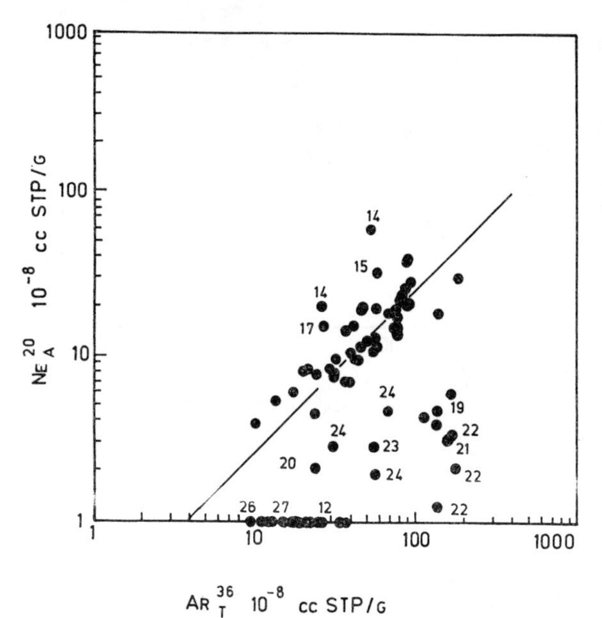

FIGURE 6 Ne_A^{20} vs. Ar_T^{36}. C1's and most C2's agree with $Ne_A^{20}/Ar_T^{36} = 0.27$. The Ne-rich chondrites Nawapali, Nogoya and Mokoia do not agree. Most C3's and C4's are either deficient in Ne_T or overabundant in Ar_T^{36} (or both). Preferential Ne_A^{20} loss is the probable explanation.

the measurements and in the evaluation of Ne_A^{20}, the correlation is reasonably good. The points that fall outside the correlation fall into two groups. The first group comprises the Ne-rich meteorites Mokoia, Nawapali, and Nogoya (dark portion). For these meteorites, the calculation of Ne_A^{20} is very uncertain. For example, in Mokoia which contains 407×10^{-8} ccSTP/g Ne_T^{20} it was calculated that all of this is Ne_B^{20}. In order to satisfy the correlation in Fig. 6 only 7×10^{-8} ccSTP/g of Ne_A^{20} are needed. A slight upward correction in point B of Fig. 4 would be sufficient. For similar reasons Ne_A^{20} in Nawapali and Nogoya may have been substantially overestimated. The second group comprises the chondrites nos. 19–24. *These are all C3's and C4's.* They are either deficient in Ne_T (hence Ne_T^{20}), or overabundant in Ar_T^{36}.

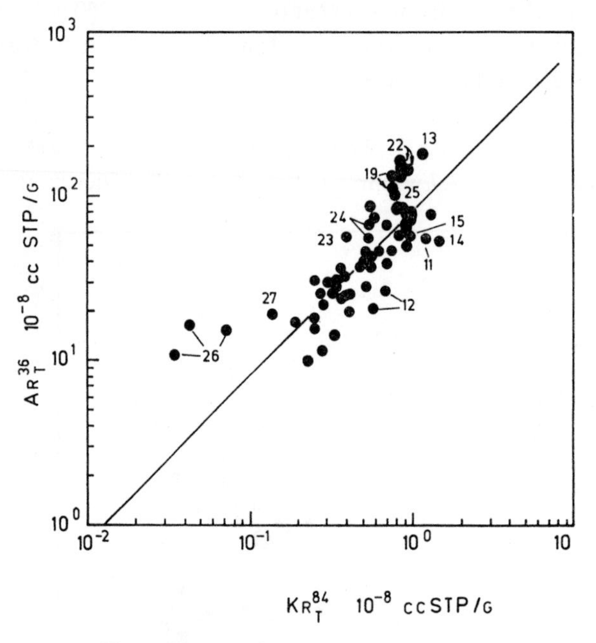

FIGURE 7 Ar_T^{36} vs. Kr_T^{84} in carbonaceous chondrites. Most C1's and C2's agree with $Ar_T^{36}/Kr_T^{84} \approx 80$. The Ne-rich meteorites Nawapali, Nogoya and Mokoia (and a Ne-rich sample of Murray) appear to be systematically poorer in Ar_T^{36} or richer in Kr_T^{84} (or both). The C3's and G4's lie generally above the distribution, i.e. these appear to be richer in Ar_T^{36}, poorer in Kr_T^{84}.

Fig. 7 shows Ar_T^{36} vs. Kr_T^{84}. The correlation is even stronger than the one between Ne_A^{20} and Ar_T^{36}: all the unnumbered points lie within $50 < Ar^{36}/Kr^{84} < 125$; the average of the ratios is 80. Again, the Ne-rich meteorites Nawapali, Nogoya, and Mokoia are located near one extremity of the distribution:

these meteorites appear to be either richer in Kr_T^{84} or poorer in Ar_T^{36}, or both than the other C2's. Also the C3's and C4's are on the whole located at one extremity, but these chondrites appear to be either richer in Ar_T^{36}, or poorer in Kr_T^{84} (or both) than the Cl's and C2's. This is interesting because the same meteorites are apparently deficient in Ne_T^{20}. This is difficult to understand in any theory of diffusion loss.

Fig. 1d shows the general relationship of Kr_T^{84} to Xe_T^{132}: Kr_T^{84} and Xe_T^{132} occur in approximately equal amounts. In earlier investigations, it was found that $Xe_T^{132} \geqq Kr_T^{84}$, but more recent work has shown that the reverse is correct. It seems reasonable to summarize the most reliable data: $Kr_T^{84}/Xe_T^{132} = 1.3 \pm 0.2$.

It has been mentioned before that the isotopic composition of Ar_T varies little compared with that of Ne_T. Fig. 8 shows a histogram of the 56 most reliable ratios (a number of results by Mazor et al. 1969 were rejected because of hydrocarbon interference in the Ar-measurements). All the ratios

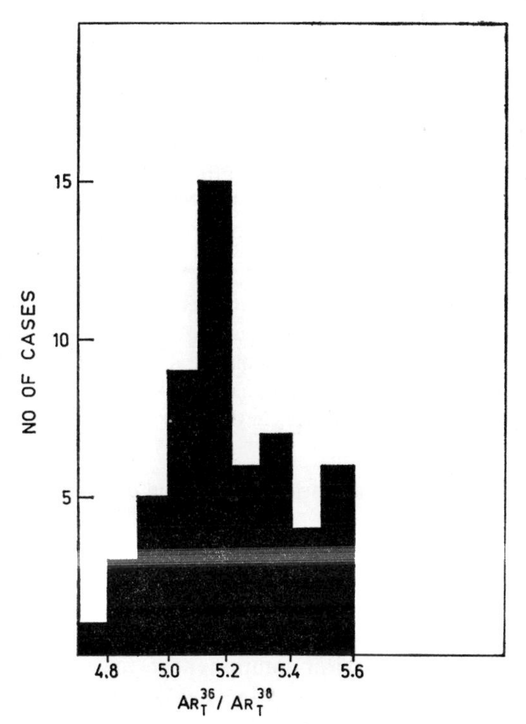

FIGURE 8 Ar_T^{36}/Ar_T^{38} in carbonaceous chondrites. The mean of 56 measurements is 5.20. This ratio seems to vary by less than 10% over the entire range of Ne_T^{20}/Ne_T^{22}.

agree within $\pm 10\%$ of the mean value, 5.20. Thus, the earlier conclusions reached by most investigators that the isotopic composition of trapped argon in carbonaceous chondrites varies by at most $\pm 10\%$ are confirmed by the more recent data.

Several authors have investigated the isotopic composition of Kr_T in carbonaceous chondrites. In Fig. 9 I have reproduced the results of Eugster et al. (1967). Marti's (1967a) results are essentially in agreement. I quote

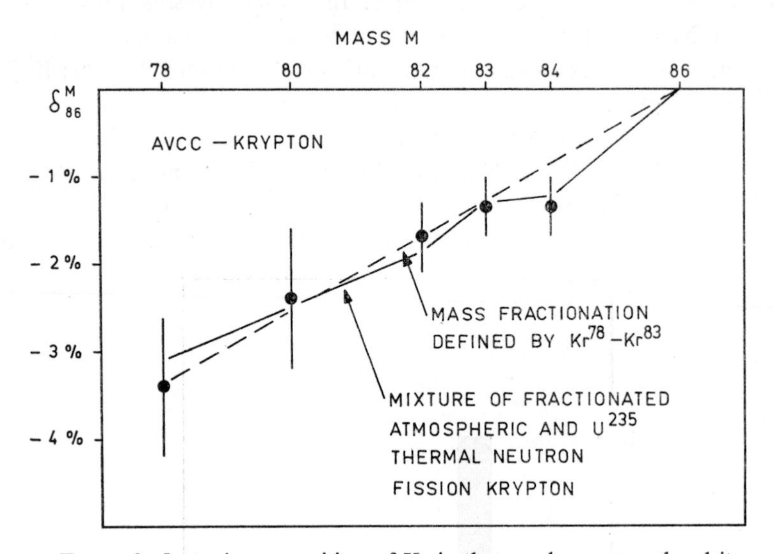

FIGURE 9 Isotopic composition of Kr in three carbonaceous chondrites, reproduced from Eugster et al., 1967. The best fit is obtained when it is assumed that Kr (in the atmosphere) is mass-fractionated (dashed line) by -0.3% per mass unit and that U^{235} thermal neutron fission or U^{238} spontaneous fission krypton is added to Kr_T in the meteorites.

Eugster et al. (1967): "The systematic trend of the δ-values

$$\left(\delta_{86}^{M} = \frac{(Kr^{M}/Kr^{86})\text{ meteorite}}{(Kr^{M}/Kr^{86})\text{ atmosphere}} - 1 \times 100\%\right)$$

with mass number suggests that a mass fractionation process is responsible for the Kr anomalies." "A fission component in meteoritic Kr alone also cannot explain the Kr anomalies since Kr^{87}/Kr^{82} is lower than the atmospheric [value]. Mass fractionation combined with a small fission component in meteoritic Kr would give the observed anomalies." (Note: mass fractionation alone is represented by the dashed curve.) "From the data alone it cannot be decided whether atmospheric or meteoritic Kr is fractionated relative to primordial Kr (i.e., Kr in the solar nebula). The Kr concentration in carbonaceous chondrites is larger than the amount of Kr present in the terrestrial atmosphere relative to the total mass of the earth, and it could be argued that the earth is only partially outgassed and that atmospheric Kr is a fraction enriched in the light isotopes."

The isotopic compositions of trapped xenon has remained an enigma, in part because of the great complexity of meteoritic xenon. A detailed discussion falls outside the scope of this review. Most carbonaceous chondrites contain xenon similar in composition to that shown in Fig. 10 (taken from Eugster

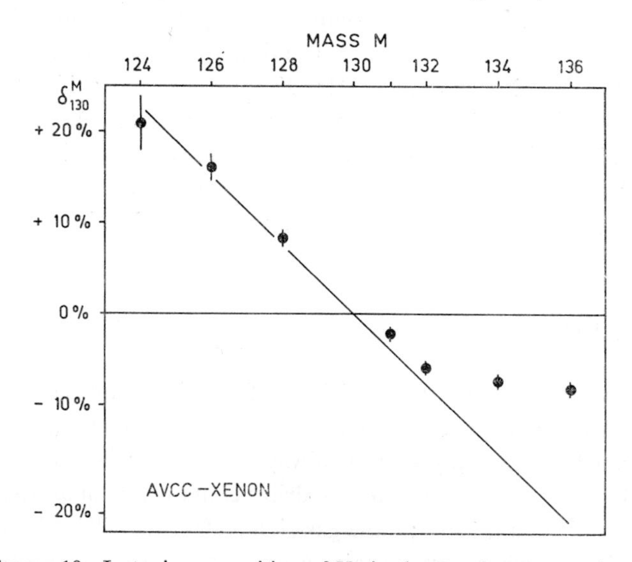

FIGURE 10 Isotopic composition of Xe in three carbonaceous chondrites (Eugster et al., 1967). The composition of Xe in most, if not all the carbonaceous chondrites is very similar, but there are significant though small variations, especially at mass numbers greater than 130. This mass spectrum presents serious difficulties. If Xe_T was merely mass-fractionated with respect to atmospheric (or the reverse; solid line), why is the fractionation factor (3.8% per mass unit) greater that that of Kr and why does it have the opposite sign? And what is the parent-nuclide of the fission component on masses 131, 132, 134, 136?. For further discussion see text.

et al., 1967); however, there are slight, but significant variations in the heavier isotopes (cf. Anders and Heymann, 1969). Reynolds (1960a), who was the first to discover these so-called "general anomalies" of meteoritic Xe, commented in one of his early papers: "On the other hand a strong mass-dependent fractionation may be responsible for most of the anomalies" (Reynolds, 1960b). Two other theories have been advanced. Kuroda (1960) and Cameron (1963) consider meteoritic Xe essentially identical to primordial, while terrestrial xenon has been modified over geolcgic time by the addition of xenon from spontaneous fission and from the solar wind. Fowler et al. (1962), on the other hand, consider terrestrial xenon identical with primordial, whereas meteoritic xenon was altered early in the evolution of the solar system by particle irradiation from the primitive sun. Eugster et al. (1967) have commented: "As can be seen in Fig. 2 (here Fig. 10), our δ-values for the light isotopes would be well represented by a mass fractionation of 3.8% per mass unit. The fission component required to bring the heavy isotopes of our AVCC Xe on the mass fractionation line in Fig. 2 would be $Xe^{131} = 25 \pm 15$, $Xe^{132} = 38 \pm 21$; $Xe^{134} = 64 \pm 10$; $Xe^{136} \equiv 100$. This again is different from all known fission spectra." And: "The Kr fractionation is much smaller (−0.3% per mass unit) than the Xe fractionation and is furthermore of opposite sign so that at least two separate fractionation processes would have to be invoked for Kr and Xe."

The Ordinary Chondrites

Most, if not all of the ordinary chondrites, (H, L, LL groups) contain small quantities of trapped argon, krypton, and xenon (Kirsten et al., 1963, were the first to conclude that Ar_T is nearly always detectable; for Kr and Xe see Zähringer, 1968). However, certain chondrites contain vast quantities of the light gases (König et al., 1961); these gas-rich meteorites will be discussed in the next section. Furthermore, certain ordinary chondrites, now classed as H3, L3, and LL3, are systematically richer in Ar_T, Kr_T, and Xe_T (first pointed out by Heymann, quoted by Wood, 1967), and occasionally contain Ne_T. Since the papers by Zähringer (1966a), Marti (1967b), and Heymann and Mazor (1968) on this subject, new data for these meteorites have become available, but these have not appreciably changed the basic conclusions. I shall therefore use the figures from the three papers. Zähringer (1966a) has arranged Ar^{36}_T according to Van Schmus and Wood's (1967) chemical-petrologic classification (Fig. 11). He concludes that in each chemical group, (except for the enstatite chondrites) Ar^{36}_T shows a decreasing trend with increasing "petrologic grade-number" (i.e. increasing degree of metamorphism, according to Van Schmus and Wood). Although Zähringer considers the carbonaceous chondrites the most striking result, this is open to question in view of the high Ar_T^{36} contents of Felix, Lancé, Ornans, Warrenton, and Vigarano (C3's and C4's) which are comparable to or even greater than those of the C1's. But the correlation among the ordinary chondrites is unmistakable and has been confirmed by other investigators. This correlation between gas-content and metamorphism immediately suggests that all chondrites, at one time, contained comparable and relatively large amounts of Ar_T^{36}, and that the distribution now seen was caused by an increasing

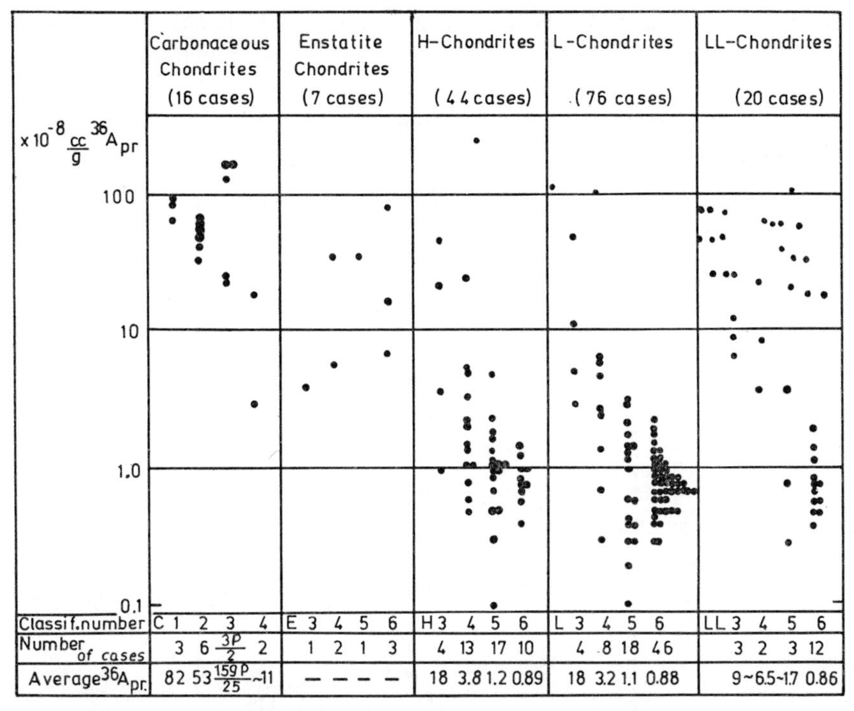

FIGURE 11 Ar_T^{36} plotted for the various chemical-petrologic types of Van Schmus and Wood (from Zähringer, 1966). According to Zähringer the trend implies that all chondrites originally contained comparable and relatively large amounts of Ar_T^{36}. Larimer and Anders (1967) have pointed out that at least part of the trend may be primary, having been established during the accretion of the parent materials of the meteorites.

degree of metamorphism, which was attended by an increasingly greater fractional loss of Ar_T^{36} (Zähringer, 1966a, 1968). However, Larimer and Anders (1967) have cautioned: "It is difficult, however, to separate the effects of accretion temperature (of the meteoritic parent materials) and metamorphism: a low accretion temperature implies a near-surface location in the parent body and hence a low degree of metamorphism. Thus the observed correlation between gas content and metamorphism (Zähringer, 1966a) does not necessarily imply a causal relationship between gas content and metamorphism. At least part of the trend may be primary, having been established during accretion."

Marti (1967a) has studied Ar_T, Kr_T, and Xe_T, and his results are shown in Fig. 12. (His study includes 7 carbonaceous chondrites and one ureilite). In this work, Marti has established an important fact, namely that both Ar_T^{36}/Xe_T^{132} and Kr_T^{84}/Xe_T^{132} in ordinary chondrites are remarkably constant over a vast range of concentrations (almost three orders of magnitude). Marti concludes: "The correlation shown in Fig. 11 requires the gases to have been removed in fixed proportions. Diffusion is likely to cause fractiona-

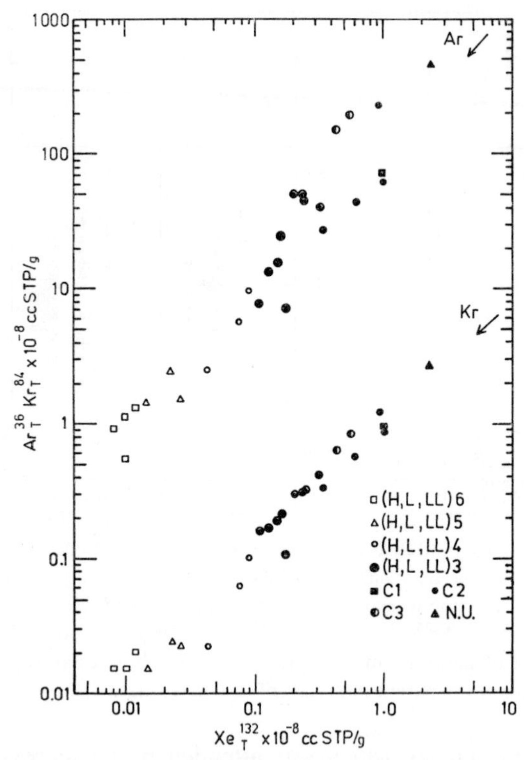

FIGURE 12 Ar_T^{36} vs. Xe_T^{132}; Kr_T^{84} vs. Xe_T^{132} (Marti, 1967a). The strong correlation over a vast range of gas contents speaks against the diffusion hypothesis, and seems to imply that chondrites *acquired* the gases in roughly the amounts now found in them.

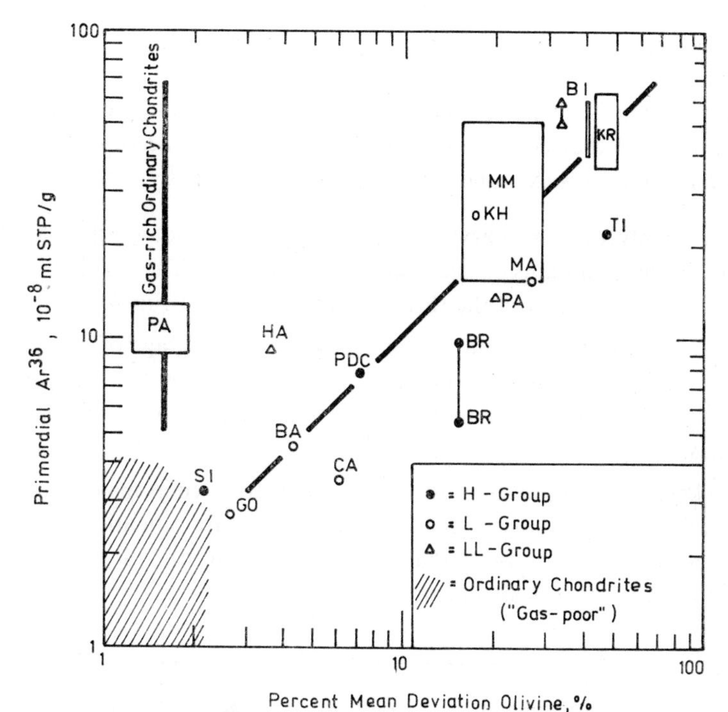

FIGURE 13 Ar_T^{36} vs. percent mean deviation of olivine composition (Heymann and Mazor, 1968). PMD measures the width of the Fe/Mg distribution in olivine grains; for low PMD values the olivines are rather uniform in composition and vice versa. Chondrites with uniform olivine have relatively low Ar_T^{36} contents, those having highly variable olivine composition are rich in Ar_T^{36}. However, the "gas-rich" chondrites do not conform to the trend; they are much richer in Ar_T^{36} than one would expect from their olivine composition.

tion of the abundance patterns. There is [only] one meteorite which shows such a fractionation."

Heymann and Mazor (1968) independently concluded that the abundance patterns of the heavy gases in the ordinary chondrites are remarkably constant. They discovered that the contents of trapped gas are roughly proportional to a parameter called "percent mean deviation of olivine composition" (Dodd et al., 1967). Fig. 13 shows this correlation for Ar_T^{36}. Perhaps the most interesting feature of this correlation is that the "gas-rich" chondrites (see next section) do not conform to the trend; these meteorites are much richer in Ar_T^{36} than one would expect from their olivine composition. Heymann and Mazor (1968) also observed that Ne_T occurs only seldomly in ordinary chondrites (except in the gas-rich variety), even when the Ar_T^{36} contents are comparable to those in the C1's and C2's.

The question arises whether the trapped gas in the ordinary chondrites is of the same kind that correlates with Ne_A in the carbonaceous chondrites (Figs. 6 and 7). Let us examine the diagnostic features:

a) Ne_A^{20}. Only one ordinary chondrite (apparently not of the gas-rich variety), Khohar, contains substantial amounts of Ne_T. The two known Ne measurements are shown in the three-component plot of Fig. 4.

b) Ar_T^{36}/Ar_T^{38}. Heymann and Mazor (1968) give an average of 9 measurements as 5.35 ± 0.15. Marti's (1967b) ratios (3) average to 5.33. In this respect Ar_T in the ordinary chondrites is identical to that in the carbonaceous chondrites.

c) The krypton isotopic composition can be interpreted as mass fractionation plus fission krypton (Eugster et al., 1967). The fractionation factor of -0.6% per mass unit is twice that for the carbonaceous chondrites, but whether this difference is significant is uncertain.

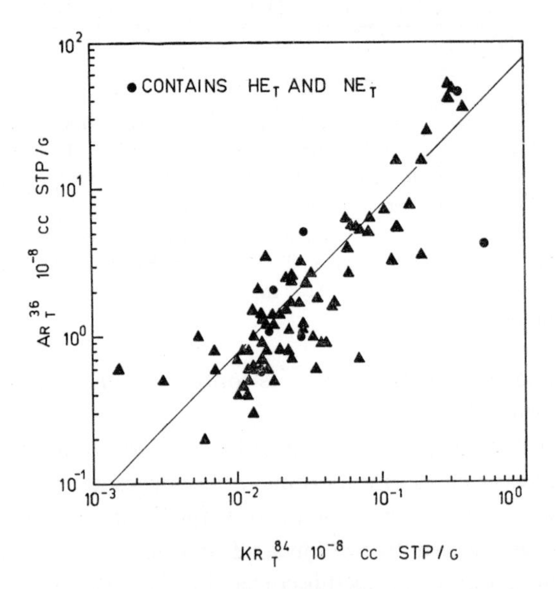

FIGURE 14 Ar_T^{36} vs. Kr_T^{84} in ordinary chondrites. The solid curve represents $Ar_T^{36}/Kr_T^{84} = 80$, taken from Fig. 7. The points scatter considerably than in Marti's representation (Fig. 12), but the trend is unmistakable. The scatter may be in part due to the difficulties in measuring accurately small amounts of Ar_T and Kr_T.

d) The xenon isotopic composition is essentially similar to that of the carbonaceous chondrites (Marti et al., 1966; Marti, 1967a; Eugster et al., 1968).

e) The Ar_T^{36}/Kr_T^{84} distribution of the H and L-group chondrites is shown in Fig. 14. The solid curve ($Ar_T^{36}/Kr_T^{84} = 80$) is taken from Fig. 7. The scatter of the data is considerably greater than in Marti's (1967b) work as noted by Zähringer (1968); nevertheless the trend is quite well established. Considering the difficulties connected with Ar_T and Kr_T measurements at low gas contents, one can argue that the distribution of the ordinary chondrites is not substantially different from that of the carbonaceous chondrites.

f) The Kr_T^{84}/Xe_T^{132} ratio varies appreciably; however when only data with Kr_T^{84}, $Xe_T^{132} > 0.1 \times 10^{-8}$ ccSTP/g are used, the average of 13 values is 1.25, in agreement with the carbonaceous chondrites.

Judging from these criteria, one may argue that the ordinary chondrites and the carbonaceous chondrites *have acquired variable amounts of one kind of trapped gas*. That the original abundance pattern was modified to some extent, or even appreciably by diffusion in a number of the chondrites is easy to accept. Thus one might suggest that any Ne_A originally present in ordinary chondrites diffused out such that it can no longer be detected in these meteorites, and that the C3's and C4's in Fig. 6 are transitional between the C1's and C2's, and the ordinary chondrites. An alternative explanation is that the C3's and C4's accreted at temperatures intermediate between the accretion temperatures of the ordinary chondrites and the C1's and C2's.

The Gas-Rich Meteorites

This is a group of meteorites which includes not only chondrites, but also achondrites (aubrites and howardites). The principal characteristic of these meteorites is that they contain vast amounts of the light trapped gases (He_T, Ne_T, and Ar_T) and that the trapped gases are often found in physically distinct portions of the meteorite such as in the dark portions of Pantar, etc. The reverse, however, is not true: at least one Pantar fragment contains dark portions which have normal gas contents (cf. Pepin and Signer, 1965). In addition to the light trapped gases, these meteorites also contain substantial amounts of Kr_T and Xe_T.

Fig. 15 shows He_T^4 vs. Ne_T^{20}. For comparison, the location of the "He-rich" carbonaceous chondrites (Fig. 2) is also shown. The most striking feature of Fig. 15 is the remarkably strong correlation of the two isotopes over a vast concentration range: most ordinary chondrites, carbonaceous chondrites, and aubrites fall in the range $200 < He_T^4/Ne_T^{20} < 500$, but the howardites seem to form a group of their own at $He_T^4/Ne_T^{20} \sim 100$ (cf.

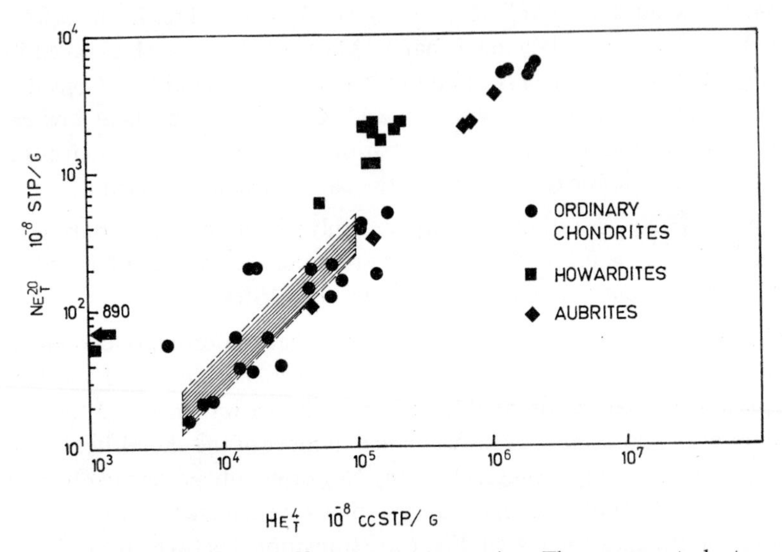

FIGURE 15 He^4 vs. Ne^{20} in gas-rich meteorites. There appear to be two populations: the ordinary chondrites, aubrites (and carbonaceous chondrites) with $He_T^4/Ne_T^{20} = 200$–500; and the howardites with $He_T^4/Ne_T^{20} \approx 100$.

Ganapathy and Anders, 1969). Preferential He_T^4 loss would have moved points to the left. While there is firm evidence that certain of these meteorites have lost some He_T^4 (see next section), it seems impossible that the entire distribution itself had come about by simultaneous He_T^4 and Ne_T^{20} loss in near-constant proportion (200–500) from meteorites that originally contained $>10^{-2}$ ccSTP/g of He_T^4. There seems to be no doubt that the correlation in Fig. 15 reflects at least grossly the concentrations in which these gases were *acquired* by the various meteorites.

We have seen tat He_T^3 in the carbonaceous chondrites is too frequently masked by He_C^3 to obtain an accurate He_T^4/He_T^3 ratio, but this is not the case for the "gas-rich" chondrites. Fig. 16 shows He_T^4 vs. He_T^3; all the points lie near $He_T^4/He_T^3 = 3000$, and there appears to be no gross systematic difference between ordinary chondrites, howardites, and aubrites in this respect. It is interesting to note that this value is similar to the one for "solar" He in the carbonaceous chondrites.

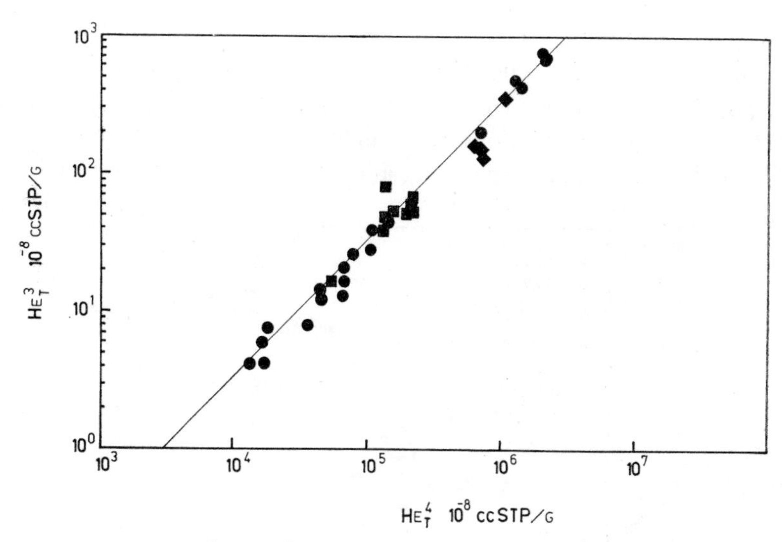

FIGURE 16 Ne_T^4 vs. Ar_T^3 in the gas-rich meteorites. Note the very strong correlation near $He_T^3/He_T^4 = 3000$.

Pepin (1967) has analyzed the neon isotopic composition of five gas-rich chondrites (Breitscheid, Fayetteville, Kapoeta, Pantar, and Pesyanoe). He concludes: "the trends shown by both total and partial gas analysis of the other four meteorites (exception: Breitscheid) are not in general directed toward the cosmogenic composition but rather toward a range of intercepts with the neon-A line". This implies that the gas-rich chondrites may contain some Ne_A, but the amounts are always relatively small: most of the points lie near the upper left hand corner in Fig. 4, i.e. near the composition of Ne_B.

Fig. 17 shows Ne_T^{20} vs. Ar_T^{36}. Again there is a remarkably good correlation between the elements, with $Ne_T^{20}/Ar_T^{26} \approx 20$. The shaded area shows the location of the C1's and C2's from Fig. 6. It is immediately obvious that the gas-rich chondrites and the carbonaceous chondrites form two distinct populations: the average Ne_A^{20}/Ar_T^{26} ratio of the latter is only 0.27. To be sure, for the points in the shaded area Ar_T^{36} is plotted against only that portion of Ne_T^{20} which is Ne_A^{20}. However, in the great majority of the cases Ne_A^{20} represents more than 50% of all the trapped neon, such that the two populations would remain distinct even if Ar_T^{36} were plotted against Ne_T^{20} for the carbonaceous chondrites. One might argue that the population with the lower Ne_T^{20}/Ar_T^{38} ratio was "derived" from the other population by diffusive loss of Ne_T (this is in essence Stauffer's original hypothesis), but we have seen that the Ne_T^{20}/Ne_T^{22} ratios speak against this.

It is interesting to plot Ne_T^{20}/Ar_T^{36} for the Ne-rich C2's (Mokoia, Murray, Nawapali, and Nogoya) in Fig. 17. It is seen that Mokoia belongs essentially to the population with $Ne_T^{20}/Ar_T^{36} \approx 20$, whereas the remaining meteorites fall between the two populations. One might thus argue that there are, indeed, two kinds of trapped argon, Ar_A and Ar_B, which, for reasons not yet

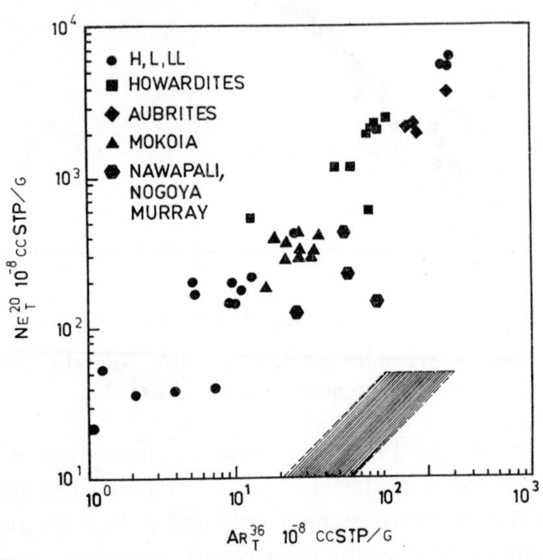

FIGURE 17 Ne_T^{20} vs. Ar_T^{36} in the gas-rich meteorites. The scatter is considerable, and there is some doubt that the meteorites form a single population. Nevertheless, the gas-rich meteorites and the carbonaceous chondrites (Cl's and C2's except Nawapali, Nogoya, and Mokoia) form two distinct groups; the former with $Ne_T^{20}/Ar_T^{36} \sim 15$; the latter with $Ne_T^{20}/Ar_T^{36} \sim 0.27$. The points for Nawapali and Nogoya (and for one Murray measurement from Chicago) seem to lie between the two populations. From this one may infer that these meteorites contain a substantial fraction of "planetary" Ar_T^{36} (or Ar^{36}).

understood appear to have virtally the same isotopic composition $Ar^{36}/Ar^{38} = 5.2 \pm 0.4$.

Regrettably, there are only few Kr_T and Xe_T data available for gas-rich chondrites. These are shown in Fig. 18. The only apparent trend is that the $Kr_T^{84} < Ar_T^{36}$ ratio is appreciably lower in the gas-rich chondrites than in carbonaceous and ordinary chondrites, perhaps by a factor 5–10. But with the limited statistics the conclusions cannot be firm. Judging from the data $Kr_T^{84} > Xe_T^{132}$, but no firm conclusions can be made regarding the value of the Kr_T^{84}/Xe_T^{132} ratio.

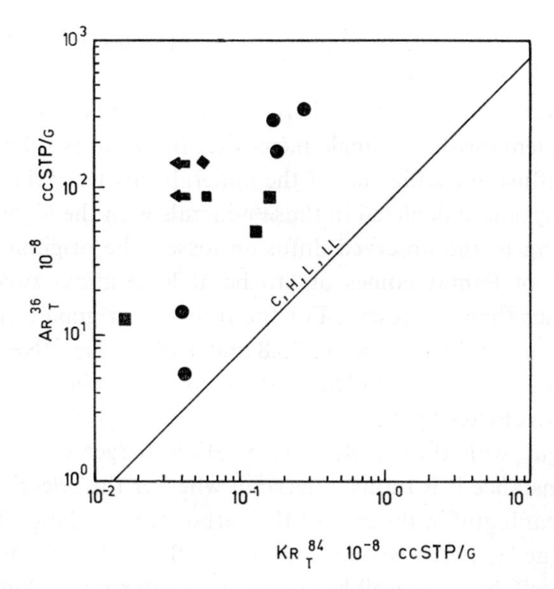

FIGURE 18 Ar_T^{36} vs. Kr_T^{84} in the gas-rich chondrites. Only few data are available, hence conclusions cannot be firm. It appears however, that gas-rich chondrites are generally poorer in Kr_T^{84} than carbonaceous and ordinary chondrites.

Separated Minerals and Temperature Runs

A brief discussion should now be given of the various attempts that have been made to locate the trapped gases within the meteorites, i.e., to find whether these are significantly concentrated in certain minerals, or at certain localities within the crystals.

Hintenberger, Wänke, and their co-workers have developed ingenious techniques which permit inert gas measurements to be made in selected minerals by the use of solvents which preferentially attack one mineral at the time (cf. Vilcsek and Wänke, 1965). In their first paper on the subject, the authors found that the gas-rich meteorite Breitscheid had suffered substantial He^4 losses, which may explain why He_T^4/Ne_T^{20} in this meteorite is only 88. In the next paper (Hintenberger, Vilcsek, and Wänke, 1965) a study was made of Breitscheid, Pantar, Kapoeta, and Murray. The authors concluded: "For the bronzite chondrites Breitscheid and Pantar, we could prove that all main mineral components contain light primordial gases (i.e. He_T and Ne_T), probably this is also true for the carbonaceous chondrite Murray and for the achondrite Kapoeta. The light primordial gases are highly concentrated in the outermost layers of the single mineral grains. The ratios for He^4/He^3 and Ne^{20}/Ne^{22} vary for different meteorites and also for different mineral components within a single meteorite. Both ratios show a correlation with the diffusion coefficients of the minerals involved; the lighter isotope being always more depleted in those minerals with the higher diffusion losses. According to the observed diffusion losses, the original primordial helium content of Pantar comes out to be at least about two orders of magnitude higher than at present. For the true and original elemental and isotopic ratios we obtained values of 13.8 and 14.0 for Ne^{20}/Ne^{22}, 2200 for He^4/He^3 and 800 for Ne^4/Ne^{20}. These ratios have been altered considerably on account of gas losses by diffusion."

One can argue with the conclusions of Hinterberger et al. on several grounds. For instance it is rather uncertain whether the selective chemical treatment is meaningful in the case of the carbonaceous chondrite Murray: the large "residue" (54% by weight, containing 38% of Ne_T) is characterized only as "silicates", but may well have been a complex mineralogical assemblage. Furthermore, the correlation between gas content, element ratios, isotope ratios, and diffusion characteristics of the minerals involved might be accidental; Pepin (1967) has pointed out that the two-component model can well account for the relationships observed among the isotopes. On the other hand, there is no doubt that certain meteorites have suffered some He_T or Ne_T loss. Also it is tacitly assumed in the diffusion hypothesis that Ne_T diffusion leads to isotope fractionation (cf. Zähringer, 1962b), but this assumption has not been proven experimentally; nor is it certain that such a fractionation, if it occurs, follows a simple $\sqrt{M}$ dependency as Zähringer has assumed.

Zähringer (1962a) was the first to attempt measurements of separated mineral phases of Kapoeta (handpicking of olivine, pyroxene, and plagioclase). From the results he concluded that most if not all of the trapped gas is present in a glass phase. König and Wlotzka (1962) measured the trapped gases in separated minerals of Pantar (olivine, orthopyroxene, troilite, silicate-troilite-metal intergrowth, magnetic and nonmagnetic fractions, etc.). From the results the authors concluded that the concentration of the trapped gases is low in the coarse-grained major minerals, but substantial in fine-grained metal and troilite.

That the trapped "solar" gas in the gas-rich meteorites is often or perhaps always concentrated near the surfaces of crystals became more firmly established by the ingenious studies on Khor Temiki by Eberhardt, Geiss,

Grögler and their coworkers (Eberhardt et al., 1965b): "The rare gas contents and isotopic compositions were determined in the Khor Temiki aubrite. This meteorite contains large amounts of trapped gases. They are only located in the fine-grained matrix and not in the larger enstatite crystals. The trapped gas contents of six grain size fractions of the matrix shows a strong anticorrelation with grain size. Measurements on separated minerals prove that this is a true grain size effect and not due to spurious enrichment of feldspar in the fine fractions. These results strongly suggest a surface effect."

On the other hand, Zähringer's (1966b) attempt to locate He_T in Fayetteville and Kapoeta by moving a polished section under an intense electron beam in a microprobe, and measuring the released He_T^4 with a mass spectrometer, seemed to contradict this finding.

Inert gas measurements in separated chondrules and matrix have been made by a number of authors, especially in connection with the isotopic anomalies in xenon. Merrihue (1966) reports: "Relative to the total meteorite (Bruderheim), chondrules are depleted in primordial xenon." Rowe (1968) reports that Mokoia chondrules contain 2.79×10^{-10} ccSTP/g of Xe^{132} as against 18.5×10^{-10} ccSTP/g in the matrix. In Chainpur, the magnetic chondrules contain 13.8×10^{-10} ccSTP/g; the matrix 37.1 $\times 10^{-10}$ ccSTP/g. Alexander and Manuel (1968) report 42.8×10^{-10} ccSTP/g of Xe^{132} in the Bjurböle chondrules; 83.0×10^{-12} ccSTP/g in the matrix, but find little difference between chondrules and matrix in Bruderheim: 1.4 vs. 1.6×10^{-12} ccSTP/g. Zähringer (1968) reports that the matrix of the C2 chondrite Essebi shows about a sixfold increase in He_T^4, Ne_T, and Ar_T over the chondrules. Although the data are still too sparse for firm conclusions, there seems to be a definite trend that the matrix is always richer in trapped gases than the chondrules.

Useful information has been obtained from experiments in which meteorite samples were outgassed at increasingly higher temperatures. If the inert gases consist of a number of components, these might be given off preferentially in sufficiently different temperature ranges such that they would become partially, or wholly resolved. Fig. 19 shows the results of one of the earliest experiments by Zähringer (1962a). The changes in He^4/He^3 and Ne^{20}/Ne^{21} imply that the trapped component comes off at lower temperatures than the cosmogenic component; on the other hand, the near-constancy of Ar^{40}/Ar^{36} above 300°C implies that Ar_R^{40} and Ar_T are similarly bound: both appear to be concentrated in the glass in Kapoeta.

Temperature-release patterns such as shown in Fig. 19 are very useful in studies of the xenon isotopic composition (cf. Reynolds, 1963), especially of the isotopic composition of trapped xenon. There is now firm evidence that Xe_F was produced in situ (cf. Reynolds, 1963; Reynolds and Turner 1964; Hohenberg et al., 1967), hence added to the genuinely trapped xenon. Fission xenon will generally increase the relative abundance of the four heaviest isotopes Xe^{131}, Xe^{132}, Xe^{134}, and Xe^{136} with respect to Xe^{130}.

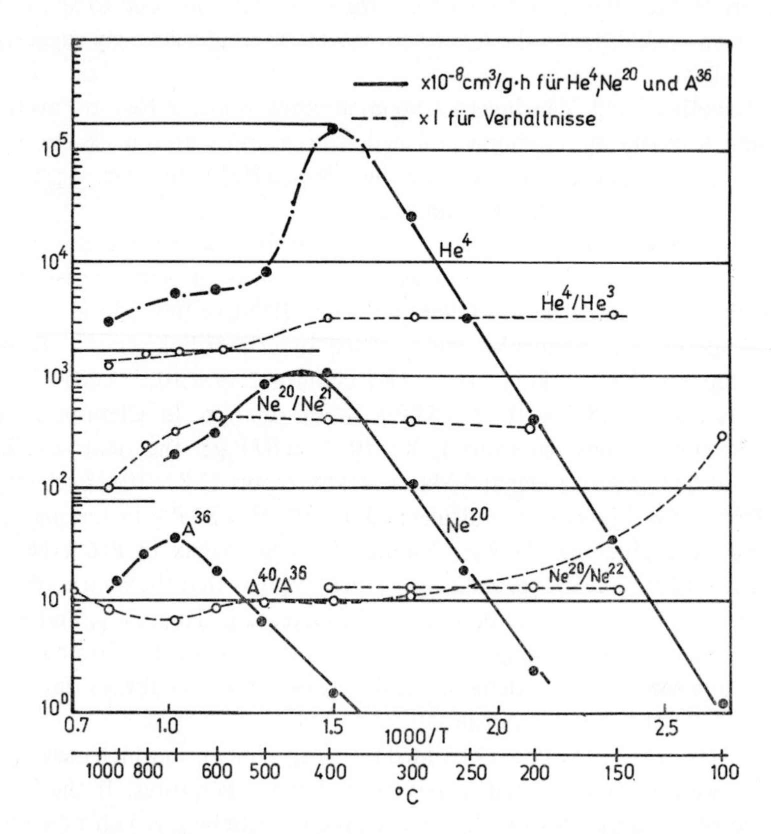

FIGURE 19 Temperature-release pattern of inert gases from the Kapoeta meteorite (Zähringer, 1962).

Hence, any gas fractions with the lowest relative abundances of these isotopes represent the closest approximation to trapped xenon. Funk et al. (1967), who have analyzed the problem in terms of a three-component mixture ("primitive" i.e. trapped, atmospheric, and fission xenon) have adopted $Xe_T^{130}/Xe_T^{132} = 0.1616 \pm 0.0007$ and $Xe_T^{133}/Xe_T^{132} = 0.3795 \pm 0.0007$.

Origin

Most theories on the origin of the trapped gases fall in one of two categories:

1) The absolute amounts and relative abundances of the gases, and the isotopic composition of Ne_T now seen in the meteorites are significantly or even substantially different from the original values. All the chondrites (or all the stony meteorites) originally contained similar amounts of trapped gases (Zähringer, 1966) in proportions similar to those now seen in Kapoeta or Pesyanoe (Zähringer, 1962a) The variations now seen have arisen from fractional diffusion of gas out of solid grains (Zähringer, 1962b).

2) The absolute amounts and relative abundances of the gases, and the isotopic composition of Ne_T now seen in the meteorites are roughly the same as the original values. The variations in relative abundances and in Ne_T^{20}/Ne_T^{22} are explained by multi-component mixtures (Signer and Suess, 1963). Diffusion of gas out of solid grains has occurred in certain meteorites, but has not substantially modified the original relationships.

Zähringer (1962a), who found that the trapped gas in Kapoeta is contained in glass, speculated that this material had solidified in the presence of large amounts of inert gases; he considered much less likely the possibility that the whole meteorite was exposed to large gas pressures at elevated temperatures.

Zähringer (1962b), in a general discussion of the trapped gases, assumed that the elements were once contained in a common reservoir, and that the observed variations have arisen in the course of the formation of the meteorites and the planets. He considered the trapping of gas in dust that had condensed in the solar nebula, and concluded that the amounts of gas thus acquired by the dust were roughly proportional to the partial pressure for each of the inert gases. Meteorites such as Pesyanoe could then not have been substantially heated on their parent object, otherwise the relative elemental abundances in this meteorite would not have remained so near the "cosmic" proportions (see Fig. 20). But not all meteorites contain the inert gases in near-cosmic proportions. For those that do not, several possibilities must be considered. The theory of Suess (1949) for the relative abundances of the inert gases in the Earth's atmosphere, which invokes selective loss of these elements from a gravitational field (after the bulk of the gas was lost convectively, leaving only 1 part in 10^7 of the cosmic abundances), cannot explain the "fractionated" (Anders, 1964) or "planetary" (Signer and Suess,

1963) abundance patterns in Fig. 20, because the predicted isotopic variation of Ar_T^{36}/Ar_T^{38} is *not* found in the meteorites. Zähringer therefore proposed a *quantized* fractionation in the meteorites such that only the He and Ne losses (by diffusion from grains) were sufficiently large for isotopic variations to

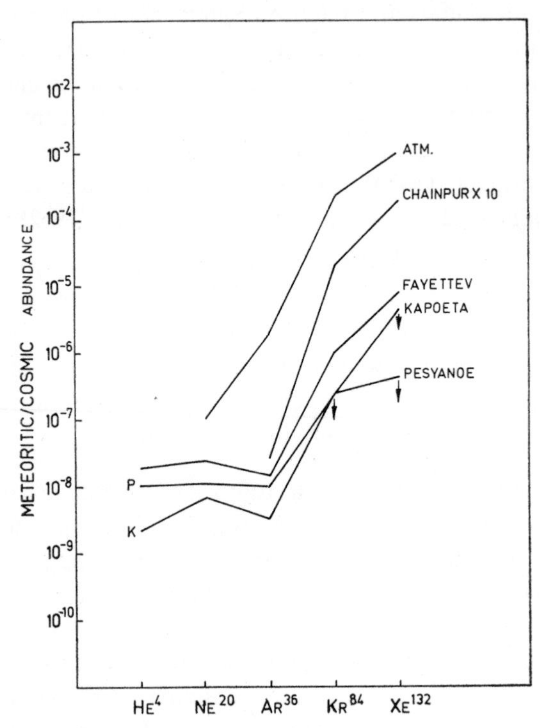

FIGURE 20 Meteoritic Abundances/Cosmic Abundances for Chainpur (LL3), Fayetteville (H), Kapoeta (howardite), and Pesyanoe (aubrite). Cosmic abundances from Cameron (1963), but (Ne^{20}/Ar^{36}) cosmic changed to 11 (Cameron, 1967). The atmospheric curve is arbitrarily adjusted to 10^{-7} for Ne^{20}; the cosmic curve is displaced from 1.0 to 10^{-10}. Near horizontal lines imply that gases occur in near-cosmic porportions ("solar"); strongly sloping lines imply fractionation ("planetary").

become detectable, whereas the losses of Ar_T, Kr_T, and Xe_T, which did occur, were too small to produce any detectable isotopic fractionations. The implications of Zähringer's hypothesis for the formation of the chondrites are these: after these meteorites acquired the trapped gases they must have gone through a period of heating such that He_T and Ne_T were preferentially lost. The temperatures could not have been much higher than 800 °C because otherwise too large a fraction of Ar_T would also have been lost. Judging from Zähringer's (1966a) recent work, one might argue that the limiting temperature could be more variable, such that the highly metamorphosed, gas-poor chondrites experienced the higher temperatures (or longer times, or both) than the less metamorphosed ones.

Zähringer's hypothesis accounts for the qualitative features of the inert gases in meteorites; however, a number of investigators have expressed serius doubt that they can explain the quantitative aspects of the distributions (cf. Pepin and Signer, 1965). A number of these objections have already been mentioned in the previous sections: e.g. the great range of Ne_T^{20}/Ne_T^{22} values in the carbonaceous chondrites for relatively small variations of Ne_T^{20}/Ar_T^{36} (Fig. 3), and the remarkably constant proportions of the elements, both in carbonaceous chondrites as well as in ordinary chondrites, over very great concentration ranges (Figs. 2, 6, 7, 12, 14, 15, and 17).

A fundamentally different theory was advanced by Suess et al. (1964), Signer (1964), and Wänke (1965) in connection with the gas-rich meteorites. These authors realized that it would require embarrassingly high partial He and Ne pressures in the gas phase to explain the amounts of He_T^4 and Ne_T^{20} now seen in meteorites such as Fayetteville. They therefore proposed low-energy particle irradiation (i.e. ion-implantation) as an alternative mechanism, in particular irradiation by the solar wind. Suess et al. (1964) pointed out that matter exposed to the solar wind will acquire the amounts of He_T and Ne_T now seen in the gas-rich meteorites *within reasonably short times*, and suggested a mechanism for the formation of the light-dark structures. The solar-wind hypothesis is now widely accepted in view of this supporting evidence:

a) the trapped gas is nearly always concentrated at the surface of crystals (Hintenberger, Vilcsek, and Wänke, 1965; Eberhardt et al., 1965b).

b) Pellas et al. (1969) have recently shown that the tracks from 1–5 MeV/A iron-group nuclei in a 101 μ thick orthopyroxene crystal from Kapoeta are about one order of magnitude denser near the surface of the crystal than at the center.

The solar-wind hypothesis raises a number of questions of its own. At which time in their history did the meteorites acquire the solar-wind emplaced gases? Where did the trapping take place? Were all the inert gases represented? Is the high Ne_T^{20}/Ne_T^{22} ratio identical to that in the sun or was it changed between the sun and the meteorites (Geiss, private communication)? And how does one account for the low Ne_T^{20}/Ne_T^{22} ratios in the carbonaceous chondrites (Fig. 4)?

Regarding the first question, I quote Wänke (1965): "Concerning the date of the solar-wind emplacement of the gases (He_T and Ne_T) in the meteoritic grains one cannot make precise statements. Since only certain, but not all of the bronzite chondrites contain the trapped gases it appears that an early date – i.e. prior to the formation of relatively large parent objects from planetary matter – is unlikely. The close proximity of gas-rich and gas-poor regions, and the 'island-like' occurrence of the light-colored, gas-free inclusions within the dark, gas-rich material seems to imply that the inclusions already existed as compact agglomerates when the emplacement of the gas (in the dark material) took place."

As far as the physical setting of the emplacement is concerned, one thing is clear: the gases were not emplaced in meteorites as we know them now, but in individual, unaggregated or only loosely aggregated grains. This could have happened on the surface of an atmosphere-free parent object (Hintenberger, Vilcsek, and Wänke, 1965) or when the parent materials of meteorites existed as dust in the solar nebula, prior to their accretion into larger bodies (Signer and Suess, 1963).

Wänke (1966b) suggested the surface of the moon as a suitable environment for the formation of the gas-rich bronzite chondrites; Mars, because of its atmosphere, must be ruled out; asteroidal objects are ruled out by Wänke on other grounds.

Suess et al. (1964) have suggested cometary nuclei as a possibility.

Anders (1964) has proposed that dust on the surface of an asteroid-size object acted as a "catcher" of solar wind. Collisions on this object would, from time to time, form shock-breccia such that the gases were essentially "shock-loaded" into the underlying rock, forming the light-dark structures. Anders' theory thus combines the essential features of the solar-wind hypothesis and the shock-hypothesis, first proposed by Fredriksson and Keil (1963). However, Suess et al. (1963) have strongly criticized the shock hypothesis.

Relatively few investigators have favored a very early implantation of solar wind, i.e., in the meteorite parent materials before the formation of the parent objects, because such theories lead to serious difficulties. The most obvious one is that the presence of appreciable gas pressures in the solar nebula would not permit the low-energy ions to penetrate far away from the sun. Heymann and Mazor (1968) have examined this possibility for the carbonaceous chondrites, but concluded that the conditions favorable for an early solar-wind emplacement are most unlikely to have prevailed when the bulk of the parent materials of the carbonaceous chondrites accreted. These authors speculated: "It seems more likely that dust, present after the dissipation of the gas acted as a 'catcher', i.e. was impregnated with solar wind, and that such gas-rich dust was accreted on the surface of already existing parent objects of meteorites where it was mixed in with surface material."

Were all the elements represented in the particle irradiation? One may argue from Figs. 15 and 17 that He^4, Ne^{20}, and Ar^{36} were present in the following approximate proportions 300:1:0.07; $He^4/He^3 \sim 3000$; $Ne^{20}/Ne^{22} \sim 12 - 13$; and $Ar^{36}/Ar^{38} \sim 5.2$. Somewhat different conditions must have prevailed for the aubrites, which have $He_T^4/Ne_T^{20} \sim 100$.

Whether Kr_T and Xe_T occurred in the solar wind cannot be firmly answered at the present time: their isotopic compositions provide no clues; the Ar_T^{36}/Kr_T^{84} relationship is not well established (Fig. 18). Pepin and Signer (1965), in reviewing this question, concluded "a few percent of the total krypton in Fayetteville would have been added with the solar rare-gas component. It is interesting to speculate that the solar wind may be fractionated to some extent, with resultant enrichment of light elements—and isotopes—relative to heavy ones. In this case Fayetteville must have lost some of its light gases after implantation, and the resemblance of their present abundances to a solar abundance pattern is coincidental. Relative abundances of the heavy rare gases in a fractionated solar wind would be very low so that virtually all krypton in the gas-rich chondrites would derive from the residual planetary gases."

The solar wind hypothesis forms a very important aspect of the two-component theories. It provides for the formation of gas-rich materials in which the relative elemental abundances and isotopic compositions are nearly constant. The vast concentration ranges in the meteorites could easily be accounted for by assuming variable exposure times, time-variations in solar wind intensity, and variable mixing ratios of exposed and unexposed material. Apparently the gas-loaded material varied also in composition and mineralogy: sometimes, as in Kapoeta, it was glass; sometimes, as in Khor Temiki, it was pyroxene and plagioclase; sometimes it consisted of the major chondritic minerals olivine, orthopyroxene, troilite, and metal. Thus it seems indeed almost certain that the solar-wind impregnation occurred on the surface of the parent objects, where collisions would produce a dust-layer of nearly the same chemical and mineralogical composition as that of the underlying rock. Collisions may be invoked to re-consolidate the gas-loaded dust, but alternative mechanisms seem equally plausible. But the most important aspect is that the solar gas became thus "superimposed" on any gas that was contained in the dust layer, or in the underlying rock with which the dust became mixed. It is most logical to assume that this gas is none other than the "planetary" component of Signer and Suess.

An alternative theory for the formation of the gas-rich meteorites was, however, advanced by Müller and Zähringer (1966), who found that carbon is enriched in the gas-rich, dark portions of Fayetteville, Kapoeta, and Pesyanoe relative to the gas-poor, light portions by factors of 2 to 3. The authors suggested that the formation of the dark portions must be closely connected with the formation of the carbonaceous chondrites (they did not suggest, however, that the dark material is simply a mixture of light material with carbonaceous chondrites as we know them.)

One may argue, however, that Müller and Zähringer's (1966) findings are not in conflict with the two-component theory. Recent data on carbon in chondrites (Moore and Lewis, 1967) show that the C-content of 0.28 % in the dark portion of Fayetteville is not uncommon among ordinary chondrites: of the 86 chondrites investigated by Moore and Lewis, 15 have C-contents comparable to or significantly greater than Fayetteville, yet only one of the 15, Weston, contains large quantities of trapped gases. Furthermore, Marti (1967b), and Heymann and Mazor (1968) have pointed out that Ar_T correlates fairly well with carbon in the ordinary chondrites. Anders and Heymann (1969) have pointed out that the correlation between Xe_T^{132} and the important trace element indium is even stronger. Thus the conclusion by Müller and Zähringer (1966) should be modified, as it appears now that the formation of most if not all of the ordinary chondrites, including the light portions of the gas-rich meteorites, is closely connected with the formation of the carbonaceous chondrites. But this is precisely what one would expect in a multi-component theory: materials rich in volatiles (inert gases, C, In), containing the gases in "planetary" proportions, i.e. $He_T^4/Ne_T^{20} \sim 200-400$; $Ne_T^{20}/Ar_T^{36} \sim 0.27$; $Kr_T^{84}/Xe_T^{132} \sim 1.3$; $Ne_T^{20}/Ne_T^{22} \sim 8.2$; and $Ar^{36}/Ar^{38} \sim 5.2$ should be formed relatively late during accretion, hence "planetary" gas should occur in most, if not all of the undifferentiated meteorites.

Why the inert gases are fractionated in the "planetary" components is not clear at the present time, nor is it clear why the Ne_T^{20}/Ne_T^{22} ratio in this component is as low as 8.2, even below that of atmospheric Ne (9.800 $\pm$ 0.0800). Furthermore, it is by no means certain that the two components "solar" and "planetary" are not mixtures themselves, or always came from the same source; the difference in "solar" He_T^4/Ne_T^{20} between the howardites on the one hand and ordinary chondrites and aubrites on the other hand seems to imply that a number of processes must have been involved in the formation of the two main components.

Adsorption has been most frequently discussed as the process that could account for the element fractionation in the "planetary" component (Signer and Suess, 1963; Pepin and Signer, 1965; Marti, 1967b; Larimer and Anders, 1967; Heymann and Mazor, 1968). Qualitatively, adsorption can account for the observed fractionation: in this respect He is more "volatile" than Ar, etc. But adsorption does not secure the gases *inside* the mineral grains, hence another process must be invoked for the firm trapping. Larimer and Anders (1967) have pointed out that equilibrium processes invariably lead to absurdly high nebular pressures: $10^3 - 10^7$ atm. for Ar, Kr, and Xe. Hence these authors suggested non-equilibrium trapping of gas during the chemical transformation of a major phase such as the formation of hydrated silicates ($\lesssim 315°K$), all in the cooling solar nebula. Dr. Geiss (private communication) has made the interesting suggestion that the "planetary" component was trapped at relatively high temperatures when certain minerals condensed in the solar nebula: gas atoms would collide with the growing crystals, and the probability that such atoms would become entrapped under a fresh layer would be roughly proportional to the "residence time" of the atom on the surface. This hypothesis, however, requires very rapid cooling of the solar nebula.

One of the greatest puzzles, that of the low Ne_T^{20}/Ne_T^{22} ratio in "planetary" gas remains unsolved. Curiously enough, relatively little attention has been given to the possibility that nuclear reactions may be accountable, probably because high-energy proton reactions on Na, Mg, Al, Si, Fe, etc. yield neon of the wrong composition, containing far too much Ne^{21} ($Ne^{20}:Ne^{21}:Ne^{22} \sim 1:1:1$). There are indications, however, that, n, α and n, 2n reactions on Na^{23} for 10–20 MeV neutrons could produce neon with Ne^{20}/Ne^{22} (via F^{20} and Na^{22}) substantially below 8.2, with little, if any production of Ne^{21} (Liskien and Paulsen, 1965; Picard and Williamson, 1965). Perhaps low-energy proton reactions should also be considered.

References

Alexander E.C. and Manuel O.K. (1969) *Geochim. Cosmochim. Acta* **33**, 298.
Anders E. (1964) *Space Sci. Rev.* **3**, 583.
Anders E. and Heymann D. (1969) *Science* **164**, 821.
Arrol W.J., Jacobi R.B. and Paneth F.A. (1942) *Nature* **149**, 245.
Bauer C.A. (1946) *Phys. Rev.* **72**, 354.
Cameron A.G.W. (1963) *Icarus* **1**, 314.
Cameron A.G.W. (1963) Lecture Notes, Yale Univ.

Cameron A. G. W. (1967) Private communication.
Dodd R. T., Van Schmus W. R. and Koffman D. M. (1967) *Geochim. Cosmochim. Acta* **31**, 921.
Eberhardt P., Eugster O. and Geiss J. (1965) *J. Geophys. Res.* **70**, 4427.
Eberhardt P., Geiss J. and Grögler N. (1965) *Tschermaks Mineral. Petrog. Mitt.* **10**, 535.
Eberhardt P., Eugster O., Geiss J. and Marti K. (1966) *Z. Naturforsch.* **21a**, 414.
Eugster O., Eberhardt P. and Geiss J. (1967) *Earth Planet. Sci. Letters* **3**, 249.
Eugster O., Eberhardt P. and Geiss J. (1968) Preprint.
Fowler W. A., Greenstein J. L. and Hoyle F. (1962) *Geophys. J.* **6**, 148.
Fredriksson K. and Keil K. (1963) *Geochim. Cosmochim. Acta* **27**, 717.
Funk H., Podosek F. and Rowe M. W. (1967) *Geochim. Cosmochim. Acta* **31**, 1721.
Ganapathy R. and Anders E. (1969) *Geochim. Cosmochim. Acta* **33**, 775.
Geiss J., and Hess D. C. (1958) *Astrophys. J.* **127**, 224.
Gentner W. and Zähringer H. (1957) *Geochim. Cosmochim. Acta* **11**, 60.
Gerling E. K. and Pavlova T. G. (1951) *Doklady. Akad. Nauk S.S.S.R.* **77**, 85.
Gerling E. K. and Rik K. G. (1955) *Doklady Akad. Nauk S.S.S.R.* **101**, 433.
Gerling E. K. and Levskii L. K. (1956) *Doklady Akad. Nauk S.S.S.R.* **110**, 750.
Heymann D. and Mazor E. (1966) *J. Geophys. Res.* **71**, 4695.
Heymann D. and Mazor E. (1967) *J. Geophys. Res.* **72**, 2704.
Heymann D. and Mazor E. (1968) *Geochim. Cosmochim. Acta* **32**, 1.
Hintenberger H., König H., Schultz L. and Wänke H. (1965) *Z. Naturforsch.* **19a**, 219.
Hintenberger, H., Vilcsek, E. and Wänke, H. (1965) *Z. Naturforsch.* **20a**, 939.
Hohenberg, C. M., Munk, M. N. and Reynolds, J. H. (1967) *J. Geophys. Res.* **72**, 3139.
Kirsten, T., Krankowsky, D. and Zähringer, J. (1963) *Geochim. Cosmochim. Acta* **27**, 261.
König, H., Keil, K., Hintenberger, H., Wlotzka, F. and Begemann, F. (1961) *Z. Naturforsch.* **16a**, 1124.
König, H. and Wlotzka, F. (1962) *Z. Naturforsch.* **17a**, 472.
Kuroda, P. K., (1960) *Nature* **187**, 4731.
Larimer, J. W. and Anders E. (1967) *Geochim. Cosmochim. Acta* **31**, 1239.
Liskien, H. and Paulsen, A. (1965) *Nuclear Phys.* **63**, 393.
Marti, K., Eberhardt, P. and Geiss, J. (1966) *Z. Naturforsch.* **21a**, 398.
Marti, K. (1967a) *Earth Planet. Sci. Letters* **2**, 193.
Marti, K. (1967b) *Earth Planet. Sci. Letters* **3**, 243.
Mazor, E., Heymann, D. and Anders, E. (1969) In preparation.
Merrihue, C. (1966) *J. Geophys. Res.*, **71**, 263.
Moore, C. B. and Lewis, C. F. (1967) *J. Geophys. Res.* **72**, 6289.
Müller, O. and Zähringer, J. (1966) *Earth Plan. Sci. Letters* **1**, 25.
Paneth, F. A., Reasbeck, P. and Mayne, K. I. (1953) *Nature* **172**, 200.
Pellas, P., Poupeau, G., Lorin, J. C., Reeves, H. and Audouze, H., Nature, in press.
Pepin, R. O. and Signer, P. (1965) *Science* **149**, 253.
Pepin, R. O. (1967) *Earth Plan. Sci. Letters* **2**, 13.
Picard, J. and Williamson, C. F. (1965) *Nuclear Phys.* **63**, 673.
Reasbeck, P. and Mayne, K. I. (1955) *Nature* **176**, 733.
Reynolds, J. H. (1960) *Phys. Rev. Letters* **4**, 351.
Reynolds, J. H. (1960) *J. Geophys. Res.* **65**, 3843.
Reynolds, J. H. (1963) *J. Geophys. Res.* **68**, 2939.
Reynolds, J. H. and Turner, G. (1964) *J. Geophys. Res.* **69**, 3263.

Rowe, M.W. (1968) *Geochim. Cosmochim. Acta* **32**, 1317.
Signer, P. and Nier, A.O. (1960) *J. Geophys. Res.* **65**, 2947.
Signer, P. and Suess, H.E. (1963) *Earth Sciences and Meteorites*, 241. North Holland.
Signer, P. (1964) *The Origin and Evolution of Atmospheres and Oceans*, 183. Wiley.
Stauffer, H. (1961) *Geochim. Cosmochim. Acta* **24**, 70.
Suess, H.E. (1949) *J. Geol.* **57**, 600.
Suess, H.E., Wänke, H. and Wlotzka, F. (1964) *Geochim. Cosmochim. Acta* **28**, 595.
Van Schmus, W.R., and Wood J.A. (1967) *Geochim. Cosmochim. Acta* **31**, 921.
Vilcsek, E. and Wänke, H. (1965) *Z. Naturforsch.* **20a**, 1282.
Wänke, H. (1965) *Z. Naturforsch.* **20a**, 946.
Wänke, H. (1966) *Fortschritte der chemischen Forschung* **7**, 322.
Wänke, H. (1966) *Z. Naturforsch.* **21a**, 93.
Wasserburg, G.J. and Hayden, R.J. (1955) *Phys. Ref.* **97**, 86.
Wood, J.A. (1967) *Icarus* **6**, 1.
Zähringer, J. and Gentner, W. (1960) *Z. Naturforsch.* **15a**, 600.
Zähringer, J. (1962) *Geochim. Cosmochim. Acta* **26**, 665.
Zähringer, J. (1962) *Z. Naturforsch.* **172a**, 460.
Zähringer, J. (1966) *Earth Planet. Sci. Letters* **1**, 379.
Zähringer, J. (1966) *Earth Planet. Sci. Letters* **1**, 20.
Zähringer, J. (1968) *Geochim. Cosmochim. Acta* **32**, 209.

References to figures

Auer, S., Braun, H.J. and Zähringer, J. (1965) *Z. Naturforsch.* **20a**, 156.
Begemann, F. (1965) *Z. Naturforsch.* **20a**, 150.
Chackett, K.F., Golden, J., Mercer, E.R., Paneth, F.A. and Reasbeck, P.R. (1950) *Geochim. Cosmochim. Acta* **1**, 3.
Dalton, J.C., Paneth, F.A., Reasbeck, P., Thomson, S.J. and Mayne, K.I. (1963) *Nature* **172**, 1168.
DuFresne, E.R. and Anders, E. (1962) *Geochim. Cosmochim. Acta* **26**, 251.
Eberhardt, P. and A. (1960) *Helv. Phys. Acta* **33**, 593.
Eberhardt, P. and Hess, D.C. (1960) *Astrophys. J.* **101**, 38.
Eberhardt, P. and A. (1961) *Z. Naturforsch.* **16a**, 236.
Eberhardt, P., Geiss, J. and Grögler, N. (1965) *J. Geophys. Res.* **70**, 4375.
Eberhardt, P., Geiss, J. and Grögler, N. (1966) *Earth Plan. Sci. Letters*, **1**, 7.
Fredriksson, K. and De Carli P. (1964) *J. Geophys. Res.* **69**, 1403.
Fredriksson, K., De Carli, P., Pepin, R.O., Reynolds, J.H. and Turner, G. (1964) *J. Geophys. Res.* **69**, 1403.
Funkhouser, J., Kirsten, T. and Schaeffer, O.A., Preprint.
Geiss, J., Hirt, B. and Oeschger, H. (1960) *Helv. Phys. Acta* **33**, 590.
Geiss, J., Oeschger, H. and Signer, P. (1960) *Z. Naturforsch.* **15a**, 1016.
Gentner, W. and Zähringer, J. (1955) *Z. Naturforsch.* **10a**, 6.
Goebel, K., Schmidlin, P. and Zähringer, J. (1959) *Z. Naturforsch.* **14a**, 996.
Hintenberger, H., König, H. and Wänke, H. (1962) *Z. Naturforsch.* **17a**, 306.
Hintenberger, H., König, H. and Wänke, H. (1962) *Z. Naturforsch.* **17a**, 92.

Hintenberger, H., Schultz, L. and Wänke, H. (1966) *Z. Naturforsch.* **21a**, 1147.
Hintenberger, H., Vilcsek, E. and Wänke, H. (1964) *Z. Naturforsch.* **19a**, 219.
Hintenberger, H., König, H., Schultz, L. and Wänke, H. (1964) *Z. Naturforsch.* **19a**, 327.
Kaiser, W. and Zähringer, J. (1964) *Z. Naturforsch.* **20a**, 963.
König, H., Keil, K. and Hintenberger, H. (1962) *Z. Naturforsch.* **17a**, 357.
Kuroda, P.K. (1961) *Geochim. Cosmochim. Acta* **24**, 40.
Kuroda, P.K. and Manuel, O.K. (1962) *J. Geophys. Res.* **67**, 4859.
Kuroda, P.K. and Crouch W.H. (1962) *J. Geophys. Res.* **67**, 4863.
Larimer, J.W. (1967) *Geochim. Cosmochim. Acta* **31**, 1215.
Merrihue, C.M., Pepin, R.O. and Reynolds, J.H. (1962) *J. Geophys. Res.* **67**, 2017.
Otting, W. and Zähringer, J. (1967) *Geochim. Cosmochim. Acta* **31**, 1049.
Pepin, R.O. (1964) *The Origin and Evolution of Atmospheres and Oceans*, 191. Wiley.
Pepin, R.O. (1968) *Origin and Distribution of the Elements*, 379. Pergamon.
Reynolds, J.H. and Lipson, J.I. (1957) *Geochim. Cosmochim. Acta* **12**, 330.
Reynolds, J.H. (1960) *Phys. Rev. Letters* **4**, 514.
Reynolds, J.H. and Turner, G. (1964) *J. Geophys. Res.* **69**, 3263.
Reynolds, J.H. (1967) *Ann. Rev. Nuc. Sci.* **17**, 253.
Rowe, M.W. (1968) *Geochim. Cosmochim. Acta* **32**, 1317.
Stauffer, H. (1961) *J. Geophys. Res.* **66**, 1513.
Stauffer, H. and Urey, H.C. (1962) *Bul. Astr. Inst. Czech.* **13**, 106.
Stauffer, H. (1961) *Geochim. Cosmochim. Acta* **24**, 70.
Suess, H.E. and Urey, H.C. (1956) *Rev. Mod. Phys.* **28**, 1956.
Tilles, D. (1962) *J. Geophys. Res.* **67**, 1687.
Vinogradov, A.P. and Zadorozhny, J.K. (1965) *Meteoritika* **26**, 77.
Wood, J.A. (1963) *Icarus* **2**, 152.
Wood, J.A. (1965) *Nature* **208**, 1085.
Zähringer, J. (1962) *Z. Naturforsch.* **17a**, 460.
Zähringer, J. (1964) *Am. Rev. Astron. and Astrophys.* **2**, 121.
Zähringer, J. (1966) *Meteoritika* **27**, 25.
Zähringer, J. and Gentner, W. (1960) *Z. Naturforsch.* **15a**, 600.

(Received 10 August 1969)

LITHIUM (3)

Walter Nichiporuk

Center for Meteorite Studies
Arizona State University
Tempe, Arizona

LITHIUM, although an element of low atomic weight, is among the least abundant on earth as well as in meteorites. This is undoubtedly related to its short half-life at the interior temperatures of stars like the sun, and arguments have been recently presented supporting the concept of lithium production during the formation of the solar system (e.g. Fowler, Greenstein, and Hoyle, 1962; Burnett, Fowler, and Hoyle, 1965; Gradsztajn, 1965). Lithium is usually determined in meteorites by emission spectrography, but the most recent determinations have been made by flame photometry, isotope dilution, neutron activation, and atomic absorption.

Suess and Urey (1956), Levin, Kozlovskaya, and Starkova (1956), and Urey (1964) have summarized the older determinations of lithium in meteorites. Suess and Urey give the amount as 5×10^{-6} g/g lithium in mean meteoritic matter, and Levin, Kozlovskaya, and Starkova give the amount as 3.2×10^{-6} g/g lithium, also in mean meteoritic matter. Urey reports two lithium values: one, 2.7×10^{-6} g/g, for the separated silicate phase of the ordinary olivine-bronzite and olivine-hypersthene chondrites; the other, 2.0×10^{-6} g/g, for the whole-rock ordinary chondrites.

Table 1 gives a list and Figure 1 shows a frequency distribution diagram of the recent determinations of the abundances of lithium in chondritic and achondritic meteorites. These determinations have been made by flame photometry (Shima and Honda, 1963 and 1967), isotope dilution (Krankowsky and Müller, 1964 and 1967; Dews, 1966; Balsiger, Geiss, Groegler, and Wyttenbach, 1968), neutron activation (Quijano-Rico and Wänke, 1969), and atomic absorption (Nichiporuk and Moore, 1970a and 1970b). In the compilation of the averages and in the plotting of the frequency

TABLE 1 Lithium in stony meteorites

Name and type*	Number of meteorites analyzed	Total number of determinations	Range 10^{-6} g/g	Ref.	Average Li 10^{-6} g/g	Atoms Li 10^6 Si**
Chondrites						
Carbonaceous C1	1	1	1.3	(1)		49.5
	1	1	1.3	(7)[a]		
Carbonaceous C2	3	3	1.4–1.6	(1)	1.5	45.5
	1	1	1.1	(6)[a]		
Carbonaceous C3	4	6	1.5–3.0	(1)	1.8	45.7
	4	4	1.0–1.6	(6)[a]		
Carbonaceous C4	1	1	0.75	(6)[a]		
Bronzite H	1	2	1.2–1.3	(2)	1.3	
	1	1	1.45	(3)		
	11	11	1.2–2.1	(1)	1.7	
	23	23	1.0–3.2	(6)[a]		
	1	3	1.8–1.8	(7)[a]		
	13	14	1.2–2.1	All	1.7	40.2
Hypersthene L	2	3	1.3–1.7	(2)	1.5	
	1	1	2.0	(3)		
	18	20	1.5–2.6	(1)	1.8	
	1	1	0.4	(8)[a]		
	9	9	0.6–1.6	(6)[a]		
	4	4	0.9–1.8	(7)[a]		
	19	24	1.3–2.6	All	1.8	39.0
Amphoterite LL	1	1	1.8	(3)		
	6	6	1.7–1.8	(1)	1.8	
	6	7	1.7–1.8	All	1.8	38.7
Enstatite I	1	1	2.3	(4)		
	2	3	1.3–2.9	(1)	1.95	
	1	1	1.1	(6)[a]		
	1	1	1.3	(7)[a]		
	2	4	1.3–2.9	All	2.1	47.3
Enstatite Intermediate	1	1	0.81	(1)		19.1
Enstatite II	2	3	0.44–0.64	(1)	0.58	12.5
	1	1	1.1	(6)[a]		
Calcium-poor achondrites						
Enstatite	1	1	0.33	(3)		
	1	1	<0.1	(5)		
	1	2	<0.1–0.33	All	~0.2	3.0

TABLE 1 (cont.)

Name and type*	Number of meteorites aualyzed	Total number of deter-minations	Range 10^{-6}g/g	Ref.	Average Li 10^{-6}g/g	Atoms Li 10^6 Si**
Hypersthene	1	1		(5)	2.6	
	1	1		(6)	2.6	
	1	5	0.44–1.1	(7)[a]		
	1	2	2.2–2.6	All	2.4	39.3
Calcium-rich achondrites						
Eucrites	3	3	5.07–7.27	(6)	6.1	106

* Classification of chondrites and carbonaceous chondrites according to Van Schmus and Wood (1967); classification of enstatite chondrites according to Anders (1964).

** Using average silicon values for chondrite groups supplied for this summary by B. Mason; individual silicon values from the literature used for the carbonaceous and enstatite chondrites.

(1) Nichiporuk and Moore (1970a)
(2) Shima and Honda (1963)
(3) Balsiger et al. (1968)
(4) Shima and Honda (1967)
(5) Nichiporuk and Moore (1970b)
(6) Quijano-Rico and Wänke (1969)
(7) Krankowsky and Müller (1964, 1967)
(8) Dews (1966)
(6)[a] Omitted from calculations of averages; values obtained only from material remaining after separation of "magnetic" fraction which was not analyzed for lithium.
(7)[a] and (8)[a] Omitted from calculations of averages; values obtained only from material remaining after leaching with acids and water.

distribution diagram, the separated non-magnetic silicate fractions of the chondrites and also the material analyzed after leaching with acids and water have not been taken into consideration because of the large divergence of these materials from the gross composition of the whole-rock samples, especially in the case of the acid- and water-leached material. Shima and Honda (1966, 1967) have stated that a major portion of lithium exists in the chondritic meteorites in an acid-soluble form.

Table 1 shows a rather uniform level of lithium in the chondrites at 40–50 atoms/10^6 Si, although in Type II enstatite chondrites the level is relatively low, at 12.5 atoms/10^6 Si. In the calcium-poor achondrites lithium

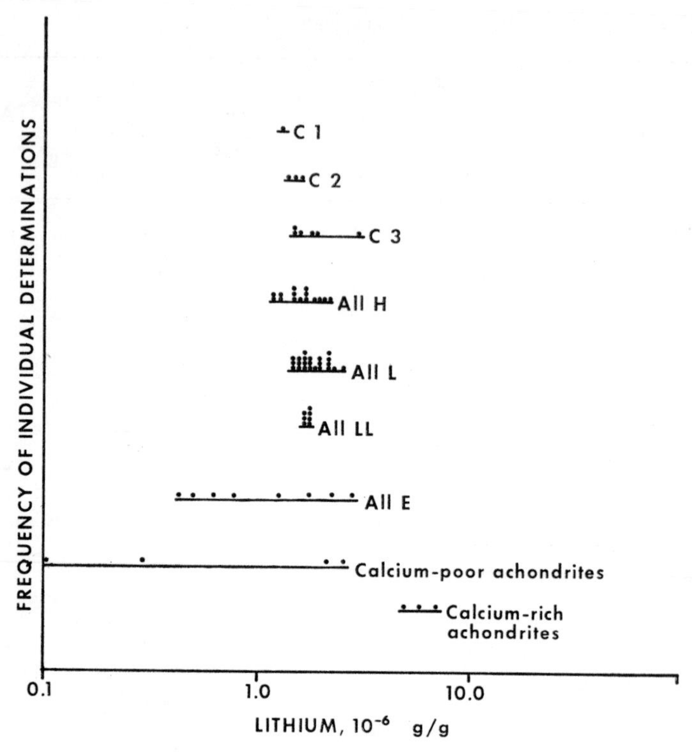

FIGURE 1 Frequency distribution of analyses for lithium in stony meteorites.

concentrations are comparable to those in the chondrites, except for the enstatite achondrites which show a marked depletion. The calcium-rich achondrites have the largest concentrations of lithium.

Lithium is not listed among the elements showing a normal depletion pattern of 1.0/0.6/0.3 in the carbonaceous chondrite groups C1/C2/C3 (Larimer and Anders, 1967). Using the data of Table 1, the atomic abundance depletion ratio of C2 to C1, and of C3 to C1 is 0.92 in each case. For the H, L, and LL chondrites as a group, the depletion relative to C1 is 0.80, but for Type II enstatite chondrites the depletion ratio is 0.25. Apparently lithium has not been fractionated among carbonaceous and ordinary chondrites but has been fractionated among enstatite chondrites.

Compared to the surface of the sun with 0.29 atoms/10^6 Si according to Aller (1961), lithium is on the average 160 times more abundant in the chondrites.

In most meteorites lithium is mainly lithophile and olivine appears to be its principal host (Shima and Honda, 1966 and 1967); however, in the enstatite chondrites chalcophile affinity of lithium cannot be ruled out (Shima and Honda, 1966 and 1967; Nichiporuk and Moore, 1970a). In the iron meteorites in which silicate and troilite inclusions are present, lithium is distinctly lithophile and concentrates in the silicate inclusions (Krankowsky and Müller, 1967). It has little or no siderophile affinity, and is seldom determined in the analyses of meteoritic iron; Fireman and Schwarzer (1957) report less than 0.01×10^{-6} g/g lithium in several iron meteorites.

Essentially equal concentrations of lithium have been observed in the light and dark silicate parts of the only gas-rich chondrite that has been investigated thus far, namely, the bronzite chondrite Pantar (Quijano-Rico and Wänke, 1969).

In addition to these studies of meteorite classes as well as of different mineralogical-chemical phases and the light-dark structures of meteorites, there are numerous determinations of lithium in the separated chondrule and matrix fractions of chondrites. The results of these determinations are summarized in Table 2.

TABLE 2 Lithium in chondrules and in matrix, in 10^{-6} g/g

Name	Type	Whole-rock meteorite	Matrix	Metal-free matrix	Chondrules	Reference
Chainpur	(LL)3	1.8			1.5	(3)
Olivenza	LL5	1.7		2.5	0.88	(4)
Bjurböle	L4				0.134	(1)
Bjurböle	L4	1.8[a]			1.1, 0.9	(2)
Bjurböle	L4		2.1, 3.4		2.3	(3)
Bjurböle	L4	2.3		3.1	1.5	(4)
Saratov	L4	1.7		1.5	1.2	(4)
Bruderheim	L6	0.4[b]			0.27, 0.089*	(1)
Tieschitz	H3	1.45			3.1	(3)
Richardton	H5	1.7		2.3	1.6	(4)
Essebi	C2	1.6[a]			1.6, 2.9	(2)

[a] Sample precleaned with 1N HCl and water
[b] Sample precleaned with HF and water
* One lath chondrule
(1) Dews (1966)
(2) Krankowsky and Müller (1967)
(3) Balsiger, Geiss, Groegler, and Wyttenbach (1968)
(4) Nichiporuk and Moore (1970a)

As found on the earth, lithium consists of two isotopes of mass 6 and 7 with a rather high value of 12.0 for the isotopic abundance ratio Li^7/Li^6. Many measurements have been made to evaluate this ratio in meteorites. The results published between 1963 and 1966 have been reviewed by Reynolds (1967). Earlier measurements were made by Ordzonikidze (1960) and the measurements made since the review include the work of Krankowsky and Müller (1967), Gradsztajn, Salome, Yaniv, and Bernas (1967), Balsiger, Geiss, Groegler, and Wyttenbach (1968), and Bernas, Gradsztajn, and Yaniv (1969). At the present time it appears uncertain whether or not there are significant variations in the Li^7/Li^6 ratio in meteorites as compared with the relatively constant values of this ratio in terrestrial rocks.

References

Aller, L.H. (1961) *The abundance of the elements.* Interscience.

Anders, E. (1964) *Space Sci. Revs.* **3**, 583.

Balsiger, H., Geiss, J., Groegler, N., and Wyttenbach, A. (1968) *Earth Planet. Sci. Letters* **5**, 17.

Bernas, R., Gradsztajn, E., and Yaniv, A. (1969) in: *Meteorite research*, ed. P. Millman, D. Reidel, Dordrecht, p. 123.

Burnett, D.A., Fowler, W.A., and Hoyle, R. (1965) *Geochim. Cosmochim. Acta* **29**, 1207.

Dews, J.R. (1966) *J. Geophys. Res.* **71**, 4011.

Fireman, E.L. and Schwarzer, D. (1957) *Geochim. Cosmochim. Acta* **11**, 252.

Fowler, W.A., Greenstein, J.L., and Hoyle, F. (1962) *Geophys. J.* **6**, 148.

Gradsztajn, E., (1965) *Ann. Phys.* (Paris) **10**, 791.

Gradsztajn, E., Salome, M., Yaniv, A. and Bernas, R. (1967) *Earth Planet. Sci. Letters* **3**, 387.

Krankowsky, D. and Müller, O. (1964) *Geochim. Cosmochim. Acta* **28**, 1625.

Krankowsky, D. and Müller, O. (1967) *Geochim. Cosmochim. Acta* **31**, 1833.

Larimer, J.W. and Anders, E. (1967) *Geochim. Cosmochim. Acta* **31**, 1239.

Levin, B.Yu., Kozlovskaya, S.V. and Starkova, A.G. (1956) *Meteoritika* **14**, 38.

Nichiporuk, W. and Moore, C.B. (1970a) *Earth Planet. Sci. Letters* **9**, 280.

Nichiporuk, W. and Moore, C.B. (1970b) Unpublished data.

Ordzonikidze, K.G. (1960) *Geokhimiya* **1**, 37.

Quijano-Rico, M. and Wänke, H. (1969) in: *Meteorite research*, ed. P. Millman, D. Reidel, Dordrecht, p. 132.

Reynolds, J.H. (1967) *Ann. Rev. Nucl. Sci.* **17**, 253.

Shima, M. and Honda, M. (1963) *J. Geophys. Res.* **68**, 2849.

Shima, M. and Honda, M. (1966) *Geochem. J.* **1**, 27.

Shima, M. and Honda, M. (1967) *Geochim. Cosmochim. Acta* **31**, 1995.

Suess, H.E. and Urey, H.C. (1956) *Revs. Mod. Phys.* **28**, 53.

Urey, H.C. (1964) *Revs. Geophys.* **2**, 1.

Van Schmus, W.R. and Wood, J.A. (1967) *Geochim. Cosmochim. Acta* **31**, 747.

(Received 4 May 1970)

BERYLLIUM (4)*

Peter R. Buseck

Departments of Geology and Chemistry
Arizona State University
Tempe, Arizona 85 281

LITTLE WORK has been done on the abundance and distribution of beryllium in meteorites. The limited available data suggests that the chondrites (6 falls and 7 finds) have a range, median, and mean beryllium content of 20 to 58 ppb, 38 ppb, and 39 ppb, respectively. The only analyzed iron meteorite (fall) has an abundance of less than 1 ppb. Beryllium appears to be strongly lithophilic.

The study of Sill and Willis (1962) is the most comprehensive investigation into the beryllium abundance in meteorites. Earlier analyses have been reported by I. Noddack and W. Noddack (1930, 1934) and Goldschmidt and Peters (1932), but they are in marked disagreement with the more recent analyses.

The analytical technique used by Sill and Willis (1962) involved preliminary crushing, mixing with Be^7 tracer, and dissolution of the sample with hydrofluoric and nitric acid. The beryllium was then extracted and determined fluorometrically with morin. Following this the sample was counted in a thallium-activated sodium iodide well counter to determine the distribution and recovery of the beryllium. The Be^7 tracer permitted evaluation of the efficiency of sample treatment. At the 95% confidence level, they accounted for 99.7% of the beryllium. They report that from the counting data they could adequately correct for the occasional 2% loss in material balance. This was done and the resulting numbers are then upper limits for the beryllium contents; they are probably very close to the true abundances. Duplicate analyses were run on all the samples, with the analyses run as much as seven months apart. For two meteorites (Norton County and

* Contribution number 48 from the Arizona State University Center for Meteorite Studies.

Potter) three replicate analyses show a range of 7 ppb. For most duplicates the range is 1 to 3 ppb.

The beryllium contents of the stony meteorites are shown in Figure 1 and Table 1. The results of the early analyses of Goldschmidt and Peters (1932) and the Noddacks (1934) indicated abundances between 0.7 and 2 ppm

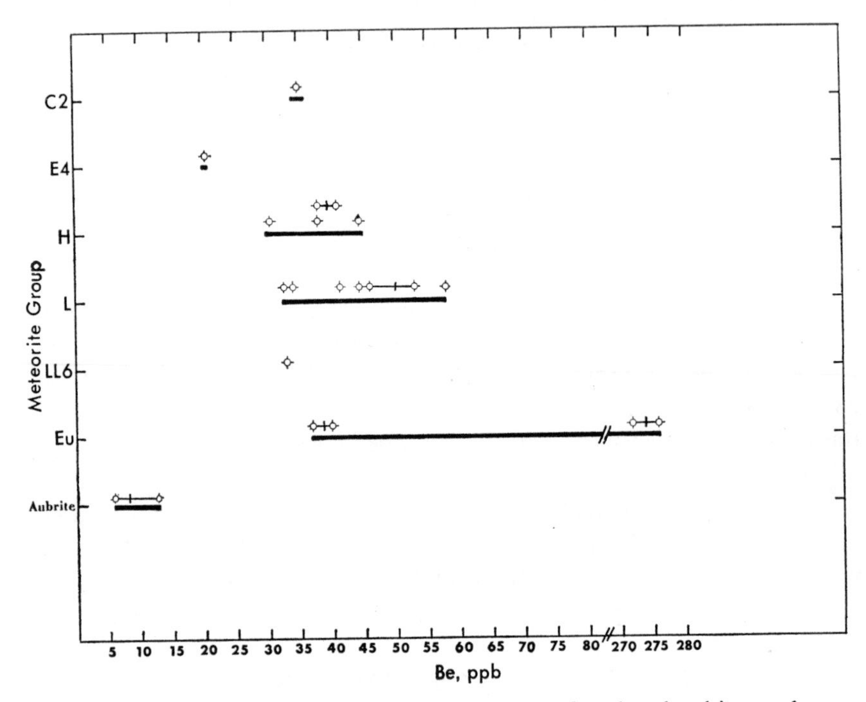

FIGURE 1 Distribution of beryllium analyses for the chondrites and achondrites. For meteorites having replicate analyses, the range is represented by a solid line connecting the extreme values, and the mean is indicated by a vertical slash.

beryllium, considerably in excess of the data in Table 1. The data of Sill and Willis (1962) are preferred and therefore they are the only ones shown. The data appear relatively homogeneous, but they are too sparse to reveal any marked abundance trends between meteorite types.

The achondritic abundances are also shown in Table 1. For the achondrites (all 3 falls) the range, median, and mean are 6 to 276 ppb, 39 ppb, and 100 ppb, respectively. In this instance there is a pronounced heterogeneity in their composition. The Pasamonte eucrite is in the same range as the chondrites, but the Sioux County eucrite is almost an order of magnitude

TABLE 1 Beryllium content of meteorites

Type	Meteorite[a]	Concentration, ppb[b]	Mean, ppb	Mean Atoms/10^6 Silicon Atoms
Chondrites				
C2	Murray	36, 34	35	0.81
E4	Abee	20, 21	20	0.36
H_	Achilles[c]	39, 37	38	0.69
	Ehole[c]	30, 31	30	0.56
H5	Plainview	41, 38	40	0.72
H6	Morland	44, 45	44	0.81
All H	—	—	38	0.69
L_	Calliham[c]	58, 58	58	0.96
L6	Bruderheim	35, 34, 33	34	0.56
	Harleton	32, 33	32	0.54
	Ladder Creek	41, 42	42	0.69
	Ness County	45, 44	44	0.74
	Potter	53, 46, 50	50	0.83
All L	—	—	43	0.72
LL6	Cherokee Springs	33	33	—
Achondrites				
Eu	Pasamonte[d]	40, 39, 37	39	0.63
	Sioux County[d]	276, 272	274	3.69
Aubrite	Norton County[d]	6, 13, 6	8	0.1
Iron[e]				
Og	Aroos[e]	<1, <1	<1	—

[a] Classification according to Van Schmus and Wood (1967) unless otherwise noted.
[b] Data from Sill and Willis (1962).
[c] Classification according to Hey (1966).
[d] Classification according to Mason (1962).
[e] Lovering (1957) found <0.8 ppm beryllium in olivine of the pallasites Admire, Albin, Brenham, and Springwater, and in the troilite phases of Brenham, Springwater, Rio Loa (H), Henbury (Om), and Toluca (Om).

higher and the Norton County aubrite is considerably lower than the mean for the chondrites.

Aroos is the only iron that has been measured for beryllium, and it contains less than 1 ppb, below the detection limit.

No mineral separates have been analyzed for beryllium. Nonetheless, the iron and stony meteorite data indicate that beryllium is lithophilic. This is strongly supported by the terrestrial occurrences of beryllium, where it always is most abundant in the oxide or silicate phases, especially in the late residual fractions of crystallized silicate magmas.

The predicted cosmic abundance of beryllium is remarkably low, presumably because it is rapidly consumed by the thermonuclear processes taking place within stellar interiors. Cameron (1968) gives a value of 0.69 atoms on a 10^6 Si atoms scale. Suess and Urey (1956) gave a value of 20, based on the older chondrite analyses. This is surely too high, as it was made previous to the recent beryllium analyses. The previous estimates may be compared to a value of 6.9 for the solar photosphere (Müller, 1968).

Acknowledgements

This report was supported in part by Grant BA-1200 from the National Science Foundation. Helpful assistance and comments were provided by Mr. Jerrald Durtsche and Dr. J. W. Larimer.

References

Cameron, A. G. W. (1968) in *Origin and Distribution of the Elements* (L. H. Ahrens, ed.). Pergamon.

Goldschmidt, V. M. and Peters, C. (1932) *Nachr. Ges. Wiss. Göttingen* **25**, 360.

Hey, M. H. (1966) *Catalogue of Meteorites* (3rd. edition). British Museum.

Lovering, J. F. (1957) *Geochim. Cosmochim. Acta* **12**, 253.

Mason, B. (1962) *Meteorites*, Wiley and Sons.

Müller, E. A. (1968) in *Origin and Distribution of the Elements* (L. H. Ahrens, ed.). Pergamon.

Noddack, I. and Noddack, W. (1930) *Naturwiss.* **18**, 757.

Noddack, I. and Noddack, W. (1934) *Svensk Kem. Tidskr.* **46**, 173.

Sill, C. W. and Willis, C. P. (1962) *Geochim. Cosmochim. Acta* **26**, 1209.

Suess, H. E. and Urey, H. C. (1956) *Rev. Mod. Phys.* **28**, 53.

Van Schmus, W. R. and Wood, J. A. (1967) *Geochim. Cosmochim. Acta* **31**, 747.

(Received 7 April 1970)

BORON (5)

P. A. Baedecker

Institute of Geophysics and Planetary Physics
University of California, Los Angeles

B, Li AND Be ARE A TRIAD of light elements which are destroyed in the hydrogen-burning zone in stars. This fact lends particular interest to their study in primitive meteorite types. In addition, many B compounds are highly volatile or mobile, and it is of interest to study its cosmochemical behavior both with respect to condensation processes in the nebula and with regard to its distribution among meteorite phases.

There is relatively little data on the concentration of B in meteorites, primarily due to the analytical difficulties involved in obtaining accurate data at the parts-per-million level. Matrix effects and difficulties with line resolution have hampered spectrographic work, and the element does not lend itself to thermal neutron-activation analysis due to the fact that the only radioactive species produced has a half-life of less than 0.1 second. Laboratory contamination is a serious problem, since most glassware has a high boron content. In recent years some success has been achieved with sensitive colorimetric and fluorimetric techniques. The most extensive investigation into the distribution of B in meteorites which has been undertaken to date is the recent work of Quijano-Rico and Wänke (1969), who determined B in a large number of chondritic and achondritic meteorites.

Goldschmidt (1954) has reviewed the early spectrographic work on the abundance of B in meteorites carried out by the Noddacks, and Goldschmidt and Peters. He reports finding 1–2 ppm of B in the Long Island and Orgueil chondrites. He failed to observe B in either the olivine or the metal phase of the Brenham pallasite. Up to 5 ppm B was observed in some irons, while others showed no detectable amount.

The most recent analytical data for B in meteorites are summarized in Table 1. The data of Harder (1961) were obtained by spectrographic

TABLE 1 B in meteorites

	No. of determinations	No. of meteorites analyzed	Concentration Range	(10^{-6} g/g) Mean	Abundance (atoms/10^6 Si atoms)	Reference
C1	1	1		~5.0	126	Harder
	4	2	5.1–7.1	5.7	144	Mills
C2	1	1		9.4	186	Quijano-Rico
C3 and C4	3	3	5.6–9.6	7.2	121	Quijano-Rico
H-group	25	25	0.14–6.0	1.0	15	Quijano-Rico
	2	2		5.0	76	Harder
	3	3	0.38–0.50	0.43	6.5	Shima
L-group	4	4	1.5–5.0	3.5	49	Harder
	10	10	0.14–1.70	0.74	10	Quijano-Rico
LL-group	1	1		1.4	19	Quijano-Rico
E4	1	1		0.77	12	Quijano-Rico
E6	1	1		0.43	5.7	Quijano-Rico
Achondrites						
Aubrites	1	1		0.42	3.9	Shima
	1	1		1.4	13	Harder
Eucrites	1	1		5.6	63	Harder
	1	1		1.3	15	Shima
	3	3	0.63–1.1	0.83	9.4	Quijano-Rico
Diogenites	1	1		0.34	3.6	Quijano-Rico
Separated phases						
H-group dark	5	5	0.19–1.57	0.54		Quijano-Rico
H-group light	5	5	0.12–0.83	0.45		Quijano-Rico
Troilite (siderite)				0.5		Harder
Olivine (pallasite)	1	1		5.0		Harder

methods. Shima (1963) and Mills (1968) both employed colorimetric methods, while Quijano-Rico and Wänke (1969) used a fluorimetric technique. The carbonaceous chondrites appear to have higher B contents than the ordinary and enstatite chondrites. Quijano-Rico and Wänke (1969) also found a correlation between the B content and petrologic grade within the H and L groups of ordinary chondrites (the most recrystallized have the lowest B concentration). This is a fractionation pattern which has been observed for a number of relatively volatile elements (e.g., In, C). The data of Quijano-Rico and Wänke for the light-dark structured H-group chondrites suggest that the dark phase may be slightly enriched in B relative to the light phase, but if so, the difference is not as great as has been observed for other trace elements (e.g., In—see Rieder and Wänke, 1969).

Wasson (1965) employed a colorimetric technique in attempting to measure the B content of iron meteorites. No B was found in five iron meteorites, which included members of the four Ga-Ge classes, and an upper limit of 0.04 ppm was set on the B concentration of the meteorites analyzed. Harder (1961) was also unable to detect B in five iron meteorites, and the metal phases of a pallasite, a mesosiderite, and one chondrite. Shima (1963) reported a concentration of 0.45 ppm for the Toluca iron, but Quijano-Rico and Wänke (1969) observed a B concentration of 0.03 for the same meteorite. They also list a value of 0.02 ppm for the Canyon Diablo iron.

From the data available, it appears that B tends to concentrate mainly in the silicate phases in meteorites. The single analysis of Harder (1961) for a troilite sample extracted from the Canyon Diablo iron (0.5 ppm) suggests that B has a greater tendency to concentrate in the sulphide phase than in the metallic phase, but clearly more data are required before definite statements can be made regarding the affinity of B for specific phases.

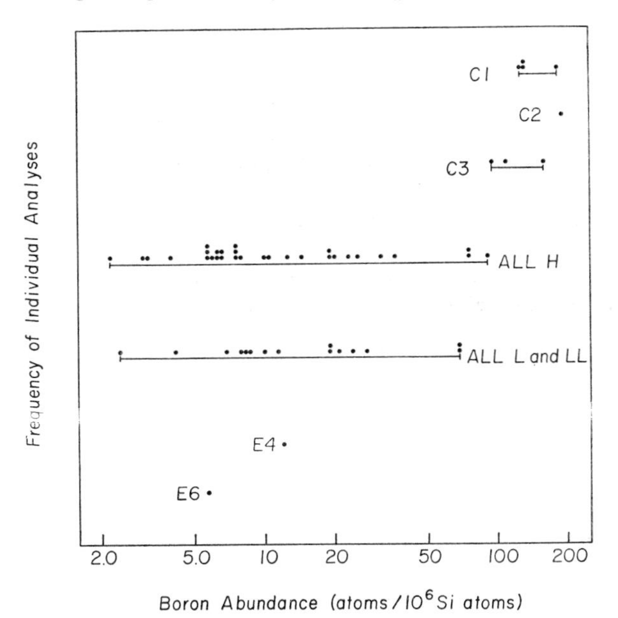

FIGURE 1 Frequency of individual analyses versus the B abundance (atoms/10^6 Si atoms) plotted on a log scale, for the various chondrite groups.

Shima (1963) has determined the isotopic composition of meteoritic B. In three chondrites, one achondrite, and one iron, she found the B^{11}/B^{10} ratio to range from 3.82 ± 0.05 to 3.96 ± 0.02. All samples of terrestrial B which were analyzed had a B^{11}/B^{10} ratio which was greater than 4.0. The ratio varied from 4.039 ± 0.012 for the G1 standard granite to 4.248 ± 0.06 for a sample of borax from the California deposit. The latter is a special case however, and the experimental value of 4.05 ± 0.05 is recommended by Shima as the terrestrial isotopic ratio. The reported meteoritic B^{11}/B^{10} ratio is therefore about 3 to 6% less than the terrestrial ratio, but must be treated with some skepticism in the light of the high reported B concentration in Toluca, and the fact that Li isotopic ratios reported by the same author during the same period have not proven reproducible in other laboratories.

Grevesse (1968) reports a solar B abundance of <14.5 relative to $Si = 10^6$ atoms (based on the solar Si abundance listed by Grevesse et al., 1968), which is considerably lower than the abundance of about 150 based on the carbonaceous chondrite data.

References

Goldschmidt, V.M. (1954) *Geochemistry*, Oxford University Press, Oxford, England.

Grevesse, N. (1968) *Solar Phys.* **5**, 159–180.

Grevesse, N., Banquet, G. and Bourgy, A. (1968) *Origin and Distribution of the Elements*, edited by L.H.Ahrens, Pergamon Press, Oxford, England, 177–182.

Harder, H. (1961) *Nachr. Akad. Wiss. Math.-Phys. Göttingen* **1**, 1.

Mills, A.A. (1968) *Nature* **220**, 1113–1114.

Quijano-Rico, M. and Wänke, H. (1969) *Meteorite Research*, edited by P.M.Millman, IAEA, Vienna, 132–145.

Shima, M. (1963) *Geochim. Cosmochim. Acta* **27**, 911–913.

Wasson, J.T. (1965) *J. Geophys. Res.* **70**, 4443–4445.

(Received 25 July 1969)

CARBON (6)

G. P. Vdovykin

V. I. Vernadsky Institute of Geochemistry and Analytical Chemistry; Committee on Meteorites USSR Academy of Sciences, Moscow

and

Carleton B. Moore

Center for Meteorite Studies Arizona State University Tempe, Arizona

THE COSMOCHEMISTRY of carbon is especially interesting because this element is found in a variety of forms in meteorites. These phases are important clues to the conditions of formation of the meteorites.

The content of total carbon in meteorites varies. Its content in carbonaceous chondrites is up to 4.8 weight per cent (Mason, 1963). The ureilites, achondrites especially enriched in carbon, contain up to 4.1 weight per cent (McCall and Cleverly, 1968). In iron meteorites carbon has been reported to vary over a large range, depending upon whether the samples taken for analysis contain the iron-nickel carbide cohenite or graphite inclusions. Buddhue (1946) in his compilation of the average compositions of iron meteorites reported a range from 0.01 average weight percent in the finest octahedrites to 0.60 average weight per cent in the fine octahedrites. Perry (1944) gave an average abundance in all iron meteorites of 0.11 weight per cent. More recent work by Moore et al. (1969) found lower carbon abundances in the iron meteorites, with averages for the different structural groups ranging from 0.007 to 0.03 weight per cent. Table 1 gives the carbon distribution in the various structural groups of iron meteorites. Moore and Lewis (1965, 1966, 1967), using total combustion and gas chromatography, have determined

TABLE 1 Carbon (wt.% as C) in iron meteorites

Name	No. of meteorites	Range	Median	Reference
H Hexahedrites	8	0.005–0.013	0.009	(1)
Octahedrites				
Ogg Coarsest	3	0.009–0.022	0.013	(1)
Og Coarse	10	0.004–0.18	0.012	(1)
Om Medium	42	0.002–0.06	0.014	(1)
Of Fine	17	0.005–0.046	0.011	(1)
Off Finest	6	0.010–0.042	0.030	(1)
Dr Ataxites	6	0.003–0.051	0.007	(1)

the contents of total carbon in many chondrites and shown that the average carbon content in enstatite chondrites is about 0.40 weight per cent, in olivine-bronzite chondrites approximately 0.10 weight per cent, and in olivine-hypersthene chondrites 0.10 weight per cent. These data and those of earlier investigations on chondrites are summarized in Table 2. Also included in Table 2 are carbon abundance data on the achondrites and stony-iron meteorites.

In Table 2 the atomic abundances have been calculated from median values of each meteorite type rather than from mean values. The carbon content distributions for each of the chondrite types has a distribution skewed towards the higher abundances. This is due to the fact that the unequilibrated ordinary chondrites in the Van Schmus and Wood (1967) group have higher carbon abundances than the equilibrated ordinary chondrites in groups 4, 5, and 6. If the unequilibrated ordinary chondrites were omitted from the calculation the median and means would be essentially identical. Carbon contents in the unequilibrated chondrites are given in Table 3. The overall distribution in different stony meteorite groups is illustrated in the accompanying figure.

In stony meteorites carbon is more or less evenly distributed between chondrules and mineral grains. In iron meteorites it is found in the form of separate inclusions or nodules measuring up to several centimeters in size.

Carbon occurs in meteorites in several different forms or phases. Of special interest are organic compounds and diamonds. The characteristics of the organic matter in meteorites shed light on the origin of this material at an early state of solar system development, and the peculiarities of the diamonds allow us to gain an idea of possible recrystallization during colli-

TABLE 2 Carbon (wt.% as C) in stony and stony-iron meteorites

Name	No. of detns.	Range	Median	Atoms/10^6 Si	References
Chondrites					
C–1	5	2.70–4.83	3.19	720,000	(3)
	4*	3.1–4.8	3.1	700,000	(4)
C–2	9	1.30–4.00	2.48	440,000	(3)
	5*	1.30–2.78	2.06	370,000	(4)
C–3	9	0.35–2.49	0.56	85,000	(3)
	4*	0.45–1.44	0.50	75,000	(4)
C–4	2	0.07–0.20	0.14	20,000	(4)
H (Bronzite)	33†	0.03–0.60	0.10	13,700	(4)
	35	0.02–0.35	0.11	15,000	(5), (7)
L (Hypersthene)	47	0.02–0.45	0.10	12 500	(4)
	43	0.03–0.53	0.09	11,000	(5), (7)
LL (Amphoterites)	25	0.02–0.44	0.12	15,000	(4)
	16	0.03–0.44	0.12	15,000	(5), (7)
E–4 (Enstatite, Type I)§	2	0.36–0.42	0.39	52,000	(4)
	4	0.36–0.56	0.39	53,000	(6)
E–5 (Enstatite, Intermediate)	1	–	0.36	47,000	(4)
	2	0.37–0.54	0.45	58,000	(6)
E–6 (Enstatite, Type II)	3	0.16–0.43	0.25	30,000	(4)
	7	0.06–0.43	0.36	43,000	(6)
Calcium-poor achondrites					
Enstatite	2	0.04–0.10	0.07	59,000	(11)
Shallowater	1	–	0.13	–	(12)
Hypersthene	1	–	0.04	3,800	(13)
Ureilites	5	1.94–4.10	3.00	370,000	(10)
Calcium-rich achondrites					
Howardites	3	0.02–0.25	0.11	11,000	(14)
Eucrites	4	0.04–0.47	0.06	6,100	(15)
Stony-irons					
Mesosiderites	2	0.06–0.10	0.08	16,000	(2)
Pallasites	10	–	0.08	–	(8)
	3	0.05–0.08	0.06	–	(9)

* Includes analyses from Ref. (3)
† Includes analyses from Ref. (5), (7)
§ Includes analyses from Ref. (6)

TABLE 3 Carbon in unequilibrated ordinary chondrites

Name	Class	Wt.% C	Reference
H Group			
Bremervorde	HU	0.22	(7)
Castalia	H5	0.27	(7)
		0.18	(7)
Clovis I	H3	0.22	(7)
Dimmitt	H (3,4)	0.13	(4)
		0.097	(7)
Grady (1937)	H3	0.10	(4)
		0.10	(7)
		0.07	(7)
Prairie Dog Creek	H3	0.35	(7)
Tieschitz	H3	0.25	(7)
L Group			
Barratta	L4	0.085	(7)
Bishunpur	L3	0.32	(4)
		0.53	(7)
Carraweena	L3	0.098	(7)
Hallingeberg	L3	0.26	(7)
Hedjaz	L3	0.27	(7)
Ioka	L3	0.12	(7)
Khohar	L3	0.32	(7)
Krymka	L3	0.27	(7)
Mezö-Madaras	L3	0.45	(7)
LL Group			
Chainpur	LL3	0.31	(4)
		0.44	(7)
Hamlet	LL (3,4)	0.12	(4)
		0.06	(4)
		0.16	(7)
Ngawi	LL3	0.39	(7)
Parnallee	LL3	0.19	(7)
		0.09	(4)
Semarkona	LL3	0.57	(7)

sions of large cosmic bodies accompanied by material transformation at superhigh pressures (Vdovykin, 1967, 1970).

Organic matter, in which carbonaceous chondrites are especially enriched (up to 7 per cent), represents a complicated mixture of compounds. Numerous reviews have been published in the last few years of investigations undertaken with the aid of various analytical methods (Vdovykin, 1967; Hayes, 1967; Nooner and Oro, 1967; Nagy, 1968; Studier et al., 1968, etc.). Compounds reported include C_{15}–C_{30} paraffins (including isoprenoids],

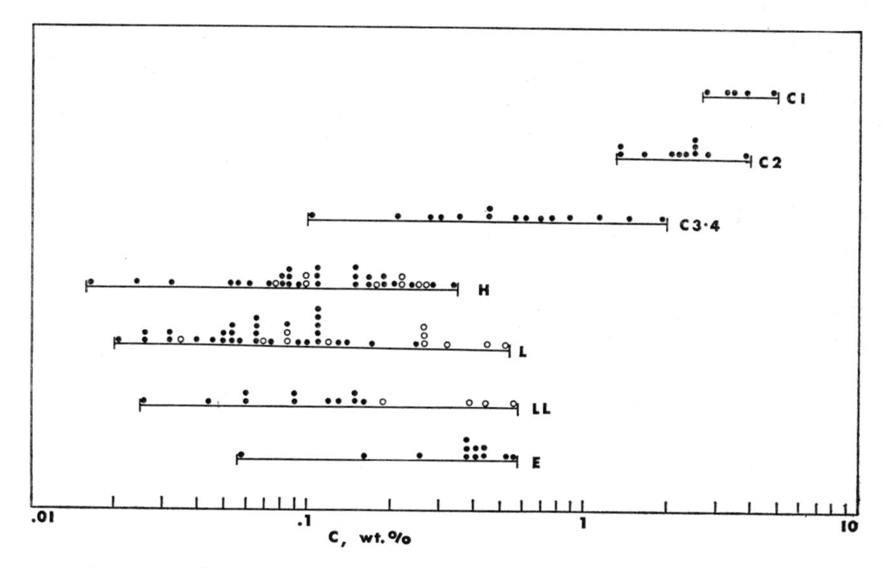

FIGURE 1 Frequency distribution of carbon in chondrites. Open circles indicate unequilibrated ordinary chondrites.

olefins, aromatic hydrocarbons, organic acids, amino acids, nitrogenous cyclic compounds, porphyrins, etc. The major part of the organic matter in meteorites is represented by high molecular weight compounds having an aromatic condensed structure (Vdovykin, 1964). The polymeric matter in the meteorite Orgueil has the following composition (in weight per cent) carbon 70.39, hydrogen 4.43, chlorine 1.22, fluorine 1.25, nitrogen 1.59, sulfur 6.91, oxygen 9.80, ash 4.58 (Hayes, 1967). The polymeric matter of carbonaceous chondrites contains free organic radicals formed during cosmic-ray irradiation; the unpaired π-electrons of these radicals are delocalized in a complicated aromatic structure (Vinogradov and Vdovykin, 1964; Duchesne et al., 1964; Vdovykin, 1967).

In some unequilibrated ordinary chondrites small amounts of organic compounds are present (Vdovykin, 1964). Traces of paraffin hydrocarbons are reported to have been detected also in the ureilites Novo Urei, Dyalpur, Goalpara (Vdovykin, 1967) and in graphite nodules from the iron meteorites

Burgavli, Yardymly (Vdovykin, 1964; 1967), Canyon Diablo, Odessa (Nooner and Oro, 1967).

Boato (1954) has shown that the total carbon of carbonaceous chondrites is heavier in the isotopic composition than the carbon of common chondrites. For various carbonaceous chondrites Boato noted that the increase in ^{13}C is parallel to the increase of the total carbon content in the meteorites. However, it has recently been shown that in carbonaceous chondrites this increase in ^{13}C is due to carbonate carbon, which is substantially heavier ($\delta C^{13} = +6$ per cent) in comparison to carbon of terrestrial carbonates (Clayton, 1963) and of organic matter in carbonaceous chondrites (Nagy, 1966; Vinogradov et al., 1967; Krouse and Modzeleski, 1969; Flory and Oro. 1969). Recent work by Belsky and Kaplan (1970) and Smith and Kaplan (1970) has

TABLE 4 Isotopic abundances of carbon-13 (as $\delta^{13}C$) for total meteorites and meteorite fractions relative to the Pee Dee belemnite standard (After Belsky and Kaplan, 1970 and Smith and Kaplan, 1970)

Meteorite	Type	Total (per mil)	Carbonate (per mil)	Soluble organic (per mil)	Insoluble organic (per mil)
Ivuna	C1	−7.5	+65.8	−24.1	−17.1
Orgueil	C1	−11.6	+70.2	−18.0	−16.9
Murray	C2	−5.6	+42.3	−5.3	−14.8
Cold Bokkeveld	C2	−7.2	+50.7	−17.8	−16.4
Erakot	C2	−7.6	+44.4	−19.1	−15.1
Mighei	C2	−10.3	+41.6	−17.8	−16.8
Mokoia	C3	−18.3	none	−27.2	−15.8
Allende	C3	−17.3	—	—	—
Karoonda	C4	−25.2	—	—	—
Abee	E4	−8.1	—	—	−6.1
Hvittis	E6	−4.2	—	—	—
Norton County	Ach	−23.9	—		—
Forest City	H5	−30.2	—	—	—
Richardton	H5	−25.8	—	—	—
Bjurbole	L4	−26.3	—	—	−24.4
Bruderheim	L6	−28.4	—	—	—

confirmed the results of the earlier investigations, and been interpreted to indicate that carbonate, residual carbon, and part of the extractable organic material are endogenous to the carbonaceous chondrites. Their ^{13}C values for a series of chondrites is given in Table 4.

Peculiarities in the distribution and composition of meteoritic organic material, such as free radicals, the absence of optically active compounds, etc., bear witness to their extraterrestrial origin. This is confirmed by experimental work on the synthesis of complicated organic molecules from simple starting compounds with the use of various energy sources (Oro, 1963; Studier et al., 1968, etc.) including irradiation with protons (Vinogradov and Vdovykin, 1969).

Carbonates are characteristic of carbonaceous chondrites only (Mason, 1963; Vdovykin, 1967), and C1 chondrites are most highly enriched in carbonates (up to 0.3 per cent). Carbonates are represented by breunnerite, dolomite, and calcite (Kvasha, 1948; Mason, 1962, 1963).

Gaseous compounds have been found in a number of carbonaceous chondrites during mass-spectrometric investigations of organic material without previous separations from the meteorites (Studier et al., 1965). As a result of heating samples of the meteorites Orgueil (C1), Murray, Cold Bokkeveld (C2) at 140 °C, CH_4, CO, CO_2, COS, etc. have been released and identified.

The presence of organic material, carbonates, and volatile compounds in carbonaceous chondrites shows that they were at least partially formed under relatively low-temperature conditions, in an environment enriched in volatile (and in particular oxygen-containing) components. Free radicals stabilized at low temperatures probably were of great importance.

Elementary carbon in meteorites is also found in different phases.

With the aid of structural methods Vdovykin (1969b) has identified in the ureilite Novo Urei the hexagonal carbon which recently was reported by El Goresy and Donnay (1968) in graphite-containing rocks of the Ries crater. This carbon modification, which has a poorly expressed hexagonal structure, has apparently formed as a result of an intensive impact.

The most abundant carbon form in meteorites is graphite. It is characteristic of common chondrites and iron meteorites. According to Vdovykin's (1967) X-ray data on the enstatite chondrite Pillistfer carbon is present only as graphite. In rare cases graphite particles are present in carbonaceous chondrites. In ureilites graphite is present in diamond-graphite intergrowths. Graphite has possibly formed in some meteorites as a result of carbonization of carbonaceous matter.

In some meteorites carbon is represented by fine inclusions of cliftonite. Cliftonite has been reported in 14 iron meteorites, the mesosiderite Vaca Muerta and the enstatite chondrite Indarch (Brett and Higgins, 1969). Cliftonite is similar in structure to graphite (Hey, 1938). Vdovykin (1967) investigated by X-ray methods cliftonite from the meteorite Canyon Diablo and found it to be, on the basis of its structure and micro-hardness, similar to the graphite of the iron meteorites Burgavli and Yardymly. Experimental investigations carried out by Brett and Higgins have shown that cliftonite is formed by the decomposition of cohenite $(Fe, Ni)_3 C$ under low pressures. This agrees with theories of the formation of meteoritic matter within relatively small cosmic bodies (Anders, 1964; Vinogradov, 1965; Urey, 1966).

Diamond is present in ureilites and in the iron meteorite Canyon Diablo. The study of diamonds in five ureilites by Vdovykin (1970) has shown that diamond is present in fine intergrowths with graphite (approximately in equal volume ratio), the sizes of such intergrowths being from 0.3 to 0.9 mm. The size of individual micro-crystals of diamond is less than 1 μ. The structure of meteoritic diamonds is similar to terrestrial diamond. An investigation using EPR and IR-spectroscopy of a pure diamond fraction isolated from the meteorite Novo Urei has shown the presence of donor nitrogen in meteoritic diamonds. The characteristics of the diamonds and of the ureilites containing them indicate that the diamonds in ureilites have formed during collisions of asteroidal bodies (Vinogradov and Vdovykin, 1963; Lipschutz 1964; Vdovykin, 1967, 1970).

In the Canyon Diablo iron, diamonds are found only in samples found on the rim of the Barringer Meteorite Crater (Nininger, 1956). Samples of the meteorite which were found on the plains adjacent to the crater lacked diamonds. Lipschutz and Anders (1961) and Heymann et al. (1966) have convincingly shown that diamonds in the meteorite were formed during impact with the Earth. Vdovykin (1969) confirmed this conclusion after studying a diamond inclusion in the Canyon Diablo iron. It consisted of diamond micromonocrystals with a lesser graphite admixture than in ureilites. The diamonds and graphite in the meteorites have a similar carbon isotopic composition (Vinogradov et al., 1967; Vdovykin, 1969a).

An especially important characteristic of meteoritic diamonds is the presence of the hexagonal phase, lonsdaleite, which is not known in terrestrial rocks but has been experimentally synthetized from graphite (Bundy and Kasper, 1967). Lonsdaleite has been identified in the ureilites Novo Urei, North Haig (Vdovykin, 1967, 1970), Goalpara, and the iron meteorite Canyon Diablo (Hanneman et al., 1967; Frondel and Marvin, 1967). It is present in the composite of diamond-graphite intergrowths and has a wurtzite-like structure. The presence of lonsdaleite in meteorites also confirms that meteoritic diamonds formed during impact.

In almost all meteorites containing cliftonite the iron-nickel carbide cohenite has been noted. In iron meteorites it is sometimes confined to nodules in association with troilite and schreibersite. The conditions of cohenite formation have been considered in detail by a number of authors. Like other mineral constituents of iron meteorites, cohenite indicates that the meteoritic matter formed at relatively low pressures (Anders, 1964; Yavnel, 1965; Brett, 1966).

During impact, meteoritic cohenite is altered. This may be seen in X-ray pictures in the form of asterism. According to Lipschutz's data (1967), the alterations in the cohenite of Canyon Diablo indicate that on impact with the Earth various samples of this meteorite were exposed to a pressure of 200–1000 kbar. Vdovykin (1969) has established that a diamond-bearing sample of Canyon Diablo had undergone a shock pressure up to 1000 kbar. This was in agreement with pressure estimations for this sample obtained according to other recrystallization signs ($>$870 kbar).

The distribution of carbon is very susceptible to the conditions of the environment. The forms of carbon are indicators of this environment, and, hence, indicators of the conditions under which meteoritic matter was formed and transformed.

References

Anders, E. (1964) *Space Sci. Rev.* **3**, 583.

Belsky, T. and Kaplan, I.R. (1970) *Geochim. Cosmochim. Acta* **34**, 257.

Boato, G. (1954) *Geochim. Cosmochim. Acta* **6**, 209.

Brett, R. (1966) *Science* **153**, 60.

Brett, R. and Higgins, G.T. (1969) *Geochim. Cosmochim. Acta* **33**, 1473.

Buddhue, P.R. (1946) *Pop. Astron.* **54**, 149.

Bundy, F.P. and Kasper, J.S. (1967) *J. Chem. Physics* **46**, 3437.

Clayton, R.N. (1963) *Science* **140**, 192.

Duchesne, J., Depireux, J. and Litt, C. (1964) *Compt. Rend. Acad. Sci. Paris* **259**, 4776.

El Goresy, A. and Donnay, G. (1968) *Science* **161**, 363.

Flory, D.A. and Oro, J. (1969) Preprint.

Frondel, C. and Marvin, U.B. (1967) *Nature* **214**, 587.

Hanneman, R.E., Strong, H.M., Bundy, F.P. (1967) *Science* **155**, 995.

Hayes, J.M. (1967) *Geochim. Cosmochim. Acta* **31**, 1395.

Hey, M.H. (1938) *Mineral. Mag.* **25**, 81.

Heymann, D., Lipschutz, M.E., Nielson, B. and Anders, E. (1966) *J. Geophys. Res.* **71**, 619.

Jarosewich, E. and Mason, B. (1969) *Geochim. Cosmochim. Acta* **33**, 411.

Krouse, H.R. and Modzeleski, V.E. (1969) Preprint.

Kvasha, L. G. (1948) *Meteoritika* **4**, 83.
Lipschutz, M. E. (1964) *Science* **143**. 1431.
Lipschutz, M. E. (1967) *Geochim. Cosmochim. Acta* **31**, 621.
Lipschutz, M. E. and Anders, E. (1961) *Geochim. Cosmochim. Acta* **24**, 83.
Mason, B. (1962) *Meteorites*. Wiley, New York.
Mason, B. (1963) *Space Sci. Rev.* **1**, 621.
McCall, G. J. H. and Cleverly W H. (1968) *Mineral. Mag.* **36**, 691.
Moore, C. B. and Lewis, C. F. (1965) *Science* **149**, 317.
Moore, C. B. and Lewis, C. F. (1966) *Earth and Planet Sci. Letters* **1**, 376.
Moore, C. B. and Lewis, C. F. (1967) *J. Geophys. Res.* **72**, 6289.
Moore, C. B., Lewis, C. F. and Nava, D. (1969) *Meteorite Research* (Ed. P. M. Millman), 738. Reidel.
Moore, C. B. et al. (1970) *Science* **167**, 495.
Nagy, B. (1966) *Proc. Nat. Acad. Sci. USA* **56**, 389.
Nagy, B. (1968) *Endeavour* **27**, 81.
Nininger, H. H. (1956) *Arizona's Meteorite Crater*. American Meteorite Museum.
Nooner, D. W. and Oro, J. (1967) *Geochim. Cosmochim. Acta* **31**, 1359.
Oro, J. (1963) *Ann. N. Y. Acad. Sci.* **108**, 464.
Otting, W. and Zähringer, J. (1967) *Geochim. Cosmochim. Acta* **31**, 1949.
Perry, S. H. (1944) *Bull. U.S. Natl. Museum.* No. 184.
Smith, J. W. and Kaplan, I. R. (1970) *Science* **167**, 1367.
Studier, M. H., Hayatsu, R. and Anders, E. (1965) *Science* **149**, 1455.
Studier, M. H., Hayatsu, R. and Anders, E. (1968) *Geochim. Cosmochim. Acta* **32**, 151.
Trofimov, A. V. (1950) *Dokl. Akad. Nauk. SSSR* **72**, 663.
Urey, H. C. (1966) *Royal Astron. Soc. Monthly Notices* **131**, 199.
Urey, H. C. and Craig, H. (1953) *Geochim. Cosmochim. Acta* **4**, 36.
Van Schmus, W. R. and Wood, J. A. (1967) *Geochim. Cosmochim. Acta* **31**, 747.
Vdovykin, G. P. (1964) *Geochimiya* **4**, 299.
Vdovykin, G. P. (1967) *Carbon matter of meteorites (organic compounds, diamonds, graphite)*. Moscow, "Nauka".
Vdovykin, G. P. (1969a) in *Advances in Organic Geochim.* **9**. Pergamon.
Vdovykin, G. P. (1969b) *Geochimiya* **9**, 1145.
Vdovykin, G. P. (1970) *Space Sci. Rev.* **10**, 483.
Vinogradov, A. P. (1956) *Pure and Appl. Chem.* **10**, No. 4.
Vinogradov, A. P., Kropotova, O. I., Vdovykin, G. F. and Grinenko, V. A. (1967) *Geochem. Internat.* **4**, 229.
Vinogradov, A. P. and Vdovykin, G. P. (1963) *Geochemistry* **8**, 743.
Vinogradov, A. P. and Vdovykin, G. P. (1964) *Geochem. Internat.* **5**, 831.
Vinogradov, A. P. and Vdovykin, G. P. (1969) *Geochimiya* **9**, 1035.
Wiik, H. B. (1969) Soc. Sci. Fennica, *Comment. Phys.-Mathem.* **34**, 135.
Wood, J. (1963) in *The Moon Meteorites and Comets* (Ed. B. M. Middlehurst and G. P. Kuiper), 337. Univ. of Chicago Press, 337.
Yavnel, A. A. (1965) *Meteoritika* **26**, 26.

References to Tables

(1) Moore et al. (1969)
(2) Patwar (Jarosewich and Mason, 1969); Clover Springs (Wiik, 1969)
(3) Mason (1963)
(4) Otting & Zähringer (1967)
(5) Moore & Lewis (1965)
(6) Moore & Lewis (1966)
(7) Moore & Lewis (1967)
(8) Wood (1963)
(9) Trofimow (1950)
(10) Wiik (1969) 3 analyses; McCall and Cleverly (1968) 2 analyses.
(11) Norton County (Moore, unpublished results); Pesyanoe (Trofimov, 1950)
(12) Moore (unpublished results)
(13) Johnstown (Moore, unpublished results)
(14) Pavlovka (Trofimov, 1950); Yurtuk (Moore et al., 1970); Frankfort (Moore, unpublished)
(15) Macibini (Urey and Craig, 1953); Sioux County, Pasamonte, Haraiya (Moore et al., 1970)

(Received 7 May 1970)

NITROGEN (7)

Carleton B. Moore

Center for Meteorite Studies
Arizona State University
Tempe, Arizona

UNTIL RECENTLY little work had been done on the nitrogen content of meteorites. Although nitrogen is one of the most abundant elements in the solar system it has been grossly depleted in the meteorites. Its most common chemical form is as atmophile molecular nitrogen and, as such, is able to diffuse into space as do other gaseous species. It is found as a major component in the mineral sinoite (Si_2N_2O) found in enstatite chondrites (Anderson, Keil, and Mason, 1964; Keil and Anderson, 1965) and in the mineral osbornite (TiN) (Bannister, 1941) found in the enstatite achondrite, Bustee. The organic material of carbonaceous chondrites contains some nitrogen-bearing molecules (Hayatsu, 1964; Studier et al., 1968; Hayatsu et al., 1968) and some water-soluble ammonium compounds, probably NH_4Cl and/or $(NH_4)_2SO_4$, (Mason, 1962). Cloez (1864) recorded 0.1% NH_4 in his analysis of Orgueil. In the low-nitrogen ordinary chondrites, stony-iron, and iron meteorites, much of the nitrogen is most likely present as dissolved interstitial atoms or molecules. Vinogradov et al. (1963) reported that nitrogen occurs in stony meteorites as ammoniacal, molecular, or possibly nitride nitrogen. No evidence of iron nitrides has been found in iron meteorites, except for a tentative identification by Buchwald (1969) in the Cacaria medium octahedrite. This meteorite has apparently been heated to about 1000°C by man and the nitrogen in the nitrides is most likely of terrestrial origin.

V. M. Goldschmidt (1923), in his table of the geochemical classification of the elements, suggested that at high temperatures and pressures, nitrogen is siderophilic in the form of nitrides. Mason (1966) raised the question of whether nitrogen in sinoite should be classified as siderophilic or lithophilic.

He pointed out that the chemical reactions which gave rise to osbornite and sinoite are of considerable significance in determining the initial physico-chemical environment of the reduced meteorites.

The nitrogen contents for meteorites in this chapter are for the most part taken from Gibson (1969) and papers resulting from this thesis. Analyses of 123 iron meteorites and 114 stony meteorites were reported. The total nitrogen contents were determined by an inert-gas fusion-gas chromatographic technique (Gibson and Moore, 1970). The values for the carbonaceous chondrites are primarily from the analyses by H. B. Wiik as reported by Mason (1963). These were determined by standard organic nitrogen analytical techniques. The few earlier analyses done by the Kjeldahl method or gas extraction agree fairly well with the more recent values. Chamberlin (1908) using gas evolution reported that stony meteorites had 32×10^{-6} g/g N and iron meteorites 38×10^{-6} g/g N. Work by Lipman (1932) using the micro-Kjeldahl method found that the nitrogen in six ordinary chondrites ranged from 16 to 65×10^{-6} g/g. Buddhue (1942) found up to 33×10^{-6} g/g of ammoniacal nitrogen in five ordinary chondrites. I. Noddack and W. Noddack (1930) reported 0.9×10^{-6} g/g nitrogen was released by heating choudrites. Nash and Baxter (1947) found less than 3×10^{-6} g/g nitrogen in eight iron meteorites. König et al. (1961) determined ammoniacal nitrogen in six chondrites and found the mean was 7.8×10^{-6} g/g. Vinogradov et al. (1963) reported a range of 8 to 38×10^{-6} g/g of ammoniacal nitrogen in

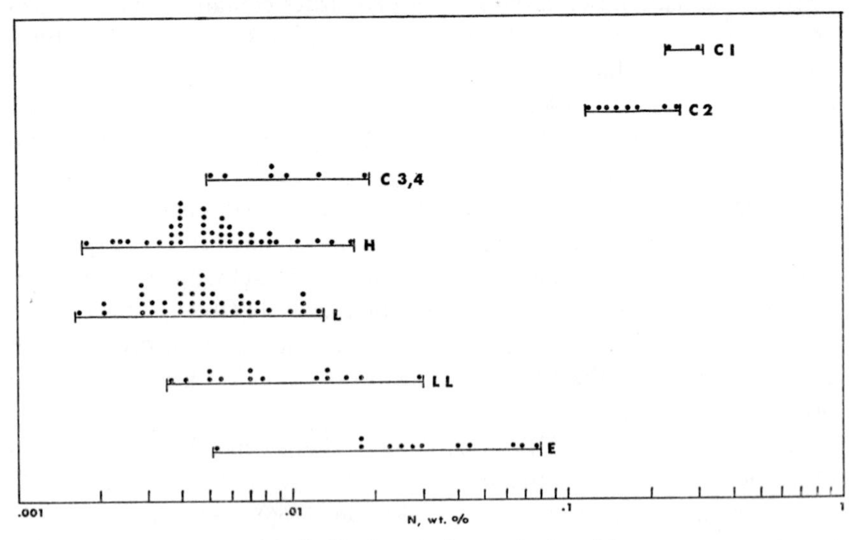

FIGURE 1 Frequency distribution of nitrogen in chondrites.

TABLE 1 Nitrogen concentrations in carbonaceous, uneguilibrated ordinary and enstatite chondrites

Meteorite	Type	Nitrogen 10^{-6} g/g	Reference
Alais	C–1	2900	1
Orgueil	C–1	2400	1
Boriskino	C–2	700	1
Erakot	C–2	2600	1
Murchison	C–2	2300	4
Al Rais	C–3	1900	1
Felix	C–3	86	3
Grosnaja	C–3	2200	1
Kainsaz	C–3	97	3
Lancé	C–3	87	3
Mokoia	C–3	100	3,5
Ornans	C–3	500	1
Renazzo	C–3	600	1
Vigarano	C–3	100	1
Vigarano	C–3	130	3
Karoonda	C–4	58	2,5
Abee	E–4	260	2,5
Indarch	E–4	430	2,5
St. Marks	E–5	180	2
St. Sauveur	E–5	220	2
Daniel's Kuil	E–6	170	2
Hvittis	E–6	650[1]	2,5
Jajh deh Kot Lalu	E–6	670[1]	2
Khairpur	E–6	54	2,5
Pillistfer	E–6	780	2
Bremervörde	H–3	51	3
Hallingeberg	L–3	75	3
Hedjaz	L–3	17	3
Manych	L–3	100	3
Chainpur	LL–3	56	3
Ngawi	LL–3	300	3
Parnallee	LL–3	70	3
Semarkona	LL–3	140	3

[1] contains Sinoite

four chondrites. Recently Goel (1970) has used neutron activation to determine nitrogen in the iron meteorites, Odessa and Canyon Diablo. Both Gibson and Goel's data are for total nitrogen, which includes dissolved molecular nitrogen. The Kjeldahl method determines only chemically active nitrogen.

Values of nitrogen contents of different classes of meteorites are given in Tables 1–3. In Table 1 nitrogen contents in individual carbonaceous chondrites, unequilibrated ordinary chondrites, and enstatite chondrites are given. Individual specimens of these meteorite types may show broad variations in their total nitrogen contents, perhaps indicating an irregular distribution of nitrogen-bearing phases. In Tables 2 and 3 are given the nitrogen contents of the major groups of stony, stony-iron and iron meteorites. Table 4 gives the nitrogen abundances in troilite nodules and the nickel-iron matrix of a selected group of iron meteorites.

TABLE 2 Nitrogen abundances in stony and stony-iron meteorites

	Number	Range	Median	Atoms/ 10^6 Si	References
Chondrites					
C–1 (Carbonaceous)	2	2400–2900	2600	49,000	Table 1
C–2 (Carbonaceous)	3	700–2600	2300	43,000	Table 1
C–3 (Carbonaceous)	9	86–2200	100	1,500	Table 1
E (Enstatite)	9	54–780	260	3,000	Table 1
H (Bronzite)	20	18–121	48	570	(3) (5)
L (Hypersthene)	28	17–109	43	480	(3) (5)
LL (Amphoterites)	11	36–298	70	740	(3) (5)
Calcium-poor achondrites					
Enstatite	1	–	44	340	(6)
Hypersthene	1	–	31	250	
Calcium-rich achondrites					
Howardites	3	40–56	56	490	(6) (7)
Eucrites	3	30–40	39	360	(6) (7)
Stony-irons					
Pallasites	–	–	–	–	–
Mesosiderites	2	59–65	62	1,000	(6)

TABLE 3 Nitrogen concentrations in iron meteorites (3)

Class	No. of meteorites analyzed	Range 10^{-6} g/g	Median 10^{-6} g/g	Mean 10^{-6} g/g
Hexahedrites	11	2.3–25.5	8.2	11.0
Coarsest Octahedrites	7	9.7–27	17.8	18.7
Coarse Octahedrites	13	2.2–75.8	22.2	29.4
Medium Octahedrites	54	2.0–131	30.6	32.8
Fine Octahedrites	21	1.0–59.9	5.6	13.9
Finest Octahedrites	4	5.7–16.4	10.0	10.5
Nickel Rich Ataxites	9	1.5–215.5	7.5	30.8
Nickel Poor Ataxites	3	21.9–61	23.0	35.3

Median Value for 123 Meteorites = 17.8 ppm Nitrogen

TABLE 4 Nitrogen content of troilite nodules (3)

Meteorite	Class	N in Troilite (ppm)	N in Metal (ppm)	Distribution Ratio Troilite/Metal
Bear Creek	Of	230	46.2	5.0
Bella Roca	Of	143	21.7	6.6
Cambria	Of	37	1.0	37.0
Canyon Diablo (cohenite-free)	Og	71	32.7	2.6
Carbo	Om			
Nodule 1		133	31.8	4.2
Nodule 2		192	31.8	6.0
Joe Wright Mt.	Om	28	37.9	0.7
Mount Stirling	Og	131	31.9	4.2
Osseo	Ogg	68	9.7	7.0
Youndegin	Og	77	22.2	3.5

References

Anderson, C.A., Keil, K. and Mason, B. (1964) *Science* **146**, 256.
Bannister, F.A. (1941) *Min. Mag.* **26**, 36.
Buchwald, V.F. (1969) Oral comm. in Gibson (1969).
Buddhue, J.D. (1942) *Pop. Astron.* **50**, 561.
Chamberlin, R.T. (1908) *Publ. No.* **106**, *Carnegie Inst.* Washington.

Cloez, S. (1864) *Compt. Rend. Acad. Sci., Paris* **59**, 37.

Gibson, E. (1969) Ph. D. Thesis, Arizona State University.

Gibson, E. and Moore, C.B. (1970) *Anal. Chem.* **42**, 461.

Goel, P.S. (1970) *Determination of nitrogen in iron meteorites* (Preprint).

Goldschmidt, V.M. (1923) *Norske Videnskaps Skrifter, Mat. Naturv. Klasse*, No. **3**, 17.

Hayatsu, R. (1964) *Science* **146**, 1291.

Hayatsu, R., Studier, M.H., Oda, A., Fuse, K., and Anders, E. (1968) *Geochim. Cosmochim. Acta* **32**, 175.

Keil, K. and Anderson, C.A. (1965) *Nature* **207**, 745.

König, H., Keil, K., Hintenberger, H., Wlotzka, F. and Begemann, F. (1961) *Z. Naturforsch.* **16a**, 1124.

Lipman, C.B. (1932) *American Mus. Novitates* No. **589**, 1.

Mason, B. (1962) *Meteorites* Wiley.

Mason, B. (1963) *Space Sci. Rev.* **1**, 621.

Mason, B. (1966) *Geochim. Cosmochim. Acta* **30**, 365.

Moore, C.B. and Gibson, E. (1969) *Science* **163**, 174.

Moore, C.B., Lewis, C.F. and Nava, D. (1969) *Meteorite Research* (Ed. P.M. Millman), (38. D. Reidel Pub. Co.

Moore, C.B., Gibson, E.K., Larimer, J.W., Lewis, C.F. and Nichiporuk, W. (1970) *Geochim. Cosmochim. Acta* (in press).

Nash, L.K. and Baxter, G.P. (1947) *J. Amer. Chem. Soc.* **69**, 2534.

Noddack, I. and Noddack, W. (1930) *Naturwiss.* **18**, 757.

Studier, M., Hayatsu, R. and Anders, E. (1968) *Geochim. Cosmochim. Acta* **32**, 151.

Vinogradov, A.P., Florenskii, K.P. and Wolynets, V.G. (1963) *Geochem.*, **905**.

References to Tables

1) H.B. Wiik in Mason B. (1963)
2) Moore C.B. et al. (1969)
3) Gibson E.K. (1969)
4) D.P. Miller from R.O. Chalmers (written communication).
5) Moore C.B. and Gibson E.K. (1969)
6) Gibson E.K. (Unpublished results)
7) Moore C.B. et al. (1970)

(Received 7 May 1970)

OXYGEN (8)

William D. Ehmann

Department of Chemistry,
University of Kentucky, Lexington, Kentucky
and Department of Chemistry,
Arizona State University, Tempe, Arizona

UNTIL RECENTLY oxygen in meteorites has been determined largely by difference after the determination of all other major components. The early data contain, therefore, all the compounded errors in the determinations of the other elements. Goldschmidt (1958) gave the atomic abundance of oxygen as 3.47×10^6 (Si $= 10^6$ atoms) in his "average" meteorite made up of 10 parts by weight silicate, 1 part troilite, and 2 parts nickel-iron. As will be seen later, this is close to the average obtained for the ordinary chondrites by use of the modern technique of non-destructive neutron activation analysis.

The direct determination of oxygen has long been a difficult analytical problem. Most of the classical methods, such as gasometry, titrimetry, and colorimetry lack sensitivity, selectivity, and speed. The technique of non-destructive 14 MeV neutron activation analysis provides a means of direct oxygen determination that is rapid, accurate (if sufficient aliquants are analyzed), and conservative of precious meteoritic material. Data on oxygen in stony meteorites by use of this technique have been recently published by Wing (1964) and Vogt and Ehmann (1965a). A summary of the combined data of both papers is presented in Table 1. The comparison of data on meteorites analyzed in both laboratories indicates good agreement within the stated error limits.

Stable oxygen consists of three isotopes—^{16}O (99.759%), ^{17}O (0.037%) and ^{18}O (0.204%). The activation procedure for the determination of oxygen is based on the ^{16}O (n, p)^{16}N reaction induced by irradiation with 14 MeV neutrons produced by a neutron generator. The analysis of oxygen by this

TABLE 1 Oxygen abundances in various classes of stony meteorites[1]

Classification[2]	Meteorites analyzed	Range of abd. %O (by weight)	Mean %O (by weight)
C1	1	–	46.0
C2	4	36.9–43.7	41.5
C3	3	34.8–38.9	36.5
C4	1	–	36.3
All C	9	34.8–46.0	*39.8*
E4	1	–	27.0
E5	1	–	36.0
E6	1	–	31.6
All E	3	27.0–36.0	*31.5*
H5	5	33.3–37.3	35.1
H6	1	–	34.4
All H	7	33.3–37.3	*35.1*
L3	1	–	38.8
L4	2	38.9–40.0	39.5
L5	3	39.2–40.8	40.1
L6	8	35.3–42.3	39.0
All L	16	35.3–42.3	*39.0*
LL6	3	38.3–39.7	*38.8*
Achondrites, Ca-rich	1	–	*42.4*
Achondrites, Ca-poor	4	42.6–48.0	*45.4*

[1] Combined activation analysis data of Vogt and Ehmann (1965a) for 39 meteorites and Wing (1964) for 9 meteorites.
[2] Classification according to Van Schmus and Wood (1967).

technique is very specific. Production of ^{16}N by the ^{15}N (n, γ) ^{16}N reaction is negligible because of the low parent isotopic abundance and its low neutron capture cross section. Only fluorine may cause a serious interference, due to the production of ^{16}N by the ^{19}F (n, α) ^{16}N reaction. For oxygen-fluorine ratios greater than 10 to 1, the presence of fluorine would produce a maximum error of 5% in the oxygen determination. Since the meteoritic oxygen-fluorine ratio is approximately 1000 to 1, the fluorine interference in the analyses of stony meteorites for oxygen is also negligible. For the same reasons, interference from ^{11}Be produced by fast neutron reactions on boron is negligible. Ordinarily scintillation spectrometry with a NaI (Tl) detector is used to measure the 7.37 second half-life ^{16}N produced in the irradiations. The principal photopeak of 6.13 MeV and its two escape peaks are counted by simply counting the entire region from approximately

4.5 to 8 MeV. The precision stated for the analyses in Table 1 is approximately ±2 to 3%. In both laboratories 3 or 4 aliquants of each meteorite were prepared (where the supply permitted) and each aliquant was analyzed 4 or 5 times to obtain the final mean abundance in a given meteorite.

The mean atomic abundances relative to Si = 10^6 atoms for the various classes of chondrites are given in Table 2. Earlier data obtained largely by difference are also listed in Table 2. The degree of oxidation, as measured

TABLE 2 Atomic abundances of oxygen in chondrites

Classification	Mean atomic abundance O (Si = 10^6)	Reference
Carbonaceous Type I	7.5×10^6	Table 1*
Carbonaceous Type II	5.3×10^6	Table 1*
Carbonaceous Type III	4.1×10^6	Table 1*
Enstatite	3.1×10^6	Table 1*
H-group	3.6×10^6	Table 1*
L-group	3.8×10^6	Table 1*
Other work		
H-group	3.39×10^6	(1)
L-group	3.48×10^6	(1)
H-group	3.41×10^6	(2)
L-group	3.47×10^6	(2)
Enstatite	2.97×10^6	(3)
Carbonaceous Type I	7.69×10^6	(4)
Carbonaceous Type II	5.51×10^6	(4)

* Using silicon data of Vogt and Ehmann (1965b).
(1) Urey and Craig (1953)
(2) Mason (1965)
(3) Mason (1966)
(4) Mason (1962–63)

by the total oxygen content of the chondrites, is greatest for the Type I carbonaceous chondrites and diminishes in the order-Type I carbonaceous, Type II carbonaceous, Type III carbonaceous, ordinary, and enstatite. All the chondrite oxygen data are less than the solar system abundance of 2.15×10^7 atoms oxygen (Si = 10^6) given by Suess and Urey (1956) and the solar value of 2.9×10^7 atoms (Si = 10^6) given by Goldberg, et al. (1950). Clearly, the chondrites must be depleted with respect to the primordial solar system abundance of oxygen. Loss of organic matter and volatile oxides is certainly the most obvious mechanism.

Vogt and Ehmann (1965a) noted that among their group of ordinary chondrites "finds" had approximately 1.8% more oxygen (39.0% vs. 37.2%) than "falls". If generally true, this is probably due to weathering effects. Deletion of data on "finds" from Table 1 would lower the L-group mean from 39.0% to 38.1%, the E-group mean from 31.5% to 29.3%, and would make negligible changes to the other group averages. Since the observed effect is small and within the stated limits of precision for the method, all available data were used in computing the group averages in Tables 1 and 2.

Addendum

A more precise 14 MeV neutron activation procedure for oxygen has recently been developed by Morgan and Ehmann (1970). Using this procedure abundances of 35.9% O in Allende (Morgan et al., 1969), 40.9% O in Murchison and 32.0% O in Lost City (Ehmann et al., 1970) have been determined. All these meteorites are recent falls.

Morgan, J.W. and Ehmann, W.D. (1970) *Anal. Chim. Acta* **49**, 287–299.

Morgan, J.W., Rebagay, T.V., Showalter, D.L., Nadkarni, R.A., Gillum, D.E., McKown, D.M. and Ehmann, W.D. (1969) *Nature* **224**, 789–791.

Ehmann, W.D., Gillum, D.E., Morgan, J.W., Nadkarni, R.A., Rebagay, T.V., Santoliquido, P.M. and Showalter, D.L. (1970) *Meteoritics* **5**, 131–136.

References

Goldberg, L., Muller, E. and Aller, L.H. (1960) *Ap. J. Suppl.* **5**, No. 45.

Goldschmidt, V.M. (1958) *Geochemistry*, Oxford University Press, London.

Mason, B. (1962–63) *Space Sci. Rev.* **1**, 621–646.

Mason, B. (1965) *Amer. Mus. Nov.* **2223**.

Mason, B. (1966) *Geochim. Cosmochim. Acta* **30**, 23–29.

Suess, H.E. and Urey, H.C. (1956) *Rev. Mod. Phys.* **28**, 53–74.

Urey, H.C. and Craig, H. (1953) *Geochim. Cosmochim. Acta* **4**, 36–82.

Van Schmus, W.R. and Wood, J.A. (1967) *Geochim. Cosmochim. Acta* **31**, 747–765.

Vogt, J.R. and Ehmann, W.D. (1965a) *Radiochim. Acta* **4**, 24–28.

Vogt, J.R. and Ehmann, W.D. (1965b) *Geochim. Cosmochim. Acta* **29**, 373–383.

Wing, J. (1964) *Anal. Chem.* **36**, 559–564.

(Received 1 February 1969)

FLUORINE (9)*

George W. Reed, Jr.

Argonne National Laboratory
Argonne, Illinois 60439

FLUORINE IS a minor element in meteorites, having a concentration range of a few to about 200 ppm by weight. Geochemically fluorine is commonly found in the phosphate mineral apatite. This is a minor mineral in the ordinary chondrites. In terrestrial apatite fluorine is the preferred substituent to chlorine. In meteorites chlorapatite is found exclusively. Recently Van Schmus (1969), using the electron microprobe, reported up to 25 ppm F in meteoritic chlorapatite. Shannon and Larsen (1925) had previously separated the phosphates, whitlockite (merrillite) and chlorapatite, from meteorites but no fluorapatite was observed. Fuchs (1962) did not identify fluorapatite in a series of optical, x-ray and microprobe studies on apatites in meteorites. Reed (1964) did not observe a positive correlation of F and Ca in chondrites. It was postulated that F could be in a major phase such as pyroxene, in a minor mineral with which F has not been identified or in an unidentified minor mineral.

Four approaches have been used in modern analyses of total F in meteorites. Fisher (1963), Reed (1964), and Reed and Jovanovic (1969) relied on different activation analysis methods; Shima (1963) determined F colorimetrically by the thorin-alizarine method, and Greenland and Lovering (1965) used emission spectrography. Fisher counted 7.4-sec N^{16} produced by (n, α) reactions on F^{19}. Large corrections were necessary because the O^{16} and N^{15} in the sample also lead to N^{16} production. Shima (1963) and Reed (1964) distilled F from the samples as H_2SiF_6 after alkali fusion. In Shima's procedure only the indigenous F was distilled. In Reed's procedure 10–50 mg F carrier added before fusion was distilled with the sample F and 1.8-hr F^{18} from the (γ, n) reaction on F^{19} was counted. The completeness of F recovery and the linearity of the absorbance were factors considered by

* Work supported by the U.S. Atomic Energy Commission.

Shima as sources of error; interfering nuclear reactions and the complexities in monitoring the photon beam were potential sources of error in Reed's procedure. Detailed considerations of sources of error are given by the investigators.

The recent data on the F contents of chondrites and achondrites are given in Tables 1 and 2. The only earlier data we are aware of are those of the Noddacks (1934) who reported 30 ppm F for a composite mixture of chondrites. Not included in the tables is a single value of 1 ppm for F in the iron meteorite Canyon Diablo (Shima, 1963). Included is an average value (2 determinations) of 44.5 ppm for silicate from the Brenham pallasite.

Fluorine concentrations vary by factors up to two within a given class of meteorites. The extent of agreement between various laboratories is illustrated in Table 3. Customarily much of the scatter in trace element contents in meteorites is ascribed to sampling on a 200–1000 mg scale. The scatter is usually small, less than 30%, for the more electropositive trace elements such as U and Ba. Highly reactive and electronegative F does not appear to parallel this behavior. The results from a given laboratory should be internally consistent and, in general, agreement between samples of the same class of meteorite is within about 50%. The bronzites are exceptions; three laboratories found a factor of two variation for this chondrite class.

The results of Shima are consistently lower than those of other workers. In view of the scatter and the relatively few meteorites measured, generalizations on the distribution of F are probably premature, although a few may be ventured. Giving more weight to the data of Greenland and Lovering and of Reed because of the number of meteorites studied, the F concentrations in the hypersthene and bronzite chondrites are about the same (~130 ppm). These are lower on the average than the type 2 enstatite and Types I and II carbonaceous chondrites. This trend is consistent with that of the highly fractionated elements. The carbonaceous chondrite Types I and II as a group average higher F contents than the ordinary chondrites. There are too few data to ascertain whether F in carbonaceous chondrites follows the pattern of mixing various amounts of unfractionated material, characterized by Type I carbonaceous chondrites, with more highly depleted material (Anders, 1964).

Among the achondrites the Ca-rich meteorites have F contents equal to or approaching those of the ordinary chondrites. The hypersthene and enstatite achondrite F contents are distinctly lower. Shima's value of 44.5 ppm F in the silicate from a pallasite indicates that the element was not highly fractionated in the silicate melt from which the olivine was derived.

TABLE 1 Fluorine contents of chondrites

Meteorite classification	No. of determinations	No. of meteorites analyzed	Range (ppm)	Mean	Atoms F 10^6 Si	Ref.
Hypersthene (L)	2	2	27.2–29.5	28.3		Shima
	8	2	146–210	178	1365	Fisher
	13	5	117–148	120[1]	950	Reed
	1	1	–	300		Sen Gupta
	–	11	68–250	148.8	1234[2]	Greenland and Lovering
Bronzite (H)	2	1	37.3–40.1	38.7		Shima
	4	1	42–97	62	495	Fisher
	6	4	89–162	122[1]	1215	Reed
	–	4	81–170	130.2	1234[2]	Greenland and Lovering
Amphoterite (LL)	2	1	60–66	63		Sen Gupta
Enstatite – 1 (Hvittis type)	3	1	113–161	122		Reed
	–	2	220–280	257		Greenland and Lovering
Enstatite – 2 (Indarch type)	4	1	122–160	139	1220	Fisher
	3	2	205–246	238[1]	1951	Reed
	2	1	250–300	275		Sen Gupta
	–	4	100–250	166		Greenland and Lovering
Carbonaceous type I	2	1	390–420	405	5750	Fisher
	2	1	150–177	158[1]	2336	Reed
	–	1	–	190	2680	Greenland and Lovering
Carbonaceous type II	2	1	210–230	220	2500	Fisher
Carbonaceous type III	2	1	64–90	66[1]	646	Reed
	–	4	166–190	172.5	1610	Greenland and Lovering

[1]) Weighted average.

[2]) Average for all ordinary chondrite falls and finds (Greenland and Lovering).

TABLE 2 Fluorine contents of achondrites

Meteorite classification	No. of deter-minations	No. of meteorites analyzed	Range (ppm)	Mean	References
Enstatite	1	1	–	10.7	Reed
	4	3	6.8–11	10.4[1]	Reed and Jovanovic
Hypersthene	2	2	12.1–12.4	12.2[1]	Reed and Jovanovic
Eucrite and Howardite	2	2	61–64	63	Reed
	8	6	23–90	60	Reed and Jovanovic
	2	1	30.6–98.2[2]	30.6	Shima
Brenham palla-site olivine	2	1	42.5–46.4	44.5	Shima

1) Includes H_2O leachable fluorine estimate.
2) Values for total and dark parts of Pasamonte respectively.

TABLE 3 Intercomparison of meteoritic F concentrations (ppm)

Meteorite	Fisher	Shima	Reed	Greenland and Lovering	Sen Gupta	Reed and Jovanovic
Mocs	147		119	160		
Holbrock	189			130		
Farmington				250	300	
Barratta	78	27.2				
Allegan			114	170		
Abee			228	280	275	
Indarch			246	220		
Hvittis			122	250		
Orgueil	405		158	190		
Lance			66	170		
Pasamonte		30.6	46			
Moore County			61			90
Norton County			10.7			10.4

References

Anders, E. (1964) "Origin, age, and composition of meteorites", *Space Sci. Rev.* **3**, 583–714.

Fisher, D.E. (1963) "The fluorine content of some chondritic meteorites", *J. Geophys. Res.* **68**, 6331.

Fuchs, L.H. (1962) "Occurrence of whitlockite in chondritic meteorites", *Science* **137**, 425.

Greenland, L. and Lovering, J.F. (1965) "Minor and trace element abundances in chondritic meteorites", *Geochim. Cosmochim. Acta* **29**, 821–858.

Noddack, I. and Noddack, W. (1934) "Die geochemischen Verteilungskoeffizienten der Elemente", *Svensk. Kem. Tidskr.* **46**, 173–201.

Reed, G.W. Jr. (1964) "Fluorine in stone meteorites", *Geochim. Cosmochim. Acta* **28**, 1729–1743.

Reed, G.W.. Jr and Jovanovic, S. (1969) "Some halogen measurements on achondrites", *Earth Planet. Sci. Lett.* **6**, 316–320.

Sen Gupta, J.G. (1968) "Determination of fluorine in silicate and phosphate rocks, micas, and stony meteorites", *Anal. Chim. Acta* **42**, 119–125.

Shannon, E.V. and Larsen, E.S. (1925) "Merrillite and chlorapatite from stony meteorites", *Am. J. Sci.* **9**, 250.

Shima, M. (1963) "Fluorine content in meteorites", *Sci. Papers I.P.C.R.* **57**, 150–154.

Van Schmus, W.R. and Ribbe, P.H. (1969) "Composition of phosphate minerals in ordinary chondrites", *Geochim. Cosmochim. Acta* **33**, 637–640.

(Received 5 May 1970)

SODIUM (11)

Gordon G. Goles

Center for Volcanology
and
Departments of Chemistry and Geology
University of Oregon
Eugene, Oregon 97403

SODIUM, CONVENTIONALLY CONSIDERED a major element in terrestrial rock analyses, is in meteorites a minor element (contents of tenths of a weight per cent) or in some cases a trace element. This fact poses problems for meteorite analysts, and introduces difficulties for reviewers of the data. Many determinations, especially those in the older literature, are not reliable. In this as in several other reviews in this monograph (Sc, Cr, Mn, and Cu) I shall rely heavily on results of instrumental neutron activation analyses by Schmitt, Goles and Smith (1971) who discuss at length analytical techniques and their relative advantages and give extensive comparisons with data of other workers. In Table 1 are presented averages of Na contents and atomic abundances as determined by Schmitt, Goles and Smith. Chondrites are classified according to Van Schmus and Wood (1967) and achondrites are identified by the code introduced by Keil (1969). Dispersions cited are estimates of the population variability, at the one-standard-deviation level. Additional important references are: Edwards and Urey (1955), Edwards (1955), Wiik (1956), Mason (1963), Craig (1964), Mason and Wiik (1964), Duke and Silver (1967), Rieder and Wänke (1969).

Some anomalies, whose probable causes may lie in analytical error, in contamination or leaching on Earth's surface, or in peculiar petrogeneses, should be noted. Among carbonaceous chondrites, Mason's (1963) value for the C1 stone *Tonk* is higher by about a factor of five than other Na contents for samples of that class, probably owing to analytical error or contamination. Sodium contents of the C2 chondrites *Nogoya* and *Murray* reported by Mason (1963) are about one-third of the average Na content for nine other members of the class. For *Murray*, duplicate analyses by Schmitt, Goles and Smith (1971) yield a mass-weighted average Na content of $(1470 \pm 80) \times 10^{-6}$ g/g, in good agreement with Mason's value of 1600×10^{-6} g/g. Edwards and Urey (1955) and Edwards (1955) found anomalously low Na and K contents in *Murray*, and Goles, Greenland and Jérome (1967) have shown that this meteorite has anomalously low halogen abundances. Either this meteorite has been strongly leached while on Earth's surface or it has had a markedly different preterrestrial history than have most members of the C2 class. Insufficient data are available to indicate whether *Nogoya* resembles *Murray* in these respects.

Edwards (1955) reported 500×10^{-6} g Na/g for olivine from the *Brenham* pallasite. Schmitt, Goles and Smith (1971) find $(92 \pm 8) \times 10^{-6}$ g Na/g for similar material and the true value may be even less (see below). Most likely, a small blank correction of roughly 400×10^{-6} g Na per analysis should be applied to the results of Edwards and of Edwards and Urey (1955); such a correction would largely obviate differences between averages of Na contents by Edwards and Urey and those given in Table 1.

TABLE 1 Sodium average contents and atomic abundances in stone meteorites

Class	Number of Specimes (samples) analysed	Na (10^{-6} g/g)	Na atoms per 10^4 Si
C 1	3 (5)	5,100 ± 400	600 ± 40
C 2	7 (13)	3,800 ± 700	350 ± 60
C 3	9 (20)	3,500 ± 500	280 ± 40
H 3	5 (8)	5,800 ± 500	420 ± 50
H 4, 5, 6	40 (56)	5,700 ± 700	410 ± 40
L 3	3 (4)	6,600 ± 700	430 ± 50
L 4, 5, 6	47 (61)	6,500 ± 800	430 ± 50
LL 3	3 (6)	6,500 ± 900	430 ± 60
LL 4, 5, 6	17 (22)	6,600 ± 800	430 ± 50
E 4	2 (4)	8,200 ± 1000	590 ± 70
E 5, 6	5 (7)	5,200 ± 800	340 ± 60
Aubrites, Ae	5 (7)	~3500	~170
Ureilites, Aop	2 (3)	300 ± 80	18 ± 5
Eucrites, Ap	20 (27)	2,800 ± 700	150 ± 40
Howardites, Aor	9 (13)	1,900 ± 1,200	110 ± 70
Nakhlites, Ado	2 (4)	3,300 ± 500	180 ± 20
Shergotty	1 (2)	11,000 ± 2,000	590 ± 11

The ureilite *Novo Urei* was studied by Ringwood (1960), who reported 3300×10^{-6} g Na/g; Schmitt, Goles and Smith (1971) find $(330 \pm 30) \times 10^{-6}$ g Na/g as an average in duplicate samples of this meteorite. Perhaps Ringwood's figure is a misprint for 330×10^{-6} g Na/g, or perhaps his analyst introduced appreciable contamination of the sample. In any case, ureilites are clearly very strongly depleted in Na, even on comparison with other achondrites.

The C1, C2 and C3 groups of chondrites are clearly distinct from each other in Na contents, and become even more so upon estimating Na atomic abundances (which in effect corrects for varied contents of volatile components in these meteorites). The Na/Si atomic ratios of the C2 class correspond rather well with those of the H, L and LL groups, which are very similar to one another, and agree with ratios in E5, 6 enstatite chondrites. The apparent agreement between Na/Si ratios in C1 and in E4 groups may well be fortuitous, since the values for the E4 class are strongly affected by the exceptionally high Na content of *Abee*, $(8600 \pm 700) \times 10^{-6}$ g Na/g. Achondrites exhibit diverse Na contents and atomic abundances, which are only in part accounted for by differences in mineralogy. Aubrites in particular are highly variable in their Na contents, so much so that a group average is almost meaningless. Diogenites, which are not listed in Table 1 owing to their absence from the collection of specimens analysed by Schmitt, Goles and Smith (1971), have Na contents of roughly 10^{-4} g Na/g, according to analyses by D.Y. Jérome and by myself.

Table 2 presents data on Na contents of mineral separates, many of which were supplied by B. H. Mason and which were analysed by instrumental activation techniques similar to those used by Stueber and Goles (1967). Pyroxene separated from meteorites which contain plagioclase may be contaminated with traces of that mineral, and the observed Na contents should be interpreted with caution. Keil (1968) reported results of microprobe analyses for Na in enstatites of E4 and E5, 6 chondrites: 0.04% for *Indarch*, 0.05% for *Kota-Kota* and for *Adhi-Kot*, 0.10% for *Abee* (all E4), and less than 0.03% for his other (E5 or E6) specimens. His results indicate that Na contents are higher in enstatite of those E chondrites rich in this element. My preliminary data on Na contents of enstatite-rich fractions (omitted from Table 2 owing to probable contamination) agree with this inference. In Table 2, there seems to be a tendency for Na contents of pyroxenes to reflect those of the whole rocks from which the pyroxenes were separated, but this trend could result from nearly constant contamination with feldspars whose Na contents are determined by compositions of the whole rocks. Present

TABLE 2 Sodium contents of pyroxene and plagioclase separates

PYROXENE (10^{-6} g Na/g)	PYROXENE (continued)
Shallowater (Ae)	*Binda* (Aor)
enstatite 268 ± 6	~orthopyroxene 310 ± 40
Johnstown (Ab)	*Shergotty*[1]
bronzite 113 ± 4	pigeonite 1570 ± 90
Tatahouine (Ab)	*Bondoc Peninsula* (M)
bronzite 62 ± 2	principally orthopyroxene 103 ± 3
Juvinas (Ap)	*Crab Orchard* (M)
pigeonite 482 ± 12	principally pigeonite 920 ± 20
Nuevo Laredo (Ap)	*Vaca Muerta* (M)
pigeonite 1370 ± 40	pigeonite and orthopyroxene 112 ± 13
	PLAGIOCLASE (% Na)
	Appley Bridge (LL 6)
	6.89 ± 0.09
	Nuevo Laredo (Ap)
	1.43 ± 0.03
	Binda (Aor)
	0.574 ± 0.010
	Shergotty[1]
	3.92 ± 0.05
	Vaca Muerta (M)
	0.701 ± 0.010

[1]) The classification of *Shergotty* is uncertain owing to numerous mineralogical and chemical peculiarities exhibited by this meteorite.

data are inadequate to disclose any variations in distribution coefficients for Na among minerals from diverse meteorite classes. A lower limit to the plagioclase-pyroxene distribution ratio of about 63 (by mass) is set by the data on mineral separates from *Vaca Muerta*, supplied by U.B.Marvin. Very pure olivine separates from pallasites have apparent Na contents of abut 50×10^{-6} g/g, but this is in part at least a measure of the ^{24}Mg (n, p) ^{24}Na interference in the method rather than a true indication of the Na content.

I shall make no attempt to review data on Na in iron meteorites, where contents of this element are very low (thus introducing severe analytical difficulties) and are strongly dependent on the distribution of silicate inclusions within the samples taken for analysis.

Acknowledgements

V. Frankum and S. S. Oxley aided me in determining data on mineral separates. I am grateful to D.Y. Jérome and B. H. Mason for advice. Analytical work was supported in part by NASA research grant NsG–319 and was done at University of California San Diego.

References

Craig, H. (1964) "Petrological and compositional relationships in meteorites", chapter 26 in *Isotopic and Cosmic Chemistry*, edited by H. Craig. S. L. Miller and G. J. Wasserburg (North-Holland, Amsterdam) 401–451.

Duke, M. B. and Silver, L. T. (1967) "Petrology of eucrites, howardites, and mesosiderites." *Geochim. Cosmochim. Acta* **31**, 1637–1666.

Edwards, G. (1955) "Sodium and potassium in meteorites." *Geochim. Cosmochim. Acta* **8**, 285–294.

Edwards, G. and Urey, H. C. (1955) "Determination of alkali metals in meteorites by a distillation process." *Geochim. Cosmochim. Acta* **7**, 154–168.

Goles, G. G., Greenland, L. P. and Jérome, D. Y. (1967) "Abundances of chlorine, bromine and iodine in meteorites." *Geochim. Cosmochim. Acta* **31**, 1771–1787.

Keil, K. (1968) "Mineralogical and chemical relationships among enstatite chondrites." *J. Geophys. Res.* **73**, 6945–6976.

Keil, K. (1969) "Meteorite composition", chapter 4 in *Handbook of Geochemistry*, edited by K. H. Wedepohl (Springer-Verlag, Berlin) 78–115.

Mason, B. (1963) "The carbonaceous chondrites." *Space Science Rev.* **1**, 621–646.

Mason, B. and Wiik, H. B. (1964) "The amphoterites and meteorites of similar composition". *Geochim. Cosmochim. Acta* **28**, 533–538.

Rieder, R. and Wänke, H. (1969) "Study of trace element abundance in meteorites by neutron activation", in *Meteorite Research*, edited by P. M. Millman (D. Reidel, Dordrecht) 74–86.

Ringwood, A. E. (1960) "The Novo Urei meteorite." *Geochim. Cosmochim. Acta* **20**, 1–4.

Schmitt, R. A., Goles, G. G. and Smith, R. H. (1971) *Elemental abundances in stone meteorites*. To be published as a NASA report; available from the author on request.

Stueber, A. M. and Goles, G. G. (1967) "Abundances of Na, Mn, Cr, Sc and Co in ultramafic rocks." *Geochim. Cosmochim. Acta* **31**, 75–93.

Van Schmus, W. R. and Wood, J. A. (1967) "A chemical-petrologic classification for the chondritic meteorites." *Geochim. Cosmochim. Acta* **31**, 747–765.

Wiik, H. B. (1956) "The chemical composition of some stony meteorites." *Geochim. Cosmochim. Acta* **9**, 279–289.

(Received 5 May 1970, revised 19 August 1970)

MAGNESIUM (12)

Brian Mason

Smithsonian Institution, Washington, D.C.

MAGNESIUM IS a major element in all classes of stony and stony-iron meteorites, and as a consequence the abundance data are very extensive. The standard procedures of analytical chemistry give reliable results when carefully applied. Michaelis et al. (1969) have shown that the older data on chondrites summarized by Urey and Craig (1953) show a greater spread of values than more recent analyses, indicating an improvement in overall quality in recent years. The figures presented in Table 1 are derived as far as possible from recent analyses of observed falls.

The data for the chondrites show that the Mg/Si ratio (atomic) is fairly uniform at 0.934 to 0.965 for the common chondrites, although the difference between the hypersthene chondrites and amphoterites on the one hand and the bronzite chondrites on the other is probably greater than the experimental errors, and hence has real significance. The differences between the common chondrites and the carbonaceous and enstatite chondrites are considerably larger than those within the common chondrites. The Mg/Si ratio for the carbonaceous chondrites is consistently slightly greater than unity, whereas for the enstatite chondrites this ratio is considerably less than unity and notably lower than for the common chondrites. The difference between the Type I and Type II enstatite chondrites is also significant. The range of values of Mg/Si ratios for the enstatite chondrites is considerably wider than for the other classes of chondrites.

The differences between the Mg/Si ratios for the different classes of chondrites were first pointed out by Urey (1961), and have been further discussed by Ahrens (1964). These differences are illustrated in a simple plot of weight per cent SiO_2 against weight per cent MgO (Fig. 1). The significance of this fractionation remains to be fully elucidated. The close approach to unity of the Mg/Si ratio for the carbonaceous and the common chondrites is con-

TABLE 1 Magnesium in meteorites

	No. of detns.	Mg, weight per cent		Atoms/ 10^3 Si	Reference
		Range	Mean		
Chondrites					
Carbonaceous, Type I	3	9.44–9.70	9.56	1061	(1)
Carbonaceous, Type II	9	10.79–14.19	11.80	1043	(1)
Carbonaceous, Type III	6	13.31–14.66	14.53	1063	(1)
Bronzite	36	13.25–14.97	14.20	965	(2)
Hypersthene	68	14.05–15.99	15.19	941	(2)
Amphoterite	14	14.48–15.75	15.29	934	(3)
Enstatite, Type I	3	10.05–11.49	10.70	727	(4)
Enstatite, Type II	5	11.62–14.01	13.31	809	(4)
Calcium-poor achondrites					
Enstatite	4	22.49–24.58	23.20	1000	(5)
Hypersthene	7	13.95–17.10	15.93	744	(6)
Chassignite	1	–	19.40	1280	(7)
Ureilites	3	22.02–23.49	22.60	1370	(8)
Calcium-rich achondrites					
Angrite	1	–	6.05	339	(9)
Nakhlite	1	–	7.25	366	(10)
Howardites	4	7.25–12.35	9.40	474	(11)
Eucrites	10	3.86–5.06	4.29	216	(12)
Stony-irons					
Pallasites (olivine)	11	25.60–29.25	28.01	1730	(12)
Mesosiderites (silicates)	9	7.35–13.61	8.95	419	(14)

(1) Wiik, in Mason (1963a)
(2) Mason (1965)
(3) Wiik, in Mason and Wiik (1964), with later analyses
(4) Mason (1966)
(5) Khor Temiki (Hey and Easton, 1967); Norton County (Wiik, 1956); Pesyanoe (Dyakonova and Kharitonova, 1960); Pena Blanca Spring (Lonsdale, 1947)
(6) Mason (1963b)
(7) Dyakonova and Kharitonova (1960)
(8) Wiik (1969)
(9) Ludwig and Tschermak (1909)
(10) Prior (1912)
(11) Bununu (Mason, 1967); Frankfort (Mason and Wiik, 1966a); Kapoeta (Mason and Wiik, 1966b); Bholgati (Jarosewich, unpublished)
(12) Mason (1967)
(13) Mason (1963c)
(14) Powell (1971)

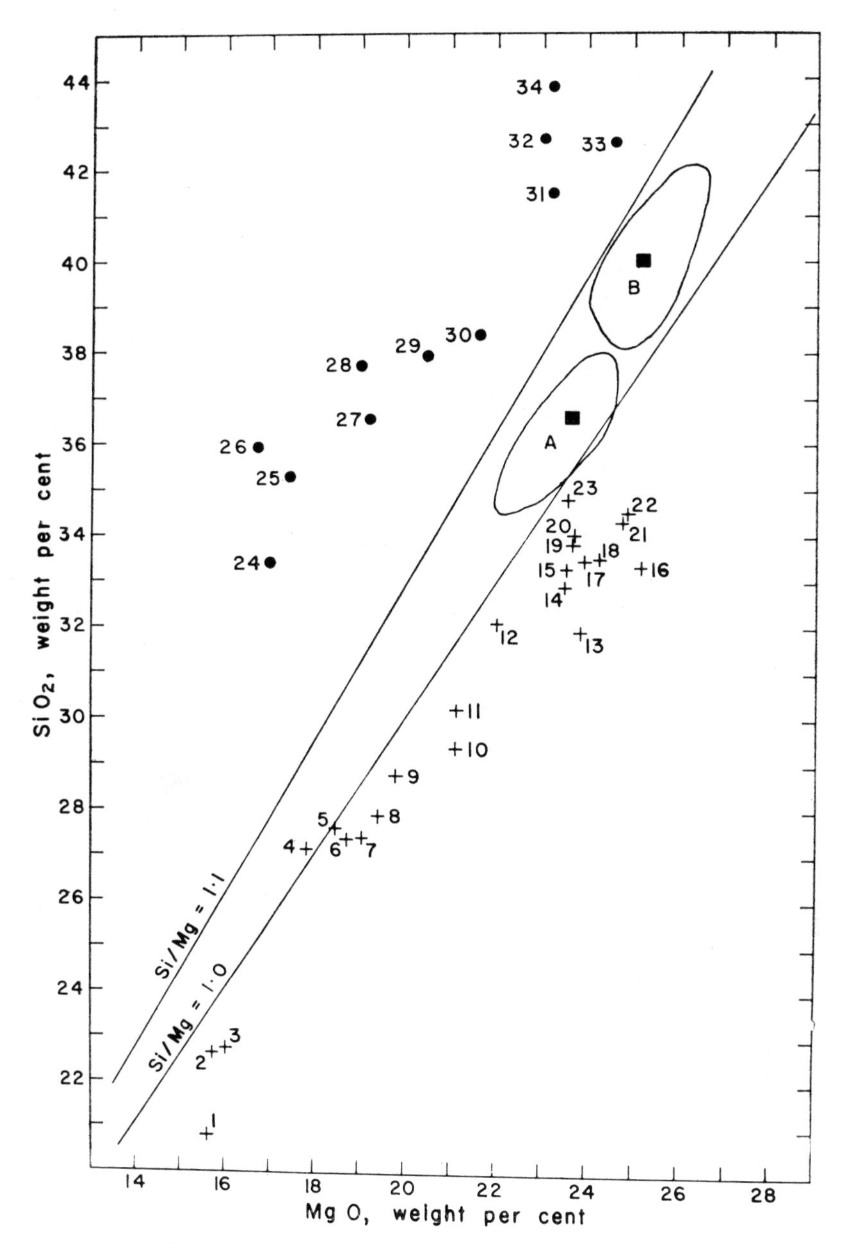

FIGURE 1 SiO_2 plotted against MgO (weight percentages) for chemical analyses of chondrites; the diagonal lines are for Si/Mg atomic ratios of 1.0 and 1.1. A is the field for 36 analyses of bronzite chondrites, B the field for 68 analyses of hypersthene chondrites, the black squares being the means for each group. The analyses of carbonaceous chondrites (1–23) and enstatite chondrites (24–34) are plotted individually.

sistent with their derivation from relatively unfractionated material that still retained the Mg/Si ratio established by nucleosynthesis. The fractionation exhibited by the enstatite chondrites may be related to the high degree of reduction shown by these meteorites, one of the features being the presence of elemental silicon in solid solution in their metal phase. Magnesium is not reduced to the elemental state in any class of meteorites. Thus the enstatite chondrites may actually be enriched in silicon (incorporated in the metal phase) rather than depleted in magnesium, relative to the common and carbonaceous chondrites.

The calcium-poor achondrites show a considerable range of Mg/Si ratios. Since the enstatite achondrites consist almost entirely of enstatite, $MgSiO_3$, this ratio should be close to unity; the exact figure in the table is coincidental, due to the balancing of the contribution of magnesium and silicon by minor minerals such as plagioclase and forsterite. The lower Mg/Si ratio for the hypersthene achondrites results from the introduction of ferrous iron partly replacing magnesium in the pyroxene. The chassignite and the ureilites show an enhancement of the Mg/Si ratio because of their high content of olivine, $(Mg, Fe)_2SiO_4$.

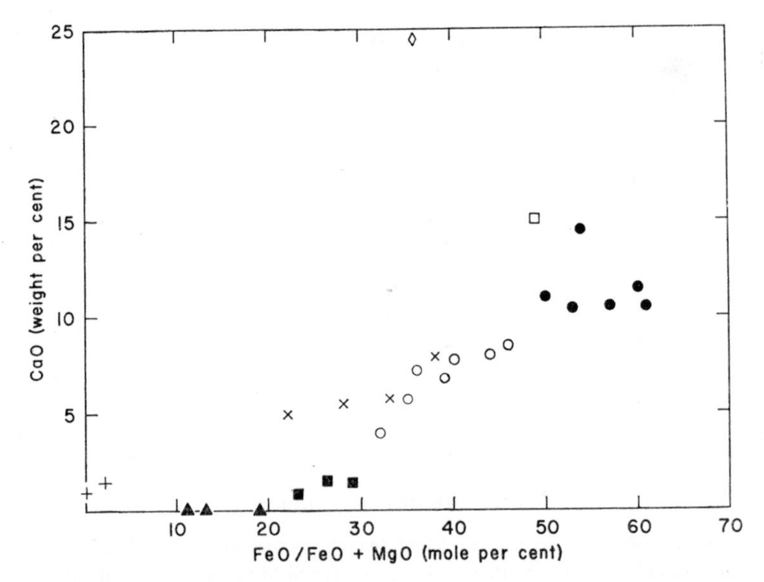

FIGURE 2 Plot of CaO (weight per cent) against FeO/FeO + MgO (mole per cent) for the achondrites and stony-irons; (+ = enstatite achondrites; ▲ = pallasites; ■ = hypersthene achondrites; × = mesosiderites; ○ = howardites; ● = eucrites; □ = nakhlite; ◇ = angrite).

The calcium-rich achondrites are notably depleted in magnesium relative to the chondrites. The howardites and the eucrites form a sequence in which the silicon and ferrous iron content remains practically constant while the magnesium content decreases (Fig. 2), hence the marked diminution in the Mg/Si ratio. Among the stony-irons, the mesosiderite silicates are essentially similar to those in the howardites, hence the similarity in Mg/Si ratios between these two classes. The silicate material in the pallasites is magnesium-rich olivine, hence the Mg/Si ratio is much higher than in other classes of meteorites and approaches 2:1.

Magnesium is usually considered to be entirely lithophilic in character; however, in the extremely reduced and sulfide-rich enstatite chondrites, a small amount of the total magnesium is present in solid solution in alabandite, (Mn, Fe)S. or as the mineral niningerite, (Mg, Fe)S. In most classes of meteorites the magnesium is combined almost entirely in the minerals olivine and/or pyroxene. The Type I and Type II carbonaceous chondrites are unique in having much of their magnesium present as the hydrated magnesium-iron silicate serpentine (or chlorite); Type I carbonaceous chondrites also contain notable amounts of hydrated magnesium sulfate.

References

Ahrens, L.H. (1964) *Geochim. Cosmochim. Acta* **28**, 411.
Dyakonova, M.I. and Kharitonova, V.Y. (1960) *Meteoritika* **18**, 48.
Hey, M.H. and Easton, A.J. (1967) *Geochim. Cosmochim. Acta* **31**, 1789.
Lonsdale, J.T. (1947) *Am. Mineral.* **32**, 354.
Ludwig, E. and Tschermak, G. (1909) *Tsch. Min. Pet. Mitt.* **28**, 109.
Mason, B. (1963a) *Space Sci. Rev.* **1**, 621.
Mason, B. (1963b) *Am. Museum Novitates*, No. **2155**.
Mason, B. (1963c) *Am. Museum Novitates*, No. **2163**.
Mason, B. (1965) *Am. Museum Novitates*, No. **2223**.
Mason, B. (1966) *Geochim. Cosmochim. Acta* **30**, 23.
Mason, B. (1967) *Geochim. Cosmochim. Acta* **31**, 107.
Mason, B. and Wiik, H.B. (1964) *Geochim. Cosmochim. Acta* **28**, 533.
Mason, B. and Wiik, H.B. (1966a) *Am. Museum Novitates*, No. 2272.
Mason, B. and Wiik, H.B. (1966b) *Am. Museum Novitates*, No. 2273.
Michaelis, H. v., Ahrens, L.H. and Willis, J.P. (1969) *Earth Planet. Sci. Lett.* **5**, 387.
Powell, B.N. (1971) *Geochim. Cosmochim. Acta* **35**, 5.
Prior, G.T. (1912) *Mineral. Mag.* **16**, 274.
Urey, H.C. (1961) *J. Geophys. Res.* **66**, 1988.
Urey, H.C. and Craig, H. (1953) *Geochim. Cosmochim. Acta* **4**, 36.
Wiik, H.B. (1956) *Geochim. Cosmochim. Acta* **9**, 279.
Wiik, H.B. (1969) *Soc. Sci. Fennica, Comm. Phys.-Math.* **34**, 135.

(Received 10 February 1970)

ALUMINUM (13)

Brian Mason

Smithsonian Institution, Washington, D.C.

ALUMINUM IS a minor constituent in all stony and stony-iron meteorites (except the pallasites, in which it is present in trace amounts only). It has therefore been determined in all complete chemical analyses of these meteorites. However, many of the data in the literature are unreliable, because the accurate determination of small amounts of aluminum (especially in the presence of much iron, as in meteorites) is extremely difficult by standard wet-chemical procedures. To obviate these difficulties, Loveland et al. (1969) have applied neutron activation analysis to the determination of this element in some 120 stony meteorites, and their results are used in Table 1, along with selected data for a few classes which they did not analyse.

Loveland et al. reported that their average aluminum abundances were lower, in general, than those previously reported, and that the individual determinations showed a much smaller dispersion around the mean value for each class than earlier determinations. For the common chondrites (bronzite, hypersthene, amphoterite) they report standard deviation dispersions of 6%, 7%, and 6% respectively, considerably less than earlier dispersion values of 25%, 28%, and 28%. Their low dispersion values indicate that, within specific chondrite classes, aluminum concentrations are remarkably uniform from one meteorite to another. This has been confirmed by Michaelis et al. (1969), who used x-ray fluorescence to analyse for aluminum and other elements in 69 meteorites.

The data for the chondrites show that the Al/Si ratio (atomic) is uniform at 0.061–0.062 for the common chondrites, whereas it is much greater for the carbonaceous chondrites (average 0.087), and somewhat lower for the enstatite chondrites (0.048). The lower ratio for the latter parallels a corresponding diminution of the Mg/Si ratio from the common chondrites to the enstatite chondrites, and can probably be ascribed to an absolute enhance-

TABLE 1 Aluminum in meteorites

	No. of detns.	Al, weight per cent		Atoms/ 10^3 Si	Reference
		Range	Mean		
Chondrites					
Carbonaceous, Type I	3	0.80–0.87	0.85	85	(1)
Carbonaceous, Type II	7	0.98–1.21	1.08	84	(1)
Carbonaceous, Type III	7	1.27–1.64	1.37	92	(1)
Bronzite	15	0.89–1.12	1.01	61	(1)
Hypersthene	23	1.00–1.31	1.10	61	(1)
Amphoterite	13	1.07–1.20	1.12	62	(1)
Enstatite, Type I	2	0.71–0.84	0.79	48	(1)
Enstatite, Type II	2	0.76–0.93	0.85	48	(1)
Calcium-poor achondrites					
Enstatite	5	0.10–1.40	0.50	20	(1)
Hypersthene	5	0.33–1.51	0.80	34	(2)
Chassignite	1	–	0.71	42	(3)
Ureilites	3	0.17–0.46	0.30	16	(4)
Calcium-rich achondrites					
Angrite	1	–	4.62	233	(5)
Nakhlite	1	–	0.92	42	(6)
Howardites	4	2.70–5.02	4.20	191	(7)
Eucrites	7	6.09–8.26	6.50	290	(1)
Stony-irons					
Pallasites (olivine)	8	0.001–0.024	0.0095	0.5	(1)
Mesosiderites (silicates)	9	3.28–6.15	4.63	211	(8)

(1) Loveland et al. (1969)
(2) Mason (1963)
(3) Dyakonova and Kharitonova (1960)
(4) Wiik (1969)
(5) Ludwig and Tschermak (1909)
(6) Prior (1912)
(7) Bununu (Mason, 1967); Frankfort (Mason and Wiik, 1966a); Kapoeta (Mason and Wiik, 1966b); Bholgati (Jarosewich, unpublished)
(8) Powell (1971)

ment of Si in the enstatite chondrites. The higher Al/Si ratios for the carbonaceous chondrites are significant. Aluminium-rich minerals have not been certainly identified in Type I and Type II carbonaceous chondrites, but Type III meteorites are known to contain aluminum-rich minerals such as spinel ($MgAl_2O_4$), anorthite ($CaAl_2Si_2O_8$), and gehlenite ($Ca_2Al_2SiO_7$) not found in other classes of chondrites. Loveland et al. point out that the Al/Na ratio increases progressively from 1.42 to 2.40 to 3.29 through the Type I, II, III carbonaceous chondrites, whereas this ratio is very uniform at about 1.45 in the ordinary chondrites.

The calcium-poor achondrites are somewhat lower in aluminum with respect to the chondrites, especially the enstatite achondrites and the ureilites; this can be correlated with their lower content of plagioclase feldspar. The calcium-rich achondrites (except the nakhlite) are notably enriched in aluminum; in the angrite this element is present in an aluminum-rich pyroxene, and in the howardites and eucrites as plagioclase. In the stony-irons the pallasite silicate is entirely olivine, which contains only traces of aluminum, hence the very low Al/Si ratio; the mesosiderites contain silicates similar in composition to the howardites, and hence show a similar Al/Si ratio.

Ahrens and Michaelis (1969) have pointed out that for all the chondrite classes, and the eucrites and howardites, the Ca/Al ratio is remarkably uniform at 1.00–1.19 (by weight), and averages 1.10

As far as is known, aluminum is entirely lithiophilic in meteorites, and lacks chalcophilic or siderophilic tendencies. In most meteorites aluminum is present almost entirely in plagioclase feldspar; meteoritic pyroxenes (except in the angrite) contain only small amounts of aluminum, and olivine only trace amounts. As mentioned above, Type III carbonaceous chondrites are noteworthy for containing aluminum-rich minerals unknown in other meteorites.

References

Ahrens, L.H. and Michaelis, H. v. (1969) *Earth Planet. Sci. Lett.* **5**, 395.
Dyakonova, M.I. and Kharitonova, V.Y. (1960) *Meteoritika* **18**, 48.
Loveland, W., Schmitt, R.A. and Fisher, D.E. (1969) *Geochim. Cosmochim. Acta* **33**, 375.
Ludwig, E. and Tschermak, G. (1909) *Tsch. Min. Pet. Mitt.* **28**, 109.
Mason, B. (1963) *Am. Museum Novitates*, No. 2155.
Mason, B. (1967) *Geochim. Cosmochim. Acta* **31**, 107.
Mason, B. and Wiik, H.B. (1966a) *Am. Museum Novitates*, No. 2272.
Mason, B. and Wiik, H.B. (1966b) *Am. Museum Novitates*, No. 2223.
Michaelis, H. v., Ahrens, L.H. and Willis, J.P. (1969) *Earth Planet. Sci. Lett.* **5**, 387.
Powell, B.N. (1971) *Geochim. Cosmochim. Acta* **35**, 5.
Prior, G.T. (1912) *Mineral. Mag.* **16**, 274.
Wiik, H.B. (1969) *Soc. Sci. Fennica, Comm. Phys.-Math.* **34**, 135.

(Received 15 February 1970)

SILICON (14)

Carleton B. Moore

Center for Meteorite Studies
Arizona State University
Tempe, Arizona

SILICON IS a major element in stony and stony-iron meteorites. The major minerals present in these meteorites are silicates, as discussed in the introductory chapter.

In classical analyses silicon is normally determined by a wet chemical gravimetric technique. Recently developed X-ray fluorescence and neutron activation non-destructive instrumental techniques have provided direct analyses for silicon. In Table 1 silicon contents have been tabulated from compilations made by Mason (1963) for carbonaceous chondrites, Mason (1965) for H and L ordinary chondrites and Mason and Wiik (1964) for LL chondrites (amphoterites). The data for enstatite chondrites have been taken from Mason (1966). Variations in the silicon contents of the chondrites are primarily due to the oxidation state and abundance of the iron present. Because of its high relatively constant abundance and the high-quality analyses, silicon is used as the basis for comparison of atomic abundances in meteorites. The abundances of individual elements are compared to 10^6 atoms of silicon. In Table 1 the mean values marked with an asterisk (*) have been used for the computations reported for other elements in this book. In addition to the wet-chemical averages, the silicon contents determined by von Michaelis et al. (1969) by X-ray fluorescence and by Vogt and Ehmann (1965) and Ehmann and Durbin (1968) by 14-MeV neutron activation are given in Table 1. The results agree quite well.

Many elements show relatively small variations with respect to the silicon abundances, but careful selection and evaluation of data indicate significant fractionation between chondrite groups. Ahrens (1964, 1965, 1967), Ahrens and von Michaelis (1968), and DuFresne and Anders (1962) have discussed

TABLE 1 Silicon (wt. % as Si) in stony and stony-iron meteorites

Name	No. of meteorites	Range	Mean	References
C–1	3	9.71–10.53	10.3*	(1)
	2	10.7–10.8	10.7	(2, 3)
	1	–	10.17	(4)
C–2	9	12.69–13.71	13.1*	(1)
	9	12.6–14.3	13.3	(2, 3)
	3	12.24–13.10	12.66	(4)
C–3	6	14.91–15.85	15.5*	(1)
	4	15.3–16.1	15.7	(2)
	4	15.50–16.29	15.84	(4)
H (Bronzite)	36	16.0–17.7	17.1*	(5)
	32	15.1–18.5	17.0	(2, 3)
	12	16.31–17.25	16.74	(4)
L (Hypersthene)	68	17.8–19.5	18.7*	(5)
	50	15.7–19.6	17.9	(2, 3)
	19	17.49–19.30	18.55	(4)
LL (Amphoterite)	12	18.1–19.3	18.8*	(6)
	9	18.7–20.0	19.2	(2, 3)
E–4 (Enstatite, Type I)	3	16.5–17.6	17.0*	(7)
	2	16.8–17.7	17.3	(2)
	2	16.30–16.65	16.45	(4)
E–5 (Enstatite, Intermed.)	2	15.6–17.1	16.4	(7)
	1	–	17.3	(3)
	1	–	17.0	(4)
E–6 (Enstatite, Type II)	4	18.0–20.5	19.5*	(7)
	7	16.1–21.7	18.5	(2, 3)
	4	17.29–22.19	19.04	(4)
Calcium-poor achondrites				
Enstatite	9	27.2–28.1	27.7*	(8)
	4	25.4–28.0	26.8	(2, 3)
Hypersthene	7	23.5–25.6	24.7*	(9)
	3	23.9–26.0	25.0	(2, 3)
Chassignite	1	–	17.5*	(10)
Ureilites	3	18.57–19.57	19.1*	(11)
	6	17.20–19.57	18.5	(18)
Calcium-rich achondrites				
Angrite	1	–	20.6*	(12)
	1	–	20.8	(3)
Nakhlite	1	–	22.9*	(13)
	1	–	23.7	(3)
Howardite	4	22.63–23.11	22.9*	(14, 15, 16)
	2	22.2–24.1	23.1	(2,3)
Eucrite	6	22.7–23.2	23.0*	(15)
	4	23.3–23.6	23.5	(2, 3)

Si/Mg, Si/Ca, Si/Al, and Si/Ti fractionation in some detail. The Si/Mg ratios in carbonaceous (1.11–1.16), ordinary (1.23–1.30), and enstatite (1.48–1.58) chondrites are significantly different.

The ratios for Si/Ca, Si/Al and Si/Ti show even greater variations between the three major chondrite groups.

In the achondrites silicon is even more abundant than in the chondrites, because of the lack of a metallic phase and the general absence of siderophilic elements. The agreement between the silicon values from wet analyses and instrumental techniques for achondrites is also very good.

Although by definition silicon is a lithophilic element, evidence also shows that under highly reducing conditions in meteorites it may also be siderophilic. Ringwood (1961) detected 2 to 5.6 atomic per cent silicon in the nickel-iron of enstatite chondrites. Detailed investigation of kamacite in the enstatite chondrites by Keil (1968) has shown that the E4, E5, and E6 chondrites have an average of 3.2, 3.3, and 1.3 weight per cent silicon in their kamacite phase.

The content of silicon in iron meteorites has proved to be a difficult quantity to measure. Early analysis such as those compiled by Farrington (1907) often reported silicon contents up to 1.2 per cent SiO_2. This was in large part due to the fact that very large sample sizes, up to 100 grams, were often taken for analysis and could contain hidden silicates. In addition, silica was dissolved from the soft soda-lime chemical glassware used at that time and hence reported in the analyses.

Modern studies of silicon in iron meteorites have been prompted by Ringwood's (1961) discovery that the metal phase of enstatite chondrites contained silicon. His primary method of analysis was based on X-ray determined lattice parameter variations. Most modern instrumental methods for the detection of trace concentrations of silicon have not proved useful for the analysis of iron meteorites. Neutron activation methods have an interference from phosphorus, and colorimetric techniques have an interference from germanium. Fredriksson and Henderson (1965) used the electron microprobe to detect 2.5 per cent silicon in the kamacite of the Horse Creek meteorite. Wasson and Wai (1970) prefer that this unusual meteorite be considered with the enstatite chondrites. More recent electron microprobe investigations include those of Wai and Wasson (1969, 1970) who reported that the concentration of silicon in iron meteorites is below the detection limit of 25×10^{-6} g/g except for a few anomalous cases. Fisher (1969) reported less than 40×10^{-6} g/g silicon in the several irons he selected for study. High silicon meteorites found by Wai and Wasson include

Tucson (0.8 per cent), Horse Creek (2.5 per cent), and Nedagolla (0.14 per cent). The advantage of the electron microprobe method is that it avoids including silicate inclusion in the sample analysed. Beaulieu and Moore (1968) used a colorimetric technique to detect 4 to 100 10^{-6} g/g silicon in iron meteorites. This work was extended (Beaulieu and Moore, 1969; and Beaulieu, 1970) to the detection of silicon in 145 iron meteorites. Meteorites containing trace amounts of magnesium and aluminum were eliminated from the calculated averages because of probable silicate contamination; of the remaining 93 samples, 8 per cent contained more than 10×10^{-6} g/g silicon, 10 per cent between 5 and 10×10^{-6} g/g silicon, and 82 per cent less than 5×10^{-6} g/g silicon. The median value was 3×10^{-6} g/g silicon. No observable correlation between silicon and iron meteorite type was detected. The meteorite Chinga with 0.04 per cent silicon was added to the list of high-silicon iron meteorites.

The isotopic composition of silicon in the Melrose chondrite was studied by Reynolds (1953) and Reynolds and Verhoogen (1953). They found that this meteoritic silicon had a δSi^{30} value of -1.0 per cent as compared with an olivine standard. Epstein and Taylor (1970) report silicon in a hypersthene achondrite and enstatite achondrite to be generally lighter than silicon in most terrestrial and lunar rocks.

References

Ahrens, L.H. (1964) *Geochim. Cosmochim. Acta* **28**, 411.
Ahrens, L.H. (1965) *Geochim. Cosmochim. Acta* **29**, 801.
Ahrens, L.H. (1967) *Geochim. Cosmochim. Acta* **31**, 861.
Ahrens, L.H. and Von Michaelis H. (1968) *Origin and Distribution of the Elements* (Ed. L.H.Ahrens), 257. Pergamon Press.
Beaulieu, P. (1970) Ph.D.Thesis, Arizona State University.
Beaulieu, P. and Moore, C.B. (1968) 31st Annual Meeting, The Meteoritical Society, Cambridge, Mass.
Beaulieu, P. and Moore, C.B. (1969) 32nd Annual Meeting, The Meteoritical Society, Houston, Texas.
Du Fresne, E.R. and Anders, E. (1963) *Geochim. Cosmochim. Acta* **26**, 1085.
Duke, M.B. and Silver, L.T. (1967) *Geochim. Cosmochim. Acta* **31**, 1637.
Dyakonova, M.I. and Kharitonova, V.Y. (1960) *Meteoritika* **18**, 66.
Ehmann, W.D. and Durbin, D.R. (1968) *Geochim. Cosmochim. Acta* **32**, 461.
Epstein, S. and Taylor, H.P., Jr., (1970) *Science* **167**, 533.
Farrington, O.C. (1907) *Field Mus. Pub., Geol. Ser.* **3**, 120.
Fisher, D.E. (1969) *Nature* **222**, 866.
Fredriksson, K. and Henderson, E.P. (1965) *Trans. Amer. Geophys. Union* **46**, 121.

Keil, K. (1968) *J. Geophys. Res.* **73**, 6945.
Ludwig, E. and Tschermak, G. (1909) *Mineralog. Petrog. Mitt.* **28**, 110.
Mason, B. (1963) *Space Sci. Rev.* **1**, 621.
Mason, B. (1963b) *Aemr. Mus. Novitates*, No.2155.
Mason, B. (1965) *Amer. Mus. Novitates*, No. 2223.
Mason, B. (1966) *Geochim. Cosmochim. Acta* **30**, 23.
Mason, B. (1966a) *Geochim. Cosmochim. Acta* **31**, 107.
Mason, B. and Wiik, H.B. (1964) *Geochim. Cosmochim. Acta* **28**, 533.
Mason, B. and Wiik, H.B. (1966b) *Am. Museum Novitates*, No. 2272.
Mason, B. and Wiik, H.B. (1966c) *Am. Museum Novitates*, No. 2273.
Prior, G.T. (1912) *Min. Mag.* **16**, 274.
Reid, A.M. and Cohen, A.J. (1967) *Geochim. Cosmochim. Acta* **31**, 661.
Reynolds, J.H. (1953) *Proc. Conf. on Nuclear Processes in Geol. Settings*, 64.
Reynolds, J.H. and Verhoogen J. (1953) *Geochim. Cosmochim. Acta* **3**, 224.
Ringwood, A.E. (1961) *Geochim. Cosmochim. Acta* **25**, 1.
Vdovykin, G.P. (1970) *Space Sci. Rev.* **10**, 483.
Vogt, J.R., and Ehmann W.D. (1965) *Geochim. Cosmochim. Acta* **29**, 373.
Von Michaelis, H., Willis, J.P., Erlank, A.J. and Ahrens, L.H. (1969) *Earth Planet. Sci. Letters* **5**, 383.
Wai, C.M. and Wasson J.T. (1969) *Geochim. Cosmochim. Acta* **33**, 1465.
Wai, C.M. and Wasson, J.T. (1970) *Geochim. Cosmochim. Acta* **34**, 408.
Wasson, J.T. and Wai, C.M. (1970) *Geochim. Cosmochim. Acta* **34**, 169.

References to Table 1

(1) Mason (1963)
(2) Vogt and Ehmann (1965)
(3) Ehmann and Durbin (1968)
(4) von Michaelis et al. (1969)
(5) Mason (1965)
(6) Mason and Wiik (1964)
(7) Mason (1966)
(8) Reid and Cohen (1967)
(9) Mason (1963b)
(10) Dyakonova and Kharitonova (1960)
(11) Wiik (Unpublished)
(12) Ludwig and Tschermak (1909)
(13) Prior (1912)
(14) Mason (1966a)
(15) Mason and Wiik (1966b)
(16 Mason and Wiik (1966c)
(17) Duke and Silver (1967)
(18) Vdovykin (1970)

(Received 2 June 1970)

PHOSPHORUS (15)

Carleton B. Moore

Center for Meteorite Studies
Arizona State University
Tempe, Arizona

PHOSPHORUS IS a minor but ubiquitous element found in all types of meteorites. It makes up about 0.1 weight per cent of the chondrites and ranges from about 0.1 to 0.4 per cent in the irons. It is usually determined in complete analyses by wet chemical colorimetric methods in both stony and iron meteorites. Phosphorus determinations have not shown a great degree of precision, and have not proven to be amenable to analysis by any particular modern instrumental technique except by X-ray fluorescence. In stones phosphorus is reported as P_2O_5, but schreibersite (iron-nickel phosphide) inclusions have been described in the metal phase of these meteorites and the actual state of the phosphorus shown by analysis is uncertain. Mason (1966) noted that although Goldschmidt classified phosphorus as siderophile in ordinary chondrites, it is present for the most part as the phosphate minerals apatite and/or whitlockite. Recent work by Fuchs (1969) has indicated that nine phosphate minerals are found in meteorites. They are whitlockite $Ca_3(PO_4)_2$; chlorapatite $Ca_5(PO_4)_3Cl$; hydroxylapatite $Ca_5(PO_4)_3OH$; sarcopside $(Fe, Mn)_3(PO_4)_2$; graftonite $(Fe, Mn)_3(PO_4)_2$; farringtonite $Mg_3(PO_4)_2$; stanfieldite $Ca_4(Mg, Fe)_5(PO_4)_6$; brianite $Na_2MgCa(PO_4)_2$ and panethite $Na_2Mg_2(PO_4)_2$.

Phosphorus has been shown to be a critical constituent in iron meteorites. Recent studies by Reed (1969), Doan and Goldstein (1969), Comerford (1969), Axon (1968) and Buchwald (1966) have discussed its occurrence as the phosphide in macroscopic (schreibersite) and microscopic (rhabdite) phases and as dissolved interstitial material, and its effect on the apparent cooling rate of these bodies. Doan and Goldstein (1969) have pointed out that total phosphorus in iron meteorites containing macroscopic schreiber-

TABLE 1 Phosphorus (wt. % as P) in stony and stony-iron meteorites

Name	No. of detns.	Range	Mean	Atoms/ 10^6 Si	References
Chondrites					
C–1	3	0.05–0.18	0.12	10,500	(1)
	1	–	0.11	9,600	(5)
	1	–	0.08	7,000	(6)
C–2	7	0.07–0.14	0.12	8,600	(1)
	3	–	0.09	6,200	(6)
C–3	5	0.09–0.17	0.14	8,200	(1)
	3	0.11–0.13	0.12	7,000	(5)
	4	0.10–0.11	0.11	6,400	(6)
	1		0.23	13,000	(3)
H (Bronzite)	36	0.00–0.23	0.11	5,800	(2)
	41	0.02–0.13	0.07	3,700	(3)
	15	0.09–0.14	0.13	6,800	(5)
	12	0.10–0.11	0.11	5,800	(6)
L (Hypersthene)	48	0.02–0.18	0.10	4,800	(3)
	68	0.00–0.30	0.12	5,800	(2)
	31	0.09–0.15	0.11	5,300	(5)
	20	0.08–0.11	0.10	4,800	(6)
LL (Amphoterites)	11	0.08–0.18	0.11	5,300	(15)
E–4 (Enstatite, Type I)	3	0.16–0.26	0.22	11,700	(4)
	2	0.20–0.21	0.20	10,600	(6)
E–5 (Enstatite, intermediate)	2	0.04–0.07	0.06	2,900	(4)
	1	–	0.19	9,100	(6)
E–6 (Enstatite, Type II)	4	0.06–0.21	0.12	5,600	(4)
	4	0.11–0.13	0.12	5,600	(6)
Calcium-poor achondrites					
Enstatite	2	0.01–0.03	0.02	650	(3)
	1	–	0.008	260	(6)
	1	–	0.003	98	(7)
Hypersthene	2	0.03–0.04	0.04	1,460	(3)
	2	0.005–0.006	0.006	220	(6)
	2	0.008–0.010	0.009	330	(7)
Chassignite	1	–	0.03	1,550	(8)
Ureilites	3	0.03–0.04	0.03	1,420	(9)
Calcium-rich achondrites					
Angrite	1	–	0.06	2,600	(3)
Nakhlite	1	–	0.05	2,000	(7)
Howardites	2	0.04–0.07	0.05	2,000	(3)
	1	–	0.005	200	(10)

TABLE 1 (cont.)

Name	No. of detns.	Range	Mean	Atoms/ 10^6 Si	References
Howardites	1	–	0.04	1,600	(6)
	1	–	0.01	400	(7)
Eucrites	7	0.005–0.08	0.05	2,000	(3)
	6	0.04–0.06	0.04	1,600	(10)
	2	0.04–0.05	0.04	1,600	(6)
	1	–	0.01	400	(7)
Stony-irons					
Mesosiderites					
Silicate	9	0.11–0.64	0.36	–	(11)
Metal	9	–	0.21	–	(16)
Total	4	0.25–0.54	0.38	30,000	(12)
Pallasites	14	–	0.16	–	(16)
Metal	4	0.05–0.37	0.16	–	(13)
	1	–	0.005	–	(14)
Olivine	1	0.08	0.08	–	(13)

site is usually underestimated. They have estimated its abundance in eleven selected meteorites by a point counting technique. Their results are compared with chemical data in Table 2. Buseck (1969) has identified a new phosphide $(Fe, Ni)_2P$ barringerite in the Ollague pallasite.

TABLE 2 Estimated total phosphorus in selected meteorites after Doan and Goldstein (1969)

Name	Ga–Ge Group	Chemical P (wt.%) (Moore et al)	Vol. % Phosphide	Estim. Total P (wt.%)
Apoala	IIIa	0.19	6	0.9
Mount Edith	IIIb		5.5	0.8
Grant	IIIb		7	1.0
Breece	IIIb		5.5	0.8
Chupaderos	IIIb		8	1.2
Tieraco Creek	IIIb		8.8	1.3
Goose Lake		0.26–0.8	6	0.9
Rodeo		0.75	6	0.9
Wallapai			7	1.0
Carlton		0.13	5	0.75
Cowra			3.5	0.50

Phosphorus in stony meteorites does not appear to show extreme variations. Table 1 gives its concentration in the various groups of stony meteorites. The iron meteorites show greater variation and grouping in their phosphorus contents. Table 3 gives the phosphorus concentration in the different iron groups in both the structural and the chemical (gallium–germanium) classifications. Results for the standard classifications show little difference between individual groups. The mean results compiled by Buddhue (1946) differ little from the new analyses by Moore et al. (1969). A comparison of phosphorus contents determined by Moore et al. (1969) within the trace

TABLE 3 Phosphorus (wt.% as P) in iron meteorites

Name	No. of meteorites	No. of analyses	Range	Mean	Reference
H Hexahedrites	34	48	–	0.29	(16)
	8	16	0.13–0.29	0.26	(17)
Octahedrites					
Ogg Coarsest	18	20	–	0.16	(16)
	4	8	0.16–0.94	0.38	(17)
Og Coarse	34	40	–	0.18	(16)
	10	20	0.04–0.61	0.20	(17)
Om Medium	92	126	–	0.18	(16)
	45	90	0.02–0.53	0.17	(17)
Of Fine	37	43	–	0.17	(16)
	17	34	0.02–0.75	0.15	(17)
Off	10	13	–	0.24	(16)
	7	14	0.02–0.30	0.11	(17)
Dr Ataxites	24	38	–	0.12	(16)
	7	14	–	0.08	(17)
Ga/Ge Group					
IIa	7		0.39–0.46	0.44	(17)
IIb	2		0.18–0.24	0.21	(17)
I	6		0.15–0.34	0.20	(17)
IIIa	13		0.09–0.36	0.15	(17)
IIIb	2		0.20–0.63	0.41	(17)
IIIab	2		0.51–0.52	0.52	(17)
IVa	10		0.02–0.16	0.05	(17)
IVb	2		0.04–0.10	0.05	(17)
IIc	1		–	0.30	(17)
IId	2		0.61–0.65	0.63	(17)

element groups defined by Wasson (1967) show that low phosphorus meteorites fall in groups IVa, IVb and IIIa, and that within group IIIa there may be a positive correlation between nickel and phosphorus.

References

Axon, H.J. (1968) *Prog. Mater. Sci.* **13**, 183.

Buchwald, V.F. (1966) *Acta Polytech. Scand. Chem. Ser.*, No. **51**.

Buddhue, J.D. (1946) *Pop. Astron.* **54**, 149.

Buseck, P.R., Moore, C.B., and Goldstein, J.I. (1967) *Geochim. Cosmochim. Acta* **31**, 1589.

Buseck, P.R. (1969) *Science* **165**, 169.

Comerford, M.F. (1969) *Meteorite Research* (Ed. P.M. Millman), 780. D. Reidel Pub. Co.

Doan, A.S., Jr. and Goldstein, J.L. (1969) *Meteorite Research* (Ed. P.M. Millman) 763. D. Reidel Pub. Co.

Duke, M.B. and Silver, L.T. (1967) *Geochim. Cosmochim. Acta* **31**, 1637.

Dyakonova, M.I. and Kharitonova, V.Y. (1960) *Meteoritika* **18**, 66.

Fuchs, L.H. (1969) *Meteorite Research* (Ed. P.M. Millman), 683. D. Reidel Pub. Co.

Greenland, L. (1964) Ph.D. Thesis, Australian National University.

Greenland, L. and Lovering, J.F. (1965) *Geochim. Cosmochim. Acta* **29**, 821.

Mason, B. (1963) *Space Sci. Rev.* **1**, 621.

Mason, B. and Wiik, H. (1964) *Geochim. Cosmochim. Acta* **28**, 533.

Mason, B. (1965) *Am. Museum Novitates*, No. 2223.

Mason, B. (1966) *Geochim. Cosmochim. Acta* **30**, 23.

Mason, B. (1966) *Geochim. Cosmochim. Acta* **30**, 365.

Moore, C.B., Lewis, C.F. and Nava, D. (1969) *Meteorite Research* (Ed. P.M. Millman), 738. D. Reidel Pub. Co.

Powell, B.N. (1969) Ph. D. Thesis, Columbia University.

Prior, G.T. (1918) *Mineral. Mag.* **18**, 151.

Reed, S.J.B. (1969) *Meteorite Research* (Ed. P.M. Millman), 749. D. Reidel Pub. Co.

Urey, H.C. and Craig, H. (1953) *Geochim. Cosmochim. Acta* **4**, 36.

Von Michaelis, H., Ahrens, L. and Willis, J.P. (1969) *Earth Planet. Sci. Letters* **5**, 387.

Wiik, H. (1956) *Geochim. Cosmochim. Acta* **9**, 279.

References to Tables

(1) H.B. Wiik (1956)

(2) B. Mason (1965)

(3) H.C. Urey and H. Craig (1953)

(4) B. Mason (1966)

(5) L. Greenland and J.F. Lovering (1965)

(6) H. Von Michaelis, L. Ahrens and J.P. Willis (1969)

(7) L. Greenland (1964)

(8) M. I. Dyakonova and V. Y. Kharitonova (1960)
(9) H. B. Wiik (unpublished)
(10) M. B. Duke and L. T. Silver (1967)
(11) B. N. Powell (1969)
(12) G. T. Prior (1918)
(13) B. Mason (1963)
(14) P. R. Buseck, C. B. Moore and J. I. Goldstein (1967)
(15) B. Mason and H. B. Wiik (1964)
(16) J. D. Buddhue (1946)
(17) C. B. Moore, C. F. Lewis and D. Nava (1969)

(Received 7 May 1970)

SULFUR (16)

Carleton B. Moore

Center for Meteorite Studies
Arizona State University
Tempe, Arizona

SULFUR IS a minor element present in all meteorites. In individual analyses of some irons, pallasites and achondrites its reported value may not be representative of the meteorite as a whole, due to a small sample size used or gross fractionation of sulfur-bearing minerals in the meteorite as a whole.

Sulfur is normally determined in stony meteorites by dissolving the specimen in an oxidizing solution to form sulfate, which after suitable separation is gravimetrically determined as $BaSO_4$. In iron meteorites and/or where present in trace amounts it may also be evolved as hydrogen sulfide in an acid dissolution or sulfur dioxide by combustion and detected by iodometric titration. Estimates of sulfur content may also be made by the modal analysis of troilite followed by the theoretical calculation of total sulfur. Sulfur is always determined in total analyses of chondrites but may be neglected in the analysis of iron meteorites.

The total content of sulfur in ordinary H, L and LL chondrites is very constant. The compilations of total sulfur in these chondrites made by Urey and Craig (1953) and Mason (1965) show very good agreement. The broad range in values for individual analyses is probably due to sampling variations. Both carbonaceous and enstatite chondrites have higher sulfur contents than the ordinary chondrites. Larimer and Anders (1967) place sulfur in their normal group of depleted elements. In this group the elements in ordinary chondrites are about 0.25 their abundance in Type I carbonaceous chondrites. The data on sulfur in different classes of stony and stony-iron meteorites are summarized in Table 1. As noted above, the data for the chondrites appear to be good but those for the other meteorites are of limited use, due to the heterogeneous distribution of sulfur-bearing phases and small sample sizes.

TABLE 1 Sulfur (wt.% as S) in stony and stony-iron meteorites

Name	No. of detns.	Range	Mean	Atoms/10^6 Si	References
Chondrites					
C–1	5	5.01–6.70	5.90	5.0×10^5	(1)
C–2	9	2.80–5.44	3.42	2.3×10^5	(1)
C–3	9	1.31–2.66	2.19	1.2×10^5	(1)
H Bronzite	38	0.92–2.63	2.07	1.1×10^5	(2)
	35	0.92–2.40	1.93	1.0×10^5	(3)
L Hypersthene	48	0.78–2.87	2.11	1.0×10^5	(2)
	67	1.26–2.95	2.22	1.0×10^5	(3)
LL Amphoterites	12	1.82–2.59	2.26	1.1×10^5	(4)
E4 (Enstatite-Type I)	3	5.65–6.12	5.85	3.0×10^5	(5)
E5 (Enstatite-Intermediate)	2	5.50–5.82	5.66	2.8×10^5	(5)
E6 (Enstatite-Type II)	6	2.62–4.44	3.32	1.5×10^5	(5)
Calcium-poor achondrites					
Enstatite	2	3.20–0.46	0.39	1.3×10^4	(2), (6)
Hypersthene	5	0.13–0.63	0.38	1.3×10^4	(2)
Ureilites	5	0.33–0.58	0.50	2.3×10^4	(7)
Calcium-rich achondrites					
Angrite	1	–	0.45	1.9×10^4	(2)
Nakhlite	1	–	0.06	2.3×10^3	(2)
Howardites	3	0.10–0.48	0.27	1.0×10^4	(2)
	1	–	0.06	2.3×10^3	(8)
Eucrites	16	0.02–0.51	0.20	7.6×10^3	(2)
	5	0.02–0.26	0.13	5.0×10^3	(8)
Stony irons					
Mesosiderites	10	0.22–4.64	1.10	8.3×10^4	(9)
Pallasites*	10*	–	0.19	–	(10)

* Reconstructed from analyses of 10 stony phase samples by ref. 10 and 14 metallic phase samples by ref. 11.

The broad range noted for sulfur in the mesosiderites shows broad-scale heterogeneity in these stony-iron meteorites.

In most meteorites sulfur is found predominately as the mineral troilite and is by definition chalcophile. Other common sulfur-bearing minerals found in many meteorites include oldhamite (CaS) and daubreelite ($FeCr_2S_4$). In Type I and II carbonaceous chondrites sulfur is found predominately as free sulfur, sulfate minerals, and in organic compounds (DuFresne and Anders, 1962; Hayes, 1967). Pentlandite [(Fe, Ni)$_9$S$_8$] is found in Type III carbonaceous chondrites.

Mason (1967) has reviewed the minerals of meteorites in some detail. Those reported in complex sulfide nodules in iron meteorites include; mackinawite, (FeS) (El Goresy, 1965), gentnerite ($Cu_8Fe_3Cr_{11}S_{18}$) (El Goresy and Ottemann, 1966), and brezinaite (Cr_3S_4) (Bunch and Fuchs, 1969). A report of stony meteorites by Ramdohr (1963) indicated the following sulfur-bearing minerals were observed; pyrite (FeS_2), sphalerite (ZnS), cubanite ($CuFe_2S_3$), chalcopyrite ($CuFeS_2$), and valleriite ($CuFeS_2$). The highly-reduced enstatite chondrites are reported to contain the following minerals: alabandite [(Mn, Fe)S] (Dawson et al., 1960), niningerite [(Mg, Fe)S] (Keil and Snetsinger, 1967), and djerfisherite ($K_3CuFe_{12}S_{14}$) (Fuchs, 1966). Keil (1968) gave detailed analyses of many of these minerals found in enstatite chondrites, and found from 0.07 to 0.24 weight per cent sulfur in schreibersite from these meteorites.

Most samples of iron meteorites taken for chemical analysis avoid the sparse randomly-distributed sulfide nodules found in these meteorites. The reported sulfur contents for iron meteorites are low, perhaps by several orders of magnitude. In Table 2 the mean concentrations compiled by Buddhue (1946) are compared with the analyses of Moore et al. (1969). The larger sulfur concentrations in the earlier analyses may be due to hidden sulfide nodules within large samples. The values reported by Moore et al. (1969) may be representative of sulfur dissolved in the metallic phases. Table 3 taken from Wood (1962) shows attempts by Henderson and Perry (1958) and Chirvinskii (1948) to estimate the true sulfur content of iron meteorites. They estimated the sulfur content by modal analyses of sulfide nodules in large slices of selected meteorites.

The isotopic abundance of sulfur in meteorites has been studied by MacNamara and Thode (1950), Vinogradov et al. (1957), Ault and Kulp (1959), Thode et al. (1961), Hulston and Thode (1965), and Kaplan and Hulston (1966). The isotopic composition in all meteorites is very constant. Most accepted analyses fall within $\delta S^{34} \pm 1.0$ per cent of the Canyon Diablo troilite standard. Kaplan and Hulston (1966) and Monster et al. (1965) have shown that the different sulfur-bearing phases from carbonaceous chondrites may vary from $\delta S^{34} - 1.7$ per cent for sulfate sulfur to $\delta S^{34} + 3.3$ per cent for elemental sulfur, but that in the bulk meteorites δS^{34} falls within ± 1 per cent of Canyon Diablo. Hulston and Thode (1965b) found evidence of cosmic-ray produced S^{36} and S^{33} in several iron meteorites.

TABLE 2 Sulfur (wt.% as S) in iron meteorites

Name	No. of meteorites	No. of analyses	Range	Mean	References
H Hexahedrites	34	48	–	0.06	(11)
	8	16	0.003–0.50	0.014*	(12)
Octahedrites					
Ogg Coarsest	18	20	–	0.02	(11)
	3	6	0.003–0.020	0.009	(12)
Og Coarse	34	40	–	0.08	(11)
	10	20	0.002–0.24	0.008†	(12)
Om Medium	92	126	–	0.09	(11)
	43	86	0.001–0.063	0.009	(12)
Of Fine	37	43	–	0.08	(11)
	18	36	0.001–0.056	0.014	(12)
Off Finest	10	13	–	0.63	(11)
	6	12	0.003–0.021	0.011	(12)
Dr Ataxites	24	38	–	0.08	(11)
	5	10	0.003–0.013	0.007	(12)

* minus one high value of 0.50 wt.% S.
† minus one high value of 0.24 wt.% S.

TABLE 3 Sulfur contents of selected iron meteorites calculated from micrometric measurements of troilite nodules in large sliced surfaces after J.A. Wood (1963)

Name	Classi-fication	Wt.% S	References
Canyon Diablo	Ogg	2.18	(13)
Osseo	Ogg	1.25	(13)
Coolac	Og	1.74	(13)
Odessa	Og	1.25	(13)
Wichita Co.	Og	1.00	(13)
Kingston	Om	0.57	(14)
Toubil River	Om	0.45	(14)
Cape York	Om	0.18	(14)
Augustinovka	Om–Of	1.49	(14)
Gibeon	Of	0.37	(14)
St. Genevieve Co.	Of	0.11	(14)

References

Ault, W.U. and Kulp, J.L. (1959) *Geochim. Cosmochim. Acta* **16**, 201.
Buddhue, J.D. (1946) *Pop. Astron.* **54**, 149.
Bunch, T.E. and Fuchs, L.H. (1969) *Am. Min.* **54**, 1509.
Chirvinskii, P.N. (1948) *Meteoritika* No. **4**, 71.
Dawson, K.R., Maxwell, J.A. and Parsons, D.E. (1960) *Geochim. Cosmochim. Acta* **21**, 127.
DuFresne, E.R. and Anders, E. (1962) *Geochim. Cosmochim. Acta* **26**, 1085.
Duke, M.B. and Silver, L.T. (1967) *Geochim. Cosmochim. Acta* **31**, 1637.
Dyakonova, M.I. and Kharitonova, V.Y. (1960) *Meteoritika* **18**, 66.
El Goresy, A. (1965) *Geochim. Cosmochim. Acta* **29**, 1131.
El Goresy, A. and Ottemann, J. (1966) *Z. Naturforsch.* **21A**, 1160.
Fuchs, L.H. (1966) *Science* **153**, 166.
Hayes, J.M. (1967) *Geochim. Cosmochim. Acta* **31**, 1395.
Henderson, E.P. and Perry, S.H. (1958) *Proc. U.S. Nat. Mus.* **107**, 339.
Hulston, J.R. and Thode, H.G. (1965) *J. Geophys. Res.* **70**, 3475.
Hulston, J.R. and Thode, H.G. (1965b) *J. Geophys. Res.* **70**, 4435.
Kaplan, I.R. and Hulston, J.R. (1966) *Geochim. Cosmochim. Acta* **30**, 479.
Keil, K. (1968) *J. Geophys. Res.* **73**, 6945.
Keil, K. and Snetsinger, K.G. (1967) *Science* **155**, 451.
Larimer, J.L. and Anders, E. (1967) *Geochim. Cosmochim. Acta* **31**, 1239.
MacNamara, J. and Thode, H.G. (1950) *Phys. Rev.* **78**, 307.
Mason, B. (1963) *Space Sci. Rev.* **1**, 621.
Mason, B. (1965) *Am. Museum Novitates*, No. 2223.
Mason, B. (1965) *Am. Museum Novitates*, No. 2223.
Mason, B. (1966) *Geochim. Cosmochim. Acta* **30**, 23.
Mason, B. (1967) *Amer. Mineralogist* **52**, 307.
Mason, B. and Wiik, H.B. (1964) *Geochim. Cosmochim. Acta* **28**, 533.
Monster, J., Anders, E. and Thode, H.G. (1965) *Geochim. Cosmochim. Acta* **29**, 773.
Moore, C.B., Lewis, C.F. and Nava, D. (1969) in *Meteorite Research* (Ed. P.M. Millman), 749. D. Reidel Pub. Co.
Powell, B.N. (1969) *Geochim. Cosmochim. Acta* **33**, 789.
Ramdohr, P. (1963) *J. Geophys. Res.* **68**, 2011.
Thode, H.G., Monster, J. and Dunford, H.B. (1961) *Geochim. Cosmochim. Acta* **25**, 159.
Urey, H. and Craig, H. (1953) *Geochim. Cosmochim. Acta* **4**, 36.
Vdovykin, G.P. (1970) *Space Sci. Rev.* **10**, 483.
Vinogradov, A.P., Chupakhin, M.S. and Grinenko, V.A. (1957) *Geochemistry* **3**, 221.
Wood, J.A. (1963) in *Moon, Meteorites and Comets* (Eds. B. Middlehurst and G.P. Kuiper), 337. Univ. of Chicago Press.

References to Tables

(1) Mason (1963)
(2) Urey and Craig (1953)
(3) Mason (1965)
(4) Mason and Wiik (1964)
(5) Mason (1966)
(6) Dyakonova and Kharitonova (1960)
(7) Vdovykin (1970)
(8) Duke and Silver (1967)
(9) Powell (1969)
(10) Wood (1963)
(11) Buddhue (1946)
(12) Moore, Lewis and Nava (1969)
(13) Henderson and Perry (1968)
(14) Chirvinskii (1948)

(Received 2 June 1970)

CHLORINE (17)*

George W. Reed, Jr.

Argonne National Laboratory
Argonne, Illinois 60439

RECENT STUDIES have been undertaken to establish the total Cl in meteorites and to identify the mineral(s) with which it is associated. Concentrations ranging from a few to several hundred ppm have been reported and association of Cl with apatite, originally reported by Shannon and Larsen (1925), has been confirmed. The site(s) of chlorine is unknown in meteorites in which apatite has not been observed, although there is evidence for the occurrence of lawrencite, $FeCl_2$.

Most modern determinations have utilized the tool of neutron activation analysis. As usually employed (von Gunten et al., 1965; Reed and Allen, 1966; Goles et al., 1967; Reed and Jovanovic, 1969) this technique requires dissolution of the sample in the presence of milligram amounts of the element to be measured, under conditions that will ensure isotopic equilibration between the irradiation-produced radioactive tracer and the added carrier. These are then radiochemically isolated and the radioactivity measured. A variation of this method has been employed by Quijano-Rico and Wänke (1969) in which the chlorine in the meteorite, acting as its own carrier, was distilled pyrohydrolytically at 1000 °C. Greenland and Lovering (1965) separated Cl by dry distillation at 1700 °C, and determined it by micro-diffusion of Cl_2 gas into a KI solution producing I_2 which was titrated.

The amount of Cl is variable within and between classes of chondrites. On an approximately one-gram sampling scale the Cl content varies greatly. Reed and Allen (1966), for instance, found Cl concentrations in Bruderheim ranging from 2.4 to 50 ppm. Goles et al. (1967) found 78 to 210 ppm Cl in Hvittis; for this same meteorite, von Gunten et al. (1965) report 199.8 and

* Work supported by the U.S. Atomic Energy Commission.

243.8 ppm, Reed and Allen (1966) report 323 ppm, and Greenland and Lovering (1965) obtained 250 ppm. A factor of two or more variation between values reported by different investigators is not unusual. The greatest spread observed in any of the studies was that quoted above for Bruderheim measured by Reed and Allen.

As pointed out by Goles et al., it is most disconcerting that similar large disagreements exist in the data for the carbonaceous chondrites. The Type I chlorine data of Reed and Allen and Greenland and Lovering lie between 200 and 300 ppm whereas Goles et al. find $\sim$100 to 800 ppm. Mighei, a type II carbonaceous chondrite, was reported to contain 350 ppm Cl by Reed and Allen and 430 and 510 ppm by Goles et al. Four type III carbonaceous chondrites measured by Greenland and Lovering averaged 348 ppm; two measured by Reed and Allen averaged 121 ppm; two measured by Goles et al. averaged 272 ppm; Quijano-Rico and Wänke report 234 ppm.

The Type I carbonaceous chondrites are thought to be the most primitive, and the least altered by fractional losses of more volatile constituents. The other two classes grade toward higher volatile element losses. Goles et al. data show this but the data of the other investigators do not support this trend.

In spite of the disagreements between laboratories, occasionally striking agreement is observed. Most results for 23 bronzites measured by Quijano-Rico and Wänke fall between 32 and 210 ppm; only Allegan gave a lower value, 7 ppm. This latter is in agreement with the 10.2 and 7.9 ppm reported by Reed and Allen.

Chlorine data for chondrites are summarized in Table 1. There seems little justification for selecting data; perhaps with more controlled sampling the spread will be reduced. If the dispersion is real on the sample scale used ($\sim$1 gm), then the most representative value for the Cl content of a given class of meteorite is that based on the weighted average. There are several meteorites for which enough determinations have been reported to warrant such averaging. They are listed in Table 2 along with sample weights as reported or as assumed by the writer. Up to 20 grams of sample were used by Greenland and Lovering; thus their values may already represent good averages and are not included in the averaging.

The chondrite, Pantar, has a light-dark structure. This type of bronzite chondrite is of especial interest since the dark part has been shown to be enriched in volatile elements such as the rare gases relative to the light phase or to other ordinary bronzite chondrites. The Cl content of the light part has been reported as 50, 52 and 56 ppm by Reed and Allen, Goles

Table 1 Chlorine contents of chondrites*

Meteorite classification	No. of determinations	No. of meteorites	Range (ppm)	Mean (ppm)	At/ 10^6Si‡	Ref.§
Hypersthene (L)	5	1	2.4–5.1	3.7†	15.7	A
	10	3	5.4–96.0	42.5	216	A
	16	5	24.6–270.9	128.2		B
	9	9	92–270	160		C
	12	5	38–130	80.4	298	D
	8	8	27–212	74.5		E
Amphoterite (LL)	2	1	57.3, 57.8	57.6		B
	1	1		190	770	D
	2	2	89, 230	160		C
	2	2	131, 121	126		E
Bronzite (H)	2	1	91–103	97		B
	3	2	0.44–10.2	6.2	33	A
	4	4	57–170	127		C
	2	1	74, 80	77	250	D
	23	23	7–210			E
Enstatite (E–4)	2	1	472.6–528.8	500.7		B
	3	2	432–770	554	2,800	A
	2	2	570–750	660	3,050	D
	1	1	570–750	900		C
	1	1		994		E
Enstatite (E–5, 6)	2	1	199.8, 243.8	221.8		B
	1	1		323	1,330	A
	2	1	78, 210	144	~660	D
	4	4	160–250	213		C
	1	1		234		E
Carbonaceous-I	3	2	210–320	257	1,750	A
	4	2	720–840	772.5	5,700	D
	1	1		290		C
Carbonaceous-II	1	1		350	2,110	A
	4	2	190–510	335	2,050	D
	1	1		108		E
	1	1		370		C
Carbonaceous-III	1	1		125	622	A
	2	2	350–360	355		C
	3	2	260–288	273	1,350	D
	3	3	248–423	309		E
Carbonaceous-IV	1	1		117	583	A
	1	1		310		C
	1	1		45		E

* Falls only.

† Average of 5 low values for Bruderheim does not include a result of 50 ppm.

‡ Abundances reported by authors, sometimes using selected data.

§ References: A = Reed and Allen, B = von Gunten *et al.*, C = Greenland and Lovering, D = Goles *et al.*, and E = Quijano-Rico and Wänke.

TABLE 2 Chlorine contents—weighted averages

Sample	Investigators	Weighted average	No. of measurements	Total weight of meteorite
Bruderheim	Reed and Allen	14.8	(5)	5.46
	Von Gunten *et al.*	102.3	(3)	1.26
	Goles *et al.*	103.3	(6)	5.15
Holbrook	Reed and Allen	27.2	(4)	4.89
	Greenland and Lovering	110	(*)	
Mocs	Von Gunten *et al.*	133.6	(5)	2.11
	Goles *et al.*	72.2	(2)	2.22
	Greenland and Lovering	220	(*)	
Hvittis	(†)	214.4	(6)	2.96
	Greenland and Lovering	250	(*)	

* These workers used up to 30 gm of some ordinary chondrites. The samples were crushed and quartered hence their results are based on a representative sample.

† Average of single and/or duplicate measurements by Von Gunten *et al.*, Reed and Allen, Goles *et al.* and Quijano-Rico and Wänke.

et al., and Quijano-Rico and Wänke, respectively. Reed and Allen found 33 ppm Cl in the dark phase whereas Quijano-Rico and Wänke found 66; in either case no significant enrichment is evident.

Chlorine contents of achondrites are summarized in Table 3. The concentrations are on the average lower than those of the chondrites.

TABLE 3 Chlorine in achondrites

Meteorite classification	No. of determinations	No. of meteorites	Range (ppm)	Mean (ppm)	At/10^6 Si	Reference
Hypersthene	1	1		13		Quijano-Rico and Wänke
Enstatite	4	1	1.9–5.6	3.8		Von Gunten *et al.*
Howardite	1	1		14.9		Reed and Jovanovic
Eucrite	2	1	26, 21.5	23.8		Reed and Jovanovic
	3	3	8–34.5	20.2		Quijano-Rico and Wänke

The only results for Cl in iron meteorites are those by Goles et al.; they report 27 and 82 ppm for two samples of the coarse octahedrite, Odessa, and 24 and 27 ppm for the very coarse octahedrite, Sardis.

The Cl in iron meteorites is thought to be associated with the mineral lawrencite, $FeCl_2$. Phosphate inclusions found in the Mt. Stirling iron meteorite by Fuchs and Olsen (1965) proved to be chlorapatite. Chlorine determined by Reed and Allen in a 100 mg sample of this apatite was about 4.3%, in agreement with the 5.3% measured by Fuchs and Olsen using the electron microprobe. Chlorapatite has now been reported in three iron meteorites: in graphite-troilite nodules in Odessa (Marshall and Keil, 1965) and Mt. Stirling (Bunch and Olsen, 1965) and in silicate inclusions in Weekeroo Station (Bunch and Olsen, 1968). It also occurs in the mesosiderite Vaca Muerta (Marvin and Klein, 1964). The Cl in Odessa reported by Goles et al. is consistent with the above observations. The Cl in Sardis (Goles et al.) might more likely be associated with lawrencite which has been blamed for the severe weathering of this meteorite.

In Fuchs' (1969) summary of the phosphate mineralogy of meteorites, those chondrites in which chlorapatite has been observed are listed. These include several hypersthene chondrites, one LL and one C–4 carbonaceous chondrite. Recent work by Van Schmus and Ribbe (1969) has added another case each to the L and LL groups, and for the first time an H-group chondrite was found that contains chlorapatite. All the chlorapatite analyses indicate a Cl content of the apatite of 5.3% with a spread of no more than 0.5%. This is compared with 6.8% for stoichiometric chlorapatite. Van Schmus and Ribbe report approximately 0.4% F and assume a few per cent OH to bring the Cl + F + OH molecular proportion to that required for stoichiometry.

The low Cl in Allegan could be considered to be consistent with the rarity of chlorapatite in bronzite chondrites were it not that most bronzite chondrites contain the same 100–200 ppm Cl found in the hypersthene chondrites. This Cl must be associated with another mineral phase, or chlorapatite, if present, is in submicron size grains which have not been detected. The phosphorus in the enstatite chondrites is in the reduced state as schreibersite, the common phosphorus phase found in iron meteorites. Thus the appreciable Cl concentration in this class of meteorites must be in other mineral phases; at least two would seem to be required to explain the large amounts of Cl that can be leached from these meteorites with water as observed by Von Gunten et al. and by Reed and Allen.

References

Berkey, E. and Fisher, D.E. (1967) "The abundance and distribution of chlorine in iron meteorites", *Geochim. Cosmochim. Acta* **31**, 1543.

Bunch, T.E. and Olsen, E. (1968) "Potassium feldspar in Weekeroo Station, Kodaikanal and Colomera iron meteorites", *Science* **160**, 1223.

Fuchs, L.H. (1969) "The phosphate mineralogy of meteorites" in *Meteorite Research*, P.M. Millman, ed. (Reidel, Dordrecht) 683–695.

Fuchs, L.H. and Olsen, E. (1965) "The occurrence of chlorapatite in the Mt. Stirling octahedrite", *Trans. Am. Geophys. Un.* **46**, 122.

Goles, G.G., Greenland, L.P. and D.Y. Jérome (1967) "Abundances of chlorine, bromine, and iodine in meteorites", *Geochim. Cosmochim. Acta* **31**, 1771–1787.

Greenland, L. and Lovering, J.J. (1965) "Minor and trace element abundances in chondrite meteorites", *Geochim. Cosmochim. Acta* **29**, 821–858.

Marvin, U.B. and C. Klein (1964) "Meteoritic zircon", *Science* **146**, 9190.

Quijano-Rico, M. and Wänke, H. (1969) "Determination of boron, lithium, and chlorine in meteorites" in *Meteorite Research*, P.M. Millman, ed. (Reidel, Dordrecht) 132–145.

Reed, G.W. and Allen, R.O. (1966) "Halogens in chondrites", *Geochim. Cosmochim. Acta* **30**, 779–800.

Reed, G.W. Jr. and Jovanovic, S. (1969) "Some halogen measurements on achondrites", *Earth Planet. Sci. Lett.* **6**, 316–320.

Shannon, E.V. and Larsen, E.S. (1925) "Merrillite and chlorapatite from stony meteorites", *Am. J. Sci.* **9**, 250.

Van Schmus, W.R. and Ribbe, P.H. (1969) "Compositions of phosphate minerals in ordinary chondrites", *Geochim. Cosmochim. Acta* **33**, 637–640.

(Received 25 September 1970)

POTASSIUM (19)

Gordon G. Goles

Center for Volcanology
and
Departments of Chemistry and Geology
University of Oregon
Eugene, Oregon 97403

POTASSIUM IS a trace element in almost all known meteoritic specimens. It is usually determined in conventional bulk analyses, often with results of marginal or no significance. Fortunately, a variety of analytical techniques has been used to estimate K contents, including isotope dilution (often in connection with ^{40}K–^{40}Ar geochronology), flame photometric, x-ray fluorescence and activation analysis methods. Consequently, data of apparently adequate accuracy are available in considerable numbers.

Potassium contents determined by direct gamma-ray spectrometry allow us to test effects of sampling errors in an illuminating way. Table 1 presents comparisons between K contents of large specimens as determined by gamma spectrometry and results of isotope dilution measurements on small specimens from the same meteorites. Where available, results obtained by other techniques are also given for comparison. Chondrite classes are identified by the code introduced by Van Schmus and Wood (1967), achondrites by that of Keil (1969).

The second meteorite listed, *Murray*, is known to have anomalous Na and halogen contents (see Na review; Goles, Greenland and Jérome, 1967). Within the context of the variable K contents of the C2 class, the values for *Murray* do not seem to be anomalous.

The H chondrite *Beardsley* has an unusually high K content. Although it was an observed fall, some specimens were subjected to a little more than a year's weathering on Earth's surface before being collected. The weathered specimens appear to have systematically lower K contents than does the

TABLE 1 Comparisons of potassium contents in stone meteorites

Meteorite (Class)	Reference (Method)	K content (10^{-6} g/g)	Comments on sample analysed
Mighei (C2)	(1) (DGS)	380 ± 30	0.750 kg.
	(2) (ID)	423	CNHM 1456; "2A", 0.2289 g.
		432	CNHM 1456; "2B".
	(3) (FP)	450	Average of 3 analyses of 1 g samples.
	(4) (XRF)	430	
	(5) (FP)	420	Determined as part of bulk analysis.
Murray (C2)	(1)	450 ± 50	1.24 kg.
	(6) (ID)	311	"I", 0.2260 g.
		312	"II", 0.2400 g.
	(7) (ID)	266	
	(3)	320	Average of 3 analyses of 1 g samples.
	(4)	260	
	(5)	330	Determined as park of bulk analysis.
Pueblito de Allende (C3, 4)	(8) (DGS)	230 to 280	NMNH 3499, 3515, 3525; four fragments totalling 2.597 kg.
	(2)	229	0.1283 g from King, Houston.
Beardsley (H5)	(1)	1250 ± 40, 1190 ± 40	"II", fresh, 0.637 kg.
		900 ± 90	"I", weathered, 0.857 kg.
		1050 ± 40	"III", weathered, 4.28 kg.
	(9) (ID)	1010	"A", weathered, 3.600 g split of 115 g sample.
		1010	"B", weathered, 1.980 g split of 115 g sample.
	(10) (ID)	1000 ± 20	1 g split of ~15 g sample.
	(11) (ID)	917 ± 13	Nininger 1349; presumably weathered.
	(12) (ID)	1232	"A", weathered, 0.2160 g.
		1197	"B", fresh, 0.2938 g.
	(3)	1024	Weathered; average of 5 analyses of 1 g samples.

TABLE 1 (cont.)

Meteorite (Class)	Reference (Method)	K content (10^{-6} g/g)	Comments on sample analysed
Beardsley (H5)	(4)	1090	
	(13) (FP)	1050	Fresh; average of 2 analyses of 1 g samples.
	(14) (NAA)	1160 ± 50	Approximately 1 g.
Forest City (H5)	(1)	780 ± 60	"I", 0.432 kg.
		720 ± 40	"II", 4.22 kg.
	(9)	827	"A", 0.792 g split of 71.762 g sample.
		835	"B", 0.792 g split of 71.762 g sample.
	(11)	820 ± 14	ANHM 2406.
	(15) (ID)	841	Approximately 1 g.
	(16) (ID)	820	"I", 0.231 g.
		765	"II", 0.038 g.
		905	"IIIA", 0.099 g.
		878	Split of "IVA"; 0.126 g.
		890	Split of "IVA"; 0.108 g.
	(17) (ID)	786.0	0.92 g sample, spiked during dissolution.
		787.9	0.50 g.
	(3)	842	Average of 4 analyses of 1 g samples.
	(4)	790	
	(5)	600, 840	Determined as parts of bulk analyses.
Plainview (H5)	(1)	840 ± 30	1.34 kg.
	(12)	919	0.2515 g.
	(15)	789	Approximately 1 g.
	(4)	740	
Richardton (H5)	(1)	840 ± 40	"I", 0.620 kg.
		690 ± 30	"II", 5.68 kg.
	(6)	676	"I", 0.3223 g.
		684	"II", 0.334 g.
	(10)	830 ± 20	1 g split of ~15 g sample.
	(11)	818 ± 12	Nininger 100.
	(12)	642	0.5875 g.
	(15)	812	Approximately 1 g.

TABLE 1 (cont.)

Meteorite (Class)	Reference (Method)	K content (10^{-6} g/g)	Comments on sample analysed
Richardton (H5)	(3)	852	Average of 6 analyses of 1 g samples.
	(4)	720	
	(5)	1100	Determined as part of a bulk analysis.
Bruderheim (L6)	(18) (DGS)	890 ± 40	2.15 kg.
	(19) (DGS)	800 ± 130	
	(15)	890	Approximately 1 g.
	(20) (ID)	896 ± 46	
	(21) (FP)	910 ± 50	Dispersion estimate from 3 replicates.
Holbrook (L6)	(1)	830 ± 40	1.12 kg.
	(10)	880 ± 20	1 g split of ~15 g sample.
	(3)	877	Average of 4 analyses of 1 g samples.
	(4)	840	
	(5)	800	Determined as part of a bulk analysis.
Mocs (L6)	(1)	900 ± 40	0.681 kg.
	(10)	880 ± 20	1 g split of ~15 g sample.
	(3)	877	Average of 4 analyses of 1 g samples.
	(5)	1,070	Determined as part of a bulk analysis.
	(14)	1040 ± 50	Approximately 1 g.
Peace River (L6)	(19)	940 ± 170	
	(7)	756	"I"
		752	"II"
		718	"III", chondrule
	(21)	870 ± 100	Dispersion estimate from 2 replicates.
	(22) (FP)	910	Determined as part of a bulk analysis.

TABLE 1 (cont.)

Meteorite (Class)	Reference (Method)	K content (10^{-6} g/g)	Comments on sample analysed
Potter (L6)	(1)	730 ± 30	"I", 1.44 kg.
		690 ± 30	"II", 1.03 kg.
	(15)	790	Approximately 1 g.
	(13)	790	1 g.
Cherokee Springs (LL5–6)	(1)	780 ± 30	"I", 2.68 kg.
		770 ± 30	"II", 5.02 kg.
	(23) (ID)	875	0.1976 g.
	(24) (FP)	900	Determined as part of a bulk analysis.
Abee (E4)	(1)	870 ± 40	2.71 kg.
	(15)	822	Approximately 1 g.
	(17)	867.8	0.826 g split of 3.2 g sample.
	(25) (ID)	867	"I", 0.1416 g.
		840	"II", 0.2223 g.
	(26) (FP)	830	Matrix.
		1,200	Fragment (clast).
Indarch (E4)	(1)	870 ± 30	1.78 kg.
	(15)	905	Approximately 1 g.
	(25)	929	"I", 0.2178 g.
		913	"II", 0.1728 g.
	(4)	800	
	(5)	900	Determined as part of a bulk analysis.
	(13)	880	1 g.
	(27) (FP)	840	1 g.
	(28) (NAA)	830 ± 20	Approximately 2 g.
Bishopville (Ae)	(1)	1260 ± 60	0.583 kg.
	(15)	199	Approximately 1 g.
	(29) (ID)	1320	"1–A"; approximately 0.5 g.
		1129, 1132	"2–A"; approximately 0.5 g.
		1132	"2–B"; approximately 0.5 g.
	(3)	830	Average of 2 analyses of 1 g samples.
	(13)	1230	Average of 4 analyses of 1 g samples.
	(27)	860	1 g.

TABLE 1 (cont.)

Meteorite (Class)	Reference (Method)	K content (10^{-6} g/g)	Comments on sample analysed
Norton Co. (Ae)	(1)	120 ± 30	1.59 kg.
	(10)	70 ± 20	1 g split of ~15 g sample.
	(15)	67	Interior; approximately 1 g.
		74	Exterior; approximately 1 g.
		68	From Moscow collection; approximately 1 g.
	(16)	69	523.3X (from Nininger Meteorite collection, Tempe); "IIA", 0.533 g.
		62	523.3X; "IIB", 0.643 g.
		53	523.3X; "IV", 0.814 g.
		107	4965E (from Nichiporuk, California Institute of Technology); "A", 1.071 g.
		107	4965E; "B", 0.428 g.
		42	Coarse enstatite from 4965E; 0.607 g.
	(30) (ID?)	230	
	(31) (ID)	23 ± 5	From Nichiporuk, CIT; "3"
		13 ± 4	From Nininger Meteorite collection, Tempe; "4"
		102 ± 2	From Nininger Meteorite collection, Tempe; "5"
		83 ± 3, 86 ± 3	From Nininger Meteorite collection, Tempe; "6"
		61 ± 3	From Nininger Meteorite collection, Tempe; "7"
	(4)	70	
	(5)	300	Determined as part of a bulk analysis.
Peña Blanca Spring (Ae)	(1)	≲120 at 3σ level	0.439 kg.
	(15)	296	Approximately 1 g.
Juvinas (Ap)	(1)	480 ± 60	0.833 kg.
	(15)	236	Approximately 1 g.
	(17)	322	0.249 g split of 4.354 g sample; spiked during dissolution.
	(3)	380	Average of 2 analyses of 1 g samples.

TABLE 1 (cont.)

Meteorite (Class)	Reference (Method)	K content (10^{-6} g/g)	Comments on sample analysed
Juvinas (Ap)	(27)	310	1 g.
	(32) (NAA)	230 ± 10	Weighted average of analyses of washed and sieved separates; about 2 g total.
		359 ± 10	Chunk; minimal handling.
Moore Co. (Ap)	(1)	150 ± 30	1.19 kg.
	(10)	210 ± 150	1 g split of ~15 g sample.
	(11)	187 ± 15	From Henderson, USNM.
	(17)	159	0.182 g.
	(19)	526	"1–B", ~0.5 g; suspected to be too high.
	(27)	210	1 g.
Nuevo Laredo (Ap)	(1)	500 ± 150	0.202 kg.
	(11)	376 ± 13	Both samples from Patterson, California Institute of Technology.
		358 ± 16	
	(17)	414	0.162 g split of 0.738 g sample.
	(33) (ID)	433	
	(13)	470	1 g; from Patterson, CIT.
Pasamonte (Ap)	(1)	380 ± 30	0.873 kg.
	(10)	430 ± 30	1 g split of Edwards' (11) sample.
	(11)	436 ± 13	"White"; split of Edwards' sample.
		413 ± 12	
	(17)	327	"197 g"; average of 3 analyses of splits from 0.35 g sample.
		276	"297y"; average of 2 analyses of a 1.08 g sample.
	(4)	330	
	(13)	570	"Grey"; 1 g.
		520	"White"; 1 g.
	(14)	380 ± 30	Approximately 1 g.
	(34) (NAA)	340 ± 50	Approximately 1 g.
Sioux Co. (Ap)	(1)	240 ± 40	0.998 kg.
	(11)	322 ± 20	Splits of sample Nininger 298.
		347 ± 12	

Table 1 (cont.)

Meteorite (Class)	Reference (Method)	K content (10^{-6} g/g)	Comments on sample analysed
Sioux Co. (Ap)	(15)	326	Approximately 1 g.
	(17)	307	Average of 2 analyses of 0.243 g split of 0.984 g sample.
		305	0.239 g split of 0.984 g sample.
	(4)	300	
	(14)	330 ± 20	Approximately 1 g.
Stannern (Ap)	(1)	760 ± 60	1.13 kg.
	(17)	657	0.090 g split of 1.401 g sample; spiked during dissolution.
	(30)	570	
	(3)	600	1 g samples, from separate specimens.
	(13)	690	
	(27)	690	1 g sample.
	(32)	548 ± 10	Weighted average of analyses of washed and sieved separates; about 2 g total.

Key to references:

(1) Rowe, Van Dilla and Anderson (1963)
(2) Kaushal and Wetherill (1970)
(3) Edwards and Urey (1955)
(4) Von Michaelis, Ahrens and Willis (1969)
(5) Wiik (1969)
(6) Krummenacher (1961)
(7) Murthy and Compston (1965)
(8) Rancitelli *et al.* (1969)
(9) Wasserburg and Hayden (1955)
(10) Geiss and Hess (1958)
(11) Gast (1960)
(12) Kaushal and Wetherill (1969)
(13) Edwards (1955)
(14) Wänke (1961)
(15) Kirsten, Krankowsky and Zähringer (1963)
(16) Bogard *et al.* (1967b)
(17) Tera *et al.* (1970)
(18) Rowe and Van Dilla (1961)

(19) Taylor (1964)
(20) J.H.Reynolds, quoted in (18)
(21) Shima and Honda (1967)
(22) Baadsgaard and Folinsbee (1964)
(23) Gopalan and Wetherill (1969)
(24) Jarosewich and Mason (1969)
(25) Gopalan and Wetherill (1970)
(26) Dawson, Maxwell and Parsons (1960)
(27) Easton and Lovering (1964)
(28) Schaeffer, Stoenner and Fireman (1965)
(29) Compston, Lovering and Vernon (1965)
(30) Vinogradov, Zadorzhnii and Knorre (1960)
(31) Burnett, Lippolt and Wasserburg (1966)
(32) Megrue (1966)
(33) Murthy, Schmitt and Rey (1970)
(34) Wänke und König (1959)

Analytical techniques: DGS – Direct gamma spectrometry
ID – Isotope dilution
FP – Flame photometry
XRF – X-ray fluorescence
NAA – Neutron activation analysis

fresh material, perhaps because of leaching of K located along grain boundaries or in water-soluble phases. (Note that the values reported by Wänke (1961; reference (14) in Table 1) seem in some cases to be too high, judging from comparisons with values by Wänke und König (1959) on samples from the same specimens.) In contrast, *Potter*, which has a terrestrial residence age of $>$20,000 years (Suess and Wänke, 1962), displays K contents which lie near the low end of the range for L chondrites but are not markedly unusual. Perhaps susceptibility to leaching is associated with the anomalously high K content of *Beardsley*. In any case, the weathered specimens of *Beardsley* provide a stringent test for sampling errors; the data suggest that effects of inhomogeneous K distributions on gram-sized samples are limited to about $\pm 5\%$ or less, and to about $\pm 20\%$ or less on 0.2 gram samples.

Such complications do not affect data on the other chondrites in Table 1. All except *Plainview* were observed falls, yet *Plainview* does not seem to have low K contents in comparison to other H chondrites. Suess and Wänke (1962) found a terrestrial age of $<$2000 years for this meteorite, which implies that many thousands of years of weathering may be required to change appreciably the K contents of at least some chondrites.

It is clear that sampling errors for chondrites must be comparable to or smaller than the analytical errors cited. It is also clear that flame photometric data are comparable in accuracy to isotopic dilution data, except sometimes when done as part of a bulk analysis (see Wiik's data for *Forest City* and *Richardton*, Jarosewich and Mason's value for *Cherokee Springs*). X-ray fluorescence and activation analysis techniques can apparently yield data of adequate accuracy, although the number of comparisons which may be made is limited.

Sampling errors are severe for aubrites (see especially data for *Bishopville*). Perhaps these brecciated achondrites contain K-rich clasts or xenocrysts. Sampling errors for eucrites are minor, if not negligible, by comparison. *Stannern* clearly has a higher K content than do the other eucrites. Jérome (personal communication) has suggested that this meteorite represents a late differentiate from the parental magma for the eucrites, on the basis of diverse lines of evidence.

Based on these comparisons, I shall treat sets of data for H, L and LL chondrites as single populations, without making distinctions among results obtained by the various techniques discussed above. I have excluded from consideration all data obtained as part of a bulk analysis, Wänke's (1961) data, and data on *Potter* (but have included data on other finds). These restrictions leave 96 analyses of H, 85 of L and 53 of LL chondrites; see Fig. 1. Mean values for the three classes of chondrites are 830×10^{-6} g K/g, $(860 \pm 12) \times 10^{-6}$ g K/g, and 830×10^{-6} g K/g. Only for the L class are the data sufficiently similar to a Gaussian distribution so that it seems valid to compute a standard error estimate. The bimodal distribution of K contents in H chondrites is almost entirely accounted for by analyses of *Beardsley* (Kirsten, Krankowsky und Zähringer, 1963, report 1025×10^{-6} g K/g for *Ekeby* and Anderson, Rowe and Urey, 1964, report 1000×10^{-6} g K/g for *Tysnes Island*). Treating results for all H chondrites other than *Beardsley* as a subset of the data, one obtains a mean value of 800×10^{-6} g K/g for "garden-variety" H chondrites (85 determinations). Because of the difference in SiO_2 contents of H and L chondrites, the "garden-variety" H and the L chondrites have exactly the same atomic abundance: 33 K atoms/10^4 Si atoms.

An indication of the hazards involved in relying on K contents determined in bulk analyses may be gained by comparing the averages above with those given by Craig (1964). Craig recommends 0.12% K for both the H and L groups, and 0.17% K for the LL group. These values are clearly too high.

Potassium contents of LL chondrites range from 199 to 2180×10^{-6} g K/g, even after excluding the value of 1.2% K found for a dark clast from *Krähenberg* by Zähringer (1968). The extreme values in this range were determined by isotope dilution (Kirsten, Krankowsky und Zähringer, 1963; Kaiser und Zähringer, 1965). Values for samples of a single meteorite, *Soko*

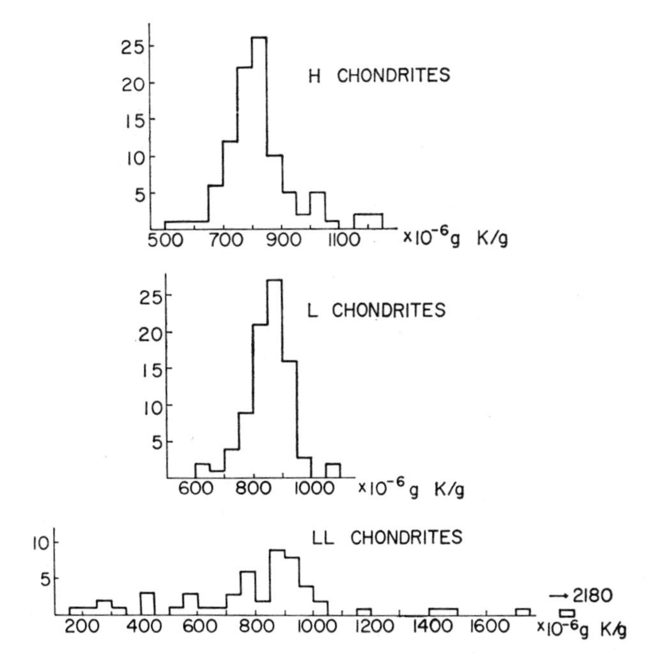

FIGURE 1 Histograms of K contents in H, L and LL chondrites. References used were those of Tables 1 and 2 plus Rowe and Manuel (1964), Funkhouser, Kirsten and Schaeffer (1967) and Gopalan and Wetherill (1968).

Banja, span nearly the entire range. There seems to be no reason to doubt that these meteorites have extremely variable K contents. The atomic abundance for LL chondrites based on the average above, about 32 K atoms/10^4 Si atoms, is very similar to those for H and L chondrites, but the meaning of an average content or atomic abundance is obscured by the variability noted. Careful searches should be made for K-rich clasts, such as those found in *Krähenberg* by Zähringer (1968) and by Kempe and Müller (1969), and the nature of the K-bearing minerals in these meteorites should be established.

For meteorites of other classes, it is not yet feasible to define mean values of K contents, either because of wide variations in reported data or because of limited information. In Table 2 are listed individual values determined by the methods discussed above. Wiik's (1969) values have been included, principally because his results on C chondrites are comprehensive and seem to be accurate in general. Results by other analysts obtained in the course of conventional bulk analyses have been excluded, although in some cases they

TABLE 2 Potassium contents in individual stone meteorites and in silicate fractions of stony-irons*

Meteorite	10^{-6} g K/g	Reference
C1		
Alais	380	(5)
	380	(32)
Ivuna	945	(2)
	570	(3)
	580	(5)
Orgueil	563, 566, 586	(2)
	347, 343, 340, 494	(3)
	560	(3)
	560	(4)
	580, 380	(5)
	310	(27)
	347, 343	(36)
	510	(37)
C2		
Al Rais	270	(5)
Cold Bokkeveld	404, 403	(2)
	430	(3)
	480	(4)
	420	(5)
	445	(15)
Erakot	362	(2)
	270	(5)
Essebi	362	(2)
	900	(5)
matrix	930	(35)
non-magnetic chondrules	290	
magnetic chondrules	270	
broken magnetic chondrules	890	
Haripura	275, 287	(2)
	600	(3)
	580	(12)
Nawapali	130	(5)
Nogoya	490	(37)
Renazzo	340	(5)

TABLE 2 (cont.)

Meteorite	10^{-6} g K/g	Reference
Santa Cruz	391	(2)
	1000	(5)
Staroye Boriskino	300	(5)
C3		
Felix	370 ± 30	(1)
	420	(3)
	370	(4)
	410	(5)
Grosnaja	449	(2)
	240	(5)
Karoonda	240	(5)
Lancé	380	(4)
	1160	(5)
	388	(7)
Leoville	300	(5)
Mokoia	295	(2)
	370	(3)
	340	(4)
	330	(5)
	311, 335	(7)
Ornans	1400	(5)
	410	(13)
	390	(37)
Warrenton	1900	(5)
	430	(13)
Vigarano	340	(4)
	480	(5)
Enstatite chondrites		
Adhi Kot (E3)	810, 1013	(25)
St. Mark's (E5)	690	(4)
	960	(5)
	757	(15)
	745, 730	(25)
Atlanta (E6)	600	(4)
	700	(5)

TABLE 2 (cont.)

Meteorite	10^{-6} g K/g	Reference
Blithfield (E6)	700	(4)
Daniel's Kuil (E6)	671	(25)
Hvittis (E6)	620	(4)
	794	(25)
Jajh deh Kot Lalu (E6)	500	(5)
	874	(25)
Khairpur (E6)	709	(25)
Pillistfer (E6)	670	(4)
Aubrites (Ae)		
Bustee	30	(13)
Cumberland Falls	140	(13)
	240	(35)
Khor Temiki	550	(35)
Shallowater	360	(13)
Staroe Pesyanoe	363	(15)
Diogenites (Ab)		
Ibbenbüren	45	(35)
Johnstown	8	(3)
	40	(4)
	10 ± 1	(14)
	90(?)	(27)
	10	(34)
	9.9 ± 0.3	(38)
bronzite separate	3.9 ± 0.28	
Shalka	20	(4)
	0.72 ± 0.06	(38)
Tatahouine	20	(15)
	1.20 ± 0.06	(38)
Goalpara (Aop)	60	(13)
	70	(27)
Angra dos Reis (Aa)	12.9 ± 1.6	(17)
	~ 50(?)	(39)
Nakhla (Ado)	1020	(13)
	1150	(27)

TABLE 2 (cont.)

Meteorite	10^{-6} g K/g	Reference
Shergotty	1500 ± 30	(10)
	1520 ± 30	
	1220	(27)
Eucrites (Ap)		
Bereba	258	(17)
	380	(35)
Chervony Kut	470	(30)
Haraiya	190 ± 5	(32)
WASP†	192 ± 5	
Jonzac	327	(17)
Howardites (Aor)		
Bholghati	166	(35)
Frankfort	200	(5)
	130 ± 20	(10)
	210	(13)
	260	(27)
Kapoeta	400	(5)
	165	(15)
Macibini	390	(15)
Malvern	400	(4)
Pavlovka	170	(13)
Petersburg	460	(13)
Yurtuk	430	(30)
Zmenj	1100	(5)
Silicates from mesosiderites		
Estherville	144 ± 7, 119 ± 6	(32)
WASP	128 ± 6, 182 ± 16	
Hainholz	70	(35)
Lowicz	192	(15)
Mt. Padbuty	200	(5)
Vaca Muerta		
anorthosite	180 ± 40	(31)

TABLE 2 (cont.)

Meteorite	10^{-6} g K/g	Reference
"*Harvard University*" lodranite silicates	168	(32)
Olivine from pallasites		
Admire	90	(13)
	52	(38)
Brenham	<15	(10)
	90	(13)
Eagle Station	0.54 ± 0.08	(38)
Finmarken	90	(13)
Krasnojarsk	2.6 ± 0.3	(38)
Marjalahti	0.77 ± 0.2	(13)
Mt. Vernon	110	(13)
Springwater	2.18 ± 0.17	(38)
Thiel Mts.	<15.3	(38)

* Data presented in Table 1 are not repeated here.
† Weighted average of separated phases; sieved, washed and passed through a Franz separator.

Key to references:

(1) to (34) as for Table 1
(35) Zähringer (1968)
(36) Compston and Vernon, cited by Easton and Lovering (1964)
(37) Ahrens, Von Michaelis and Fesq (1969)
(38) Megrue (1968)
(39) Smales et al. (1970).
Note that data reported by Megrue (refs. 32 and 38) were determined by E.F. Norton.

may be sufficiently accurate to be useful. At present, data on K contents in these classes are best employed in pointing out problems to be solved by future work of a more systematic nature.

Potassium contents in C1 stones are variable; they appear to have a bi-modal distribution. There is little justification for choosing a value from the observed range to represent the cosmic abundance of K. For these who enjoy playing that game, the average of about 430×10^{-6} g K/g is equivalent to an atomic abundance of about 30 K atoms/10^4 Si atoms, similar to those observed for the H, L and LL groups. Note the contrast with the distribution of Na, which clearly has a higher atomic abundance in C1 stones than in H, L and LL chondrites (see chapter on Na).

Potassium contents in C2 chondrites range over a factor of about 8. A rationale for this marked variability is suggested by Zähringer's (1968) data on *Essebi:* there are at least two very different sets of chondrules in this meteorite, with the broken ones closely resembling the matrix in their high K contents. Perhaps some of the K-rich "chondrules" are really clasts similar to those inferred to be present in aubrites and observed in LL chondrites. In any case, inhomogeneous admixture of minor amounts of K-rich material with the bulk, K-poor constituents of these chondrites would account for the observed variations in K contents.

Potassium contents in C3 chondrites also range over a factor of about 8, if Wiik's value for Lancé is valid.

Enstatite chondrites appear to display a slight systematic decrease in K contents with increasing petrographic grade (Van Schmus and Wood, 1967; Keil, 1968). *Abee* (Table 1) has an exceptionally high Na content (see chapter on Na) but a "normal" K content. Variations in K contents between matrix and clasts in this meteorite should be investigated further.

Among achondrites, the variability in K contents of aubrites has already been mentioned. Diogenites have very low K contents, as expected from their mineralogy. Some of the higher values for diogenites in Table 2 may reflect contamination or inadequate correction for blanks. Note, however, that Megrue (1968; reference (38)) employed a rather stringent washing procedure which was alleged to remove surface contamination, but could also have removed indigenous K in labile sites. Especially in the brecciated diogenites (all but *Tatahouine*), there may be appreciable amounts of indigenous K located in their matrices as xenolithic microclasts or xenocrysts.

The low K contents of the ureilite *Goalpara* are consistent with the low Na contents observed in this meteorite by Schmitt, Goles and Smith (1971). The Na/K ratio for *Goalpara*, 3.1 (g/g), is much lower than those for ordinary chondrites which average around 7.5.

The high K contents of *Nakhla* and *Shergotty* contrast with those of almost all other Ca-rich achondrites (Wiik's value for the howardite *Zmenj* should be checked). Eucrites and howardites have been classified in Table 2 according to the model developed by Jérome (1970) in which the latter are considered to be polymict mélanges with eucritic and diogenitic fragments as the principal components. Admixture of diogenitic material to eucritic fragments would markedly decrease K in the resulting mélanges compared to eucrites. However, there is apparently a chondritic component in howardites, which would tend to increase observed K contents. Since extant eucrites vary so much in K contents, it is difficult to disentangle these effects.

Duke and Silver (1967) have pointed out that the silicates in mesosiderites resemble howardites in many respects (see also Prior, 1916, 1918). Data in Table 2 do not contradict this observation.

Potassium contents in olivine separates from pallasites are subject to severe errors related to contamination and limited sensitivity. Geiss and Hess (1958) discuss problems of determining contamination from reagents, concluding that at these levels it is extraordinarily difficult to make adequate estimates of blanks. Megrue (1968) washed his samples with acid and triply-distilled water to remove surface contamination. (He did not wash the *Thiel Mts.* sample.) Even if he had leached indigenous K from his samples, his results are likely to be most nearly representative of the olivine itself. It would be interesting to see if K-rich inclusions may be observed in *Admire* olivine with a microprobe.

Some K contents have been determined in mineral separates from stone meteorites or with a microprobe (e.g., Baadsgaard, Campbell and Folinsbee, 1961; Compston, Lovering and Vernon, 1965; Schaeffer, Stoenner and Fireman, 1965; Bogard et al., 1967a; Gopalan and Wetherill, 1970). It is clear that K is enriched in oligoclase when present, in plagioclase when K-feldspars are apparently absent, and in Ca-rich fractions of the *Indarch* enstatite chondrite. Van Schmus (1968) has shown that most of the potassium in ordinary chondrites is contained in the plagioclase feldspar; he records 0.73–1.13% K in H-group feldspar, 0.45–1.01% in L-group feldspar, and 0.13–0.77% in LL-group feldspar.

Potassium feldspars, some of very impressive size, have been found in a number of iron meteorites (Bogard et al., 1967b; Bunch and Olsen, 1968; Wasserburg, Sanz and Bence, 1968). These silicate phases, along with the clasts studied by Kempe and Müller (1969), may in part comprise the K-rich component inferred above to be present in several diverse meteorite classes. Investigations of this possibility might have important implications for hypotheses of meteorite genesis and interrelations.

As with Na, there is no point in reviewing K contents of iron meteorites owing to their strong dependence on the distribution of silicate inclusions within the samples taken for analysis.

Acknowledgements

Preparation of this review was supported in part by NASA grant NGL 38-003-010. I am indebted to the Horace H. Rackham School of Graduate Studies, The University of Michigan, for an appointment as a visiting scholar which enabled me to use library facilities at Ann Arbor.

References

Ahrens, L. H., Von Michaelis, H. and Fesq, H. W. (1969) "The composition of the stony meteorites (IV) Some analytical data on Orgueil, Nogoya, Ornans and Ngawi." *Earth Planet. Sci. Letters* **6**, 285–288.

Anderson, E. C., Rowe, M. W. and Urey, H. C. (1964) "Potassium and aluminum 26 contents of three bronzite chondrites." *J. Geophys. Res.* **69**, 564–565.

Baadsgaard, H., Campbell, F. A. and Folinsbee, R. E. (1961) "The Bruderheim meteorite." *J. Geophys. Res.* **66**, 3574–3577.

Baadsgaard, H. and Folinsbee, R. E. (1964) "Peace River meteorite." *J. Geophys. Res.* **69**, 4197–4200.

Bogard, D. D., Burnett, D. S., Eberhardt, P. and Wasserburg, G. J. (1967a) "^{87}Rb–^{87}Sr isochron and ^{40}K–^{40}Ar ages of the Norton County achondrite". *Earth Planet. Sci. Letters* **3**, 179–189.

Bogard, D. D., Burnett, D., Eberhardt, P. and Wasserburg, G. J. (1967b) "^{40}Ar–^{40}K ages of silicate inclusions in iron meteorites." *Earth Planet Sci. Letters* **3**, 275–283.

Bunch, T. E. and Olsen, E. (1968) "Potassium feldspar in Weekeroo Station, Kodaikanal and Colomera iron meteorites." *Science* **160**, 1223–1275.

Burnett, D. S., Lippolt, H. J. and Wasserburg, G. J. (1966) "The relative isotopic abundances of K^{40} in terrestrial and meteoritic samples." *J. Geophys. Res.* **71**, 1249–1269.

Compston, W., Lovering, J. F. and Vernon, M. J. (1965) "The rubidium-strontium age of the Bishopville aubrite and its component enstatite and feldspar." *Geochim. Cosmochim. Acta* **29**, 1085–1099.

Craig, H. (1964) "Petrological and compositional relationships in meteorites", chapter 26 in *Isotopic and Cosmic Chemistry*, edited by H. Craig, S. L. Miller and G. J. Wasserburg (North-Holland, Amsterdam) 401–451.

Dawson, K. R., Maxwell, J. A. and Parsons, D. E. (1960) "A description of the meteorite which fell near Abee, Alberta, Canada." *Geochim. Cosmochim. Acta* **21**, 127–144.

Duke, M. B. and Silver, L. T. (1967) "Petrology of eucrites, howardites, and mesosiderites." *Geochim. Cosmochim. Acta* **31**, 1637–1666.

Easton, A. J. and Lovering, J. F. (1954) "Determination of small quantities of potassium and sodium in stony meteoritic material, rocks and minerals." *Anal. Chim. Acta* **30**, 543–548.

Edwards, G. (1955) "Sodium and potassium in meteorites". *Geochim. Cosmochim. Acta* **8**, 285–294.

Edwards, G. and Urey, H. C. (1955) "Determination of alkali metals in meteorites by a distillation process." *Geochim. Cosmochim. Acta* **7**, 154–168.

Funkhouser, J., Kirsten, T. and Schaeffer, O.A. (1967) "Light and heavy rare gases in four fragments of the St. Severin meteorite." *Earth Planet. Sci. Letters* **2**, 185–190.

Gast, P.W. (1960) "Alkali metals in stone meteorites." *Geochim. Cosmochim. Acta* **14**, 1–4.

Geiss, J. and Hess, D.C. (1958) "Argon-potassium ages and the isotopic composition of argon from meteorites." *Astrophys. J.* **127**, 224–236.

Goles, G.G., Greenland, L.P. and Jérome, D.Y. (1967) "Abundances of chlorine, bromine and iodine in meteorites." *Geochim. Cosmochim. Acta* **31**, 1771–1787.

Gopalan, K. and Wetherill, G.W. (1968) "Rubidium-strontium age of hypersthene (L) chondrites." *J. Geophys. Res.* **73**, 7133–7136.

Gopalan, K., and Wetherill G.W. (1969) "Rubidium-strontium age of amphoterite (LL) chondrites." *J. Geophys. Res.* **74**, 4349–4358.

Gopalan, K. and Wetherill, G.W. (1970) "Rubidium-strontium studies on enstatite chondrites: whole meteorite and mineral isochrons." *J. Geophys. Res.* **75**, 3457–3467.

Jarosewich, E. and Mason, B. (1969) "Chemical analyses with notes on one mesosiderite and seven chondrites." *Geochim. Cosmochim. Acta* **33**, 411–416.

Jérome, D.Y. (1970) *Composition and origin of some achondritic meteorites.* Ph. D. dissertation. Univ. of Oregon.

Kaiser, W. and Zähringer, J. (1965) "Kalium-Analysen von Amphoterit-Chondriten und deren K.A. Alter." *Z. Naturforsch.* **20a**, 973–965.

Kaushal, S.K. and Wetherill, G.W. (1969) "Rb^{87}-Sr^{87} age of bronzite (H group) chondrites." *J. Geophys. Res.* **74**, 2717–2726.

Kaushal, S.K. and Wetherill, G.W. (1970) "Rubidium 87-strontrium 87 age of carbonaceous chondrites." *J. Geophys. Res.* **75**, 463–468.

Keil, K. (1968) "Mineralogical and chemical relationships among enstatite chondrites." *J. Geophys. Res.* **73**, 6945–6976.

Keil, K. (1969) "Meteorite composition", chapter 4 in *Handbook of Geochemistry*, edited by K.H. Wedepohl (Springer-Verlag, Berlin) 78–115.

Kempe, W. and Müller, O. (1969) "The stony meteorite Krähenberg." In *Meteorite Research* (editor P.M. Millman, pub. by D. Reidel, Dordrecht, Holland), 418–428.

Kirsten, T., Krankowsky, D. und Zähringer, J. (1963) "Edelgas- und Kalium-Bestimmungen an einer größeren Zahl von Steinmeteoriten." *Geochim. Cosmochim. Acta* **27**, 13–42.

Krummenacher, D. (1961) "Le dosage du potassium des météorites par dilution isotopique, et la détermination de leur âge par la méthode K/A." *Helv. Chim. Acta* **44**, 1054–1057.

Megrue, G.H. (1966) "Rare-gas chronology of calcium-rich achondrites." *J. Geophys. Res.* **71**, 4021–4027.

Megrue, G.H. (1968) "Rare-gas chronology of hypersthene achondrites and pallasites." *J. Geophys. Res.* **73**, 2027–2033.

Murthy, V.R. and Compston, W. (1965) "Rb-Sr ages of chondrules and carbonaceous chondrites." *J. Geophys. Res.* **70**, 5297–5307.

Murthy, V.R., Schmitt, R.A. and Rey, P. (1970) "Rubidium-strontium age and elemental and isotopic abundance of some trace elements in lunar samples." *Science* **167**, 476–479.

Prior, G.T. (1916) "On the genetic relationship and classification of meteorites". *Mineral. Mag.* **18**, 26–44.

Prior, G.T. (1918) "On the mesosiderite-grahamite group of meteorites: with analyses of Vaca Muerta, Hainholz, Simondium, and Powder Mill Creek." *Mineral. Mag.* **18**, 151–172.

Rancitelli, L.A., Perkins, R.W., Cooper, J.A., Kaye, J.H. and Wogman, N.A. (1969) "Radionuclide composition of the Allende meteorite from non-destructive gamma-ray spectrometric analysis." *Science* **166**, 1269–1272.

Rowe, M.W. and Manuel, O.K. (1964) "γ radioactivity in the Fayetteville meteorite." *J. Geophys. Res.* **69**, 1944–1945.

Rowe, M.W. and Van Dilla, M.A. (1961) "On the radioactivity of the Bruderheim chondrite." *J. Geophys. Res.* **66**, 3553–3556.

Rowe, M.W., Van Dilla, M.A. and Anderson, E.C. (1963) "On the radioactivity of stone meteorites." *Geochim. Cosmochim. Acta* **27**, 983–1001.

Schaeffer, O.A., Stoenner, R.W. and Fireman, E.L. (1965) "Rare gas isotope contents and K-Ar ages of mineral concentrates from the Indarch meteorite." *J. Geophys. Res.* **70**, 209–213.

Schmitt, R.A., Goles, G.G. and Smith, R.H. (1971) *Elemental abundances in stone meteorites.* To be published as a NASA report; available from the author on request.

Shima, M. and Honda, M. (1967) "Distributions of alkali, alkaline earth and rare earth elements in component minerals of chondrites." *Geochim. Cosmochim. Acta* **31**, 1995–2006.

Smales, A.A., Mapper, D., Webb, M.S.W., Webster, RK. and Wilson, J.D. (1970) "Elemental composition of lunar surface material." *Science* **167**, 509–512.

Suess, H.E. and Wänke, H. (1962) "Radiocarbon content and terrestrial age of twelve stony meteorites and one iron meteorite." *Geochim. Cosmochim. Acta* **26**, 475–480.

Taylor, H.W. (1964) "Gamma radiation emitted by the Peace River chondrite." *J. Geophys. Res.* **69**, 4194–4196.

Tera, F., Eugster, O., Burnett, D.S. and Wasserburg, G.J. (1970) "Comparative study of Li, Na, K, Rb, Cs, Ca, Sr and Ba abundances in achondrites and in Apollo 11 lunar samples." *Geochim. Cosmochim. Acta, Suppl. I*, 1637–1657.

Van Schmus, W.R. (1968) "The composition and structural state of feldspar from chondritic meteorites." *Geochim. Cosmochim. Acta* **32**, 1327–1342.

Van Schmus, W.R. and Wood, J.A. (1967) "A chemical-petrologic classification for the chondritic meteorites." *Geochim. Cosmochim. Acta* **31**, 747–765.

Vinogradov, A.P., Zadorozhnii, I.K. and Knorre, K.G. (1969) "On argon in meteorites (in Russian)." *Meteoritika* **18**, 92–99.

Von Michaelis, H., Ahrens, L.H. and Willis, J.P. (1969) "The composition of stony meteorites. (II) The analytical data and an assessment of their quality." *Earth Planet. Sci. Letters* **5**, 387–394.

Wänke, H. (1961) "Über den Kaliumgehalt der Chondrite, Achondrite und Siderite". *Z. Naturforsch.* **16a**, 127–130.

Wänke, H. and König, H. (1959) "Eine neue Methode zur Kalium-Argon-Alterbestimmung und ihre Anwendung auf Steinmeteorite." *Z. Naturforsch.* **14a**, 860–866.

Wasserburg, G.J., Lang, H.G. and Bence, A.E. (1968) "Potassium-feldspar phenocrysts in the surface of Colomera, an iron meteorite." *Science* **161**, 684–687.

Wiik, H.B. (1969) "On regular discontinuities in the composition of meteorites." *Comm. Phys.-Math.* **34**, 135–145.

Zähringer, J. (1968) "Rare gases in stony meteorites." *Geochim. Cosmochim. Acta* **32**, 209–237.

(Received 24 August 1970)

CALCIUM (20)

Brian Mason

Smithsonian Institution, Washington, D.C.

CALCIUM IS a minor element in most classes of stony and stony-iron meteorites, being present at about the 1% level; exceptions are the calcium-rich achondrites and mesosiderites, which contain considerably greater amounts, and the pallasites, in which this element is present in trace amounts only. Calcium is determined in all complete analyses of these meteorites, so the data on this element in the literature are very extensive; however, the quality is not good, as has been demonstrated by Michaelis et al. (1969). Classical wet-chemical analysis for calcium at the 1% level, especially in the presence of much magnesium (as in meteorites), may produce poor results unless great care is taken. An extensive series of x-ray fluorescence analyses by Nichiporuk et al. (1967) and Michaelis et al. (1969) has shown very uniform calcium concentrations within the specific chondrite classes; their results are extremely consistent, and those of Michaelis et al. are used in Table 1, except for the carbonaceous chondrites and the amphoterites, for which a broader cover of reliable analyses is obtainable from the literature.

The Ca/Si ratios are remarkably uniform and distinctive for the three major groups of chondrites, as follows: carbonaceous, 0.073; ordinary (hypersthene, bronzite, amphoterite), 0.048; enstatite, 0.036. For the classes within these major groups the differences in this ratio are small and probably not significant. In the Type III carbonaceous chondrites the enrichment in calcium is manifested by the presence of calcium-rich minerals such as anorthite ($CaAl_2Si_2O_8$) and gehlenite ($Ca_2Al_2SiO_7$) not found in other classes of chondrites.

The calcium-poor achondrites, as their name implies, are notably depleted in calcium in relation to the chondrites, whereas the calcium-rich achondrites show a notable enrichment; the individual calcium values for many meteorites in these groups are illustrated under magnesium, Fig. 2. The two meteor-

TABLE 1 Calcium in meteorites

	No. of detns.	Ca, weight per cent		Atoms/ 10^6 Si	Reference
		Range	Mean		
Chondrites					
Carbonaceous, Type I	3	0.87–1.34	1.06	72,100	(1)
Carbonaceous, Type II	9	1.11–1.63	1.34	71,900	(1)
Carbonaceous, Type III	6	1.40–1.87	1.70	74,300	(1)
Bronzite	12	1.15–1.22	1.19	49,800	(2)
Hypersthene	19	1.22–1.35	1.28	48,400	(2)
Amphoterite	9	1.07–1.43	1.25	46,400	(3)
Enstatite, Type I	2	0.81–0.87	0.84	35,700	(2)
Enstatite, Type II	3	0.85–0.98	0.90	35,500	(2)
Calcium-poor achondrites					
Enstatite	4	0.47–1.16	0.80	20,900	(4)
Hypersthene	6	0.54–1.88	1.02	28.900	(5)
Chassignite	1	–	0.37	14,800	(6)
Ureilites	3	0.31–0.97	0.62	22,800	(7)
Calcium-rich achondrites					
Angrite	1	–	17.50	595,000	(8)
Nakhlite	1	–	10.85	332,000	(9)
Howardites	4	2.87–5.78	4.47	137,000	(10)
Eucrites	10	7.15–8.20	7.65	234,000	(11)
Stony-irons					
Pallasites (olivine)	5	0.0020–0.0075	0.0046	174	(12)
Mesosiderites (silicates)	9	2.07–5.66	4.05	115,000	(13)

(1) Wiik, in Mason (1963a)
(2) Michaelis et al. (1969)
(3) Wiik, in Mason and Wiik (1964), with later analyses
(4) Khor Temiki (Hey and Easton, 1967); Norton County (Wiik, 1956); Pesyanoe (Dyakonova and Kharitonova, 1960); Pena Blanca Spring (Lonsdale, 1947)
(5) Mason (1963b)
(6) Dyakonova and Kharitonova (1960)
(7) Wiik (1969)
(8) Ludwig and Tschermak (1909)
(9) Prior (1912)
(10) Bununu (Mason, 1967); Frankfort (Mason and Wiik, 1966a); Kapoeta (Mason and Wiik, 1966b); Bholgati (Jarosewich, unpublished)
(11) Mason (1967)
(12) Buseck and Goldstein (1969)
(13) Powell (1971)

ites richest in calcium are the angrite Angra dos Reis and the nakhlite Nakhla; in both of these the calcium is present as a calcium-rich pyroxene. In the other meteorites of these groups most of the calcium is present as calcic plagioclase. In the stony-irons, the pallasites contain only traces of calcium, in solid solution in olivine; the mesosiderite silicates resemble those of the howardites, and have comparable calcium contents.

Calcium is usually considered to be entirely lithiophilic in character; however, in the extremely reduced and sulfide-rich enstatite chondrites and enstatite achondrites, some of the calcium is present as the sulfide oldhamite. In the common chondrites the calcium is distributed over a number of minerals – as the calcium phosphates chlorapatite and/or whitlockite, as the pyroxene diopside, and in solid solution in orthopyroxene and sodic plagioclase. As mentioned above, the Type III carbonaceous chondrites are characterized by the presence of some calcium-rich minerals; the distribution of calcium in the Type I and II carbonaceous chondrites is not well known, but gypsum, calcite, and dolomite have been recorded from some meteorites in these classes.

References

Buseck, P.R. and Goldstein, J.I. (1969) *Bull. Geol. Soc. Am.* **80**, 2141.
Dyakonova, M.I. and Kharitonova, V.Y. (1960) *Meteoritika* **18**, 48.
Hey, M.H. and Easton, A.J. (1967) *Geochim. Cosmochim. Acta* **31**, 1789.
Lonsdale, J.T. (1947) *Am. Mineral.* **32**, 354.
Ludwig, E. and Tschermak, G. (1909) *Tsch. Min. Pet. Mitt.* **28**, 109.
Mason, B. (1963a) *Space Sci. Rev.* **1**, 621.
Mason, B. (1963b) *Am. Museum Novitates*, No. 2155.
Mason, B. (1967) *Geochim. Cosmochim. Acta* **31**, 107.
Mason, B. and Wiik, H.B. (1964) *Geochim. Cosmochim. Acta* **28**, 533.
Mason, B. and Wiik, H.B. (1966a) *Am. Museum Novitates*, No. 2272.
Mason, B. and Wiik, H.B. (1966b) *A. Museum Novitates*, No. 2273.
Michaelis, H. v., Ahrens, L.H. and Willis, J.P. (1969) *Earth Planet. Sci. Lett.* **5**, 387.
Nichiporuk, W., Chodos, A., Helin, E. and Brown, H. (1967) *Geochim. Cosmochim. Acta* **31**, 1911.
Powell, B.N. (1971) *Geochim. Cosmochim. Acta* **35**, 5.
Prior, G.T. (1912) *Mineral. Mag.* **16**, 274.
Wiik, H.B. (1956) *Geochim. Cosmochim. Acta* **9**, 279.
Wiik, H.B. (1969) *Soc. Sci. Fennica, Comm. Phys.-Math.* **34**, 135.

(Received 20 February 1970)

SCANDIUM (21)

Gordon G. Goles

Center for Volcanology
and
Departments of Chemistry and Geology
University of Oregon
Eugene, Oregon 97403

SCANDIUM IS a dispersed trace element. In mineral systems like those of meteorites, it tends to be found in pyroxenes. Emission spectrographic techniques are commonly used for the estimation of this element, although during the past decade many activation analysis determinations have been made. An excellent example of the application of the first technique to the study of meteorites is the work of Greenland and Lovering (1965), who determined Sc in 50 chondrites. Schmitt, Goles and Smith (1971) report instrumental neutron activation analysis results for Sc in 180 stone meteorites. Intercomparisons among all sets of results for Sc in chondrites demonstrate very good general agreement, and imply that the precision of Greenland and Lovering's results is about ±40% (relative standard deviation). Since instrumental neutron activation analyses for Sc in stone meteorites seem to be precise to ±10% or better, I shall here rely on the results of Schmitt, Goles and Smith. Additional important references are: Bate, Potratz and Huizenga (1960; seven stone meteorites), Kemp and Smales (1960; five stone meteorites), Greenland (1963; eighteen stone meteorites in addition to those listed by Greenland and Lovering, 1965).

Table 1 presents Sc average contents and atomic abundances, largely from the work of Schmitt, Goles and Smith (1971). Chondrites are classified according to Van Schmus and Wood (1967) and achondrites are identified by the code introduced by Keil (1969).

Dispersions cited are estimates of single standard deviations of the populations; analytical errors contribute very little to most of these values. Atomic

TABLE 1 Scandium average contents and atomic abundances in stone meteorites

Meteorite class	Number of specimens (samples) analysed	Average contents, 10^{-6} g Sc/g	Atomic abundances Sc/10^6 Si
C1	3 (4)	5.1 ± 1.7	31 ± 10
C2	7 (11)	7.4 ± 1.5	35 ± 7
C3	9 (15)	9 ± 2	36 ± 9
H3	5 (8)	7.2 ± 1.1	26 ± 4
H4, 5, 6	38 (54)	7.6 ± 1.5	28 ± 6
L3	3 (4)	7.4 ± 1.0	25 ± 3
L4, 5, 6	44 (58)	8.2 ± 1.2	28 ± 4
LL3	3 (5)	8.5 ± 1.2	28 ± 4
LL4, 5, 6	17 (21)	7.8 ± 1.1	26 ± 4
E4	2 (3)	7.2 ± 1.0	26 ± 4
E5, 6	5 (6)	7 ± 2	22 ± 8
Aubrites, Ae	5 (7)	6.4 ± 1.8	16 ± 4
Diogenites, Ab*	2 (2)	11 ± 2	26 ± 5
Ureilites, Aop	2 (3)	7.9 ± 1.0	26 ± 3
Eucrites, Ap	20 (27)	26 ± 7	68 ± 8
Howardites, Aor	9 (13)	21 ± 6	59 ± 17
Nakhlites, Ado	2 (3)	48 ± 4	131 ± 11
Shergotty	1 (2)	30 ± 4	80 ± 11

* Private communication from D.Y. Jérome

abundances (effectively, Sc/Si ratios) of all classes of chondritic meteorites are indistinguishable from one another on a rigorous statistical basis. In contrast, Sc and Si are clearly incoherent among the achondrites. Ureilites closely resemble chondrites in Sc contents and atomic abundances, as indeed Ringwood's (1960) speculations concerning the origin of *Novo Urei* would imply they should, but otherwise there appears to be little that is systematic in the distribution of Sc among achondrites. No coherence with Fe is indicated, but perhaps a Sc–Ca coherence may be discerned.

Data on Sc contents in mineral separates are given in Table 2. Most of these separates were generously provided by B. H. Mason, although a few were hand-cobbed by me. U. B. Marvin kindly supplied the separates from *Vaca Muerta*. All were analysed by instrumental activation analysis techniques similar to those of Stueber and Goles (1967). Data on Na and Fe in pigeonite and anorthite from *Nuevo Laredo* demonstrate clearly that these

TABLE 2 Scandium contents (10^{-6} g/g) in mineral separates

Meteorite	Mineral and Sc content
PYROXENE	
Coolidge (C3)	clinobronzite 19.3 ± 0.3
Clovis (H3)	clinobronzite 10.3 ± 0.3
Miller (H5)	bronzite 8.01 ± 0.19
Richardton (H5)	bronzite 9.67 ± 0.19
Holbrook (L6)	hypersthene 10.8 ± 0.2
Shaw (L6)	~90% orthopyroxene, ~10% chrome diopside 18.4 ± 0.2
Chainpur (LL3)	pigeonite 10.51 ± 0.15
Appley Bridge (LL6)	orthopyroxene, Fs_{25} 15.5 ± 0.2
Kota-Kota (E4)	principally clinoenstatite 7.01 ± 0.14
Jajh deh Kot Lalu (E6)	principally orthoenstatite 11.4 ± 0.2
Shallowater (Ae)	enstatite 5.49 ± 0.08
Johnstown (Ab)	bronzite 14.4 ± 0.2
Tatahouine (Ab)	bronzite 13.4 ± 0.4
Juvinas (Ap)	pigeonite 37.2 ± 0.7
Nuevo Laredo (Ap)	pigeonite 44.8 ± 0.8
Pasamonte (Ap)	pigeonite 53.3 ± 0.7
Binda (Aor)	orthopyroxene 23.7 ± 0.3
*Shergotty**	pigeonite 70.7 ± 0.9
Bondoc Peninsula (M)	principally orthopyroxene 7.8 ± 0.8
Crab Orchard (M)	principally pigeonite 20.0 ± 0.3
Vaca Muerta (M)	pigeonite and orthopyroxene 16.0 ± 0.3
OLIVINE	
Argonia (P)	2.08 ± 0.16
Eagle Station (P)	2.33 ± 0.18
Glorieta Mt. (P)	2.85 ± 0.07
Mt. Dyrring (P)	1.54 ± 0.12
Springwater (P)	1.43 ± 0.07
Mt. Padbury (M)	1.46 ± 0.04
PLAGIOCLASE	
Appley Bridge (LL6)	1.77 ± 0.03
Nuevo Laredo (Ap)	1.35 ± 0.03
Binda (Aor)	2.69 ± 0.04
*Shergotty**	4.48 ± 0.06
Vaca Muerta (M)	0.401 ± 0.012
OPAQUE MINERALS	
Allegan (H5)	chromite 7.1 ± 0.6
Richardton (H5)	chromite 4.6 ± 0.2
Mt. Dyrring (P)	chromite 6 ± 2
McKinney (L4)	troilite 8.4 ± 1.1

* The classification of *Shergotty* is uncertain owing to numerous mineralogical and chemical peculiarities exhibited by this meteorite.

separates are cross-contaminated to roughly the level of one part in twenty, so that anorthite from *Nuevo Laredo* most likely contains essentially no Sc. A similar calculation for *Vaca Muerta* separates suggests that there is some Sc-bearing, Fe-free mineral(s) in the "anorthite" sample, but this may be, for example, a phosphate mineral rather than anorthite.

Calculations based on published norms imply that at least three meteorites for which both pyroxene and whole-rock Sc contents are known have another important host mineral for this element. About half of the Sc in *Miller* and in *Holbrook*, one-third of that in *Richardton*, is located in their pyroxenes, assuming that no gross sampling errors have been made. The location of the remainder is not well understood. It seems likely that Sc is captured to some degree by Fe-bearing minerals; see data on olivines and opaque minerals in Table 2. Still, it is very implausible that these should account for one-half to two-thirds of the Sc present in the three chondrites mentioned above. Mason and Wiik (1961a, 1961b, 1963) suggest that phosphate minerals might be found in these chondrites, and such minerals may well act as hosts for much of the Sc (and, presumably, rare earths) present.

All samples of olivine from pallasites for which data is given in Table 2 were selected with great care and appear to be quite clean upon microscopic examination. Hamaguchi *et alii* (1965) report 1.7×10^{-6} g Sc/g for "silicate phase" from *Admire*, which agrees very well with my data, but they do not discuss the purity of their separate. Among opaque minerals listed, the *Mt. Dyrring* chromite sample is a single crystal of mass 0.635 mg with apparently euhedral faces, and is by far the purest of the three chromites. Bulk compositions of the two other chromites imply that they are contaminated with about 20% *(Allegan)* and 40% *(Richardton)* of non-Cr-bearing phases. It is barely conceivable that all of the Sc observed in these two samples is attributable to the contaminants. The troilite sample, of mass 6.5 mg, approximates ideal Fe content; it was drilled out of a small nodule exposed on a sawn surface of *McKinney* and is probably quite pure.

Scandium thus has marked potential as a probe for investigating paragenetic relationships (or demonstrating their absence) among meteoritic minerals. It is present in trace amounts and consequently is free from saturation effects. It is essentially absent from plagioclase but seems to be widely distributed in the other phases. One would expect that well-defined Sc partition coefficients, coupled with adequate experimental data, could be used to set limits on the (P, T) conditions under which meteorites originated. The major drawback at present is technical: one must either devise very selective and efficient mineral separation procedures, which would yield useful material for activation analyses, or extend one or another of the microprobe techniques so that Sc can be determined at the level of 10^{-6} g/g.

Acknowledgements

V. Frankum and S. Oxley assisted with the analytical work on mineral separates here reported, which was done with support in part from NASA grant NsG–319 at University of California San Diego.

References

Bate, G. L., Potratz, H. A. and Huizenga, J. R. (1960) "Scandium, chromium, and europium in stone meteorites by simultaneous neutron activation analysis." *Geochim. Cosmochim. Acta* **18**, 101–107.

Greenland, L. P. (1963) Ph. D. dissertation, Australian National University.

Greenland, L. P. and Lovering, J. F. (1965) "Minor and trace element abundances in chondritic meteorites." *Geochim. Cosmochim. Acta* **29**, 821–858.

Hamaguchi, H., Watanabe, T., Onuma, N., Tomura, K. and Kuroda, R. (1965) "Neutron activation analysis of scandium." *Analyt. Chim. Acta* **33**, 13–20.

Keil, K. (1969) "Meteorite composition", chapter 4 in *Handbook of Geochemistry*, edited by K. H. Wedepohl (Springer-Verlag, Berlin) 78–115.

Kemp, D. M. and Smales, A. A. (1960) "The determination of scandium in rocks and meteorites by neutron-activation analysis." *Analyt. Chim. Acta* **23**, 410–418.

Mason, B. and Wiik, H. B. (1961a) "The Miller, Arkansas, chondrite." *Geochim. Cosmochim. Acta* **21**, 266–271.

Mason, B. and Wiik, H. B. (1961b) "The Holbrook, Arizona, chondrite." *Geochim. Cosmochim. Acta* **21**, 276–283

Mason, B. and Wiik, H. B. (1963) "The composition of the Richardton, Estacado, and Knyahinya meteorites." *Amer. Mus. Nov.*, No. **2154**, 1–18.

Ringwood, A. E. (1960) "The Novo Urei meteorite." *Geochim. Cosmochim. Acta* **20**, 1–4.

Schmitt, R. A., Goles, G. G. and Smith, R. H. (1971) *Elemental abundances in stone meteorites.* To be published as a NASA report; available from the author on request.

Stueber, A. M. and Goles, G. G. (1967) "Abundances of Na, Mn, Cr, Sc and Co in ultramafic rocks." *Geochim. Cosmochim. Acta* **31**, 75–93.

Van Schmus, W. R. and Wood, J. A. (1967) "A chemical-petrologic classification for the chondritic meteorites." *Geochim. Cosmochim. Acta* **31**, 747–765.

(Received 5 May 1970, revised 19 August 1970)

TITANIUM (22)

Brian Mason

Smithsonian Institution, Washington, D.C.

TITANIUM IS a minor element in stony and stony-iron meteorites, usually in the range of 500–5000 ppm by weight. It is normally determined by standard colorimetric methods in complete analyses of these meteorites, but was frequently omitted in older analyses. In a survey of selected analyses of 36 bronzite chondrites and 68 hypersthene chondrites, the range of TiO_2 was 0.02–0.42 per cent, with a mean of 0.15 per cent (900 ppm Ti) for each class (Mason, 1905). This range is not consistent with the uniformity of mineralogical composition of the common chondrites, and Moore and Brown (1962), by spectrographical analysis of a large number of chondrites, found a rather constant titanium content. For 19 bronzite chondrite falls the mean was 620 ppm Ti, equivalent to 0.103 per cent TiO_2; for 23 hypersthene chondrite falls the mean was 660 ppm Ti, equivalent to 0.110 per cent TiO_2. These figures are in agreement with recent wet-chemical analyses of ordinary chondrites.

The data on titanium in different classes of stony and stony-iron meteorites are summarized in the accompanying table. These determinations have been made by standard wet-chemical procedures (usually colorimetric), except for the figures for the bronzite and hypersthene chondrites, which are spectrographic. This table shows a fairly uniform level of titanium in the chondrites at 2000–3000 atoms/10^6 Si, although there is some evidence for a moderate degree of enrichment in the Type II and Type III carbonaceous chondrites. In the calcium-poor achondrites titanium contents are comparable to those in the chondrites, except for the enstatite achondrites, which show a marked depletion; in these meteorites the titanium resides almost entirely in the troilite, which is small in amount and irregularly distributed. The calcium-rich achondrites show a marked enrichment in titanium, both absolutely and relative to silicon. The unique meteorite Angra dos Reis shows extreme enrichment; this meteorite consists almost entirely of a calcium-rich aluminous pyroxene, in which the titanium is present in atomic

Table 1 Titanium in meteorites

	No. of detns.	Ti (ppm) Range	Ti (ppm) Mean	Atoms /10^6 Si	References
Chondrites					
Carbonaceous, Type I	3	400–450	430	2,400	(1)
Carbonaceous, Type II	9	480–700	540	2,900	(1)
Carbonaceous, Type III	6	600–1400	870	3,600	(1)
Bronzite	19	510–780	620	2,100	(2)
Hypersthene	23	460–810	660	2,100	(2)
Amphoterites	11	540–1300	840	2,600	(3)
Enstatite, Type I	3	400–800	570	2,000	(4)
Enstatite, Type II	7	300–900	620	1,900	(4)
Calcium-poor achondrites					
Enstatite	4	160–480	340	750	(5)
Hypersthene	2	1100–1300	1,200	2,800	(6)
Chassignite	1	–	960	3,200	(7)
Ureilites	3	720–1100	880	2,700	(8)
Calcium-rich achondrites					
Angrite	1	–	14,000	40,000	(9)
Nakhlite	1	–	2,300	5,800	(10)
Howardites	4	660–2800	2,200	5,500	(11)
Howardites	4	660–2800	2,200	5,500	(11)
Eucrites	5	3400–5800	4,600	12,000	(12)
Stony irons					
Pallasites (olivine)	4	10–60	25	78	(13)
Mesosiderites	1	–	960	4,700	(14)

(1) Wiik, in Mason (1963a)
(2) Moore and Brown (1962)
(3) Wiik, in Mason and Wiik (1964), with later analyses
(4) Mason (1966), with later analyses
(5) Khor Temiki (Hey and Easton, 1967); Norton County (Wiik, 1956); Pesyanoe (Dyakonova and Kharitonova, 1960); Pena Blanca Spring (Lonsdale, 1947)
(6) Mason (1963b)
(7) Dyakonova and Kharitonova (1960)
(8) Wiik (unpublished)
(9) Ludwig and Tschermak (1909)
(10) Prior (1912)
(11) Bununu (Mason, 1967); Frankfort (Mason and Wiik, 1966a); Kapoeta (Mason and Wiik, 1966b); Bholgati (Jarosewich, unpublished)
(12) Duke and Silver (1967)
(13) Lovering (1957)
(14) Jarosewich and Mason (1969)

substitution, probably as the component $CaTiAl_2O_6$. Among the stony-irons the pallasites are notably deficient in titanium, because this element does not readily enter olivine, the silicate mineral; the mesosiderites contain silicates similar to the howardites, and the titanium content is comparable in these two classes.

The chondritic abundances are compatible to that in the solar atmosphere, 1500 atoms Ti/10^6 Si, according to Aller (1961).

In most meteorites titanium is mainly lithophile, but also has moderate chalcophile affinity; in enstatite chondrites and enstatite achondrites, however, it is almost completely chalcophile. Titanium is considered to have little or no siderophile tendency, and is seldom looked for in analyses of meteoritic iron; Moss (pers. comm.) reports it below the limit of detection (<5 ppm) in several irons.

Titanium may be present in several phases in a single meteorite. The only meteoritic minerals in which this element is an essential constituent are ilmenite ($FeTiO_3$), rutile (TiO_2), and osbornite (TiN). Ilmenite is not uncommon as an accessory mineral; Ramdohr (1963) reports... "It was observed in more than 50 per cent of all the stony meteorites examined. It appears to be absent in some of the strongly reduced chondrites and in carbonaceous chondrites. Usually it occurs in very small quantities, only a few grains being found in sections of normal size." Rutile is a rare accessory mineral (in some meteorites as exsolution lamellae in ilmenite and chromite) in a few chondrites and some mesosiderites (Buseck and Keil, 1966). Osbornite was described many years ago from the Bustee enstatite achondrite, and may occur in other enstatite achondrites and enstatite chondrites in trace amounts. However, most of the titanium in meteorites is dispersed in the more abundant minerals in substitution for the major elements. Olivine contains little titanium, probably in the 10–100 ppm range. Orthopyroxene contains about 1000 ppm (except for enstatite from enstatite chondrites and enstatite achondrites, which contains less than 100 ppm), whereas the clinopyroxenes diopside and pigeonite can take up considerably larger amounts, usually up to about 5000 ppm; the clinopyroxene from the Angra dos Reis meteorite is exceptional with 15,000 ppm. Plagioclase may contain a little titanium, but no measurements in meteoritic plagioclase have been reported; in terrestrial plagioclase the maximum appears to be about 400 ppm. Chromite ($FeCr_2O_4$) is an accessory mineral in stony meteorites, usually in the 0.1–0.5 per cent range, but it contains significant amounts of titanium; in the common chondrites it contains 2–4 per cent TiO_2 (Snetsinger et al., 1967). In the enstatite chondrites and enstatite achondrites almost all the titanium is present in solid solution in troilite; usually about 0.5 per cent in enstatite chondrites (Keil, 1968), but as much as 4.1 per cent was recorded in troilite from the Norton County enstatite achondrite (Keil and Fredriksson, 1963), and up to about 10 per cent in this mineral from the Khor Temiki enstatite achondrite (Keil, 1969). In the common chondrites Moss et al. (1967) found 192, 100 and 132 ppm in troilite from three hypersthene chondrites, and 398 ppm in troilite from a bronzite chondrite. Titanium has not been recorded from other sulfide minerals except daubreelite ($FeCr_2S_4$), in which Keil (1968) found up to 1500 ppm.

Titanium is clearly a dispersed element in stony meteorites, but the major amount is probably contained in the pyroxenes, except in the enstatite chondrites and enstatite achondrites.

No studies have been reported on the isotopic composition of titanium in meteorites.

References

Aller, L.H. (1961) *The abundance of the elements*. Interscience.
Buseck, P.R. and Keil, K. (1966) *Am. Mineral.* **51**, 1506.
Duke, M.B. and Silver, L.T. (1967) *Geochim. Cosmochim. Acta* **31**, 1637.
Dyakonova, M.I. and Kharitonova, V.Y. (1960) *Meteoritika* **18**, 48.
Hey, M.H. and Easton, A.J. (1967) *Geochim. Cosmochim. Acta* **31**, 1789.
Jarosewich, E. and Mason, B. (1969) *Geochim. Cosmochim. Acta* **33**, 411.
Keil, K. (1968) *J. Geophys. Res.* **73**, 6945.
Keil, K. (1969) *Earth Planet. Sci. Letters* **5**.
Keil, K. and Fredriksson, K. (1963) *Geochim. Cosmochim. Acta* **27**, 939.
Lonsdale, J.T. (1947) *Am. Mineral.* **32**, 354.
Lovering, J.F. (1957) *Geochim. Cosmochim. Acta* **12**, 253.
Ludwig, E. and Tschermak, G. (1909) *Tsch. Min. Pet. Mitt.* **28**, 109.
Mason, B. (1963a) *Space Sci. Rev.* **1**, 621.
Mason, B. (1963b) *Am. Museum Novitates*, No. 2155.
Mason, B. (1965) *Am. Museum Novitates*, No. 2223.
Mason, B. (1966) *Geochim. Cosmochim. Acta* **30**, 23.
Mason, B. (1967) *Geochim. Cosmochim. Acta* **31**, 107.
Mason, B. and Wiik, H.B. (1964) *Geochim. Cosmochim. Acta* **28**, 533.
Mason, B. and Wiik, H.B. (1966a) *Am. Museum Novitates*, No. 2272.
Mason, B. and Wiik, H.B. (1966b) *Am. Museum Novitates*, No. 2273.
Moore, C.B. and Brown, H. (1962) *Geochim. Cosmochim. Acta* **26**, 495.
Moss, A.A., Hey, M.H., Elliott, C.J. and Easton, A.J. (1967) *Mineral. Mag.* **36**, 101.
Prior, G.T. (1912) *Mineral. Mag.* **16**, 274.
Ramdohr, P. (1963) *J. Geophys. Res.* **68**, 2011.
Snetsinger, K.G., Keil, K. and Bunch, T.E. (1967) *Am. Mineral.* **52**, 1322.
Wiik, H.B. (1956) *Geochim. Cosmochim. Acta* **9**, 279.

(Received 10 January 1969)

VANADIUM (23)

Walter Nichiporuk

Center for Meteorite Studies
Arizona State University
Tempe, Arizona

VANADIUM IS a frequently determined trace lithophile element in meteorites, and numerous analyses are available. Although it is usually determined in the meteorites by emission spectrography, it has also been determined by neutron activation analysis, both "classical" and instrumental, and by mass spectrometry. In a summary of the older analyses for vanadium in meteorites, Goldschmidt (1958) gave 50×10^{-6} g V/g in the silicate phase of the chondritic stony meteorites, $1{,}500 \times 10^{-6}$ g V/g in the troilite phase of iron meteorites, and 5×10^{-6} g V/g in the metal phase of iron meteories. By weighting these phases in the proportion of 10 parts silicate, 1 part troilite, and 2 parts metallic nickel-iron, Goldschmidt obtained an average of 100×10^{-6} g V /g in mean meteoritic matter. The Noddacks (1934) used the chondritic stony meteorites as a proper average of meteoritic materials and reported a mean of 91×10^{-6} g V/g. Either of these averages, that of the Noddacks or of Goldschmidt, is not very far from the truth, although each has to be decreased slightly in the light of recent data.

A summary of the recent determinations of vanadium in 119 stony and four stony-iron meteorites is given in Table 1 and the distribution of a total of 152 individual vanadium analyses is indicated graphically in Figure 1. These determinations have been made by emission spectrography (Lovering, 1957; Greenland and Lovering, 1965; Nichiporuk and Bingham, 1970), except the data on one carbonaceous and four hypersthene chondrites which are by "classical" neutron activation analysis (Kemp and Smales, 1960). The vanadium data are somewhat variable in quality and replicate analyses of several chondrites show a considerable spread, so that the averages are uncertain to about 20 per cent. In the case of one enstatite chondrite (Abee),

TABLE 1 Vanadium in stony meteorites

Name	Type*	Number of meteorites analyzed	Total number of determinations	Range, 10^{-6} g/g†	Mean V 10^{-6} g/g	Atomic V abundance Si = 10^6 atoms[a]
Chondrites						
Carbonaceous	C1	1	2	41–57	49	254
Carbonaceous	C2	2	2	56–71	64	270
Carbonaceous	C3	6	11	50–117	88	312
Bronzite	H4–H6, H	32	35	44–88	61	196
Hypersthene	L3–H6, L	49	61	45–94	65	193
Amphoterite	LL4–LL6	7	11	53–93	74	217
Enstatite	E4	2	3	55–163	83	268
				(55–56)[b]	(56)[b]	(181[b])
Enstatite	E5–E6	5	6	49–151	82	232
				(49–100)[c]	(65)[c]	(183)[c]
Calcium-poor achondrites						
Enstatite		4	4	14–21	17	34
Hypersthene		2	2	61–242	152	339
Calcium-rich achondrites						
Nakhlites (Lafayette)		1	1		169	407
Howardites		1	1		86	207
Eucrites		7	7	39–156	75	180
Stony irons						
Pallasites (Olivine)		4	6	10–14	15	

* Classification of chondrites according to Van Schmus and Wood (1967).

† Based on the data of Kemp and Smales (1960), Greenland and Lovering (1965), Nichiporuk and Bingham (1970), and Lovering (1957).

[a] Using average silicon values for stony meteorite groups supplied for the purposes of this summary by B. Mason.

[b] Omitting Greenland and Lovering's high vanadium value of 163×10^{-9} g/g for Abee, which these authors suspect might be due to an unrepresentative sample.

[c] Omitting Greenland and Lovering's high vanadium value of 151×10^{-9} g/g for St. Mark's for the same reason as in the preceding footnote.

the spread of a factor of 3 indicated an uncertainty in the average result of about 50 per cent. For this meteorite it may require more detailed study of the mineralogical composition and vanadium content to say with a high degree of certainty whether the analyses are poorly done or the meteorite is unusual.

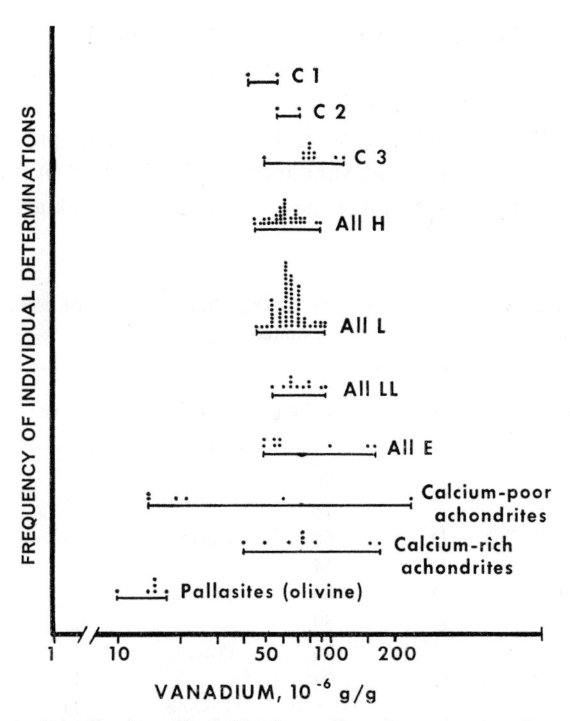

FIGURE 1 Distribution of individual vanadium determinations in stony and stony-iron meteorites.

The summary in Table 1 shows a fairly uniform abundance of vanadium in the chondrites at 200–300 atoms/10^6 Si, although there is some evidence for a moderate degree of enrichment in the C2 and C3 chondrites. The vanadium abundance in the C1 chondrites is in agreement with mean chondritic abundance of 220 atoms/10^6 Si reported by Suess and Urey (1956), but is somewhat larger than the abundance of 160 atoms/10^6 Si in the solar atmosphere according to Aller (1961).

Vanadium in achondrites is much more variable than in the chondrites. In the calcium-poor achondrites vanadium is markedly depleted in the enstatite types and markedly enriched in the hypersthene types relative to the chondrites. There is no apparent enrichment of vanadium in the calcium-rich achondrites, except the meteorite Lafayette in which there is an appreciable enrichment. In the pallasites (olivine), much like in the enstatite achondrites, vanadium is markedly deficient relative to the chondrites.

Vanadium has been studied in the separated light and dark parts of brecciated stony meteorites by Mazor and Anders (1967) and by Nichiporuk

(1970). Of a total of eleven such meteorites studied only one, Cumberland Falls (Merrill, 1920), was a brecciated mixture of dissimilar component fragments and only two, Jodzie (Mazor and Anders) and Fayetteville (Müller and Zähringer, 1966), were known to contain primordial gases mainly in their dark parts. The vanadium levels observed in the brecciated stony meteorites by Mazor and Anders and by Nichiporuk are as shown in Table 2.

TABLE 2 Vanadium in the dark (D) and light (L) parts of brecciated stony meteorites

Name*	Sample	V, 10^{-6} g/g	Name	Sample	V, 10^{-6} g/g
Arriba	D	70	Merua	D	57
	L	76		L	58
Cumberland	D	68	Pasamonte	D	114
Falls	L	14		L	97
Fayetteville	D	66	Plainview	D	55
	L	59		L	67
Jelica	D	76	Potter	D	67
	L	79		L	58
Jodzie*	D	90–120	Texline	D	65
	L	130		L	65
Kelly	D	62			
	L	60			

* Spectrographic determinations on Jodzie by Norman Suhr, Pennsylvania State University, as reported by Mazor and Anders (1967). Spectrographic determinations on all other meteorites by Elizabeth Bingham, California Institute of Technology, as reported by Nichiporuk (1970).

The data on vanadium in specific iron meteorites are listed in Table 3. These determinations have been made by instrumental neutron activation analysis (Linn et al., 1968), by mass spectrometry (Stauffer and Honda, 1961), and by "classical" neutron activation analysis (Kemp and Smales, 1960). For the meteorite Yardymly (Aroos) the mass-spectrometric result is a value after correction for the cosmic-ray produced vanadium and represents the primordial vanadium content of the meteorite. All results in the table are upper limits only, since the vanadium determination in iron meteorites at these levels becomes very difficult even with mass-spectrometric and neutron activation techniques. Nonetheless, the results do indi-

TABLE 3 Vanadium in iron meteorites

Name	Class	V, 10^{-6} g/g	Reference
Canyon Diablo	Og	<0.2	(1)
Mount Stirling	Og	<0.22	(3)
Yardymly (Aroos)	Og	0.2	(2)
Youndegin	Og	<0.10	(3)
Carbo	Om	<0.07	(3)
Bear Creek	Of	<0.23	(3)
Bella Roca	Of	<0.20	(3)
Carlton	Of	<0.2	(1)

(1) Kemp and Smales (1960)
(2) Stauffer and Honda (1961)
(3) Linn, Moore and Schmitt (1968)

cate that the vanadium levels in meteoritic iron are very low indeed, much lower than the 5×10^{-6} g/g limit selected by Goldschmidt (1958) in his calculation of the average composition of meteorites.

In addition to these studies of main meteorite series, there is a number of analyses of accessory mineral phases of meteorites. Results are reported by Lovering (1957), Nichiporuk and Chodos (1959) and Linn et al. (1968) for the troilite phase* and by Yavnel (1950) and Bunch et al. (1967) for the chromite. The only troilite from stony meteorites was separated from the bronzite chondrite Cullison and found to contain a limiting amount of 13×10^{-6} g V/g (Nichiporuk and Chodos). The observed concentrations of vanadium in the troilite from the Brenham and Springwater pallasites are 8×10^{-6} g/g and 9×10^{-6} g/g respectively, according to Lovering, and 24×10^{-6} g/g in the troilite from Brenham, according to Nichiporuk and Chodos. Much greater variations exist in the vanadium content of the troilite from iron meteorites. Studies by Lovering showed a variation between 10×10^{-6} g V/g and 320×10^{-6} g V/g, and those by Nichiporuk and Chodos a variation between less than 13×10^{-6} g V/g and 415×10^{-6} g V/g. The most recent studies of vanadium in the troilites from iron meteorites have been made by Linn et al. They did not find any troilite inclusions with vanadium contents smaller than 0.6×10^{-6} g/g or greater than 48×10^{-6} g/g.

Chromite is an accessory mineral in stony meteorites, usually in the range of 0.1–0.6 per cent by weight, but it contains significant amounts of vanadium;

* Often impure, containing one or more of the following minerals: daubreelite, nickel-iron, chromite, schreibersite, graphite.

in the bronzite and hypersthene chondrites it contains 0.36 to 0.82 per cent V_2O_3 (Bunch et al., 1967). A small inclusion of chromite extracted from the Sikhote-Alin iron meteorite by Yavnel contained about 0.3 per cent by weight vanadium.

TABLE 4 Vanadium isotopic ratios in meteoritic and terrestrial samples as determined by Balsiger et al. (1969)

Sample	Chondrite class	$V^{50}/V^{51} \times 10^{-3}$
Meteoritic		
Abee	E4	2.441 ± 0.018
Bruderheim	L6	2.452 ± 0.014
Dimmitt	H (3, 4)	2.425 ± 0.027
Mezö-Madaras	L3	2.453 ± 0.022
Orgueil	C1	2.414 ± 0.066
Terrestrial		
Diabase W-1		2.444 ± 0.032
V-reagent I		2.453 ± 0.016
V-reagent II		2.464 ± 0.015

A number of results are reported by Balsiger et al. (1969) for the V^{50}/V^{51} abundance ratio in chondritic meteorites of different selected subclasses. The results are shown in Table 4 and compared with those obtained for terrestrial diabase W-1 and reagent vanadium. It can be seen that for all these samples the vanadium isotopic ratio exhibits rather constant values.

References

Aller, L.H. (1961) *The abundance of the elements*. Interscience.
Balsiger, H., Geiss, J., and Lipschutz, M.E. (1969) *Earth Planet. Sci. Letters* **6**, 117.
Bunch, T.E., Keil, K. and Snetsinger, K.G. (1967) *Geochim. Cosmochim. Acta* **31**, 1569.
Goldschmidt, V.M. (1958) *Geochemistry*. Edited by A. Muir. Clarendon Press.
Greenland, L. and Lovering, J.F. (1965) *Geochim. Cosmochim. Acta* **29**, 821.
Kemp, D.M. and Smales, A.A. (1960) *Anal. Chim. Acta* **23**, 397.
Linn, T.A., Jr., Moore, C.B. and Schmitt, R.A. (1968) *Geochim. Cosmochim. Acta* **32**, 561.
Lovering, J.F. (1957) *Geochim. Cosmochim. Acta* **12**, 253.
Mazor, E. and Anders, E. (1967) *Geochim. Cosmochim. Acta* **31**, 1441.
Merrill, G.P. (1920) *Proc. U.S. National Mus.* **57**, 97.

Müller, O. and Zähringer, J. (1966) *Earth Plan. et Sci. Letters* **1**, 25.
Nichiporuk, W. (1970) Paper in preparation.
Nichiporuk, W. and Bingham, E. (1970) Paper in preparation.
Nichiporuk, W. and Chodos, A.A. (1959) *J. Geophys. Res.* **64**, 2451.
Noddack, I. and Noddack, W. (1934) *Svensk Kem. Tidskr.* **46**, 173.
Stauffer, H. and Honda, M. (1961) *J. Geophys. Res.* **66**, 3584.
Suess, H.E. and Urey, H.C. (1956) *Revs. Mod. Phys.* **28**, 53.
Van Schmus, W.R. and Wood, J.A. (1967) *Geochim. Cosmochim. Acta* **31**, 747.
Yavnel, A.A. (1950) *Meteoritika* **8**, 134.

(Received 30 August 1969, revised 8 June 1970)

CHROMIUM (24)

Gordon G. Goles

Center for Volcanology
and
Departments of Chemistry and Geology
University of Oregon
Eugene, Oregon 97403

THE MINOR ELEMENT chromium is determined in the conventional scheme of analysis of stone meteorites, so that there are many reported data. Individual results vary over a wide range, although the results of analyses selected for their presumed high quality (such as those by H. B. Wiik) indicate that the true variability of Cr contents may be much less than suggested by the total corpus of data. Severe analytical difficulties may arise in the conventional approach from the presence of the refractory spinel chromite in many stone meteorites; dissolution of chromite, which contains on the average about one-third of the Cr in ordinary chondrites, requires great care. Consequently, it is preferable where possible to use results of Cr analyses by procedures which do not involve dissolution of the sample, such as instrumental activation analyses or x-ray fluorescence. I shall here rely principally on data from instrumental neutron activation analyses for Cr by Schmitt, Goles and Smith (1971) and by myself.

Schmitt, Goles and Smith (1971) discuss comparisons of their results with those determined by other workers and other methods. Briefly, Cr contents determined by Wiik (1956; see also Mason, 1963a, 1965) agree very well with those determined by Schmitt, Goles and Smith, if one excludes the value given for *Ochansk* by Wiik (1956) which is probably a misprint. Greenland (1963) determined Cr by an emission spectrographic technique in numerous meteorites, of which 22 chondrites and 5 achondrites were also analysed by Schmitt, Goles and Smith. Agreement is good, although the precision of the emission spectrographic results seems to be markedly poorer than that of the activation analyses. Duke and Silver (1967) report the results of analyses by Maynes and by Blake of five eucrites, each of which has been analysed by Schmitt, Goles and Smith. Agreement of Cr results is very good. Nichiporuk et al. (1967) have determined Cr in 57 chondrites of the CH and CL groups via x-ray fluorescence; their results seem to be systematically too low by about 10 to 15%. Yates, Tackett and Moore (1968) have used an x-ray fluorescence technique to determine Cr in 35 chondrites and find contents in very good agreement with those of Schmitt, Goles and Smith.

Table 1 presents average contents and atomic abundances of Cr in stone meteorites, largely from the work of Schmitt, Goles and Smith (1971). Chondrites are classified according to Van Schmus and Wood (1967) and achondrites are identified by the code introduced by Keil (1969). Dispersions cited are estimates of the population variability (one standard deviation), and are not directly related to analytical uncertainties although they may reflect in part non-representative sampling. Despite the sizeable dispersions,

TABLE 1 Chromium average contents and atomic abundances in stone meteorites

Class	Number of specimens (samples) analysed	Cr, 10^{-6} g/g	Cr atoms per 10^4 Si
C1	3 (4)	2400 ± 300	127 ± 16
C2	7 (11)	3100 ± 300	124 ± 12
C3	9 (15)	3500 ± 400	123 ± 14
H3	5 (8)	3400 ± 500	106 ± 15
H4, 5, 6	40 (56)	3400 ± 400	109 ± 12
L3	3 (4)	3750 ± 180	109 ± 5
L4, 5, 6	47 (61)	3800 ± 400	109 ± 11
LL3	3 (5)	3640 ± 190	104 ± 5
LL4, 5, 6	17 (21)	3700 ± 400	106 ± 11
E4	2 (3)	3210 ± 110	101 ± 4
E5, 6	5 (6)	3300 ± 500	96 ± 14
Aubrites, Ae	5 (7)	500 ± 200	10 ± 5
*Diogenites, Ab	3 (3)	6800 ± 1600	150 ± 40
Ureilites, Aop	2 (3)	4900 ± 400	138 ± 10
Eucrites, Ap	20 (27)	2300 ± 900	50 ± 20
Howardites, Aor	9 (13)	4600 ± 1300	110 ± 30
Nakhlites, Ado	2 (3)	1680 ± 130	40 ± 4
Shergotty	1 (2)	700 ± 300	15 ± 6

* D.Y. Jerome, personal communication

there is a hint that carbonaceous chondrites may have higher Cr atomic abundances than do the other classes of chondrites. Nevertheless, Cr and Si are clearly coherent to a remarkable degree in chondrites, an observation which is quite puzzling when considered in the light of what is known about the phases in which Cr is located in these meteorites (Keil, 1962, 1968; Mason, 1963a). Chromium is a dispersed element in the carbonaceous chondrites, is concentrated in chromite (as mentioned above) in the H, L and LL groups, and resides in the very different mineral daubréelite in enstatite chondrites. If chondrites were chance agglomerates of previously-existing minerals like those now present, as implied by many hypotheses of their origin, one might expect to see much less coherence between Cr and Si than is indicated by the data of Table 1.

In contrast, Cr and Si are incoherent in achondrites, most of which have lower Cr atomic abundances than do the chondrites. Chromium contents and atomic abundances of ureilites support Ringwood's (1960) suggestion that a small amount of a low-melting fraction has been lost from a parent material like carbonaceous chondrites to form these meteorites. The high Cr contents and atomic abundances of diogenites are related to the "noteworthy amount of chromium combined in the pyroxene" (Mason, 1963b) of these achondrites.

Data on Cr in mineral separates in Table 2 were obtained by an instrumental activation analysis technique similar to that used by Stueber and Goles (1967). B. H. Mason provided most of the separates; those from *Vaca Muerta* were supplied by U. B. Marvin, and I hand-cobbed the olivine separates. The plagioclase separates are contaminated with pyroxene to a greater or lesser degree, as is clearly indicated by their Fe contents. Upon making crude corrections for this cross-contamination, their apparent Cr contents go to zero. Observed Cr contents are reported here because they serve to place useful limits on the extent of pyroxene cross-contamination of these plagioclase separates, which are discussed in others of the chapters I have written for this volume. The olivine separates are very clean, although minute traces of chromite might be present, and much of the observed Cr is probably present in solid solution in the olivine lattice.

Chromium contents in pyroxenes seem to reflect, sometimes in striking ways, diverse paragenetic and perhaps petrogenetic events. Compare, for example, the Cr content of enstatite from the *Kota-Kota* (E4) chondrite with those of enstatites from *Jajh deh Kot Lalu* (E6) and *Shallowater* (Ae) stones, which presumably reflect the equilibration of enstatite with Cr-bearing sulfide phases in the latter two specimens but not (or to a markedly lesser

TABLE 2 Chromium contents (10^{-6} g/g) in mineral separates

PYROXENE	
Coolidge (C3)	
clinobronzite	4060 ± 60
Clovis (H3)	
clinobronzite	5140 ± 70
Miller (H5)	
bronzite	4550 ± 90
Richardton (H5)	
bronzite	5840 ± 110
Holbrook (L6)	
hypersthene	3930 ± 80
Shaw (L6)	
~90% orthopyroxene, ~10% chrome diopside	7060 ± 130
Chainpur (LL3)	
pigeonite	5060 ± 90
Appley Bridge (LL6)	
orthopyroxene, Fs_{25}	2810 ± 50
Kota-Kota (E4)	
principally clinoenstatite	2270 ± 30
Jajh deh Kot Lalu (E6)	
principally orthoenstatite	47 ± 11
Shallowater (Ae)	
enstatite	5.6 ± 1.2
Johnstown (Ab)	
bronzite	5000 ± 90
Tatahouine (Ab)	
bronzite	5040 ± 90
Juvinas (Ap)	
pigeonite	3290 ± 70
Nuevo Laredo (Ap)	
pigeonite	2940 ± 60
Pasamonte (Ap)	
pigeonite	3440 ± 60
Binda (Aor)	
principally orthopyroxene	5760 ± 100
*Shergotty**	
pigeonite	1690 ± 30
Bondoc Peninsula (M)	
principally orthopyroxene	5800 ± 100
Crab Orchard (M)	
principally pigeonite	6640 ± 120
Vaca Muerta (M)	
pigeonite & orthopyroxene	3820 ± 70
OLIVINE	
Argonia (P)	339 ± 17
Eagle Station (P)	211 ± 18
Glorieta Mt. (P)	430 ± 10
Mt. Dyrring (P)	284 ± 14
Springwater (P)	184 ± 8
PLAGIOCLASE	
Appley Bridge (LL6)	1084 ± 19
Nuevo Laredo (Ap)	123 ± 3
Binda (Aor)	495 ± 9
*Shergotty**	53.7 ± 1.3
Vaca Muerta (M)	73.0 ± 1.3

* The classification of *Shergotty* is uncertain owing to numerous mineralogical and chemical peculiarities exhibited by this meteorite.

degree) in the former (see also Keil, 1968). It would be very interesting to follow up the fragmentary evidence for systematic differences in Cr content of pyroxenes in chondrites according to their class and petrographic type. Similarly, Cr contents of pyroxenes from achondrites might be useful in suggesting genetic relationships among these meteorites and certainly are helpful in classifying them (cf. *Shergotty* and other Ca-rich achondrites). Chromium contents in bronzites from the two diogenites analysed, *Johnstown* and *Tatahouine*, are very similar and agree well with the value of about

5500×10^{-6} g/g found by Haramura in a split of the same separate from *Johnstown* (Mason, 1963b; Mason also calls attention to the compositional uniformity of pyroxenes in this class of achondrites).

Although not strictly pertinent to this review, one should note the excellent studies of opaque minerals in stony meteorites by Ramdohr (1963) and of chromite in chondrites by Bunch, Keil and Snetsinger (1968).

Chromium in iron meteorites is concentrated in inclusions of diverse mineralogy and it is rarely possible either to avoid minute inclusions completely or to sample them in a truly representative way. Consequently, the results of extensive and, apparently, analytically-reliable studies of Cr contents in bulk specimens of iron meteorites by Lovering et al. (1957) and by Smales, Mapper and Fouché (1967) are difficult to interpret. I have summarized these results, as ranges of Cr contents, in Table 3 but the reader should

TABLE 3 Chromium contents (10^{-6} g/g) in metal from iron meteorites

LOVERING ET AL. (1957)		SMALES, MAPPER AND FOUCHÉ (1967)	
Ga-Ge groups I	<1–5.3	Structural groups H	28–169
II	2.6–101	Of	49–355
III	<1–100	Om (a)	5.1–7.7
VI	<1–112	(b)	18–63
		Og	8.3–14.2
		Ogg	12.5–29
WAI AND WASSON (1970)			
Boguslavka H	<20		
Clark County Om	260		
Guffey D	450		
Rafrüti D	<30		
Tucson D	2200 (2600, kamacite; 2500, taenite)*		
Santiago Papasquiero D	500		
Babb's Mill D	260		
N'Goureyma D	<60		
Nedagolla D	2600		

* From Bunch and Fuchs (1969)

beware of taking them too seriously, especially the higher values. A powerful tool for studying the distribution of Cr in phases of iron meteorites is the electron-beam microprobe; results obtained via this technique by Bunch and Fuchs (1969) and by Wai and Wasson (1970) are also given in Table 3. Unfortunately the technique at present has limited utility owing to lack of sensitivity. For crude mass-balance calculations, the grand average of 37×10^{-6} g Cr/g "metal" found by Lovering et al. might be considered preferable to the mean of all results of 115×10^{-6} g Cr/g "metal" reported by Smales, Mapper and Fouché. Investigation of possible systematic differences between structural classes or Ga–Ge groups among the iron meteorites must await additional data outlining the true distribution of Cr.

Nichiporuk and Chodos (1959) determined Cr in a series of troilite nodules from iron meteorites, and found the remarkable range of 181×10^{-6} g Cr/g to 5.57% Cr. They suggest that their observed contents are affected by the presence of inclusions of chromite or daubréelite, an inference foreshadowed by the work of Dyakonova (1958) on Cr in *Sikhote-Alin*. This would seem to be a very clear case where analyses should be made by microprobe techniques.

Lovering et al. (1957) analysed metal phases of pallasites, and demonstrated that they were poor in Cr. The Cr contents reported, ranging from 8.3×10^{-6} g/g for *Admire* to less than 1×10^{-6} g/g for five other pallasites, could be accounted for by contamination of their samples with chromite to the extent of about one part in 10^5 or less. Their value of 52×10^{-6} g Cr/g metal from the *Pinnaroo* mesosiderite may imply that their sample was contaminated with non-metallic Cr-bearing phases.

Acknowledgements

V. Frankum and S. Oxley assisted with the analytical work on mineral separates here reported, which was done with support in part from NASA grant NsG–319 at University of California San Diego.

References

Bunch, T.E. and Fuchs, L.H. (1969) "A new mineral: Brezinaite, Cr_3S_4, and the *Tucson* meteorite." *Amer. Mineral.* **54**, 1509–1518.

Bunch, T.E., Keil, K. and Snetsinger, K.G. (1967) "Chromite composition in relation to chemistry and texture of ordinary chondrites." *Geochim. Cosmochim. Acta* **31**, 1569–1582.

Duke, M.B. and Silver, L.T. (1967) "Petrology of eucrites, howardites, and mesosiderites." *Geochim. Cosmochim. Acta* **31**, 1637–1666.

Dyakonova, M.I. (1958) "Khimicheskii sostav Sikhote-Alinskogo meteorita." *Akad. Nauk. S.S.S.R., Meteoritika* **16**, 42–48.

Greenland, L.P. (1963) Ph. D. dissertation, Australian National University.

Keil, K. (1962) "On the phase composition of meteorites." *J. Geophys. Res.* **67**, 4055.

Keil, K. (1968) "Mineralogical and chemical relationships among enstatite chondrites." *J. Geophys. Res.* **73**, 6945–6976.

Keil, K. (1969) "Meteorite composition", chapter 4 in *Handbook of Geochemistry*, edited by K.H. Wedepohl (Springer-Verlag, Berlin) 78–115.

Lovering, J.F., Nichiporuk, W., Chodos, A. and Brown, H. (1957) "The distribution of gallium. germanium, cobalt, chromium, and copper in iron and stony-iron meteorites in relation to nickel content and structure." *Geochim. Cosmochim. Acta* **11**, 263–278.

Mason, B. (1963a) "The carbonaceous chondrites." *Space Science Rev.* **1**, 621–646.

Mason, B. (1963b) "The hypersthene achondrites." *Amer. Mus. Nov.*, no. 2155, 1–13.

Mason, B. (1966) "The enstatite chondrites", *Geochim. Cosmochim. Acta* **30**, 23–29.

Nichiporuk, W. and Chodos, A.A. (1959) "The concentration of vanadium, chromium, iron, cobalt, nickel, copper, zinc and arsenic in the meteoritic iron sulphide nodules." *J. Geophys. Res.* **64**, 2451–2463.

Nichiporuk, W., Chodos, A., Helin, E. and Brown, H. (1967) "Determination of iron, nickel, cobalt, calcium, chromium and manganese in stony meteorites by X-ray fluorescence." *Geochim. Cosmochim. Acta* **31**, 1911–1930.

Ramdohr, P. (1963) "The opaque minerals in stony meteorites." *J. Geophys. Res.* **68**, 2011–2036.

Ringwood, A.E. (1960) "The Novo Urei meteorite", *Geochim. Cosmochim. Acta* **20**, 1–4.

Schmitt, R.A., Goles, G.G. and Smith, R.H. (1971) *Elemental abundances in stone meteorites.* To be published as a NASA report; available from the author on request.

Smales, A.A., Mapper, D. and Fouché, K.F. (1967) "The distribution of some trace elements in iron meteorites, as determined by neutron activation." *Geochim. Cosmochim. Acta* **31**, 673–720.

Stueber, A.M. and Goles, G.G. (1967) "Abundances of Na, Mn, Cr, Sc and Co in ultramafic rocks." *Geochim. Cosmochim. Acta* **31**, 75.

Van Schmus, W.R. and Wood, J.A. (1967) "A chemical-petrologic classification for the chondritic meteorites." *Geochim. Cosmochim. Acta* **31**, 747.

Wai, C.M. and Wasson, J.T. (1970) "Silicon in the Nedagolla ataxite and the relationship between Si and Cr in reduced iron meteorites." *Geochim. Cosmochim. Acta* **34**, 408–410.

Wiik, H.B. (1956) "The chemical composition of some stony meteorites." *Geochim. Cosmochim. Acta* **9**, 279.

Yates, A.M., Tackett, S.L. and Moore, C.B. (1968) "Chromium and manganese in chonrites." *Chem. Geol.* **3**, 313–322.

(Received 5 May 1970, revised 19 August 1970)

Keil, K. (1968) "Mineralogical and chemical relationships among enstatite chondrites." *J. Geophys. Res.*, **73**, 6945–6976.

Keil, K. (1969) "Meteorite composition." Chapter 4 in *Handbook of Geochemistry*, edited by K. H. Wedepohl (Springer-Verlag, Berlin) 78–115.

[illegible] (1957) "The distribution of [illegible] in iron and stony-iron meteorites [illegible]." *Geochim. Cosmochim. Acta* **11**, 285–[illegible]

[illegible] "The carbonaceous chondrites." *Space Science Rev.*, **1**, [illegible]

[illegible]

[illegible]

[illegible] and [illegible] in the [illegible] iron-silicate nodules." *J. Geophys. Res.* [illegible]

[illegible]

[illegible] (1962) [illegible]

[illegible]

MANGANESE (25)

Gordon G. Goles

Center for Volcanology
and
Departments of Chemistry and Geology
University of Oregon
Eugene, Oregon 97403

THIS MINOR TRANSITION ELEMENT is usually determined in bulk analyses of stone meteorites, although the results often have only marginal significance, judging from the fact that the range of reported abundances is wide even within a given class. As in the case of Cr, analyses selected for their presumed high quality exhibit much less variability. Probably the most effective technique for determination of Mn in large numbers of meteorites is instrumental neutron activation analysis. Schmitt, Goles and Smith (1971) have used this technique to determine Mn in 206 samples of 143 chondritic specimens and 55 samples of 39 achondrites; I shall rely principally on their results here. Comparisons of Mn data of Schmitt, Goles and Smith with those of Wiik (Mason 1963a), Greenland and Lovering (1965), Nichiporuk et al. (1967), Höfler and Sorantin (1967), Schaudy, Kiesl and Hecht (1967, 1968), Yates, Tackett and Moore (1968), Von Michaelis, Ahrens and Willis (1969), and Ahrens, Von Michaelis and Fesq (1969) demonstrate good to excellent agreement among these authors, who used diverse techniques. While many of the Mn values of Moore and Brown (1962) agree well with those determined by authors listed above, others seem to be beyond reasonable limits of sampling errors.

Manganese average contents and atomic abundances in classes of stone meteorites, from Schmitt, Goles and Smith (1971), are presented in Table 1. Chondrites are classified according to Van Schmus and Wood (1967) and achondrites are identified by the code introduced by Keil (1969). Dispersions cited are estimates of the population variablity (one standard deviation),

TABLE 1 Manganese average contents and atomic abundances in stone meteorites

Meteorite class	Number of specimens (samples) analysed	Average contents, 10^{-6} g Mn/g	Atomic abundances, Mn/10^6 Si
C1	3 (5)	1900 ± 300	92 ± 13
C2	7 (13)	1630 ± 60	62 ± 2
C3	9 (20)	1490 ± 120	49 ± 4
H3	5 (8)	2250 ± 140	68 ± 5
H4, 5, 6	40 (55)	2260 ± 170	67 ± 5
L3	3 (4)	2400 ± 200	65 ± 5
L4, 5, 6	48 (61)	2460 ± 150	67 ± 4
LL3	3 (6)	2500 ± 300	67 ± 8
LL4, 5, 6	17 (22)	2560 ± 100	70 ± 3
E4	2 (4)	3200 ± 1000	100 ± 30
E5, 6	5 (7)	1800 ± 400	50 ± 12
Aubrites, Ae	5 (7)	1400 ± 600	28 ± 12
Ureilites, Aop	2 (3)	2890 ± 90	77 ± 3
Eucrites, Ap	20 (27)	3900 ± 600	85 ± 14
Howardites, Aor	9 (13)	3900 ± 400	88 ± 8
Nakhlites, Ado	2 (4)	3700 ± 150	82 ± 3
Shergotty	1 (1)	1540 ± (50)	34

and are not directly related to analytical uncertainties, although they may in part reflect non-representative sampling.

Manganese contents and atomic abundances systematically decrease in the sequence of C1, 2, 3 classes. In contrast, Mn and Si are closely coherent in ordinary chondrites, without regard to petrographic grades. Dispersions are small for C2, C3, H, L and LL chondrites but large for C1 and E classes. Perhaps this distinction is related to differences in mineralogy; Mn is located in olivine and pyroxene in the ordinary and C2, 3 chondrites but is present largely in minor sulfides in the enstatite chondrites. Its location in C1 stones is not well understood, but if this element is concentrated in some minor, inhomogeneously-distributed mineral such as magnetite in these meteorites, its variability could readily be explained.

Aubrites display great variability in Mn contents, consistent with the brecciated, polymict character of these achondrites. If the Mn content of enstatite from *Shallowater* (Table 2) is typical of these stones, essentially all of the Mn resides in minor phases. Ureilites have Mn contents and atomic abundances slightly greater than do ordinary chondrites, which is consistent with Ringwood's (1960) suggestion for their origin.

TABLE 2 Manganese contents (10^{-6} g/g) in mineral spearates

Meteorite / mineral	Mn
PYROXENE	
Coolidge (C3)	
clinobronzite	1700 ± 30
Clovis (H3)	
clinobronzite	3500 ± 70
Miller (H5)	
bronzite	3820 ± 70
Richardton (H5)	
bronzite	4370 ± 150
Holbrook (L6)	
hypersthene	3460 ± 140
Shaw (L6)	
~90% orthopyroxene	3100 ± 50
~10% chrome diopside	
Chainpur (LL3)	
pigeonite	2810 ± 50
Appley Bridge (LL6)	
orthopyroxene, Fs_{25}	3050 ± 100
Kota-Kota (E4)	
principally clinoenstatite	970 ± 20
Jajh deh Kot Lalu (E6)	
principally orthoenstatite	48 ± 12
Shallowater (Ae)	
enstatite	13.4 ± 0.5
Johnstown (Ab)	
bronzite	3790 ± 90
Tatahouine (Ab)	
bronzite	3770 ± 70
Juvinas (Ap)	
pigeonite	7330 ± 120
Nuevo Laredo (Ap)	
pigeonite	6270 ± 110
Pasamonte (Ap)	
pigeonite	6230 ± 100
Binda (Aor)	
orthopyroxene	5000 ± 80
*Shergotty**	
pigeonite	5410 ± 90
Bondoc Peninsula (M)	
principally orthopyroxene	2930 ± 60
Crab Orchard (M)	
principally pigeonite	4400 ± 70
Vaca Muerta (M)	
pigeonite & orthopyroxene	4110 ± 60
OLIVINE	
Argonia (P)	2050 ± 40
Eagle Station (P)	1180 ± 70
Glorieta Mt. (P)	2200 ± 50
Mt. Dyrring (P)	2220 ± 50
Springwater (P)	2270 ± 50
Mt. Padbury (M)	1550 ± 30
Determined by microprobe (Buseck and Goldstein 1969):	
Eagle Station (P)	1250
Marjalahti (P)	2000
Ollague (P)	2000
Pavlodar (P)	2400
Springwater (P)	2300
OPAQUE MINERALS	
Allegan (H5)	
chromite	3730 ± 70
Mt. Dyrring (P)	
chromite	4530 ± 60
McKinney (L4)	
troilite	198 ± 4

* The classification of *Shergotty* is uncertain owing to numerous mineralogical and chemical peculiarities exhibited by this meteorite.

Eucrites, howardites and nakhlites are very much alike in their average Mn contents and atomic abundances. The relatively wide dispersions are probably indicative of sampling problems rather than being a fundamental petrogenetic characteristic—these meteorites have many coarse grains in them and Mn is strongly concentrated in their pyroxenes (Table 2). *Shergotty* seems to have a very low Mn content, but additional samples should be analysed to confirm this feature.

Manganese contents in mineral separates are given in Table 2. See chapters on Sc and Cr for brief discussions of the sources and probable purities of these separates. Data on Mn in plagioclase separates are not reported, owing to contamination with pyroxenes, which seems adequate to account at least to the first order for all observed Mn in these separates.

Engel and Engel (1962) have observed that Mn contents of hornblendes and pyroxenes decrease with increasing metamorphic grade. Although data are sparse, there may be a tendency for Mn contents in pyroxenes from ordinary chondrites to *increase* with petrographic grade, suggesting that these meteorites are not related to one another by metamorphic processes in the usual sense. A striking contrast in Mn contents, which may be attributed to equilibration between enstatite and sulfides under reducing conditions, may be seen in comparing *Kota-Kota* and *Jajh deh Kot Lalu* pyroxenes. As Keil (1968) pointed out, differences in textural, mineralogical and chemical features among enstatite chondrites are not necessarily related solely to thermometamorphism. Like those of Cr, Mn contents would be useful in ranking enstatite chondrites and achondrites according to their degree of equilibration.

As was noted for Cr, Mn contents in bronzites from the two diogenites analysed, *Johnstown* and *Tatahouine*, are indistinguishable and agree very well with the value determined by Haramura on a split of the same separate from *Johnstown* (Mason 1963b).

Manganese contents in eucritic pyroxenes are variable and might prove to be a guide to the position of the parent rock in the presumed magmatic sequence which relates these meteorites (Duke and Silver 1967). Howardites appear to be mélanges of eucritic and diogenitic clastic materials (Jérome 1970), and silicates in mesosiderites seem to be closely related to howardites (Prior 1916, 1918). Data on Mn in pyroxene separates from *Binda*, *Crab Orchard* and *Vaca Muerta* agree with these hypotheses, but pyroxene from *Bondoc Peninsula* has an anomalously low Mn content.

There is excellent agreement between Mn contents of pallasitic olivines determined by me, using instrumental neutron activation analysis, and those determined by Buseck and Goldstein (1969) using an electron-beam microprobe. Olivine from *Eagle Station*, the only example of Prior's (1920) second class of pallasites in Table 2, clearly is much poorer in Mn than are the other olivines, which have constant Mn contents within analytical precision.

Such variations might yield information on petrogenetic relations (see also Craig 1964).

Manganese enters chromite to a substantial degree, but is nearly excluded from troilite. These observations agree with those of Bunch, Keil and Olsen (1970), summarized in Table 3 along with much additional information on Mn distributions in minerals of silicate inclusions in iron meteorites. Note

TABLE 3 Manganese contents (%) in minerals of silicate inclusions of iron meteorites as determined by electron-beam microprobe (Bunch, Keil and Olsen 1970)

	Odessa type	Copiapo type
Clinopyroxenes	0.15	0.17
Orthopyroxenes	0.25	0.24
Olivines	0.22	0.22
Chromite	1.5–2.6	2.06–2.6
Troilite	0.16	$<$0.02–2.6
Alabandite	55.0	–
Chlorapatite	0.03–0.11	0.11
Whitlockite	$\sim$0.03	–

that I have given in Table 3 a summary only of data on the two common kinds of inclusions, the so-called *Odessa* and *Copiapo* types. Bunch, Keil and Olsen argue that silicates in these types of inclusions are reasonably well-equilibrated with one another, so that one might use their data to approximate distribution coefficients for Mn among meteoritic minerals.

Manganese contents of carefully-selected metal phases from iron meteorites have been determined by Bauer and Schaudy (1970). They are generally low, ranging from 9.8 to 22 $\times 10^{-6}$ g Mn/g; metal from the *Estherville* mesosiderite has 75 $\times 10^{-6}$ g Mn/g (Cor less, if some silicates were in fact present in the sample taken for analysis). *San Cristóbal*, an iron with numerous silicate inclusions, exhibits high and variable Mn contents. It seems likely that bulk analyses of iron meteorites for Mn would be dominated by the fraction of that element present in inclusions, unless great care is taken in selecting samples.

Acknowledgements

V. Frankum and S. Oxley assisted with the analytical work on mineral separates here reported, which was done with support in part from NASA grant NSG–319 at University of California San Diego.

References

Ahrens, L.H., Von Michaelis, H. and Fesq, H.W. (1969) "The composition of the stony meteorites (IV). Some analytical data on Orgueil, Nogoya, Ornans and Ngawi." *Earth Planet. Sci. Letters* **6**, 285–288.

Bauer, R. and Schaudy, R. (1970) "Activation analytical determination of elements in meteorites, 3. Determination of manganese, sodium, gallium, germanium, copper and gold in 21 iron meteorites and 2 mesosiderites." *Chem. Geol.* **6**, 119–131.

Bunch, T.E., Keil, K. and Olsen, E. (1970) "Mineralogy and petrology of silicate inclusions in iron meteorites." *Contr. Mineral. and Petrol.* **25**, 297–340.

Buseck, P.R. and Goldstein, J.I. (1969) "Olivine compositions and cooling rates of pallasitic meteorites." *Geol. Soc. Am. Bull.* **80**, 2141–2158.

Craig, H. (1964) "Petrological and compositional relationships in meteorites." chapter 26 in *Isotopic and Cosmic Chemistry*, edited by H. Craig, S.L. Miller and G.J. Wasserburg (North-Holland, Amsterdam) 401–415.

Duke, M.B. and Silver, L.T. (1967) "Petrology of eucrites, howardites, and mesosiderites." *Geochim. Cosmochim. Acta* **31**, 1637–1666.

Engel, A.E.J. and Engel, C.G. (1962) "Hornblendes formed during progressive metamorphism of amphibolites, Northwest Adirondack Mountains, New York." *Geol. Soc. Am. Bull.* **73**, 1499–1514.

Greenland, L.P. and Lovering, J.F. (1965) "Minor and trace element abundances in chondritic meteorites." *Geochim. Cosmochim. Acta* **29**, 821–858.

Höfler, H. and Sorantin, H. (1967) "Application of nondestructive activation analysis to meteorites: Determination of aluminium, vanadium, manganese and gold in stony and iron meteorites." *Chem. Geol.* **2**, 279–287.

Jérome, D.Y. (1970) *Composition and origin of some achondritic meteorites* (Ph. D. dissertation, University of Oregon, Eugene).

Keil, K. (1968) "Mineralogical and chemical relationships among enstatite chondrites." *J. Geophys. Res.* **73**, 6945–6976.

Keil, K. (1969) "Meteorite composition". chapter 4 in *Handbook of Geochemistry*, edited by K.H. Wedepohl (Springer-Verlag, Berlin) 78–115.

Mason, B. (1963a) "The carbonaceous chondrites." *Space Science Rev.* **1**, 621–646.

Mason, B. (1963b) "The hypersthene achondrites." *Amer. Mus. Nov.*, No. 2155, 1–13.

Moore, C.B. and Brown, H. (1962) "The distribution of manganese and titanium in stony meteorites." *Geochim. Cosmochim. Acta* **26**, 495–502.

Nichiporuk, W., Chodos, A., Helin, E. and Brown, H. (1967) "Determination of iron, nickel, cobalt, calcium, chromium and manganese in stony meteorites by X-ray fluorescence." *Geochim. Cosmochim. Acta* **31**, 1911–1930.

Prior, G.T. (1916) "On the genetic relationship and classification of meteorites." *Mineral. Mag.* **18**, 26–44.

Prior, G.T. (1918) "On the mesosiderite-grahamite group of meteorites: with analyses of Vaca Muerta, Hainholz, Simondium, and Powder Mill Creek." *Mineral. Mag.* **18**, 151–172.

Prior, G.T. (1920) "The classification of meteorites." *Mineral. Mag.* **19**, 51–63.

Ringwood, A.E. (1960) "The Novo Urei meteorite." *Geochim. Cosmochim. Acta* **20**, 1–4.

Schaudy, R., Kiesl, W. and Hecht, F. (1967) "Activation analytical determination of elements in meteorites." *Chem. Geol.* **2**, 279–287.

Schaudy, Kiesl, W. and Hecht, F. (1968) "Activation analytical determination of elements in meteorites, 2. Determination of manganese, sodium, gallium, copper, gold and chromium in 21 meteorites." *Chem. Geol.* **3**, 307–312.

Schmitt, R.A., Goles, G.G. and Smith, R.H. (1971) *Elemental abundances in stone meteorites.* To be published as a NASA report; available from the author on request.

Van Schmus, W.R. and Wood, J.A. (1967) "A chemical-petrological classification for the chondritic meteorites." *Geochim. Cosmochim. Acta* **31**, 747–765.

Von Michaelis, H., Ahrens, L.H. and Willis, J.P. (1969) "The composition of stony meteorites. (II) The analytical data and an assessment of their quality." *Earth Planet. Sci. Letters* **5**, 387–394.

Yates, A.M., Tackett, S.L. and Moore, C.B. (1968) "Chromium and manganese in chondrites." *Chem. Geol.* **3**, 313–322.

(Received 14 September 1970)

IRON (26)

Brian Mason

Smithsonian Institution, Washington, D.C.

IRON IS a major element in all classes of meteorites except the enstatite achondrites, and as a consequence the abundance data are very extensive. The standard procedures of analytical chemistry give reliable results for total iron when carefully applied. Michaelis et al. (1969) have shown that the older data on chondrites summarized by Urey and Craig (1953) show a greater spread of values than more recent analyses, indicating an improvement in overall quality in recent years. The figures presented in Table 1 are derived as far as possible from recent analyses of observed falls.

As discussed in the introductory chapter, the individual classes of chondrites can be distinguished by their total iron content, and by the relative proportion of iron in oxidic compounds (mainly ferromagnesian silicates) and in nickel-iron and troilite (Fig. 1, Introduction). Urey and Craig (1953) utilized the total iron content to divide the chondrites into H (high iron) and L (low iron) groups. Bronzite chondrites are all H, averaging 27.6%, whereas hypersthene chondrites are all L, averaging 21.8%; the amphoterites are sometimes considered a subclass of hypersthene chondrites, denoted LL, and averaging 20.0%. Carbonaceous chondrites seem to be all H, since on a volatile-free basis they contain 25–27%, and their Fe/Si ratio is similar to or somewhat greater than that for the bronzite chondrites. Enstatite chondrites show a wide range of total iron contents, and designation as H or L types is of doubtful utility.

Urey (1961) pointed out the significance of the Fe/Si ratio in chondrites, both for the recognition of different classes and subclasses, and as an indication of chemical fractionations between different classes. The data in Table 1 show small but consistent differences between the three types of carbonaceous chondrites, the Fe/Si ratio being 0.901 for Type I, 0.841 for Type II, and 0.816 for Type III. The latter figure is essentially identical with that for the bronzite chondrites (0.812). The lower iron content of the

14 Mason (1495)

TABLE 1 Iron in meteorites

	No. of detns.	Fe, weight per cent		Atoms 10^3 Si	Reference
		Range	Mean		
Chondrites					
Carbonaceous, Type I	3	17.76–19.01	18.40	901	(1)
Carbonaceous, Type II	9	20.85–23.78	21.90	841	(1)
Carbonaceous, Type III	6	24.04–25.94	25.15	816	(1)
Bronzite	36	24.57–30.88	27.61	812	(2)
Hypersthene	60	20.15–23.61	21.81	577	(2)
Amphoterite	13	18.56–21.30	20.03	536	(3)
Enstatite, Type I	4	30.35–35.02	32.96	975	(4)
Enstatite, Type II	5	22.17–29.03	25.46	657	(4)
Calcium-poor achondrites					
Enstatite	4	0.47–1.55	1.02	19	(5)
Hypersthene	7	11.67–16.56	13.49	275	(6)
Chassignite	1	–	20.61	593	(7)
Ureilites	3	11.31–16.40	14.45	381	(8)
Calcium-rich achondrites					
Angrite	1	–	7.47	182	(9)
Nakhlite	1	–	16.60	365	(10)
Howardites	4	13.69–14.09	13.93	306	(11)
Eucrites	9	12.19–15.77	14.42	315	(12)
Stony-irons					
Pallasites	23	29.82–68.61	52.21	2596	(13)
Mesosiderites	5	43.97–51.76	48.18	2109	(14)

(1) Wiik, in Mason (1963a)
(2) Mason (1965)
(3) Wiik, in Mason and Wiik (1964), with later analyses
(4) Mason (1966)
(5) Khor Temiki (Hey and Easton, 1967); Norton County (Wiik, 1956); Pesyanoe (Dyakonova and Kharitonova, 1960); Pena Blanca Spring (Lonsdale, 1947)
(6) Mason (1963b)
(7) Dyakonova and Kharitonova (1960)
(8) Wiik (1969)
(9) Ludwig and Tschermak (1909)
(10) Prior (1912)
(11) Bununu (Mason, 1967); Frankfort (Mason and Wiik, 1966a); Kapoeta (Mason and Wiik, 1966b); Bholgati (Jarosewich, unpublished)
(12) Mason (1967)
(13) Chirvinsky (1949)
(14) Patwar (Jarosewich and Mason, 1969); Vaca Muerta, Hainholz (Prior, 1918); Clover Springs (Wiik, 1969); Mt. Padbury (McGall, 1966)

hypersthene chondrites and the amphoterites is reflected in the Fe/Si ratios of 0.577 and 0.536 respectively. Enstatite chondrites show a wide range of Fe/Si ratios, the most iron-rich ones having ratios slightly exceeding 1.

The interpretation of this range of Fe/Si ratios in terms of iron-silicate fractionation in the chondrites has been cogently discussed by Anders (1964). He came to the following conclusions:

1) The metal-silicate fractionation in chondrites involved loss of metal from primordial matter with Fe/Si $\approx$ 0.8–1.0.

2) At the time of fractionation, the material had gone through a high-temperature stage and contained individual metal and silicate grains. The metal phase was more highly reduced than the present metal phase in chondrites.

3) The fractionation probably occurred while the material was in a dispersed state.

At the time he wrote, a serious objection to his postulate of primordial matter with Fe/Si $\approx$ 0.8–1.0 was the apparent low abundance of iron in the Sun—Fe/Si = 0.12, according to Goldberg et al. (1960). This seemed to require that primordial matter was low in iron, and that high-iron chondrites must have been enriched in this element. However, this apparent impasse has recently been resolved by a reevaluation of the spectrographic data for the solar abundance of iron (Garz et al., 1969). This reevaluation has led to an increase in the figure for the solar abundance of iron by a factor of $\sim$10 compared to that given by Goldberg et al., the Fe/Si ratio now being given as 1.0. Thus it now appears that the high-iron chondrites, specifically the Type I carbonaceous chondrites, are closely akin to original solar matter.

Relative to the chondrites, all the achondrites (except the unique chassignite) show marked depletion in iron and correspondingly lower Fe/Si ratios. To a considerable extent, this is due to marked impoverishment of these meteorites in nickel-iron and troilite, which are present in very small amounts (except in the ureilites). The enstatite achondrites are extremely depleted: they consist essentially of almost iron-free enstatite ($MgSiO_3$).

The stony-irons consist of mixtures, in somewhat variable amounts, of achondritic silicates and metal phase. They average about 50% silicates, 50% nickel-iron, and this is reflected in high Fe/Si ratios.

The iron meteorites consist essentially of nickel-iron, with minor to trace amounts of accessory minerals such as troilite, schreibersite, and graphite. Samples for analysis are generally selected to avoid the accessory minerals as far as possible, and the resulting data thus correspond more closely to the metal phase than to the overall composition of the meteorite. The metal phase is essentially a three-component system Fe-Ni-Co; the cobalt content is uniformly low, ranging from 0.3–1.0%, so the iron and nickel contents are inversely related. The range of iron content in iron meteorites can thus be derived directly from a plot of nickel contents, such as Fig. 1 in the nickel chapter.

Buddhue (1946) compiled the analytical data for the different classes of iron meteorites, and calculated average compositions, after eliminating those analyses that appeared unreliable. His results are as follows:

Class	Number of analyses	Fe, weight per cent
Hexahedrites	48	93.59
Coarsest octahedrites	20	92.33
Coarse octahedrites	40	91.22
Medium octahedrites	126	90.67
Fine octahedrites	43	90.53
Finest octahedrites	13	86.75
Nickel-rich ataxites	38	79.63
All irons	428	89.70

(The total number of analyses for "All irons" is greater than the sum of the analyses for the individual classes, because in the overall average Buddhue included analyses of metal from stony-irons and chondrites; however, the averages for the metal from these groups are very close to the overall average.)

In stony and stony-iron meteorites the principal iron-bearing minerals are nickel-iron, troilite, and the ferromagnesian silicates olivine and pyroxene. Small amounts of the accessory minerals chromite ($FeCr_2O_4$) and ilmenite ($FeTiO_3$) are usually present. The Type I and Type II carbonaceous chondrites are unique in having much of their iron present as the hydrated magnesium-iron silicate serpentine (or chlorite), and in having magnetite (Fe_3O_4) and pentlandite ($(Fe, Ni)_9S_8$) as accessory minerals.

References

Anders, E. (1964) *Space Sci. Rev.* **3**, 583.
Buddhue, J.D. (1946) *Pop. Astron.* **54**, 149.
Chirvinsky, P.N. (1949) 4.*Meteoritika* **6**, 5

Dyakonova, M.I. and Kharitonova, V.Y. (1960) *Meteoritika* **18**, 48.
Garz, T., Kock, M., Richter, J., Baschek, B., Holweger, H. and Unsöld, A. (1969) *Nature*, **223**, 1254.
Goldberg, L., Müller, A.E. and Aller, L.H. (1960) *Astrophys. J. Suppl.* **5**, 1.
Hey, M.H. and Easton, A.J. (1967) *Geochim. Cosmochim. Acta* **31**, 1789.
Jarosewich, E. and Mason, B. (1969) *Geochim. Cosmochim. Acta* **33**, 411.
Lonsdale, J.T. (1947) *Am. Mineral.* **32**, 354.
Ludwig, E. and Tschermak, G. (1909) *Tsch. Min. Pet. Mitt.* **28**, 109.
McCall, G.J.H. (1966) *Mineral. Mag.* **35**, 1029.
Mason, B. (1963a) *Space Sci. Rev.* **1**, 621.
Mason, B. (1963b) *Am. Museum Novitates*, No. 2155.
Mason, B. (1965) *Am. Museum Novitates*, No. 2223.
Mason, B. (1966) *Geochim. Cosmochim. Acta* **30**, 23.
Mason, B. (1967) *Geochim. Cosmochim. Acta* **31**, 107.
Mason, B. and Wiik, H.B. (1964) *Geochim. Cosmochim. Acta* **28**, 533.
Mason, B. and Wiik, H.B. (1966a) *Am. Museum Novitates*, No. 2272.
Mason, B. and Wiik, H.B. (1966b) *Am. Museum Novitates*, No. 2273.
Michaelis, H. v., Ahrens, L.H. and Willis, J.P. (1969) *Earth Planet. Sci. Lett.* **5**, 387.
Prior, G.T. (1912) *Mineral. Mag.* **16**, 274.
Prior, G.T. (1918) *Mineral. Mag.* **18**, 151.
Urey, H.C. (1961) *J. Geophys. Res.* **66**, 1988.
Urey, H.C. and Craig, H. (1953) *Geochim. Cosmochim. Acta* **4**, 36.
Wiik, H.B. (1956) *Geochim. Cosmochim. Acta* **9**, 279.
Wiik, H.B. (1969) *Soc. Sci. Fennica, Comm. Phys.-Math.* **34**, 135.

(Received 27 July 1970)

COBALT (27)

Carleton B. Moore

Center for Meteorite Studies
Arizona State University
Tempe, Arizona

COBALT IS a predominantly siderophilic element whose chemical distribution in meteorites is closely related to that of nickel. In chondritic, stony-iron, and iron meteorites, it is a minor element usually determined in most modern chemical analyses. It has been detected using wet-chemical, colorimetric, emission spectrographic, neutron activation, and x-ray spectrographic methods of analysis. Recent compilations of cobalt abundances by Nichiporuk et al. (1967) and Schmitt et al. (1970) show a range of 260–1010 $\times 10^{-6}$ g/g in ordinary chondrites.

In the stony meteorites, cobalt is found mainly in the metallic phase, and reported as Co, except in the metal-free carbonaceous chondrites, where it is reported as CoO. The fractionation of cobalt between the bronzite (H), hypersthene (L), and amphoterite (LL) classification groups is quite distinct, just as it is for nickel. Variations within each group are most probably primarily due to sampling inhomogeneity rather than to intrinsic differences in cobalt content. The cobalt contents of enstatite chondrites and all three groups of carbonaceous chondrites are similar to those of the bronzite (H) group chondrites. Data by Schmitt et al. (1970) and Mason (1966) indicate the cobalt contents of the enstatite chondrites generally follow the trend of iron concentration. Keil (1968), using electron microprobe analysis, has shown that the cobalt concentrations in the metal phase of the enstatite chondrites ranges from 0.37 to 0.77 weight per cent and in schreibersite from 0.19 to 0.39 weight per cent. These values are similar to the cobalt concentrations in iron meteorites. The abundance of cobalt in the non-metallic phase of chondrites is not well known, due to its low abundance and the difficulty of making clean phase separations.

TABLE 1 Cobalt (wt.% as Co) in stony and stony-iron meteorites

Name	No. of detns.	Range	Mean	Atoms/10^6 Si	References
Chondrites					
C–1	1	–	530	2400	(4)
	2	–	470	2200	(5)
	1	–	530	2400	(6)
	4	500–520	510	2400	(7)
	4	460–550	480	2200	(8)
C–2	2	592–610	600	2200	(4)
	6	470–630	590	2100	(5)
	8	500–600	570	2100	(8)
	11	460–700	530	1900	(7)
C–3	4	620–720	650	2000	(4)
	3	550–720	650	2000	(6)
	9	440–820	650	2000	(7)
	15	480–740	620	1900	(8)
H Bronzite	27	730–1,010	860	2400	(4)
	8	710–1,150	930	2600	(6)
	61	440–1,260	810	2300	(8)
L Hypersthene	30	260–620	500	1300	(4)
	51	440–780	570	1500	(6)
	60	290–870	560	1400	(8)
LL Amphoterites	10	200–1,100	560	1400	(9)
	25	250–870	450	1100	(8)
E Enstatite	3	670–1,000	840	2200	(4)
	5	440–1,280	890	2400	(6)
	9	680–1,000	830	2200	(8)
Calcium-poor achondrites					
Enstatite	2	4.6–5.8	5.2	9	(4)
	6	2.1–57	4*	7	(8)
Hypersthene	2	18–19	19	37	(10)
Ureilites	3	80–139	114	290	(8)
Calcium-rich achondrites					
Nakhlite	4	37–47	43	90	(4), (8)
Howardites	12	3–54	19	40	(8)
Eucrites	27	2–22	2	4	(8)
Stony irons					
Mesosiderites	4	–	2800	11,500	(11)
Pallasites (iron)	9	4700–6000	5500	–	(2)

* Median

The reported concentrations of cobalt in the achondrites are difficult to evaluate because of the sampling difficulties in these meteorites. The cobalt-containing metallic inclusions are sparse and randomly distributed. In the stony-iron meteorites the high metal phase contents account for the high cobalt abundances. In Table 1 the cobalt contents for the metallic phase of the pallasites are taken from the analyses of Lovering et al. (1957), while the cobalt contents for the mesosiderites are from a theoretical composite reported by Wood (1963).

Cobalt has been shown by Nichiporuk et al. (1967) to be present in the range of 18–6000 10^{-6} g/g in troilite nodules from twelve iron meteorites, one pallasite and one chondrite. In mineral separates from the Sikhote-Alin iron Yavnel (1950) found 0.47% cobalt in the kamacite, 0.03% in the schreibersite, 0.01% in the troilite, and 0.01% in the chromite.

TABLE 2 Cobalt (wt.% as Co) in iron meteorites

Name	No. of meteorites	No. of analyses	Range	Mean	References
H Hexahedrites	34	48	–	0.66	(1)
	13	–	0.45–0.56	0.48	(2)
	8	16	0.39–0.45	0.34	(3)
Octahedrites					
Ogg Coarsest	18	20	–	0.50	(1)
	6	–	0.45–0.51	0.50	(2)
	4	8	0.42–0.48	0.46	(3)
Og Coarse	34	40	–	0.54	(1)
	14	–	0.42–0.55	0.49	(2)
	10	20	0.34–0.50	0.46	(3)
Om Medium	92	126	–	0.59	(1)
	21	–	0.45–0.61	0.52	(2)
	45	90	0.32–0.61	0.45	(3)
Of Fine	37	43	–	0.57	(1)
	10	–	0.38–0.60	0.40	(2)
	16	32	0.34–0.65	0.45	(3)
Off Finest	10	13	–	0.61	(1)
	8	–	0.44–0.70	0.57	(2)
	6	12	0.39–1.02	0.58	(3)
Dr Ataxites	24	38	–	1.01	(1)
	7	–	0.54–0.74	0.65	(2)
	6	12	0.43–0.75	0.63	(3)

Cobalt is ubiquitous in all iron meteorites and has been shown to increase with increasing nickel content (Moore et al., 1969).

In eighty-eight iron meteorites, Lovering et al. (1957) reported an average cobalt concentration of 0.51 %. They detected no significant difference between their 4 Ga–Ge classification groups. Their cobalt values ranged from 0.38 to 0.74 % and they noted a positive correlation between cobalt and nickel. The range for nine pallasites was 0.47–0.60 % with an average of 0.55 %. Since the structure of iron meteorites is a function of nickel content, there is also some correlation between cobalt and structure. Moore et al. (1969) in an investigation of 100 iron meteorites, found a range of values from 0.32 to 1.02 %. They showed a good positive correlation between cobalt and nickel, which gave a least-squares fit that did not pass through the origin but rather had the relationship Co = 0.0197 Ni + 0.308. Such a curve indicates a preferential concentration of cobalt relative to nickel. Nickel contents of iron meteorites in the structural classification groups taken from Buddhue (1946), Lovering et al. (1957) and Moore et al. (1969) are given in Table 2. In Table 3 the cobalt concentrations in the iron meteorites analyzed by Moore et al. (1969) are reported according to the Ga-Ge chemical classification groups.

TABLE 3 Cobalt in iron meteorites data from Moore et al. (1)

Ga-Ge group	No. of meteorites	Concentration (10^{-6} g/g) range	Mean
I	7	0.42–0.49	0.45
IIa	7	0.39–0.46	0.43
IIb	2	–	0.48
IIc	1	–	0.53
IId	2	0.61–0.66	0.63
IIIa	13	0.45–0.52	0.48
IIIb	4	0.47–0.57	0.51
IVa	10	0.34–0.42	0.39
IVb	2	0.70–0.71	0.71

References

Buddhue, J.D. (1946) *Pop. Astron.* **54**, 149.
Greenland, L. (1964) Ph. D. Thesis, Australian National University.
Greenland, L. and Lovering, J.F. (1965) *Geochim. Cosmochim. Acta* **29**, 821.

Keil, K. (1968) *J. Geophys. Research.* **73**, 6945.
Lovering, J.F., Nichiporuk, W., Chodos, A., and Brown, H. (1957) *Geochim. Cosmochim. Acta* **11**, 263.
Mason, B. (1963) *Space Sci. Rev.* **1**, 621.
Mason, B. (1966) *Geochim. Cosmochim. Acta* **30**, 23.
Mason, B. and Wiik, H. (1964) *Geochim. Cosmochim. Acta* **28**, 533.
Moore, C.B., Lewis, C.F. and Nava, D. (1969) *Meteorite Research* (Ed. P.M.Millman), 738. D.Reidel Pub. Co.
Nichiporuk, W., Chodos, A., Helin, E. and Brown, H. (1967) *Geochim. Cosmochim. Acta* **31**, 1911.
Schmitt, R.A., Goles, G.G. and Smith, R.H. (1970) Unpublished data.
Wiik, H.B. (1956) *Geochim. Cosmochim. Acta* **9**, 279.
Wood, J. (1963) in *The Moon Meteorites and Comets*" (Ed. B.M.Middlehurst and G.P.Kuiper), 337. Univ. of Chicago Press.
Yavnel, A.A. (1950) *Meteoritika* **8**, 134.

References for Tables

(1) Buddhue (1946)
(2) Lovering (1957)
(3) Moore et al. (1969)
(4) Nichiporuk (1967)
(5) Wiik (1956)
(6) Greenland and Lovering (1965)
(7) Mason (1963)
(8) Schmitt (1970)
(9) Mason and Wiik (1964)
(10) Greenland (1964)
(11) Wood (1963)

(Received 7 May 1970)

NICKEL (28)

Carleton B. Moore

Center for Meteorite Studies
Arizona State University
Tempe, Arizona

NICKEL IS present in all meteorites, although in the analyses of some achondrites it may not be reported due to the fact that the small samples taken for analysis do not include metallic nickel-iron. Nickel is distinctly siderophilic, and thus is present almost entirely in the metal phase. In those carbonaceous chondrites with no metal present, nickel is chalcophilic, occurring in the nickel-iron sulfide pentlandite. The fractionation pattern of nickel in chondrites is similar to the other siderophilic elements, including iron, cobalt, phosphorus, and the noble metals. Its concentration increases in the carbonaceous chondrites, going from about 1.0 weight per cent in Type I to 1.3 weight per cent in Type II and 1.4 weight per cent in Type III carbonaceous chondrites. Like iron, it is enriched in H group ordinary chondrites with about 1.7 weight per cent, as compared to approximately 1.2 and 0.9 weight per cent in L and LL ordinary chondrites. The enstatite chondrites are similar to the H group chondrites in their range of nickel contents.

Nickel is always included in complete chemical analyses of chondrites. It is usually detected by the gravimetric analysis of nickel precipitated by dimethylglyoxime. Variations of nickel contents within individual meteorite groups are probably due for the most part to sampling inhomogeneity and incomplete extraction of the metal phase in sequential analyses. The chondrite analyses given in Table 1 include analyses of carbonaceous chondrites by H. B. Wiik as reported by Mason (1963) and selected superior analyses of ordinary chondrites compiled by Mason (1965). A similar compilation made in 1953 by Urey and Craig produced similar results. Included for comparison are two modern studies of nickel in chondrites done by a spectrophotometric method by Greenland and Lovering (1965) and X-ray fluorescence spec-

TABLE 1 Nickel (wt. % as Ni) in stony and stony-iron meteorites

Name	No. of meteorites	Range	Mean	Atoms/10^6 Si	References
Chondrites					
C–1	5	0.75–1.09	0.98	46,000	(1)
	1	–	1.11	52,000	(2)
	1	–	1.15	54,000	(3)
C–2	9	1.17–1.34	1.23	45,000	(1)
	2	1.26–1.29	1.28	47,000	(2)
C–3	9	1.23–1.50	1.33	41,000	(1)
	4	1.36–1.51	1.42	44,000	(2)
	3	1.35–1.51	1.45	45,000	(3)
H (Bronzite)	36	1.21–1.93	1.63	46,000	(4)
	7	1.71–2.06	1.91	53,000	(3)
	27	1.38–1.99	1.70	48,000	(2)
L (Hypersthene)	67	0.39–1.48	1.10	28,000	(4)
	29	0.98–1.57	1.27	32,000	(3)
	29	0.67–1.41	1.16	30,000	(2)
LL (Amphoterite)	12	0.69–1.28	0.91	23,000	(5)
E–4 (Enstatite, Type I)	3	1.66–1.95	1.81	51,000	(6)
E–5 (Enstatite, Intermed.)	2	1.62–1.81	1.71	44,000	(6)
E–6 (Enstatite, Type II)	6	1.11–1.96	1.53	38,000	(6)
Calcium-poor achondrites					
Enstatite	3	0.0065–0.026	0.016	280	(2)
Hypersthene	2	0.0013–0.014	0.0075	150	(2)
	2	0.040–0.093	0.067	1,300	(9)
Chassignite	1	–	0.06	1,600	(7)
Ureilites	6	0.09–0.23	0.13	3,300	(8)
Calcium-rich achondrites					
Nakhlite	1	–	0.099	2,100	(9)
	1	–	0.011	230	(2)
Howardite	1	–	0.0013	27	(2)
Eucrites	6	0.0009–0.002	0.0013	27	(2)
Stony-irons					
Mesosiderites					
Total	4	–	4.39	180,000	(10)
Total	8	3.2–4.8	4.2	170,000	(11)
Metal	9	7.6–8.8	8.2	–	(11), (16)
Pallasites					
Olivine	12	0.004–0.007	0.0053	–	(12)
Metal	23	7.9–16.4	10.5	–	(12)
Total	10	–	4.66	–	(10)

troscopy by Nichiporuk et al. (1967). The contents of the LL group chondrites or amphoterites were taken from the review of Mason and Wiik (1964). Mason's (1966) review of the enstatite chondrites compiled data for this group.

As indicated above the nickel contents reported for the achondrites are usually suspect, due to the sporadic distribution of the nickel-bearing metallic phase in these meteorites. In many, if not most cases, the samples analyzed probably contain no metallic material and the results reported may be indicative of the nickel contents of the combined silicate phases.

Because of their high proportion of nickel-iron both the mesosiderites and pallasites have high nickel contents. The values reported in Table 1 include those compiled and calculated by Wood (1963), together with more recent analyses for mesosiderites by Powell (1969) and for pallasites by Buseck and Goldstein (1969). It is interesting to note the nickel contents reported by Buseck and Goldstein (1969) for the olivine phase of pallasites are similar to many of the nickel determinations in achondrites.

The nickel content of several mineral phases has been determined by wet chemical, instrumental and microprobe analysis. Nichiporuk and Chodos (1969) determined the nickel in sulfide nodules from iron, stony-iron, and chondritic meteorites. The range of values was from 0.017 per cent to 6.61 per cent, and most likely indicates the presence of mineral phases other than troilite in the nodules. Ramdohr (1963) has recognized exsolution lamellae of pentlandite in troilite in some chondrites. Keil and Fredriksson (1963) reported 2.8 per cent nickel in troilite from the Norton County achondrite. Kamacite from enstatite chondrites has been reported by Keil (1968) and Wasson and Wai (1970) to contain from 6.4 per cent to 8.5 per cent nickel. Wasson and Wai (1970) report that kamacite grains in enstatite achondrites contain from 3.8 per eent to 6.4 per cent nickel. Schreibersite has a high nickel content. Goldstein and Ogilvie (1963) report from 23.4 per cent to 37.2 per cent nickel in schreibersite in iron meteorites, and 44 per cent in rhabdites from the same meteorites. In enstatite chondrites Wasson and Wai (1970) and Keil (1968) report schreibersite with from 7.1 per cent to 31.8 per cent nickel. They and Reed (1968) also report the presence of the nickel-rich mineral perryite with 75.5 per cent to 82 percent nickel in the enstatite chondrites. Lovering (1964) reported nickel contents ranging from 0.60 per cent to 3.14 per cent in cohenite from iron meteorites. Buseck and Goldstein's (1969) reported values of from 40 to 70 $\times 10^{-6}$ g/g nickel in the olivine in pallasites are lower than earlier results reported by Lovering (1957); 150 to 350 $\times 10^{-6}$ g/g. Unpublished results of electron microprobe analyses by Fredriksson reported by Mason (1966) indicate that olivine from chondrites contains 100 to 200 $\times 10^{-6}$ g/g nickel and that the olivine in the Marjalahti pallasite has 50 to 150 $\times 10^{-6}$ g/g nickel. Duke (1965) reported less than 3 $\times 10^{-6}$ g/g nickel in pyroxene from the eucrites, 6–20 $\times 10^{-6}$ g/g in pyroxene from the hypersthene achondrites, and 30 to 50 $\times 10^{-6}$ g/g in olivine + pyroxene from chondrites.

Reliable analyses of iron meteorites indicate that the nickel content may range from about 4.1 per cent to 62 per cent. The vast majority of iron meteorites contain between 5 per cent and 10 per cent nickel. Those with the very lowest nickel contents appear to be hexahedrites made up primarily of kamacite which has lost nickel to adjacent schreibersite inclusions. The highest nickel value is for the Oktibbeha County meteorite. The original mass of this iron was very small and it has been difficult to verify its meteoritic origin. There are, however, a moderate number of nickel-rich ataxites

TABLE 2 Nickel (wt. % as Ni) in structural classes of iron meteorites

Name	No. of meteorites	No. of analyses	Range	Mean	References
H Hexahedrites	34	48	–	5.57	(13)
	8	16	5.50–6.11	5.68	(14)
	29	–	4.13–5.86	5.52	(15)
Octahedrites					
Ogg Coarsest	18	20	–	6.54	(13)
	4	8	5.57–6.71	6.35	(14)
	13	–	5.68–6.54	6.30	(15)
Og Coarse	34	40	–	7.39	(13)
	10	20	6.15–8.67	7.28	(14)
	35	–	6.57–8.80	7.10	(15)
Om Medium	92	126	–	8.22	(13)
	45	90	7.01–10.29	8.15	(14)
	97	–	6.26–10.84	8.20	(15)
Of Fine	37	43	–	9.00	(13)
	18	36	7.48–13.43	8.80	(14)
	36	–	7.50–10.50	9.20	(15)
Off Finest	10	13	–	11.65	(13)
	7	14	7.66–15.80	9.30	(14)
	19	–	9.09–16.7	11.8	(15)
Dr Ataxites	24	38	–	18.85	(13)
	7	14	9.68–18.17	14.96	(14)
	30	–	9.30–36.1	16.5	(15)

with nickel contents between 20 per cent and 36 per cent. Table 2 gives the nickel contents of the various structural groups and Table 3 gives the nickel content of groups classified by their chemical composition, primarily gallium and germanium contents. The data listed are from two sets of recent analytical data. Those by Moore et al. (1969) were determined by wet

TABLE 3 Nickel contents (wt. %) of iron meteorites in chemical classification groups

Name	No. of meteorites	Range	Mean	References
I	58	6.38–8.64	7.51	(15)
	7	6.68–8.18	7.62	(14)
IIa	30	5.27–5.86	5.57	(15)
	8	5.50–6.11	5.70	(14)
IIb	11	5.68–6.43	6.06	(15)
	3	5.60–6.71	6.24	(14)
IIc	8	9.72–11.50	10.61	(15)
	1	–	9.81	(14)
IId	9	9.96–11.30	10.63	(15)
	2	10.28–10.57	10.44	(14)
IIIa	62	7.29–9.10	8.20	(15)
	13	7.39–8.43	7.90	(14)
IIIab	13	8.30–9.28	8.79	(15)
IIIb	20	9.20–10.70	9.95	(15)
	4	8.57–9.95	9.27	(14)

chemistry and those by Wasson (1967a, 1967b, 1969) by atomic absorption spectroscopy. Difficulties may arise in any technique that does not completely dissolve the sample, since small amounts of insoluble schreibersite will have a high nickel content and thus decrease the apparent total nickel. Moore et al. (1969) have discussed the validity of older analyses of iron meteorites. There are many reliable individual analyses reported in the literature but it is often difficult to make a reliable value judgement for statistical purposes. Many of these analyses were compiled by Buddhue (1946) and are reported in Table 2. Figure 1 gives a histogram of the overall nickel concentrations. This type of histogram was first utilized by Yavnel (1958) and has been updated by Boustead (1964) reported by Axon (1967).

The distribution of nickel in the metal phases of meteorites has been studied extensively. Nichiporuk (1958) separated kamacite and taenite from 2 hexahedrites and 6 octahedrites and found that nickel in kamacite ranged

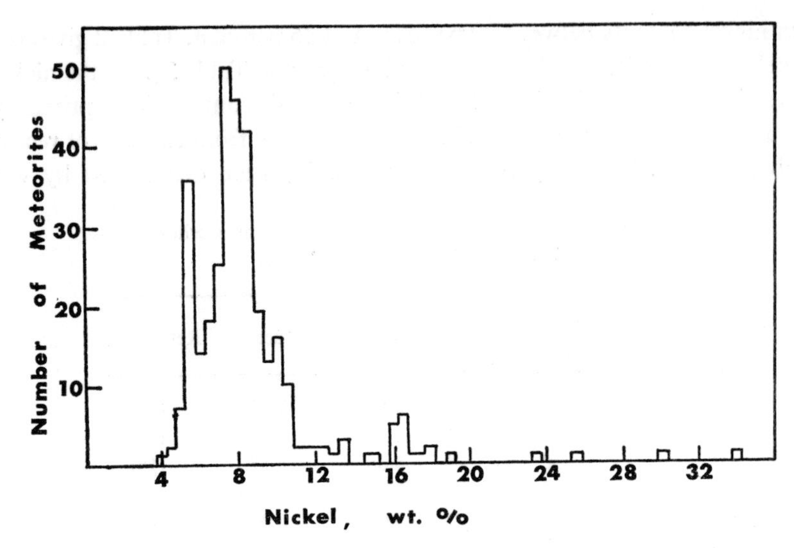

FIGURE 1 Histogram of the nickel content of analyzed iron meteorites plotted at $\frac{1}{2}$% Ni intervals. After Axon (1968) and Yavnel (1958)

from 3.3 per cent to 5.8 per cent while the taenite contained 31 to 54 per cent nickel. Electron microprobe analysis of the iron meteorites indicate great difficulties in a wet chemical approach to the problem. These studies confirmed earlier qualitative conclusions (Perry, 1944; Uhlig, 1954) that there are nickel concentration gradients in taenite bands. This work originally undertaken by Yavnel et al. (1958) and followed up by Agrell et al. (1963) and Goldstein (1965) showed the now well known M shaped nickel profile across taenite plates. On the basis of this work Wood (1964) and Goldstein and Ogilvie (1965) have developed methods to explain the observed concentration gradients and calculate meteorite cooling rates. These studies together with more recent ones have been reviewed by Goldstein (1969), Axon (1967), and Wood (1968). The nickel content in both kamacite and taenite is a function of bulk composition and cooling rate. The highest nickel concentrations are found at the kamacite-taenite interface.

References

Agrell, S.O., Long, J.V.P. and Ogilvie, R.E. (1963) *Nature* **198**, 749.
Axon, H.J. (1968) *Prog. Mater. Sci.* **13**, 183.
Boustead, J. (1964) Thesis, Manchester University.

Buddhue, J.D. (1946) *Pop. Astron.* **54**, 149.
Buseck, P.R. and Goldstein, J.I. (1969) *Geol. Soc. Amer. Bull.* **80**, 2141.
Duke, M.B. (1965) *J. Geophys. Res.* **70**, 1523.
Dyakonova, M.I. and Kharitonova, V.Y. (1960) *Meteoritika* **18**, 66.
Goldstein, J.I. (1965) *J. Geophys. Res.* **70**, 6223.
Goldstein, J.I. (1969) *Meteorite Research* (Ed. P.M. Millman), 722. D. Reidel Pub. Co.
Goldstein, J.I. and Ogilvie, R.E. (1963) *Geochim. Cosmochim. Acta* **27**, 623.
Goldstein, J.I. and Ogilvie, R.E. (1965) *Geochim. Cosmochim. Acta* **29**, 893.
Greenland, L. (1964) Ph.D. Thesis, Australian National University.
Greenland, L. and Lovering, J.F. (1965) *Geochim. Cosmochim. Acta* **29**, 821.
Keil, K. (1968) *J. Geophys. Research* **73**, 6945.
Keil, K. and Fredriksson, K. (1963) *Geochim. Cosmochim. Acta* **27**, 939.
Lovering, J.F. (1957) *Geochim. Cosmochim. Acta* **12**, 253.
Lovering, J.F. (1964) *Geochim. Cosmochim. Acta* **28**, 1745.
Mason, B. *Space Sci. Rev.* **1**, 621.
Mason, B. (1965) *Am. Museum. Novitates*, No. 2223.
Mason, B. (1966) *Geochim. Cosmochim. Acta* **30**, 365.
Mason, B. and Wiik, H.B. (1964) *Geochim. Cosmochim. Acta* **28**, 533.
Moore, C.B., Lewis, C.F. and Nava, D. (1969) *Meteorite Research*, Ed. P.M. Millman, 683. D. Reidel Pub. Co.
Nichiporuk, W. (1958) *Geochim. Cosmochim. Acta* **13**, 233.
Nichiporuk, W. and Chodos, A.A. (1959) *J. Geophys. Res.* **64**, 2451.
Nichiporuk, W., Chodos, A.A., Helin, E. and Brown, H. (1967) *Geochim. Cosmochim. Acta* **31**, 1911.
Perry, S.H. (1944) *U.S. Nat. Mus. Bull.*, **184**.
Powell, B.N. (1969) Ph. D. Thesis, Columbia University.
Powell, B.N. (1969b) *Geochim. Cosmochim. Acta* **33**, 789.
Ramdohr, P. (1963) *J. Geophys. Res.* **68**, 2011.
Reed, S.J.B. (1968) *Mineral. Mag.* **36**, 850.
Uhlig, H.H. (1954) *Geochim. Cosmochim. Acta* **6**, 282.
Urey, H. and Craig, H. (1953) *Geochim. Cosmochim. Acta* **4**, 36.
Vdovykin, G.P. (1970) *Space Sci. Rev.* **10**, 483.
Wasson, J.T. (1967a) *Geochim. Cosmochim. Acta* **31**, 161.
Wasson, J.T. (1967b) *Geochim. Cosmochim. Acta* **31**, 2065.
Wasson, J.T. (1969) *Geochim. Cosmochim. Acta* **33**, 859.
Wasson, J.T. and Wai, C.M. (1970) *Geochim. Cosmochim. Acta* **34**, 169.
Wood, J.A. (1963) in *The Moon Meteorites and Comets* (Ed. B.M. Middlehurst and G.P. Kuiper), 337. Univ. of Chicago Press.
Wood, J.A. (1964) *Icarus* **3**, 429.
Wood, J.A. (1968) *Meteorites and the Origin of Planets*, McGraw-Hill, New York, 117.
Yavnel, A.A. (1958) *Meteoritika* **15**, 115.
Yavnel, A.A., Borovski, I.B., Illin, N.P., and Marchukova, I.D. (1958) *Doklady Akad. Nauk. S.S.S.R.* **123**, 256.

References to Tables

(1) Mason (1963)
(2) Nichiporuk et al. (1967)
(3) Greenland and Lovering (1965)
(4) Mason (1965)
(5) Mason and Wiik (1964)
(6) Mason 1966)
(7) Dyakonova et al. (1960)
(8) Vdovykin (1970)
(9) Greenland (1964)
(10) Wood (1962)
(11) Powell (1969a)
(12) Buseck and Goldstein (1969)
(13) Buddhue (1946)
(14) Moore et al. (1969)
(15) Wasson (1967a, 1967b, 1969)
(16) Powell (1969b)

(Received 2 June 1970)

COPPER (29)

Gordon G. Goles

Center for Volcanology
and
Departments of Chemistry and Geology
University of Oregon
Eugene, Oregon 97403

SEVERAL STUDIES have been made of contents of this trace element in stone meteorites. Smales, Mapper andWood (1957) used radiochemical activation analysis to determine Cu in six ordinary chondrites, an aubrite and a mesosiderite. Greenland and Lovering (1965) analysed fifty chondrites for Cu by emission spectrography, and Greenland and Goles (1965) determined Cu in twenty chondrites by activation analysis. Schmitt, Goles and Smith (1971) used instrumental neutron activation analysis to determine Cu in one-hundred-six chondrites and twenty-five achondrites. They compare their results with those of the workers mentioned above and find satisfactory agreement. Values for Cu in chondrites by Schaudy, Kiesl and Hecht (1967, 1968) are clearly too high, for reasons which are not understood. Wiik (1969) summarizes his Cu determinations, many of which may also be too high.

In Table 1 are listed average Cu contents and atomic abundances from Schmitt, Goles and Smith (1971). Dispersions cited are estimates of the population variability at the single standard deviation level. They may in some cases be slightly larger than the true variability owing to contributions from analytical errors, judging from the data of Wong (1969) on a few H and L chondrites. Chondrites are classified according to Van Schmus and Wood (1967) and achondrites are identified by the code introduced by Keil (1969).

As was noted for several other elements (e.g., Na, Mn), Cu average contents and atomic abundances decrease in the sequence C1, 2, 3. Both average

TABLE 1 Copper average contents and atomic abundances in stone meteorites

Meteorite Class	Number of specimens (samples) analysed	Cu, 10^{-6} g/g	Cu atoms per 10^6 Si
C 1	3 (4)	127 ± 14	540 ± 60
C 2	7 (12)	116 ± 12	390 ± 40
C3	10 (19)	108 ± 17	300 ± 50
H3	4 (7)	95 ± 9	240 ± 30
H4, 5, 6	24 (35)	90 ± 20	230 ± 50
L3	3 (4)	80 ± 16	190 ± 40
L4, 5, 6	32 (41)	94 ± 16	220 ± 40
LL3	2 (4)	90 ± 30	210 ± 80
LL4, 5, 6	15 (18)	80 ± 40	190 ± 90
E4	2 (3)	193 ± 12	500 ± 30
E5, 6	4 (14)	110 ± 60	260 ± 140
Aubrites, Ae	5 (7)	13 ± 6	22 ± 13
Ureilites, Aop	2 (3)	11 ± 1	25 ± 2
Eucrites, Ap	7 (9)	6 ± 5	10 ± 8
Howardites, Aor	5 (5)	3 ± 3	6 ± 6
Nakhlites, Ado	2 (3)	12 ± 2	23 ± 4

contents and atomic abundances are essentially constant in all classes of "ordinary" chondrites, whether of low or of intermediate and high petrographic grade. The atomic abundances, at least, for "ordinary" chondrites are lower than those for any of the C groups. Copper contents and atomic abundances in the two groups of E chondrites, however, differ by about a factor of two. Without better understanding of the reasons for these variations (or their absence), it does not seem feasible to try to define the primordial or "cosmic" abundance of Cu to better than about ±50% relative precision.

All classes of achondrites are markedly poorer in Cu than are the chondrites. In this sense, the complementary differentiate of the achondrites may be the iron meteorites and metal phases of stony-irons. Lovering et al. (1957), Cobb (1967), Smales, Mapper and Fouché (1967), Moore, Lewis and Nava (1969), and Bauer and Schaudy (1970) have determined Cu in many samples of meteoritic metal phases, and on the whole agree well. Copper contents generally lie in the range from tens to a few hundreds times 10^{-6} g/g, with a few extreme outliers in both directions. The overall average for iron meteorites given by Smales, Mapper and Fouché is 172×10^{-6} g Cu/g, and their averages for structural classes among hexahedrites and octahedrites range from 121 to 141 $\times 10^{-6}$ g Cu/g, in good agreement with those of Cobb which range from 126 to 175 $\times 10^{-6}$ g Cu/g. Copper contents of pallasitic metal phases are similar to those of iron meteorites.

Copper contents of various mineral separates are given in Table 2. These values were determined by instrumental neutron activation analysis; see chapters on Sc and Cr for brief discussions of sources and probable purities of the separates.

TABLE 2 Copper contents (10^{-6} g/g) in mineral separates

PYROXENE	
Coolidge (C3)	
clinobronzite	30 ± 2
Clovis (H3)	
clinobronzite	50 ± 20
Miller (H5)	
bronzite	6 ± 2
Richardton (H5)	
bronzite	15 ± 2
Holbrook (L6)	
hypersthene	13 ± 3
Chainpur (LL3)	
pigeonite	9.5 ± 1.6
Appley Bridge (LL6)	
orthopyroxene, Fs_{25}	6 ± 2
Kota-Kota (E4)	
principally clinoenstatite	90 ± 20
Jajh deh Kot Lalu (E6)	
principally orthoenstatite	5 ± 3
Shallowater (Ae)	
enstatite	2.7 ± 0.5
Johnstown (Ab)	
bronzite	3.0 ± 0.5
Tatahouine (Ab)	
bronzite	≤1.7
Juvinas (Ap)	
pigeonite	2.4 ± 0.4
Nuevo Laredo (Ap)	
pigeonite	1.9 ± 1.3
Binda (Aor)	
orthopyroxene	2.9 ± 0.7
*Shergotty**	
pigeonite	13.0 ± 1.5
Bondoc Peninsula (M)	
principally orthopyroxene	8.4 ± 1.5
Crab Orchard (M)	
principally pigeonite	4.8 ± 1.3
OLIVINE	
Argonia (P)	1.9 ± 0.6
Eagle Station (P)	7 ± 3
Glorieta Mt. (P)	1.5 ± 0.8
Mt. Dyrring (P)	7 ± 2
Springwater (P)	1.2 ± 0.8
Mt. Padbury (M)	1.9 ± 0.2
OPAQUE MINERALS	
Allegan (H5)	
chromite	7.9 ± 1.1
Richardton (H5)	
chromite	12 ± 5
Mt. Dyrring (P)	
chromite	22 ± 9
McKinney (L4)	
troilite	580 ± 11

* The classification of *Shergotty* is uncertain owing to numerous mineralogical and chemical peculiarities exhibited by this meteorite.

The method is subject to interferences from Na. Although corrections can be made for these interferences, I have not reported Cu contents for plagioclase separates where the apparent values are probably too high in all cases. Copper is strongly concentrated in troilite, but even so a mass balance calculation suggests that most of the Cu in "ordinary" chondrites of intermediate and high petrographic grade resides in grains of native Cu like those observed by Ramdohr (1963). Copper contents in *Coolidge* (C3), *Clovis* (H3) and *Kota-Kota* (E4) pyroxenes are notably higher than in other silicates. This trend may be related to equilibration with sulfides, as seems to be the case for Mn in enstatite separates (see chapter on Mn). If so, it is curious that pyroxene from the LL3 chondrite *Chainpur* should have a relatively low Cu content. Perhaps the concept of "low petrographic grade" bears different implications for the LL chondrites than for the other classes.

Acknowledgements

V. Frankum and S. Oxley assisted with the analytical work on mineral separates here reported, which was done with support in part from NASA grant NSG–319 at University of California San Diego.

References

Bauer, R. and Schaudy, R. (1970) "Activation analytical determination of elements in meteorites, 3. Determination of manganese, sodium, gallium, germanium, copper and gold in 21 iron meteorites and 2 mesosiderites." *Chem. Geol.* **6**, 119–131.

Cobb, J. C. (1967) "A trace-element study of iron meteorites." *J. Geophys. Res.* **72**, 1239–1341.

Greenland, L. and Goles, G. G. (1965) "Copper and zinc abundances in chondritic meteorites." *Geochim. Cosmochim. Acta* **29**, 1285–1292.

Greenland, L. and Lovering, J. F. (1965) "Minor and trace element abundances in chondritic meteorites." *Geochim. Cosmochim. Acta* **29**, 821–858.

Keil, K. (1969) "Meteorite composition", chapter 4 in *Handbook of Geochemistry*, edited by K. H. Wedepohl (Springer-Verlag, Berlin) 78–115.

Lovering, J. F., Nichiporuk, W., Chodos, A. and Brown, H. (1957) "The distribution of gallium, germanium, cobalt, chromium, and copper in iron and stony-iron meteorites in relation to nickel content and structure." *Geochim. Cosmochim. Acta* **11**, 263–278.

Moore, C. B., Lewis, C. F. and Nava, D. (1969) "Superior analyses of iron meteorites", in *Meteorite Research*, edited by P. M. Millman (Springer-Verlag, New York) 738–748.

Ramdohr, P. (1963) "The opaque minerals in stony meteorites." *J. Geophys. Res.* **68**, 2011–2036.

Schaudy, R., Kiesl, W. and Hecht, F. (1967) "Activation analytical determination of elements in meteorites." *Chem. Geol.* **2**. 279–287.

Schaudy, R., Kiesl, W. and Hecht, F. (1968) "Activation analytical determination of elements in meteorites, 2. Determination of manganese, sodium, gallium, copper, gold and chromium in 21 meteorites." *Chem. Geol.* **3**, 307–312.

Schmitt, R.A., Goles, G.G. and Smith, R.H. (1971) *Elemental abundances in stone meteorites*. To be published as a NASA report; available from the author on request.

Smales, A.A., Mapper, D. and Fouché, K.F. (1967) "The distribution of some trace elements in iron meteorites, as determined by neutron activation." *Geochim. Cosmochim. Acta* **31**, 673–720.

Smales, A.A., Mapper, D. and Wood, A.J. (1957) "The determination, by radioactivation, of small quantities of nickel, cobalt and copper in rocks, marine sediments and meteorites." *Analyst* **82**, 75–88.

Van Schmus, W.R. and Wood, J.A. (1967) "A chemical-petrologic classification for the chondritic meteorites." *Geochim. Cosmochim. Acta* **31**, 747–765.

Wiik, H.B. (1969) "On regular discontinuities in the composition of meteorites." *Comm. Phys.-Math.* **34**, 135–145.

Wong, D. (1969) *Copper homogeneity in chondritic meteorites*. M.S. dissertation, Oregon State University.

(Recevied 14 September 1970)

ZINC (30)

Carleton B. Moore

Center for Meteorite Studies
Arizona State University
Tempe, Arizona

ZINC IS FOUND in meteorites as a trace element. Ordinary chondrites contain about 50 μg/g. Nava (1968) has shown that in the Leedey chondrite zinc is evenly distributed throughout the meteorite, but earlier work (Greenland and Lovering, 1965; Nishimura and Sandell, 1964) indicates a more uneven distribution. Zinc appears to be predominantly lithophilic in character but under reducing conditions may be chalcophilic. Ramdohr (1963) has reported the zinc sulfide mineral sphalerite in the enstatite chondrites Pillistfer and Hvittis. Keil (1962) reported zincian daubreelite with a range of 0.04 to 5.2% zinc in the enstatite chondrites. El Goresy (1965, 1967) reported sphalerite in the sulfide nodules and kamacite of the Odessa iron meteorite.

Zinc has been determined by a number of analytical techniques. Nichiporuk and Chodos (1959) used X-ray fluorescence for the analysis of zinc in sulfide (troilite) nodules of twelve iron and two silicate meteorites. It was found that zinc varied greatly from nodule to nodule. In 1962 Nishimura and Sandell used a colorimetric method to determine zinc separated by an anion exchange technique from iron and stony meteorites. Greenland (1964) also used a colorimetric technique to investigate the fractionation of zinc in chondritic meteorites. A more extensive study of the distribution of zinc in meteorites was reported by Nishimura and Sandell (1964). Their report reviewed most of the earlier studies of zinc in meteorites. Their work confirmed the fact that zinc is predominately found in the silicate phases within chondrites. Greenland and Lovering (1965) investigated the zinc contents of fifty meteorites, including representatives of all chondrite groups. They showed that in terms of cosmic abundance and distribution zinc may be included in the group of strongly fractionated chalcophile elements.

Table 1 Zinc (10^{-6} g/g) in stony and stony-iron meteorites

Name	No. of meteorites	Range	Mean	Atoms/ 10^6 Si	References
Chondrites					
C–1	2	330–450	390	1600	(4), (5), (6), (7)
C–2	2	–	180	590	(4), (6), (7)
C–3	6	105–130	120	330	(4), (5), (6), (7)
H (Bronzite)	8	28–89	51	130	(3), (4), (5), (6,) (7)
L (Hypersthene)	15	8–102*	58	130	(3), (4), (5), (6), (7)
E–3	2	350–460	400	1000	(3), (4), (5), (6), (7)
E–4	1	–	85	200	(7)
E–6	4	8–28	17	37	(3), (4), (5), (6), (7)
Calcium-poor achondrites					
Enstatite	2	5–25	15	24	(7), (8)
Hypersthene	2	3–63	34	60	(8)
Calcium-rich achondrites					
Nakhlite	1	–	84	160	(7), (8)
Eucrites	3	2–11	5	10	(3), (8)
Howardite	1	–	37	75	(8)
Stony-irons					
Pallasites (metal)	1	–	<1	–	(3)

* Unequilibrated L–3 chondrite Khohar (Zn = 157 10^{-6} g/g) omitted from mean.

TABLE 2 Zinc (10^{-6} g/g) in iron meteorites

Name	No. of meteorites	Range	Mean	Reference
H Hexahedrites	7	0.48–<1	(<1 median)	(1)
	9	12–16	14	(2)
Octahedrites				
Ogg Coarsest	3	27–37	31	(1)
	4*	14–45	29	(2)
Og Coarse	2	12–42	25	(1)
	22	14–59	30	(2)
Om Medium	20	<1–32	1.5 (median)	(1)
	14	11–39	18	(2)
Of Fine	6	<1–21	<1 (median)	(1)
	10	12–17	14	(2)
Off Finest	5	13–61	25	(2)
Dr Ataxites	8	11–53	18	(2)

* Does not include Youndegin iron meteorite with 425 10^{-6} g/g Zn.

TABLE 3 Zinc contents in sulfide nodules

Meteorites	Class	Zn, 10^{-6} g/g	Reference
Coahuila	H	99	(9)
Hex River Mtns.	H	178	(2)
Indian Valley	H	<50	(9)
Sikhote-Alin	H-Ogg	<50	(9)
Osseo	Ogg	20	(2)
Youndegin	Ogg	13	(2)
		13.2	(2)
Mt. Stirling	Ogg	12.7	(2)
		21.9	(2)
Bendego	Og	20.2	(2)
Camp Verde	Og	21.6	(2)
Canyon Diablo*	Og	<50	(9)
Canyon Diablo (1)†	Og	211	(9)
Odessa	Og	<50	(9)
Odessa	Og	0.33	(9)
Bella Roca	Om	14.6	(2)
		13.6	(2)
Cambria	Om	12.7	(2)
		12.6	(2)
Henbury	Om	353	(9)
Henbury (1)	Om	122	(9)
Joe Wright Mtn.	Om	14.6	(2)
		14.4	(2)
Toluca	Om	521	(9)
Toluca (1)	Om	<50	(9)
Toluca (2)	Om	83	(9)
Toluca (3)	Om	<50	(9)
Ballinoo	Of	<50	(9)
Bear Creek	Of	<50	(9)
Cambria	Of	<50	(9)
Duchesne	Of	16.2	(2)
		101	(9)
Duchesne (1)	Of	77	(9)
Duchesne (2)	Of	<50	(9)
Moonbi	Of	242	(9)
Brenham	P	63	(9)
Cullison	C	61	(9)

* The results are based on the portion of the sample free of metallic inclusions.

† Numbers denote different nodules from individual meteorites.

In 1965 Greenland and Goles used neutron activation to determine additional zinc contents of chondrites. They were able to show that the zinc abundances for enstatite chondrites fell into a high or a low group. The high zinc enstatite chondrites had concentrations similar to the carbonaceous chondrites while the low zinc enstatite chondrites had abundances more like the depleted ordinary chondrites. Mason (1966) in his review of the enstatite chondrites pointed out that zinc has a higher content in the iron-rich enstatite chondrites and a lower content in those with a lower iron content. It is interesting to note that sphalerite has been reported in the low zinc enstatite chondrites but not in those with a high zinc content. Greenland (1967) reported additional zinc analyses obtained by neutron activation using radiochemical separations. Smales, Mapper and Fouché (1967) also utilized a neutron activation technique to determine zinc in 67 iron meteorites.

Atomic absorption spectrophotometry has been shown by Nava (1968) to be a sensitive method for the analysis of low concentrations of zinc. He used this technique to determine the zinc contents of five stony meteorites and seventy-five iron meteorites. In addition, zinc was determined from sulfide inclusions from eleven iron meteorites. The data of both Smales et al. (1967) and Nava (1968) indicate that zinc has a very low abundance in most iron meteorites except for those of the coarse and coarsest structural groups or gallium-germanium group I. Nava's (1968) data for the low-zinc irons are generally in the range of 12 to 18 μg/g while those of Smales et al. (1967) are often listed less than 1 μg/g. Unfortunately, not enough independent analyses on iron meteorites have been done to resolve this problem.

The distribution of zinc in stony and stony-iron meteorites is given in Table 1. Table 2 contains zinc data for the iron meteorites and Table 3 reports the zinc contents in sulfide nodules from the investigations of Nichiporuk and Chodos (1959) and Nava (1968).

References

El Goresy, A. (1965) *Geochim. Cosmochim. Acta* **29**, 1131.
El Goresy, A. (1967) *Geochim. Cosmochim. Acta* **31**, 1667.
Greenland, L. (1963) *J. Geophys. Res.* **64**, 2451.
Greenland, L. (1964) Ph.D. Thesis, Australian National University, Canberra.
Greenland, L. (1967) *Geochim. Cosmochim. Acta* **31**, 849.
Greenland, L. and Goles, G.G. (1965) *Geochim. Cosmochim. Acta* **29**, 1285.
Greenland, L. and Lovering, J.F. (1965) *Geochim. Cosmochim. Acta* **29**, 821.
Keil, K. (1962) *J. Geophys. Res.* **67**, 4055.
Mason, B. (1966) *Geochim. Cosmochim. Acta* **30**, 23.

Nava, D. (1968) Ph. D. Thesis, Arizona State University, Tempe, Arizona.
Nichiporuk, W. and Chodos, A. (1959) *J. Geophys. Res.* **64**, 2451.
Nishimura, M. and Sandell, E.B. (1962) *Anal. Chim. Acta* **26**, 242.
Nishimura, M. and Sandell, E.B. (1964) *Geochim. Cosmochim. Acta* **28**, 1055.
Ramdohr, P. (1963) *J. Geophys. Res.* **68**, 2011.
Smales, A.A., Mapper, D. and Fouché, K.F. (1967) *Geochim. Cosmochim. Acta* **31**, 673.

References to Tables

(1) Smales A.A., Mapper D. and Fouché K.F. (1967)
(2) Nava (1968)
(3) Nishimura M. and Sandell E.B. (1964)
(4) Greenland L. and Goles G.G. (1965)
(5) Greenland L. and Lovering J.R. (1965)
(6) Greenland L. (1967)
(7) Nishimura M. and Nasu T. (1964)
(8) Greenland (1964)
(9) Nichiporuk W. and Chodos A. (1959)

(Received 7 May 1970)

GALLIUM (31)

P. A. Baedecker and **J. T. Wasson**

Department of Chemistry and Institute of Geophysics and Planetary Physics, University of California, Los Angeles, California 90024

A LARGE NUMBER of accurate Ga determinations have been reported for meteorites of the various classes, and especially for iron meteorites and chondrites. Interest in the Ga concentrations of iron meteorites attaches to the fact that whereas Ga varies from 0.2 to 100 ppm, the concentration range is much smaller (generally less than a factor of 2) within any group of irons which appear to be genetically related. Gallium concentrations in chondrites are relatively constant within each class, although Tandon and Wasson (1968) have shown a correlation between Ga and certain elements mainly concentrated in the metallic phases of L-group chondrites. As pointed out by Anders (1964), Ga is one of a large number of elements which is fractionated in a systematic fashion between the petrologic types of C-group chondrites. The geochemical affinity of Ga varies according to the oxidation state of the meteorite. In reduced meteorites, such as the enstatite chondrites, it is mainly found in the magnetic separates, whereas in more oxidized objects an appreciable fraction is found in the non-magnetic fraction.

The earliest quantitative determinations of the concentrations of Ga in meteorites were the spectrographic studies of I. and W. Noddack (1930, 1934), Goldschmidt and Peters (1931), and Goldschmidt (1937). The Noddacks (1934) reported an average chondritic abundance of 4.6 ppm. Goldschmidt (1937) reported an average concentration for iron meteorites of 8 ppm. The advent of neutron-activation analysis as a highly sensitive analytical tool has led to the acquisition of a great deal of data for Ga in chondrites and iron meteorites. One of the earliest applications of this technique to the determination of trace elements in meteorites was the work of Brown and Goldberg (1949), who determined Ga and Pd in a number of iron meteorites.

TABLE 1 Gallium abundances in chondrites

Meteorite type	No. of determinations	No. of meteorites analyzed	Concentration Range	(10^{-6} g/g) Mean	Abundance (atoms/10^6 Si atoms)	Reference
Carbonaceous						
C1	2	2	9.2–10.0	9.6	38	Fouché
	3	2	12–17	14	54	Greenland
C2	2	2	8.1–8.2	8.2	25	Fouché
	2	2	9.6–10.4	10	31	Greenland
	1	1		7.8	24	Rieder
C3	2	2	6.5–6.8	6.7	17	Fouché
	7	6	6.0–11.5	9.1	24	Greenland
	3	3	5.3–6.8	5.9	15	Rieder
H-Group	7	7	4.9–5.8	5.3	13	Fouché
	2	2	4.0–6.2	5.1	12	Greenland
	9	9	4.2–6.8	5.4	13	Onishi
	9	7	4.9–5.8	5.4	13	Rieder
	12	4	2.5–5.4	4.1	10	Schaudy
L-Group	7	7	4.7–8.6	5.6	12	Fouché
	2	2	4.0–6.6	5.2	11	Greenland
	9	9	4.4–6.0	5.2	11	Onishi
	39	13	3.0–8.7	4.9	11	Schaudy
	59	26	3.6–6.2	5.2	11	Tandon
Enstatite						
E4	5	2	17–22	17	41	Fouché
	3	2	17–32	22	53	Greenland
E5 and E6	5	2	11–13	12	25	Fouché
	6	4	12–17	15	30	Greenland
	1	1		12	25	Rieder

The data which have been reported for the chondrites are summarized in Table 1, and graphically in Figure 1. All of the chondrite data were obtained by neutron-activation analysis, except for the data of Onishi and Sandell (1955), who employed a photometric technique in the analysis of 19 chondrite finds. The data from different researchers are generally in good agreement, except for the results of Greenland (1965) for the carbonaceous and enstatite chondrites, which are high in comparison with the other activation-analysis data.

Gallium concentrations and abundances (relative to Si) are shown in Table 1. The Ga abundances decrease in the order C1 > C2 > C3 > H

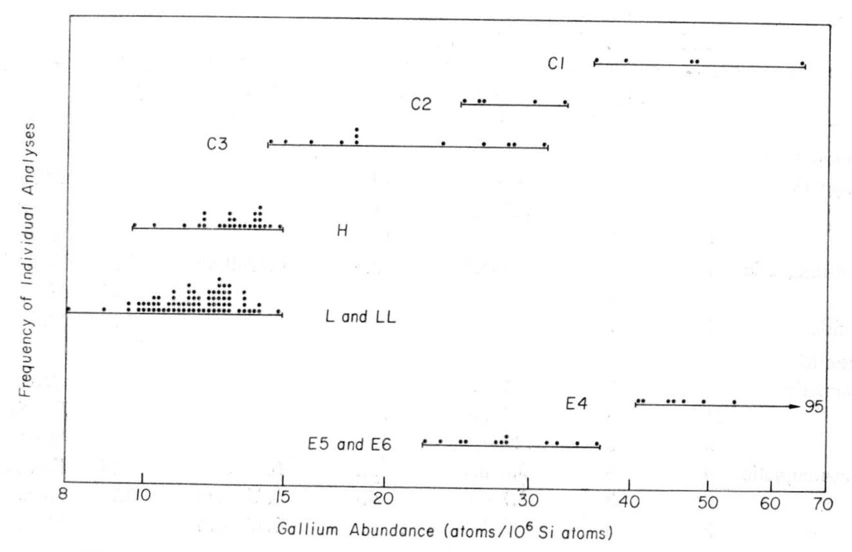

FIGURE 1 Frequency of individual analyses versus the gallium abundance (atoms/10^6 Si atoms) plotted on a log scale, for the various chondrite groups.

> L. Similarly, the E4 chondrites are enriched relative to the E5 and E6 chondrites.

Tandon and Wasson (1968) have determined Ga in a petrologic suite of L-group chondrites (as defined by the classification scheme of Van Schmus and Wood (1968)). Data for more volatile elements such as In and the rare gases show a strong correlation between trace element content and the petrologic type in the order L3 (least metamorphosed) >L4 > L5 > L6 (most metamorphosed). However, Tandon and Wasson were unable to observe any similar correlation for Ga, within experimental error.

Fouché and Smales (1967) and Rieder and Wänke (1969) have determined Ga in the magnetically separated fractions of ordinary and enstatite chondrites, and Rieder and Wänke (1969) have determined Ga in the separated light and dark portions of five H-group chondrites. These data are summarized in Table 2. The Ga data for the ordinary chondrites show appreciable concentrations in both the magnetic and non-magnetic fractions. The enstatite chondrites formed under very strongly reducing conditions, and most of the Ga is found in the magnetic portions of these meteorites. Moss et al. (1967) have also determined Ga in separated fractions from five chondrites, by a colorimetric method. These workers performed a magnetic separation of the powdered material, followed by selective dissolution of the metal and sulfide minerals. Their data are also summarized in Table 2.

TABLE 2 Ga concentrations in separated fractions of meteorites

Material	No. of determinations	No. of specimens analyzed	Concentration (10^{-6} g/g) Range	Mean	Fraction/metal ratio Range	Mean	Reference
H-chondrites							
magnetic	4	3	11–15	13			Cobb
	8	8	6.2–17	11			Fouché
	1	1		16			Moss
non-magnetic	7	7	2.5–5.5	3.8	0.15–0.89	0.39	Fouché
	1	1		2.4		0.16	Moss
sulfide	1	1		9		0.56	Moss
L-chondrites							
magnetic	3	3	10–15	1.3			Cobb
	11	9	3.4–26	11			Fouché
	3	3	12–19	15			Moss
non-magnetic	9	9	3.8–8.9	5.3	0.15–2.1	0.74	Fouché
	3	3	3.7–4.8	4.1	0.21–0.32	0.27	Moss
sulfide	3	3	2–14	6	0.13–0.74	0.40	Moss
E4 chondrites							
magnetic	1	1		64			Cobb
	3	2	61–71	66			Fouché
non-magnetic	3	2	1.0–5.2	3.7	0.014–0.080	0.058	Fouché
E5 and E6 chondrites							
magnetic	3	2	52–60	56			Fouché
	1	1		69			Moss
	1	1		41			Rieder
non-magnetic	2	2	0.25–0.48	0.37	0.0042–0.0092	0.0067	Fouché
	1	1		0.7		0.01	Moss
	1	1		0.3		0.0073	Rieder
Pallasite							
metal	3	3	19–28	23			Cobb
	2	2	18–21	20			Goldberg
	1	1		14			Smales
	36	17	4.5–26	19			Wasson
silicate	4	4		<2	<0.007–<0.14		Lovering
Mesosiderite							
metal	1	1		7.0			Cobb
	4	1		4.0			Schaudy
	2	1		8.6			Wasson
Siderophyre							
silicate	3	1		0.4			Schaudy
Group I irons							
troilite	2	2	0.3–0.4	0.4	0.005–0.006	0.006	Goldberg
H-group							
dark	5	5	4.6–5.6	5.2			Rieder
light	8	5	3.7–6.3	5.4			Rieder

The data which have been obtained for the metallic fraction of pallasites and mesosiderites are also presented in Table 2, along with data for two samples of troilite from a group I iron meteorite. These data demonstrate a distinct but minor tendency of Ga to appear in the troilite. There is no significant difference in the Ga concentrations of the light and dark portions of the H-group chondrites analyzed by Rieder and Wänke.

Gallium (along with Ge) has been a particularly important element in elucidating groups of possibly genetically related iron meteorites. Goldberg, Uchiyama, and Brown (1951) determined Ga in 45 iron meteorites by neutron activation, and found the concentrations to vary between 0.43 and 96 ppm. They observed that the iron meteorites appeared to fall into three distinct groups based on their Ga contents. The Ga classes thus defined could be roughly related to the structure of the various iron meteorites. Lovering, Nichiporuk, Chodos, and Brown (1957) determined Ga and Ge in 88 irons by a spectrographic technique. They found that Ge showed even greater variations than did Ga. They also found that the Ge and Ga concentrations were strongly correlated, and were able to distinguish four "Ga-Ge groups" of iron meteorites. The recent studies of Wasson (1967, 1968, 1969ab), and Wasson and Kimberlin (1967) have resolved additional "chemical groups" of iron meteorites. These workers have determined Ni, Ga, Ge and Ir in about half of the known iron meteorites (approximately 320). They have retained some aspects of the classification scheme of Lovering et al. (1957), but find that the Ga-Ge groups recognized by these workers can be resolved into smaller groups based on more accurate analyses for Ga and Ge, as well as determinations of Ni and Ir. The bases for recognizing chemical groups according to Wasson (1967) are as follows:

1. The range of concentration of a given element among various members of a group are small compared to the range encountered in the meteorites as a whole.

2. The concentrations of any two elements plotted against one another show a smooth variation, and in most cases will show a strong positive or negative correlation.

3. The structures of the members of a group will be quite similar, both in the content and arrangement of major and minor phases.

The data which have been obtained by various workers for the Ga concentrations in iron meteorites are summarized in Table 3. Mean values and ranges of concentration are reported for the nine chemical groups as defined

TABLE 3 Gallium concentrations in iron meteorites

Group	No. of determinations	No. of meteorites analyzed	Concentration (10^{-6} g/g) Range	Mean	Reference
I	9	7	43–100	79	Cobb
	61	8	55–96	76	Goldberg
	13	13	50–93	79	Lovering
	10	10	63–95	78	Smales
	107	51	54–100	79	Wasson
IIA	10	7	50–74	63	Cobb
	16	5	59–66	63	Goldberg
	14	14	40–66	51	Lovering
	7	7	51–59	54	Smales
	65	31	57–65	60	Wasson
IIB	3	2	52–61	55	Cobb
	8	2	45–57	51	Goldberg
	2	2	35–53	44	Lovering
	2	2	26–37	32	Smales
	23	11	46–59	53	Wasson
IIC	2	1		12	Cobb
	3	1		38	Goldberg
	2	2	19–44	32	Lovering
	2	2	33–40	37	Smales
	13	6	37–39	38	Wasson
IID	2	1	53–67	60	Cobb
	1	1		74	Lovering
	1	1		65	Smales
	14	6	70–83	77	Wasson
IIIA	8	6	13–24	20	Cobb
	43	10	17–21	20	Goldberg
	14	14	9–19	16	Lovering
	6	6	15–23	19	Smales
	110	54	17–22	20	Wasson
IIIB	11	3	18–20	19	Goldberg
	6	6	8–15	13	Lovering
	2	2	20–21	20	Smales
	38	18	16–21	18	Wasson
IVA	4	3	2.0–3.3	2.9	Cobb
	22	6	1.8–2.5	2.2	Goldberg
	7	7	1.7–2.8	<2.1	Lovering
	9	9	1.7–2.3	2.1	Smales
	59	28	1.6–2.5	2.1	Wasson

TABLE 3 (cont.)

Group	No. of determinations	No. of meteorites analyzed	Concentration (10^{-6} g/g) Range	Mean	Reference
IVB	1	1		0.68	Cobb
	3	1		0.43	Goldberg
	2	2	1.0–<2	<1.5	Lovering
	1	1		0.18	Smales
	19	9	0.17–0.31	0.22	Wasson
Anom.	18	12			Cobb
	24	7			Goldberg
	27	27			Lovering
	30	30			Smales
	187	89			Wasson

by Wasson and co-workers. Those meteorites which could not be unambiguously assigned to one of the nine groups were listed as anomalous, and no attempt was made to summarize the data for these meteorites. The number of anomalous meteorites analyzed by each worker is listed at the end of the table. Additional work may show that many of these meteorites may be members of as yet unresolved chemical groups, but the data now available do not warrant expanding the classification scheme to include them. Figure 2 is a plot of Ge versus Ga and versus Ni in iron meteorites, based on the work of Wasson and co-workers. The data points for the irons which have been classified as belonging to a resolved chemical group have not been plotted, but instead the areas on the diagram within which the analyses for these meteorites are seen to fall are represented by closed lines. The data for the remaining 30% of the irons which cannot be classified as belonging to one of the nine chemical groups, are plotted as dots on the diagram.

Nichiporuk (1958) has determined Ga in the kamacite and taenite phases of iron meteorites by means of selective dissolution. In all meteorites analyzed, Ga was observed to be enriched in the nickel-rich taenite phase.

Inghram et al. (1948) have determined the isotopic composition of meteoritic Ga. Within experimental error (0.2%) there was no difference between the Ga^{69}/Ga^{71} ratio for the Canyon Diablo iron, and that of a sample of commercial Ga.

The only analysis which has been reported in the literature on the abundance of Ga in an achondrite is the single determination of 3.4 ppm for the Stannern eucrite by Schaudy et al. (1967).

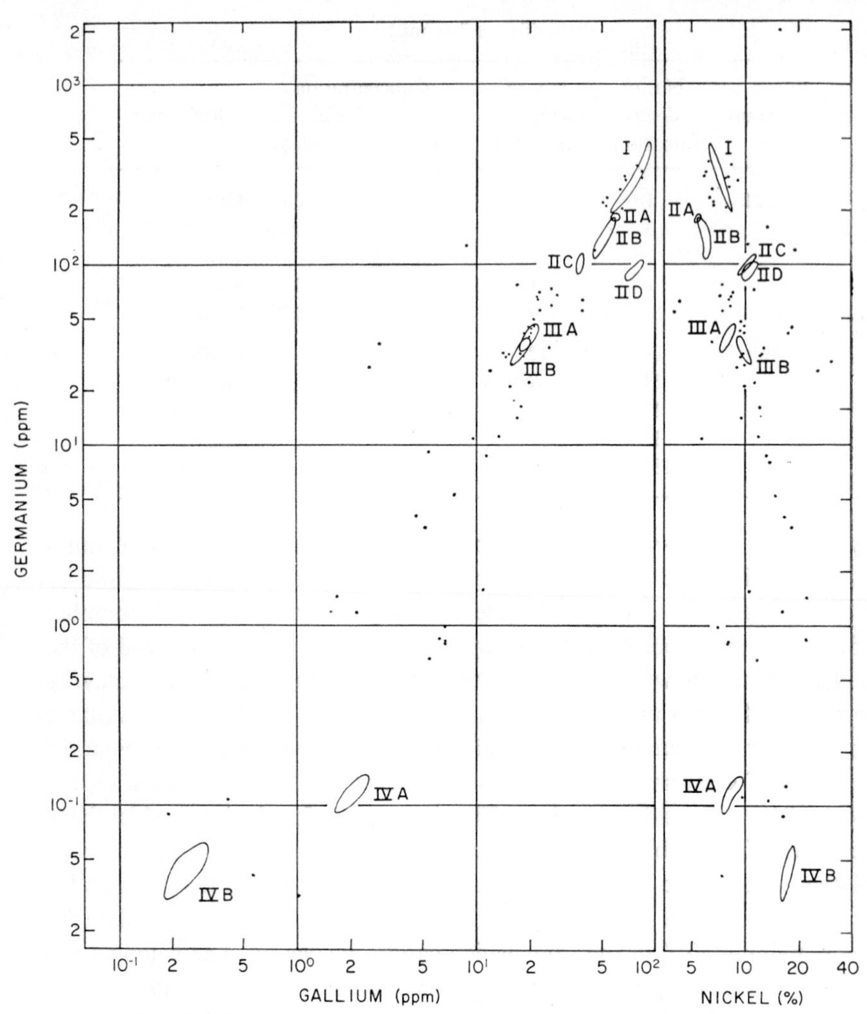

FIGURE 2 Plots of Ga concentration versus Ge concentration and versus Ni concentration for about 200 iron meteorites. Locations of eight chemical groups are outlined and designated by Roman numerals and letters. No points have been plotted for group members, and some anomalous irons plot within the outlined areas.

Müller (1968) lists the solar abundance of Ga to be log $N_{Ga} = 2.72$ (log $N_H = 12.00$). Using a Si abundance of log $N_{Si} = 7.24$ (which Müller suggests is a questionable value) the solar abundance of Ga relative to Si $= 10^6$ is 30. This is in reasonable agreement with the C1 chondrite abundance of about 40–50.

References

Anders, E. (1964) *Space Sci. Rev.* **3**, 583–714.
Brown, H. and Goldberg, E. (1949) *Science* **109**, 347–353.
Cobb, J.C. (1967) *J. Geophys. Res.* **72**, 1329–1341.
Cobb, J.C. and Moron, G. (1965) *J. Geophys. Res.* **70**, 5309–5311.
Fouché, K.F. and Smales, A.A. (1967) *Chem. Geol.* **2**, 5–33.
Goldberg, E., Uchiyama, A. and Brown, H. (1951) *Geochim. Cosmochim. Acta* **2**. 1–25.
Goldschmidt, V.M. (1937) *Skr. Norske Vidensk. Akad.* Oslo 1, No. 4, 85.
Goldschmidt, V.M. and Peters, C. (1931) *Nachr. Akad. Wiss. Göttingen Math.-Phys. Kl.*, 165.
Greenland, L. (1965) *J. Geophys. Res.* **70**, 3813–3817.
Inghram, M.G., Hess, D.C., Brown, H.S. and Goldberg, E. (1948) *Phys.Rev.* **74**, 343–344.
Lovering, J.F. (1957) *Geochim. Cosmochim. Acta* **12,** 253–261.
Lovering, J.F., Nichiporuk, W., Chodos, A., and Brown, H. (1957) *Geochim. Cosmochim. Acta* **11**, 263–278.
Moss, A.A., Hey, M.H., Elliot, C.J. and Easton, A.J. (1967) *Mineral. Mag.* **36**, 101–119.
Müller, E. (1958) *Origin and Distribution of the Elements*, L.H. Ahrens, editor, Pergamon Press, Oxford, 155–176.
Nichiporuk, W. (1958) *Geochim. Cosmochim. Acta* **13**, 233–247.
Noddack, I. and Noddack, W. (1930) *Naturwissenschaften* **18**, 757.
Noddack, I. and Noddack, W. (1934) *Svensk. Kem. Tid.* **46**, 173.
Onishi, H. and Sandell, E.B. (1955) *Geochim. Cosmochim. Acta* **8**, 78–82.
Rieder, R. and Wänke, H. (1969) *Meteorite Research* (ed. P.M. Millman) 75–86.
Schaudy, R., Kiesl, W. and Hecht, F. (1967) *Chem. Geol.* **2**, 279–287, (1968) *Chem. Geol.* **3**, 307–312.
Smales, A.A., Mapper, D. and Fouché, K.F. (1967) *Geochim. Cosmochim. Acta* **31**, 673–720.
Tandon, S.N. and Wasson, J.T. (1968) *Geochim. Cosmochim. Acta* **32**, 1087–1109.
Van Schmus, W.R. and Wood, J.A. (1967) *Geochim. Cosmochim. Acta* **31**, 747–765.
Wasson, J.T. (1967) *Geochim. Cosmochim. Acta* **31**, 161–180.
Wasson, J.T. (1968) *J. Geophys. Res.* **73**, 3207–3211.
Wasson, J.T. (1969a) *Geochim. Cosmochim. Acta* **33**, 859–876.
Wasson, J.T. (1969b) unpublished data.
Wasson, J.T. and Goldstein, J.I. (1968) *Geochim. Cosmochim. Acta* **32**, 329–339.
Wasson, J.T. and Kimberlin, J. (1967) *Geochim. Cosmochim. Acta* **31**, 2065–2093.
Wasson, J.T. and Sedwick, S.P. (1968) *Nature* **222**, 22–24.

References for Tables

Cobb (1967), *Cobb and Moran* (1965)
Fouché and Smales (1967)
Goldberg et al. (1951)

Greenland (1965)
Lovering et al. (1957), *Lovering* (1957)
Onishi and Sandell (1955)
Rieder and Wänke (1969)
Schaudy et al. (1967, 1968)
Smales et al. (1967)
Tandon and Wasson (1968)
Wasson (1967, 1968, 1969ab), *Wasson and Goldstein* (1968), *Wasson and Kimberlin* (1967), *Wasson and Sedwick* (1968)

(Received 15 October 1969)

GERMANIUM (32)

P. A. Baedecker and **J. T. Wasson**

Department of Chemistry and Institute of Geophyscis and Planetary Physics, University of California, Los Angeles, California 90024

GERMANIUM IS FOUND at trace levels in all meteorite classes, almost exclusively in the kamacite and taenite phases. It is the most valuable of several elements used for the classification of iron meteorites, chiefly because the Ge concentration range is at most a factor of 7 (and generally less than 1.5) within a genetic group of iron meteorites, while the range throughout the iron meteorites as a whole is a factor of 10^6. In chondrites the range of abundances (relative to Si) is about a factor of 4, the highest concentrations being found in volatile-rich meteorites such as C1 and E4 chondrites. Very few Ge data are available for achondrites or the stony portion of stony-irons.

Goldschmidt (1954) has reviewed the early (prior to 1940) determinations of Ge in meteorites, which were carried out by spectrographic techniques. In more recent years, much more extensive investigations of Ge in meteorites have been reported.

The data which have been obtained for germanium in chondrites are summarized in Table 1 and Fig. 1. Most of the investigators employed spectrophotometric techniques. However the data of Fouché and Smales (1967), Rieder and Wänke (1969), and Tandon and Wasson (1968) were obtained by neutron-activation analysis. In addition to the analyses listed, Burkser et al. (1962) have determined Ge in five chondrites and one achondrite by a colorimetric technique. However, there is a large discrepancy between their results and those of the other investigators. The results obtained by various workers are generally in good agreement, but there are some inconsistencies. For example, Shima's (1964) results are lower than those of the other investigators for the type II carbonaceous, H-group, and

TABLE 1 Germanium Abundances in Chondrites

Meteorite type	No. of determinations	No. of meteorites analyzed	Concentration range	(10^{-6} g/g) Mean	Abundance (Si = 10^6)	Reference
Carbonaceous	2	2	32–35	34	130	Fouché
C1	1	1		37	140	Greenland
Carbonaceous	2	2	25.0–25.4	25	75	Fouché
C2	1	1		27	70	Rieder
	3	1	16–19	17	51	Shima
Carbonaceous	2	2	18–19	19	46	Fouché
C3	3	3	8.3–18	13	32	Greenland
	3	3	17–28	22	55	Rieder
H-Group	3	3	9.3–13	11	25	Cohen
	1	1		12	27	El Wardani
	7	7	11–13	13	29	Fouché
	8	8	10–13	12	27	Greenland
	10	10	9.6–12	11	25	Onishi
	8	6	10–16	13	30	Rieder
	11	8	4–11	7.7	17	Shima
L-Group	1	1		7.3	15	Cohen
	6	4	6.6–9.6	8.8	18	El Wardani
	9	9	4.6–12	8.7	18	Fouché
	28	28	6.6–12	9.3	19	Greenland
	10	8	9.0–12	10	21	Onishi
	9	7	5.9–10	8.1	17	Shima
	59	26	6.2–17	11	22	Tandon
LL-Group	1	1		7.6	16	Cohen
	3	3	6.9–12	9.0	19	Greenland
Enstatite	5	2	42–54	47	107	Fouché
E4	2	2	46–48	48	110	Greenland
	2	1	29	29	67	Shima
Enstatite	5	2	22–29	25	50	Fouché
E5 and E6	3	3	1–48	39	77	Greenland
	1	1		19	37	Rieder

Type I enstatite chondrites, and disagree in both directions with neutron-activation determinations on iron meteorites.

The fractionation of Ge in chondritic meteorites was first pointed out by Greenland (1963). This fractionation can be seen in Table 1, where the

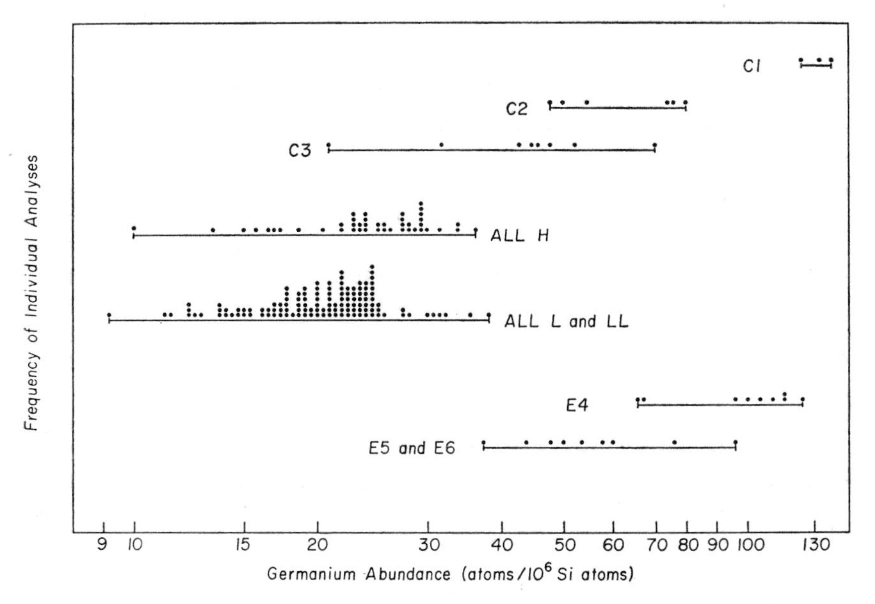

FIGURE 1 Frequency of individual analyses versus the germanium abundance (atoms/10^6 Si atoms) plotted on a log scale, for the various chondrite groups.

abundance decreases in the order C1 > C2 > C3 > H = L. The abundance of Ge in the E4 chondrites (relative to Si) is approximately the same as the C1 abundance. The E5 and E6 chondrites have abundances which average about a factor of 2 lower. This pattern is observed for a number of trace elements, and is attributed by Larimer and Anders (1967) to the high volatility of certain compounds of these elements.

Tandon and Wasson (1968) have determined the Ge concentrations in a petrologic suite (as defined by Van Schmus and Wood (1967)) of L-group chondrites. They find no correlation between germanium concentration and petrologic grade, although such a correlation has been observed for several other elements such as In and C.

Germanium is strongly concentrated in the kamacite and taenite of most meteorites, as is illustrated in Table 2 by the data of Fouché and Smales (1967), Rieder and Wänke (1969), and Shima (1964) for the separated magnetic and non-magnetic portions of ordinary and enstatite chondrites. The data for the non-magnetic fraction should not be accepted as an accurate indication of the concentration of Ge in the silicate fraction, however, due to the difficulty of obtaining silicate material which is completely free of

TABLE 2 Germanium abundances in separated fractions of meteorites

Meteorite type	No. of determinations	No. of meteorites analyzed	Concentration (10^{-6} g/g)		Ratio (Phase/Metal)		Reference
			range	mean	Range	mean	
H-Group							
magnetic	8	8	62–67	64			Fouché
	1	1		65			Moss
	1	1		33			Shima
non-magnetic	7	7	0.13–0.92	0.48	0.0022–0.015	0.0076	Fouché
	1	1		<0.5		<0.008	Moss
	1	1		0.17		0.0051	Shima
sulphide	1	1		<8		<0.12	Moss
L-Group							
magnetic	11	9	92–168	130			Fouché
	3	3	73–127	107			Moss
non-magnetic	9	9	0.22–3.9	1.7	0.0023–0.030	0.013	Fouché
	3	3		<0.5		<0.005	Moss
sulphide	3	3	<0.1–<2		<0.0008–<0.03		Moss
E4							
magnetic	3	2	170–120	190			Fouché
non-magnetic	3	2	1.4–2.5	2.1	0.008–0.012	0.011	Fouché
E5–6							
magnetic	3	2	110–140	120			Fouché
	1	1		123			Moss
	1	1		52			Rieder
non-magnetic	2	2	0.47–0.77	0.62	0.0034–0.0063	0.0048	Fouché
	1	1		2.5		0.020	Moss
	1	1		5.9		0.11	Rieder
Pallasite							
metal	4	2	29–57	43			Shima
	3	3	11–51	36			Smales
	38	17	11–75	46			Wasson
silicate	4	4		<20			Lovering
	3	2	0.63–0.86	0.78	0.015–0.021	0.018	Shima
troilite	2	2	6–8	8	0.14–0.19	0.17	Lovering
	2	2	12–17	14	0.31–0.38	0.35	Shima
Mesosiderite							
metal	1	1		51			Wasson
Irons							
troilite	2	2	6–10	8	0.035–0.044	0.040	Lovering
	1	1		122		0.30	Shima
	2	2	19–32	26	0.53–0.089	0.071	Smales
H-Group							
dark	8	5	10–13	12			Rieder
light	5	5	10–18	14			Rieder

metal. Nevertheless, the data do demonstrate the strong tendency of germanium to favor the metallic phases.

Moss et al. (1967) determined Ge colorimetrically in separated fractions of five chondrites. Their technique involved a preliminary magnetic separation, followed by selective chemical dissolution of the metal and sulfide minerals. Their data are summarized in Table 2.

The data for the separated pallasite phases, presented in Table 2, also indicate a strong tendency of germanium to concentrate in the metal phases, as well as a slight tendency to favor the troilite. The data of Shima (1964) and Lovering (1957) suggest a troilite-to-metal concentration ratio in the range of 0.14 to 0.38 for the pallasites. Shima's (1964) value for the concentration of Ge in the Odessa troilite suggests a similar ratio for the irons, but Lovering (1957) and Smales et al. (1958) obtained ratios of less than 0.1 for troilite inclusions in iron meteorites.

Yavnel' (1950) studied the distribution of a number of elements, including Ge, between various mineral fractions of the Sikhote-Alin iron. He found approximately 300 ppm Ge in the kamacite, <30 ppm in schreibersite, ~ 30 ppm in troilite, and 100 ppm in the cohenite, using a spectrographic technique. Nichiporuk (1958) found Ge to be enriched in the taenite relative to the kamacite phase of iron meteorites. The two phases were separated by selective dissolution, and Ge was determined by emission spectrography. Goldstein (1967) studied the Ge distribution in iron meteorites by electron-probe microanalysis. He found that germanium is concentrated almost entirely in the metallic phases, and was unable to detect any Ge in the schreibersite, troilite, or cohenite inclusions, above the detectability limit of the probe, which was 40 ppm. His data indicate upper limits on the schreibersite/metal, troilite/metal and cohenite/metal concentration ratios of 0.025, 0.05, and 0.13, based on the meteorites having the highest Ge metal concentrations in each class. Goldstein reported a Ge concentration ratio of about 2.5 at the taenite-kamacite interface in the remarkable Butler iron meteorite, which contains 2000 ppm Ge (Wasson, 1966). Goldstein's Ni concentrations in Butler taenite are about 8 times higher than those in the kamacite at the interface between the two phases. Thus, Ge fractionates less strongly than Ni between the two phases. The taenite/kamacite ratios at the phase interfaces in other meteorites studied by Goldstein are in the range of about 1.5–4, and can be considered constant within experimental error.

Also presented in Table 2 are the results of Rieder and Wänke (1969) who have analyzed the separated light and dark portions of five H-group chondrites having a light-dark (polymict-brecciated) structure. They found

TABLE 3 Germanium concentrations in iron meteorites

Ga–Ge group	No. of determinations	No. of meteorites analyzed	Concentration (10^{-6} g/g) Range	Mean	Reference
I	3	3	270–520	380	El Wardani
	14	14	160–410	300	Lovering
	20	5	210–470	350	Shima
	10	10	220–520	340	Smales
	107	51	190–520	330	Wasson
IIA	14	14	130–170	150	Lovering
	3	3	89–130	110	Shima
	6	6	170–190	180	Smales
	61	29	170–190	180	Wasson
IIB	2	2	73–140	110	Lovering
	2	1	120–140	130	Shima
	2	2	94–180	190	Smales
	23	11	110–180	140	Wasson
IIC	2	2	77–91	84	Lovering
	2	2	99–100	100	Smales
	13	6	88–110	98	Wasson
IID	1	1		76	Lovering
	1	1		90	Smales
	9	4	83–93	89	Wasson
IIIA	14	14	29–53	38	Lovering
	2	1	12–32	22	Shima
	8	8	36–44	40	Smales
	111	53	33–46	39	Wasson
IIIB	7	7	25–34	30	Lovering
	2	1	56–72	64	Shima
	4	4	33–38	35	Smales
	36	17	28–38	33	Wasson
IVA	1	1		<1.0	El Wardani
	7	7		<1.0	Lovering
	12	12	0.12–0.30	0.18	Smales
	59	28	0.092–0.15	0.12	Wasson
IVB	1	1		<1.0	El Wardani
	2	2		<1.0	Lovering
	1	1		0.04	Smales
	19	9	0.03–0.064	0.046	Wasson
Anomalous	2	2			El Wardani
	28	28			Lovering
	3	3			Shima
	30	30			Smales
	187	89			Wasson

no difference in the germanium contents of the two portions within their experimental error.

As has been discussed in the chapter on Ga, Ge is a particularly important element for identifying different chemical groups of related iron meteorites. Lovering et al. (1957), having analyzed 88 iron meteorites for Ge and Ga by a spectrographic technique, found that iron meteorites could be classified as belonging to one of four groups, based on their Ge and Ga concentrations. In subsequent work Wasson (1967, 1968, 1969ab), Wasson and Kimberlin (1967) and Wasson and Goldstein (1968) have determined Ga, Ge, Ir, and Ni in about 320 irons by neutron activation, and have resolved nine "chemical groups". The data which have been obtained by various workers for the Ge concentrations in iron meteorites are summarized in Table 3. Mean values and ranges of concentration are reported for the nine groups as defined by Wasson and co-workers. Those meteorites which could not be unambiguously assigned to one of the nine groups were listed as anomalous. The data of Wasson and coworkers and Smales et al. (1957) were obtained by activation analysis. Lovering et al. (1957) employed spectrographic methods, while Shima (1964) and El Wardani (1957) employed spectrophotometric techniques. The activation analysis results are in good agreement, with the results of Smales et al. falling within the ranges defined by Wasson for the various chemical groups (with the exception of group IVA). The results of Shima (1964) are in many cases inconsistent with the work of Wasson and Smales. The results by Lovering and coworkers appear to be too low in most cases, and for a given chemical group, scatter over a greater range than that indicated by the activation-analysis data.

Plots of Ge versus Ga and Ni in iron meteorites are presented as Fig. 2 in the chapter on Ga. The areas on the diagram defined on the basis of data for the resolved chemical groups are represented by closed lines, with the individual points not shown. The data for the remaining 30% of the irons which cannot be classified as belonging to one of the nine chemical groups are plotted as dots on the diagram.

Cohen (1960) determined the concentration of Ge in the Norton County aubrite to be 0.21 ppm. The non-magnetic portion of the same meteorite was found to have 0.09 ppm Ge. These represent the only data which have appeared in the literature for the concentration of Ge in achondrites.

Shima (1963) measured the isotopic composition of germanium in eight iron meteorites by solid-source mass spectrometry. No difference in the isotopic compositions of meteoritic and terrestrial Ge was detected.

The stellar Ge abundance is given by Müller (1968) to be 18 atoms/10^6 Si atoms, which is revised downward considerably from the value of 62 atoms/ 10^6 Si atoms reported by Goldberg, Muller and Aller (1960). Both results are much smaller than the value of about 130 atoms/10^6 Si atoms reported for the C1 chondrites and given above in Table 1.

References

Burkser, E.S., Lazebnik, K.I. and Alekseeva, K.N. (1962) *Meteoritika* **22**, 94–96.
Cohen, A.J. (1960) *Intern. Geol. Cong. XXI, Norden*, pt. 1, 30–39.
El Wardani, S.A. (1957) *Geochim. Cosmochim. Acta* **13**, 5–19.
Fouché, K.F. and Smales, A.A. (1967) *Chem. Geol.* **2**, 5–33.
Goldberg, L., Müller, E.A., and Aller, L.H. (1960) *Astrophys. J.* **5**, 1–138.
Goldschmidt, V.M. (1954) *Geochemistry*, Oxford University Press, Oxford, England.
Goldstein, J.I. (1967) *J. Geophys. Res.* **72**, 4689–4696.
Greenland, L. (1963) *J. Geophys. Res.* **68**, 6507–6514.
Greenland, L. and Lovering, J.F. (1965) *Geochim. Cosmochim. Acta* **29**, 821–858.
Larimer, J. and Anders, E. (1967) *Geochim. Cosmochim. Acta* **31**, 1239–1270.
Lovering, J.F. (1957) *Geochim. Cosmochim. Acta* **12**, 253–261.
Lovering, J.F., Nichiporuk, W., Chodos, A. and Brown, H. (1957) *Geochim. Cosmochim. Acta* **11**, 263–278.
Moss, A.A., Hey, M.H., Elliot, C.J. and Easton, A.J. (1967) *Mineral. Mag.* **36**, 101–119.
Müller, E.A. (1968) *Origin and Distribution of the Elements* (1968) Edited by L.H.Ahrens, pp. 155–76. Pergamon Press, Oxford, England.
Nichiporuk, W. (1958) *Geochim. Cosmochim. Acta* **13**, 233–247.
Onishi, H. (1956) *Bull. Chem. Soc. Japan* **29**, 686–694.
Rieder, R. and Wänke, H. (1969) *Meteorite Research* (ed. P.M.Millman), 75–86.
Shima, M. (1963) *J. Geophys. Res.* **68**, 4289–4292.
Shima, M., (1964) *Geochim. Cosmochim. Acta* **28**, 517–532.
Smales, A.A., Mapper, D., Morgan, J.W., Webster, R.K. and Wood, A.J. (1958) *Proc. 2nd. Int. Conf. Peaceful Uses Atomic Energy*, U.N. Geneva **2**, 242–248.
Smales, A.A., Mapper, D., and Fouché, K.F. (1967) *Geochim. Cosmochim. Acta* **31**, 673–720.
Tandon, S.N. and Wasson, J.T. (1968) *Geochim. Cosmochim. Acta* **32**, 1087–1109.
Van Schmus, W.R. and Wood, J.A. (1967) *Geochim. Cosmochim. Acta* **31**, 747–765.
Wasson, J.T. (1966) *Science* **153**, 976–978.
Wasson, J.T. (1967) *Geochim. Cosmochim. Acta* **31**, 161–180.
Wasson, J.T. (1968) *J. Geophys. Res.* **73**, 3207–3211.
Wasson, J.T. (1969a) *Geochim. Cosmochim. Acta* **33**, 859–876.
Wasson, J.T. (1969b) unpublished data.
Wasson, J.T. and Goldstein, J.I. (1968) *Geochim. Cosmochim. Acta* **32**, 329–339.
Wasson, J.T. and Kimberlin, J. (1967) *Geochim. Cosmochim. Acta* **31**, 2065–2093.
Wasson, J.T. and Sedwick, S.P. (1969) *Nature* **222**, 22–24.
Yavnel', A.A. (1950) *Meteoritika* **8**, 134–148.

References for Tables

Cohen (1960)
El Wardani (1957)
Fouché and Smales (1967)
Greenland (1963), *Greenland and Lovering* (1965)
Lovering (1957), *Lovering et al.* (1957)
Moss et al. (1967)
Onishi (1956)
Rieder and Wänke (1969)
Shima (1964)
Smales et al. (1958), Smales *et al.* (1967)
Tandon and Wasson (1968)
Wasson (1967, 1968, 1969ab), *Wasson and Kimberlin* (1967), *Wasson and Goldstein* (1968), *Wasson and Sedwick* (1969)

(Received 15 October 1969)

ARSENIC (33)

Michael E. Lipschutz

Departments of Chemistry and Geosciences
Purdue University
Lafayette, Indiana 47907

ARSENIC IS PRESENT as a trace element in meteorites in amounts ranging from 0.02–30 $\times 10^{-6}$ g/g. While a few determinations have been made by spectrophotometry (Onishi and Sandell, 1955), X-ray fluorescence (Nichiporuk and Chodos, 1959) or spark source mass spectrometry (Berkey and Morrision, 1968) the overwhelming number have been made by techniques based upon neutron activation analysis. These techniques include: "classical" neutron activation in which the irradiated sample is completely dissolved and the arsenic fraction extensively purified radiochemically before counting (Esson et al., 1965; Fouché and Smales, 1967; Hamaguchi et al., 1961, 1969; Smales et al., 1958, 1967); "instrumental" neutron activation in which no chemistry is performed and arsenic is determined by direct counting of the irradiated meteorite (Cobb, 1967); and combinations of these approaches (Case et al., 1971; Kiesl, 1967; Kiesl et al., 1967; Kiesl and Hecht, 1969).

In meteorites, arsenic is a dispersed element and seems about equally siderophile and chalcophile in geochemical behavior, although the former may predominate. Onishi and Sandell (1955) have determined the arsenic contents in separated metal, sulfide and silicate portions of two composites (each consisting of 7 H- and L-group finds) to be 11, 8, 0.4 and 13, 11, 0.2 $\times 10^{-6}$ g/g, respectively. The geochemical behavior of arsenic is also demonstrated by its generally high concentration in iron meteorites and by the analyses of: whole-rock samples and separated silicate and/or metal portions of 4 individual chondritic finds (Onishi and Sandell, 1955); separated magnetic and non-magnetic portions of 7 H-, 9 L- and 4 E-group chondrites—all but 3 of which are falls (Fouché and Smales, 1967); metal and silicate portions of one pallasite (Hamaguchi et al., 1961) and metal and non-metallic (doubtless including sulfide) portions of the siderophyre Steinbach (Kiesl and Hecht, 1969). Troilite inclusions from iron meteorites appear to be quite variable in their arsenic content. Smales et al. (1958) established upper limits of 0.03 and 0.07×10^{-6} g/g in an inclusion from Cañon Diablo, while Nichiporuk and Chodos (1959) found 140 and 270×10^{-6} g/g in inclusions from Odessa and Henbury, respectively, and upper limits of 50×10^{-6} g/g in 20 other inclusions from 12 irons (including Cañon Diablo, Odessa and Henbury), 1 pallasite and an H-group chondrite. Kiesl and Hecht (1969) reported 0.28×10^{-6} g/g in troilite from the Mt. Joy hexahedrite an upper limit of $<5 \times 10^{-9}$ g/g in troilite from two octahedrites and 11×10^{-6} g/g in a troilite-iron inclusion from the H-group chondrite Mocs. Since stable arsenic is monoisotopic there have been no studies of its isotopic composition.

The data for 12 carbonaceous, 47 ordinary (including 20 finds) and 4 enstatite chondrites are summarized in Table 1 and illustrated in Figure 1. In general, replicate analyses of the carbonaceous chondrites are in good agreement and indicate that the arsenic concentrations in these (and in the ordinary chondrites as well) are rather similar (Figure 1). Although the data suggest that the arsenic concentrations in ordinary chondrites decrease in the order H–L–LL (Table 1), as would be expected from its geochemical be-

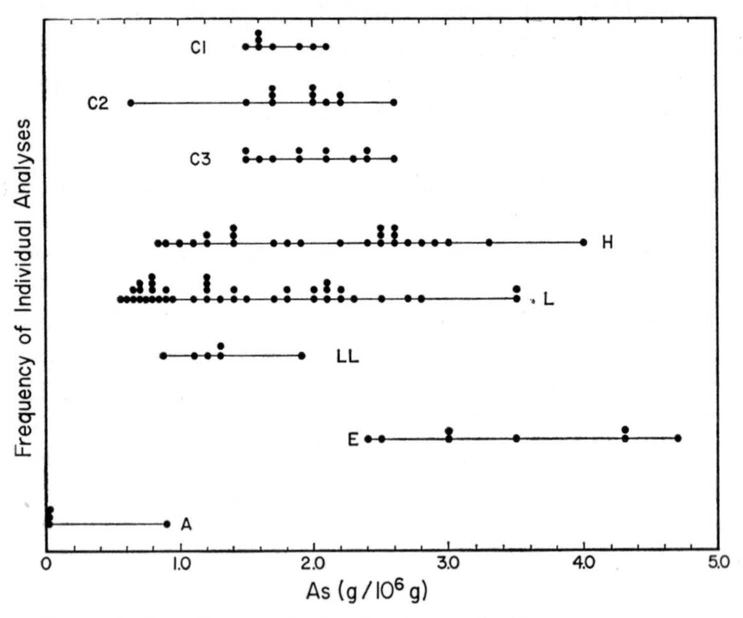

FIGURE 1 Arsenic concentrations in stony meteorites.

TABLE 1 Arsenic in stony meteorites

Type	Number analyzed	No. of detns.	As contents Range (10^{-6} g/g)	(ref.)	Mean (10^{-6} g/g)	(atoms/ 10^6 Si atoms)
C1	2	3	1.6–2.0	(1)	1.7	
	2	2	1.9–2.1	(2)	2.0	
	1	3	1.5–1.7	(3)	1.6	
	2	8	1.5–2.1	All	1.8	6.6
C2	2	2	2.0	(1)	2.0	
	2	2	2.1–2.2	(2)	2.2	
	4	7	0.64–2.6	(3)	1.8	
	1	1	1.7	(9)	1.7	
	4	12	0.64–2.6	All	1.9	5.4
C3	4	4	1.5–2.4	(1)	1.8	
	2	2	2.3–2.4	(2)	2.4	
	2	4	1.6–2.6	(3)	2.0	
	2	2	1.9–2.1	(9)	2.0	
	6	12	1.5–2.6	All	2.0	4.8
H	6	7	0.84–1.9	(1)	1.3	
	6	6	2.5–3.3	(2)	2.6	
	1[a]	1	1.4	(3)	1.4	
	7	7	0.9–4.0	(4)	2.5	
	1	2	1.4	(5)	1.4	
	1	3	2.2–3.0	(6)	2.6	
	17[a]	26	0.84–4.0	All	2.1	4.6
L	6[b]	6	0.67–1.3	(1)	0.97	
	10	10	0.92–3.5	(2)	1.6	
	1	1	0.61	(3)	0.6	
	7	7	0.8–3.5	(4)	1.9	
	3	7	0.57–0.79	(5)	0.7	
	1	3	1.4–2.2	(6)	1.8	
	5	5	0.80–2.8	(7)	2.1	
	2	2	2.1–2.2	(8)	2.1	
	25[b]	41	0.57–3.5	All	1.5	3.0
LL	4	4	1.1–1.3	(1)	1.2	
	1	1	0.87	(3)	0.9	
	1	1	1.9	(7)	1.9	
	5	6	0.87–1.9	All	1.3	2.6
E4	2	2	4.3–4.7	(2)	4.5	
	2	4	2.5–4.3	(3)	3.3	
	2	6	2.5–4.7	All	3.7	8.2

TABLE 1 (cont.)

Type	Number analyzed	No. of detns.	Range (10^{-6} g/g)	As contents (ref.)	Mean (10^{-6} g/g)	(atoms/ 10^6 Si atoms)
E6	2	2	2.4–3.0	(2)	2.7	
	2	2	2.4–3.0	All	2.7	5.2
Eu	1	3	0.02	(6)	0.02	
	1	1	0.91	(7)	0.9	
	2	4	0.02–0.91	All	0.2	0.4

[a] Not including a determination for Tieschitz which seemed suspiciously high to the original authors.

[b] Not including determinations for Holbrook and Mezö-Madaras which seemed suspiciously high to the original authors.

(1) Case et al. (1971)
(2) Fouché and Smales (1967)
(3) Hamaguchi et al. (1969)
(4) Onishi and Sandell (1955)
(5) Esson et al. (1965)
(6) Hamaguchi et al. (1961)
(7) Kiesel et al. (1967)
(8) Kiesel (1967)
(9) Kiesl and Hecht (1969)

havior, the ranges for each group overlap to such an extent that this tendency is not statistically significant. Furthermore, the relationships among the ordinary chondrites are clouded by a possible systematic difference between the results of Esson et al. (1961), Hamaguchi et al. (1969), and Case et al. (1971) which, in general, agree and are about a factor of 2 lower than those of the five other studies listed in Table 1. This difference is indicated not only by the H-, L-, LL- and E-group means for each investigation (Table 1) but also by the analyses of individual meteorites where, in 9 out of 10 cases, the same tendency is evident. Despite this problem the arsenic concentrations within each of the ordinary chondrite groups appear to be independent of chemical-petrologic type and of the degree of equilibration of the unequilibrated ordinary chondrites (Case et al., 1971).

The mean abundances for the C1 and ordinary chondrites (Table 1) show arsenic to be an undepleted element (Larimer and Anders, 1967) and the analyses of the C1 chondrites suggest a mean cosmic abundance value of

6.6 atoms/10^6 Si atoms. Neither Aller (1961) nor Müller (1968) list solar photospheric values for this element.

As noted, the possible systematic difference associated with the ordinary chondrite data may exist for the E-group data as well. Nevertheless the arsenic contents in these chondrites seem higher than those in the carbonaceous and ordinary chondrites, as expected from the highly reduced nature of the enstatite chondrites and their high metal and sulfur contents. The limited data suggest that despite the overlap in the ranges for each type (Table 1), the arsenic contents in Type I (E4) enstatite chondrites are somewhat higher than those in Type II (E6), again as might be expected from their mineralogical and chemical composition.

The data for iron meteorites are summarized in Table 2 and illustrated in Figure 2. In 9 of 13 cases, replicate analyses of the same meteorite have been in good agreement. In Smithonia, the only meteorite studied by spark source

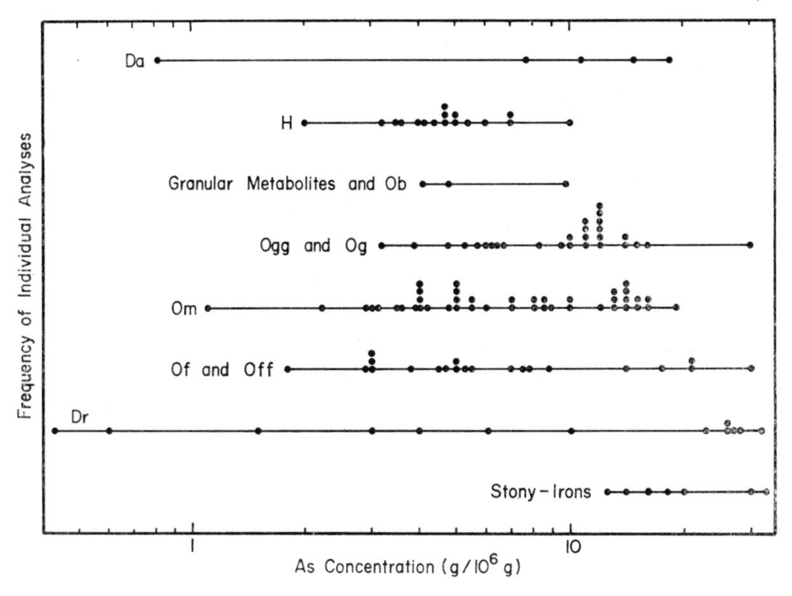

FIGURE 2 Arsenic concentrations in iron and stony-iron meteorites. For convenience, Ogg and Og have been grouped as have Of and Off.

mass spectrometry, the variability of the results (a factor of 5) may be due to experimental difficulties (Berkey and Morrison, 1968). In the remaining cases (Canyon Diablo, Gibeon and Henbury) the extreme results vary by factors of 2–5 which, in view of the structural and chemical variations known for

TABLE 2 Arsenic in iron and stony-iron meteorites

Type	Number analyzed	No. of detns.	As content Range (10^{-6} g/g)	(ref.)	Mean (10^{-6} g/g)
Da	5[a, b]	5	0.81–18	(1)	10
H	6	6	3.5–5.4	(1)	4.4
	3[b]	4	3.2–7	(2)	5.1
	3	3	4.1–7.0	(3)	5.3
	1	4	2–10	(4)	5
	10	17	2–10	All	5.0
Granular metabolite*	2	2	4.1–4.8	(1)	4.4
Ob	1	1	9.9	(1)	9.9
Ogg	5	5	5.3–11	(1)	8.3
	1	1	6.7	(2)	6.7
	1	1	30	(3)	30
	7	7	5.3–30	All	11
Og	8	8	4.8–14	(1)	10
	4	5	5.7–15	(2)	11
	1	1	8.3	(3)	8.3
	1	1	16	(5)	16
	1	3	12–14	(6)	13
	1	2	3.2–3.9	(7)	3.6
	13	20	3.2–16	All	10
Om	19	19	1.1–19	(1)	8.2
	10[b]	13	2.2–14	(2)	6.8
	5	5	3.9–16	(3)	8.7
	1	1	2.9	(5)	2.9
	1	3[c]	14–15	(6)	14
	1	2	8.0–8.2	(7)	8.1
	32	43	1.1–16	All	8.1
Of	10	11	1.8–21	(1)	7.0
	1	1[b]	4.7	(2)	4.7
	4	4	5.0–14	(3)	8.1
	14	16	1.8–21	All	7.1
Off	4	4	3.0–21	(1)	9.7
	1	1	30	(3)	30
	5	5	3.0–30	All	14
Dr	6	6	0.43–31	(1)	12
	3[b]	5	3–28	(2)	17

TABLE 2 (cont.)

Type	Number analyzed	No. of detns.	As content: Range (10^{-6} g/g)	(ref.)	Mean (10^{-6} g/g)
	1	1	0.6	(3)	0.6
	1	1	27	(5)	27
	10	13	0.43–31	All	14
P†	1	1	20	(1)	20
	2	2	14–30	(2)	22
	1	1	33	(3)	33
	1	1[d]	18	(6)	18
	5	5	14–33	All	23
M	1	1	12	(2)	12
‡	1	1	16	(2)	16

[a] Includes Santa Rosa which may be a metabolite.

[b] Not including upper limits for Nedagolla (Da), Clark Co. (Om), Charlotte (Of) and Deep Springs (Dr) for which other actual determinations reported lie below these limits and Boguslavka and Braunau (H), for which there are no other measurements (Cobb, 1967).

[c] Not including a determination for the same meteorite (Henbury) considered suspiciously high by the original authors.

[d] Hamaguchi et al. (1961) also obtained a value of 0.02 ppm for the silicate phase of this meteorite (Admire).

* It has been demonstrated (Jain and Lipschutz, 1968) that granular metabolites can be produced from shock-loaded octahedrites by extended annealing below the α–γ transformation temperature. These samples may well be heat-altered, shock-loaded octahedrites.

† Metal phase only.

‡ This meteorite (Pitts) is classified as an octahedrite with silicate inclusions.

(1) Smales et al. (1967)
(2) Cobb (1967)
(3) Smales et al. (1958)
(4) Berkey and Morrison (1968)
(5) Kiesl et al. (1967)
(6) Hamaguchi et al. (1961)
(7) Onishi and Sandell (1955)

Canyon Diablo (Lipschutz, 1965; Wasson, 1968), suggests that arsenic may be variable in large meteorites. For meteorites of a given structural class the arsenic concentrations are quite variable (Table 2), and are not directly correlated with nickel or the meteorites' Ga–Ge or chemical groups. However, the data (Table 2) do suggest that the arsenic contents of hexahedrites are less variable and are lower than those of most other meteorite groups and that meteorites of nickel-rich groups are also somewhat enriched in arsenic. In Smales et al.'s study (1967), a number of other elements (Ag, Cr, Cu, Ga, Ge, In, Mo, Pd and Zn) were determined in the same samples. Smales et al. (1967) report that arsenic is strongly correlated with Sb and Pd, weakly correlated with Cu and strongly correlated in an inverse manner with Cr.

The only achondrites in which arsenic has been determined are eucrites (Table 1). Despite the difference in the results it seems that in keeping with their mineralogy these meteorites have lower arsenic contents than any other meteoritic group. Stony-irons (and especially the metal portion of the pallasites), on the other hand, seem to be relatively rich in arsenic (Table 2).

Acknowledgement

This research was supported in part by the U.S. National Science Foundation, grant GA-1474.

References

Aller, L.H. (1961) *The Abundance of the Elements*. Interscience.

Berkey, E. and Morrison, G.H. (1968) *Origin and Distribution of the Elements* (L.H. Ahrens, ed.) 345–357. Pergamon.

Case, D.R., Laul, J.C., Pelly, I.Z., Wechter, M.A., Schmidt-Bleek, F. and Lipschutz, M.E. (1971) In preparation.

Cobb, J.C. (1967) *J. Geophys. Res.* **72**, 1329.

Esson, J., Stevens, R.H. and Vincent, E.A. (1965) *Mineral. Mag.* **35**, 88.

Fouché, K.F. and Smales, A.A. (1967) *Chem. Geol.* **2**, 105.

Hamaguchi, H., Nakai, T. and Endo, T. (1961) *Nippon Kagaku Zasshi* **82**, 1485.

Hamaguchi, H., Onuma, N., Hirao, Y., Yokoyama, H., Bando, S. and Furukawa, M. (1969) *Geochim. Cosmochim. Acta* **33**, 507.

Jain, A.V. and Lipschutz, M.E. (1968) *Nature* **220**, 139.

Kiesl, W. (1967) *Z. Anal. Chem.* **227**, 13.

Kiesl, W. and Hecht, F. (1969) *Meteorite Research* (P.M. Millman, ed.) 67–74, Reidel.

Kiesl, W., Seitner, H., Kluger, F. and Hecht F. (1967) *Monatsh. Chem.* **98**, 972.

Larimer, J.W. and Anders, E. (1967) *Geochim. Cosmochim. Acta* **31**, 1239.

Lipschutz, M. (1965) *Nature* **208**, 636.

Müller, E. (1968) *Origin and Distribution of the Elements* (L.H. Ahrens, ed.) 155–176. Pergamon.

Nichiporuk, W. and Chodos, A.A. (1959) *J. Geophys. Res.* **64**, 2451.
Onishi, H. and Sandell, E.B. (1955) *Geochim. Cosmochim. Acta* **7**, 1.
Smales, A.A., Mapper, D., Morgan, J.W., Webster, R.K. and Wood, A.J. (1958) *Proceedings of the Second International Conference on the Peaceful Uses of Atomic Energy* **2**, 242–248.
Smales, A.A., Mapper, D. and Fouché, K.F. (1967) *Geochim. Cosmochim. Acta* **31**, 673.
Wasson, J.T. (1968) *J. Geophys. Res.* **73**, 3207.

(Received 22 October 1969, revised 8 June 1970)

SELENIUM (34)

Ithamar Z. Pelly* and **Michael E. Lipschutz†**

Department of Chemistry
Purdue University
Lafayette, Indiana 47907

SELENIUM HAS BEEN DETERMINED in 91 meteorites (not including upper limits) and is present in these in trace quantities ranging from 0.002–34 $\times 10^{-6}$ g/g. Byers (1938) determined selenium, and DuFresne (1960) both selenium and Te, by spectrophotometric techniques. All other determinations have been made by techniques based upon neutron activation and have been conducted in conjunction with measurement of other trace elements: Te—Schindewolf (1960); Te, In—Akaiwa (1966); Te, other elements—Turner (1965), Greenland (1967), Case et al., (1971); other elements—Kiesl (1967), Kiesl et al. (1967), Seitner et al. (1968), Kiesl and Hecht (1969), Dörfler and Kiesl (1970). There have been no determinations of the isotopic composition of meteoritic selenium.

In meteorites selenium is a dispersed element and, as on Earth, is chalcophile in geochemical behavior. The selenium contents in separated metal and stony (doubtless including sulfide) portions of the siderophyre Steinbach are 0.04 and 13 $\times 10^{-6}$ g/g respectively (Kiesl and Hecht, 1969). Byers (1938) reported 23 $\times 10^{-6}$ g/g in troilite (FeS) from Canyon Diablo, and Kiesl and Hecht (1969) a range of 130–300 $\times 10^{-6}$ g/g in troilite from a hexahedrite and two octahedrites. The selenium content of a troilite-metal grain from the H-group chondrite Mocs was reported as 52 $\times 10^{-6}$ g/g (Kiesl and Hecht, 1969) compared with a range of 7.3–11 $\times 10^{-6}$ g/g from whole-rock analyses of the same chondrite (Greenland, 1967; Kiesl, 1967; Kiesl et al., 1967). Based upon these results and the nearly constant selenium-sulfur ratios in chondrites (see below) troilite appears to be the host mineral for selenium in meteorites. Here it should be noted that the experimental technique of Kiesl

* Now at Dept. of Geology, Negev University, Beersheba, Israel
† Also Department of Geosciences, Purdue

and co-workers involved addition of the irradiated meteorite specimen and inert carrier to a distillation apparatus containing concentrated H_2SO_4, followed by volatilization at ~200°C of selenium and other elements as the halides (cf. Kiesl, 1967; Kiesl et al., 1967). Under these conditions it is probable that not all of the meteorite sample is dissolved; yet, as will be seen, the selenium results obtained by Kiesl and co-workers generally agree with those obtained by other workers using more classical activation techniques on the same or similar stony meteorites. This agreement suggests, therefore, that probably all of the selenium is sited in minerals, such as troilite, soluble in conc. H_2SO_4 or in hydrogen halide—H_2SO_4 solutions at elevated temperature. Whether any chondritic selenium, like much of the Te (cf. Pelly and Lipschutz, this volume), is soluble in water has not been determined.

The data of 11 irons (10 being finds), one pallasite find, two mesosiderites (including one find), two achondrites (including one find), and 12 carbonaceous, 57 ordinary (including 15 finds) and 6 enstatite chondrites are sum-

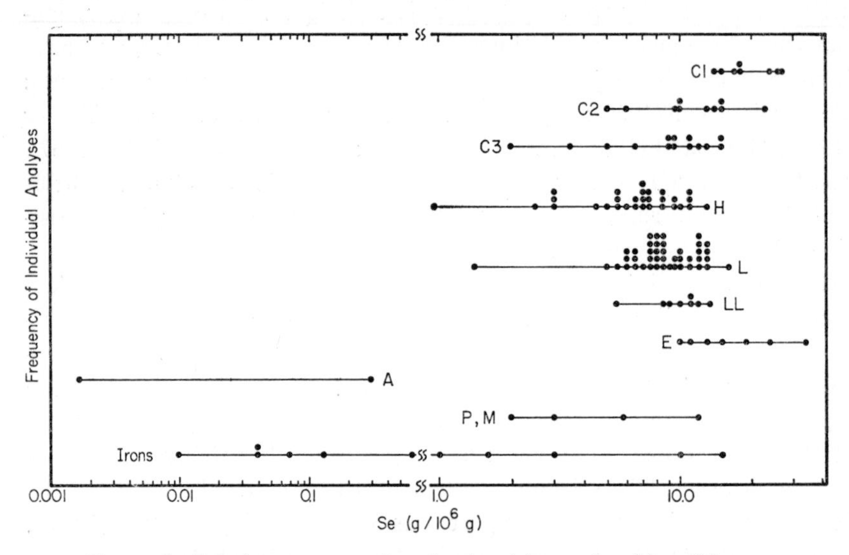

FIGURE 1 Selenium concentrations in chondrites, achondrites (A), mesosiderites (M), pallasites (P) and iron meteorites.

marized in Table 1 and illustrated in Figure 1. Replicate analyses of the same meteorite by different investigators differ by as much as a factor of 4, although when such replicate analyses are carried out by the same investigator the difference is usually much less. Though some of these discrepancies might

TABLE 1 Selenium in meteorites

Type	Number analyzed	No. of analyses	Se contents Range (10^{-6} g/g)	[ref.]	Mean (10^{-6} g/g)	Mean (atoms/10^6 Si atoms)
			chondrites			
C1	1	2	14–18	(1)	16	
	2	2	26–27	(2)	26	
	2	4	15–24	(3)	18	
	2	8	14–27	All	20	69
C2	1	2	5.0–6.1[b]	(1)	*5.6*	
	3	3	9.6–14	(2)	11	
	3	4	9.8–15	(3)	13	
	1	1	23	(4)	23	
	4	10	(5.0–23)	All	(12)	(33)
	4	8	9.6–23	All[a]	14	38
C3	1	2	2.0–3.6[b]	(1)	*2.8*	
	4	4	6.5–11	(2)	9.4	
	6	6	5.2–15	(3)	10	
	2	2	13–15	(4)	14	
	6	14	(2.0–15)	All	(9.4)	(22)
	6	12	5.2–15	All[a]	10	23
H	3	3	4.8–6.3	(2)	5.7	
	6	7	7.0–9.5	(3)	7.9	
	11	11	1.0–9.7	(5)	5.5	
	2	2	8.6–11	(6)	9.8	
	3	3	3–13[c]	(7)	9.0	
	2	2	11	(8)	11	
	1	2	3.2–4.5	(9)	3.8	
	24	30	1.0–13	All	7.0	15
L	4	4	6.4–8.2	(2)	7.1	
	6	6	5.9–12	(3)	8.4	
	12	14	1.4–13	(5)	8.7	
	2	2	6.5–13	(6)	9.8	
	7	7	7.7–16	(8)	11	
	1	1	4.8	(9)	4.8	
	2	2	7.8–8.5	(10)	8.2	
	5	7	7.1–12	(11)	9.0	
	26	43	1.4–16	All	8.9	17
LL	1	1	9.0	(2)	9.0	
	4	4	8.4–14	(3)	11	

TABLE 1 (cont.)

Type	Number analyzed	No. of analyses	Se contents Range (10^{-6} g/g)	[ref.]	Mean (10^{-6} g/g)	Mean (atoms/10^6 Si atoms)
			chondrites			
LL	1	1	11	(5)	11	
	1	1	10	(8)	10	
	1	1	5.6	(11)	5.6	
	7	8	5.6–14	All	10	19
E4	1	1	4[b]	(1)	*4*	
	2	2	11–34	(2)	22	
	1	1	15	(5)	15	
	2	4	(4–34)	All	(16)	(33)
	2	3	11–34	All[a]	20	42
E6	4	4	10–24	(2)	16	29
			achondrites			
Eu	1	1	0.0016[d]	(6)	0.002	
	1	1	0.5	(11)	0.5	
	2	2	0.0016–0.5	All	0.25	0.4
			stony-irons			
M	2	2	3–12	(7)	7.5	
P	1	1	5.8	(6)	5.8	
	1	1	2	(7)	2	
	1	2	2–5.8	All	4	
			irons			
H	1	1	0.01[e]	(8)	0.01	
O	4	4	1–15[b, f]	(7)	*7.3*	
	4	4	0.04–1.6[f]	(8)	0.59	
	1	1	0.07	(12)	0.07	
	9	9	(0.04–15)	All	(3.5)	
	5	5	0.04–1.6	All[a]	0.49	
Dr	1	1	0.13	(11)	0.13	

be ascribed to sample inhomogeneity, most can be accounted for by noting that the C2, C3, and E4 (but not C1) results of Akaiwa (1966) are a factor of 2–3 lower than those of other investigators of the same meteorites, and lie outside of the remaining range for the respective group. Turner's (1965) results are accompanied by large experimental uncertainties and might be systematically low, although DuFresne (1960) or Greenland (1967) report comparable results for the two meteorites Turner studied. Byers' (1938) results for iron meteorites are 1–2 orders of magnitude higher than those of Kiesl and co-workers and are similar to the selenium contents of ordinary chondrites. Apart from Byers' (1938) results, there exists no evidence of selenium possessing appreciable siderophile geochemical character, and his results therefore are suspect.

Within each group of ordinary chondrites, the selenium contents vary by about a factor of 10 (Table 1, Figure 1). This variation appears to reflect real compositional differences between the chondrites of each group, since duplicate analyses of 13 ordinary chondrites by different investigators seldom differ by as much as a factor of 2. DuFresne (1960) noted a seeming difference between the selenium contents of 4 chondritic finds and 20 chondritic falls—the ranges being 1.0–2.9 and 5.3–13 $\times 10^{-6}$ g/g, respectively. Results for 11

(1) Akaiwa (1966)
(2) Greenland (1967)
(3) Case et al. (1971)
(4) Kiesl and Hecht (1969)
(5) DuFresne (1960)
(6) Schindewolf (1960)
(7) Byers (1938)
(8) Seitner et al. (1968)
(9) Turner (1965)
(10) Kiesl (1967)
(11) Kiesl et al. (1967)
(12) Dörfler and Kiesl (1970)

[a] Doubtful data omitted.
[b] The quality of these data is uncertain (see text).
[c] These data do not include negative results for two other similar chondrites.
[d] An upper limit of 0.007 $\times 10^{-6}$ g/g for Johnstown, a diogenite, is not included.
[e] The datum is for Braunau. Upper limits of 0.01 $\times 10^{-6}$ g/g for two samples of Coahuila and Mt. Joy (Seitner et al., 1968) and a negative result for Negrillos (Byers, 1938) are omitted.
[f] Negative results for four other octahedrites (Byers, 1938) and upper limits of 0.01 $\times 10^{-6}$ g/g for five more octahedrites (Kiesl et al., 1967; Seitner et al., 1968) have been omitted.

other chondritic finds generally range from 7.1–11 × 10^{-6} g/g (Greenland, 1967; Kiesl et al., 1967; Seitner et al., 1967; Case et al., 1971) with but one exception of 3 × 10^{-6} g/g (Byers, 1938). It would seem therefore that there is no systematic difference between the selenium contents of chondritic finds and falls. Although it may appear that the mean contents for the ordinary chondrite groups increase in the order H–L–LL (Table 1) the ranges for each group are so similar (Figure 1, Table 1) that this trend cannot be statistically significant. The selenium concentrations within each chondritic group are independent of chemical-petrologic type and of the degree of equilibration of the unequilibrated ordinary chondrites (Case et al., 1971). There appears to be no significant difference between the selenium contents of the co-existing light and dark portions of Pantar (Turner, 1965) or Fayetteville (Case et al., 1971).

If the results of Akaiwa (1966) are ignored, the selenium concentration ranges within each type of carbonaceous chondrite vary by factors of 2–3 (Figure 1, Table 1), which may reflect real sample variations or systematic differences in analytical technique. The mean abundances for C1 (Table 1) and all ordinary chondrites, 16 atoms/10^6 Si atoms, show selenium to be a normally-depleted element (Larimer and Anders, 1967), and the mean abundance in C1 is about twice that in C2 and three times that in C3. The C1 data suggest a mean cosmic abundance value of 69 atoms/10^6 Si atoms, which is similar to abundances estimated previously. Neither Aller (1961) nor Müller (1968) list solar photospheric values for selenium.

The selenium concentration ranges of E4 and E6 chondrites are similar to those of the carbonaceous chondrites. The concentration range for E6 is included within that for E4 although it appears that the mean content in E4 (ignoring the datum of Akaiwa, 1966) is somewhat higher than that in E6. In terms of Larimer and Anders' (1967) notation the depletion factors of Types I and II are, respectively, 0.60 and 0.42.

Based solely on their own results it has been suggested that, in chondrites, selenium is fractionated from its congeners, S and Te, (DuFresne, 1960) and from Te but not from S (Greenland, 1967). In explanation of the latter Greenland (1967) cites Sindeeva (1964) as having shown empirically that at chondritic Te levels even limited anionic isomorphism between Te and selenium does not occur. Consideration of all of the selenium, Te (Pelly and and Lipschutz, this volume), and S (Moore, this volume) data suggests that these chemically similar elements may not be fractionated from each other. If all doubtful selenium and Te data are eliminated, the Se/S, Te/S, and Se/Te ratios are nearly constant in all chondritic groups but E6, where the ratios involving Te are markedly different (Table 2). The Se/Te and Te/S ratios also suggest that the Te data for type C3 may be open to question, a conclusion reached on other grounds by Pelly and Lipschutz (ibid.). Clearly, additional reliable Te data are needed for groups C3 and particularly, E6 before the existence of the hypothetical selenium–tellurium fractionation in chondrites is accepted.

TABLE 2 Selenium, sulfur and tellurium atomic ratios in chondrites

Type	C1	C2	C3	H	L	LL	E4	E6
Se/S × 10^5	14	17	19	15	17	17	14	19
Se/Te	12	13	21	11	14	12	12	56
Te/S × 10^5	1.2	1.3	0.92	1.4	1.2	1.4	1.2	0.35

As expected from their generally low sulfide contents, achondrites contain much smaller amounts of selenium than do chondrites. The two achondritic data differ by more than two orders of magnitude (Table 1), the lower value being derived from a find, Nuevo Laredo. If only the higher datum is considered, the Se/S ratio, 16×10^{-5}, is similar to typical chondritic ratios. The only selenium data for the mesosiderites was obtained by Byers (1938). These values are similar to chondritic ones, and their mean yields a Se/S ratio of 28×10^{-5} which, in view of the limited data, is not inconsistent with the chondritic ratios. What little data are available (Table 1) suggest that selenium is about equally abundant in pallasites and mesosiderites, and is probably less abundant in iron meteorites. Even considering only the data of Kiesl and co-workers (Table 1), the selenium contents of the iron meteorites vary greatly, as would be expected from the known inhomogeneous distribution of troilite, the probable host mineral, in irons.

Acknowledgement

This research was supported in part by the U.S.National Science Foundation, grant GA–1474.

References

Akaiwa, H. (1966) *J. Geophys. Res.* **71**, 1919.
Aller, L.H. (1961) *The Abundance of the Elements*. Interscience.
Byers, H.G. (1938) *Ind. Eng. Chem. News Ed.* **16**, 459.

Case, D.R., Laul, J.C., Pelly, I.Z., Wechter, M.A., Schmidt-Bleek, F. and Lipschutz, M.E. (1971) in preparation.

Dörfler, G. and Kiesl, W. (1969) *The Kayakent meteorite*, preprint.

DuFresne, A. (1960) *Geochim. Cosmochim. Acta* **20**, 141.

Greenland, L. (1967) *Geochim. Cosmochim. Acta* **31**, 849.

Kiesl, W. (1967) *Z. Anal. Chem.* **227**, 13.

Kiesl, W. and Hecht, F. (1969) *Meteorite Research* (P.M. Millman, ed.), 67–74. Reidel.

Kiesl, W., Seitner, H., Kluger, F. and Hecht, F. (1967) *Monatsh. Chem.* **98**, 972.

Larimer, J.W. and Anders, E. (1967) *Geochim. Cosmochim. Acta* **31**, 1239.

Müller, E. (1968) *Origin and Distribution of the Elements* (L.H. Ahrens, ed.). 155–176 Pergamon.

Schindewolf, U. (1960) *Geochim. Cosmochim. Acta* **19**, 134.

Seitner, H., Kiesl, W., Kluger, F. and Hecht, F. (1968) *Wet chemical analysis and determination of trace elements by neutron activation in meteorites*, preprint.

Sindeeva, N.D. (1964) *Mineralogy and Types of Deposits of Selenium and Tellurium.* Wiley.

Turner, G. (1965) *J. Geophys. Res.* **70**, 5433.

(Received 4 June 1970)

BROMINE (35)*

George W. Reed Jr.

Argonne National Laboratory
Argonne, Illinois 60439

BROMINE IS a trace element in meteorites; its concentrations range from about 20 ppb to a few ppm. The amount appears to be highly variable within a given class of meteorites and to some extent in samples (1 gm scale) from the same meteorite. In spite of the spread in the data, the various classes of meteorites have distinctly different average Br concentrations. Bromine is fractionated between the ordinary ([Br] $\sim$ few tenths ppm) and the carbonaceous and enstatite chondrites ([Br] $\sim$ few ppm). The achondritic data are more sparse but falls in the range of the ordinary chondrites with no significant differences between the classes. The Br in iron meteorites tends to be lower ([Br] $<$ 0.1 ppm) than that in stones.

Early wet-chemical Br analyses by von Fellenberg and Lunde (1926), Selivanov (1940), and Behne (1953) are probably less reliable than those based on the more modern activation analysis technique. All recent analyses reported were based on the measurement of radioactive Br^{80m} and/or Br^{82} isolated from samples irradiated with neutrons. The radiochemical procedures used in all the investigations were similar. Counting techniques varied from reliance on purely β decay, or γ-ray spectrometry to γ–γ and β–γ coincidence spectrometry.

Wyttenbach et al. (1965) measured Br radioactivities derived from both stable isotopes, Br^{79} and Br^{81}, and were able to establish that meteoritic Br has the same isotopic composition as terrestrial Br within the 2% precision of the measurements.

Comparison of the results of replicate determinations on a single chondritic meteorite is helpful in placing interlaboratory measurements in perspective. Wyttenbach et al., (1965) report values of 35 and 69 ppb for Mocs

* Work supported by the U.S. Atomic Energy Commission.

TABLE 1 Bromine in chondrites

Meteorite classification	No. of determinations	No. of meteorites	Range (ppm)	Mean (ppm)	At/10^6 Si	Ref. §
Hypersthene (L)	7	5	0.035–1.56	0.51		A
	10	10	0.12–0.39*	0.23		B
	9	3	0.52–0.0225*†	0.15	0.28	C
	8	5	0.026–0.32	0.17	0.33	D
	9	4	0.019–0.18*	0.09	0.14	E
Hypersthene (Holbrook)*	1			1.00		B
	3		1.0–2.1	1.50	2.8	C
	2		0.60–0.80	0.72	1.3	E
Bronzite (H)	1	1		0.17		A
	9	8	0.8–0.78	0.26		B
	5	3	0.08–0.20‡	0.15	0.33	C
	5	4	0.12–1.46†	0.59	1.2	D
	3	2	0.012–0.083†	0.047	0.10	E
Amphoterite (LL3)	1	1		0.87	1.6	E
(LL6)	1	1		0.18		A
Enstatite (E4)	2	2	3.5, 6.5	5.0	10.4	C
	2	1	1.59, 1.90	1.75		D
	2	2	3.34, 4.8	4.07	8.3	E
(E5, 6)	1	1		1.04		A
	1	1		1.5	2.7	C
	1	1		1.54		D
	1	1		0.95	1.7	E
Carbonaceous I	2	2	2.0, 11.0	6.5	21.3	C
	1	1		5.72	18.6	D
	4	2	3.34–5.1	4.34	15.5	E
Carbonaceous II	1	1		3.5	8.2	C
	4	3	0.81–4.77	2.54	6.1	D
	4	2	0.25+–3.9	1.87	8.8+	E
Carbonaceous III	2	2	0.5, 2.6	1.5	3.4	C
	4	2	1.73–3.25	2.45	5.5	D
	3	2	1.2–1.7	1.5	3.2	E

(the latter number for a less veined sample). Two samples of Bruderheim give 0.97 and 1.56 ppm. Filby (1965) obtained 0.78 and 0.33 ppm for Pultusk. Reed and Allen (1966) reported values of 0.052 to 0.225 ppm in five measurements on Bruderheim, about a factor two variation for Harleton and Holbrook and less variability for two New Concord and three Allegan samples. Lieberman and Ehmann (1967) found less than 50% variation in 8 of 9 ordinary chondrites but the carbonaceous chondrite Cold Bokkeveld yielded 0.81 and 1.87 ppm. Goles et al. (1967) also obtained duplication within 50% for 5 of 6 meteorites and a range of 0.109 to 0.18 ppm from six measurements on Bruderheim.

Another possible source of variation is the water solubility of the Br in meteorites. Wyttenbach et al. (1965) observed 31% and 88% H_2O extractable Br in two meteorites. Reed and Allen (1966) reported that over one half the Br in 11 of 16 chondrites was leachable. In most of the other samples about 30% of the Br was leached. Thus a sample cut with a water-cooled saw or rinsed by rain or ground water can give a spurious result for total Br. Wyttenbach et al. (1965) noted that in their two samples the extractable Br and Cl were correlated. Reed and Allen (1966) observed a soluble-Br/insoluble-Br ratio of ~4 in ordinary and C–1 carbonaceous chondrites. The water-extractable Br/Cl ratio was of the order 0.1 and 0.01 for ordinary and for carbonaceous and enstatite chondrites, respectively.

The Br contents of stone meteorites are summarized in Tables 1 and 2. Those of iron meteorites are given in Table 3. The hypersthene chondrite data of Wyttenbach et al. are spread evenly over the range 0.035–1.56 ppm; these data are averaged. Holbrook data by Filby and Ball (1.00 ppm), Goles et al. (0.72 ppm) and Reed and Allen (~1.5 ppm) are listed separately. The Br data for the "find", Plainview, 0.75 (Reed and Allen, 1966), 22

* There appears to be a consensus that Holbrook Br concentrations are significantly different from most other hypersthene chondrites. Mocs may have less Br than the average on the basis of the agreement between results by Wyttenbach *et al.*, and Goles *et al.*

† Data on the "find" Plainview are not included in the table.

‡ Pantar – dark is not included

§ References: A = Wyttenbach *et al.*
B = Filby and Ball.
C = Reed and Allen.
D = Liebermann and Ehmann.
E = Goles *et al.*

\+ The low type II carbonaceous chondrite values were obtained for Murray. Some samples of this meteorite were collected later and may have had Br leached out. The At Br/10^6 Si does not include Murray data.

TABLE 2 Bromine in achondrites

Meteorite classification	No. of determin- ations	No. of meteorites	Range (ppm)	Mean (ppm)	Ref.*
Eucrite	2	2	0.08, 0.21	0.15	B
	2	1	0.38, 0.39	0.39	D
	2	1	0.53, 56	0.54	F
Howardite	1	1		0.21	B
	2	1	0.18, 0.60	0.39	D
	1	1		0.36	F
Hypersthene (diogenite)	2	2	0.07, 0.19	0.13	B
	2	1	0.059, 0.16	0.11	D
Enstatite (aubrite)	2	2	0.16, 0.23	0.20	B
	1	1		0.067	A
	2	1	0.074, 0.084	0.077	D

* References: A → E same as in Table 1.
F = Reed and Jovanovic

TABLE 3 Bromine in iron meteorites

Meteorite	and classification	Range (ppm)	Mean (ppm)	Ref.*
Odessa	Og	0.12, 0.18	0.15	E
Sardis	Ogg	0.058, 0.019	0.039	E
Cape York	Om		0.04	B
Cape York	Om		0.04	B
Gibeon	Of		0.07	B
Morradal	D		0.07	B
Corrizatillo	?		0.12	B
Mt Stirling (Chlorapatite)	Og		40.0	C

* See footnote to Table 1.

(Goles et al. 1967) and 0.22 (Lieberman and Ehmann, 1967) are not tabulated.

The sampling for the iron meteorites (Table 3) and the achondrites (Table 2) is not extensive; however, there is a reasonable agreement between laboratories for the Br content of achondrites.

The dark parts of ordinary chondrites having dark-light structure and chondrites of "low metamorphic grade" have been found to be enriched in rare gases relative to the light fraction or "higher grade chondrites". The dark phase of Pantar appears to have more Br than the light phase, 0.74 ppm dark *vs* 0.20 ppm light. (Reed and Allen, 1966) or 0.083 ppm light (Goles et al. 1967). The high Br in Tysnes (Filby and Ball, 1965) may be related to its classification as an H4 meteorite (Van Schmus and Wood, 1967); similarly this may account for Goles et al. (1967) value of 0.87 ppm for Br in Chainpur (LL3) *vs* a value of 0.18 ppm for the LL6 chondrite Benton (Wyttenbach et al. 1965). A similar trend is noted between Abee and Indarch (E4) and Hvittis (E6), 4.1 *vs* 0.95 ppm (Goles et al. 1967) or 4.5 *vs* 1.5 (Reed and Allen, 1966). The E6 meteorite, Pillistfer, measured by Lieberman and Ehmann (1967) is consistent with the above although their value for Abee is lower than others reported. The high Br contents of Holbrook, Forest City and Dhurmsala are not consistent with their L6 classification.

Br is considered to be a dispersed element. The large variations in Br contents in meteorites suggests that it is probably in minor, inhomogeneously distributed phases (Dodd, 1969). In the case of the ordinary chondrites the possibility exists that Br is associated with apatite. Reed and Allen (1966) found 40 ppm Br in chlorapatite from the Mt. Stirling iron meteorite. Assuming a similar content in the apatite in ordinary chondrites which contain a few tenths per cent of this mineral one estimates that a Br content of about 0.1 ppm can be accounted for; this is in agreement with concentrations given in Table 1.

References

Behne, W. (1953) "Untersuchungen zur Geochemie des Chlor und Brom", *Geochim. Cosmochim. Acta* **3**, 186–214.

Dodd, R.T. (1969) "Metamorphism of the ordinary chondrites: A review", *Geochim. Cosmochim. Acta* **33**, 161–203.

Filby, R.H. (1964), "The determination of bromine in rocks by neutron activation analysis", *Analytica Chim. Acta* **31**, 434.

Filby, R.H. and T.K.Ball (1965) "Zinc and bromine in some meteorites by neutron activation analysis", *International Conf. on Modern Trends in Activation Anal.*, College Station, Texas.

Goles, G.C., Greenland, L.P. and Jérome, D.Y. (1967), "Abundances of chlorine, bromine, and iodine in meteorites", *Geochim. Cosmochim. Acta* **31**, 1771.

Lieberman, K.W. and Ehmann, W.D. (1967), "Determination of bromine in stony meteorites by neutron activation", *J. Geophys. Res.* **72**, 6279.

Reed, G.W. and Allen, R.O. (1966) "Halogens in chondrites", *Geochim. Cosmochim. Acta* **30**, 779–800.

Reed, G.W. and Jovanovic, S. (1969) "Some halogen measurements on achondrites", *Earth Planet. Sci. Lett.* **6**, 316–320.

Selivanov, S. (1940) (1957) "Cl and Br in massive crystalline rocks", *DAN USSR*, **28**, 809, Quoted by A.P. Vinogradov, AERE Translation 797.

Van Schmus, W.R. and Wood, J.A. (1967) "A chemical-petrological classification for the chondritic meteorites", *Geochim. Cosmochim. Acta* **31**, 747–765.

Von Fellenberg, T. and Lunde, G. (1926) "Studies on the occurrence of iodine in nature, 10, a contribution to the geochemistry of iodine", *Biochem. Z.* **175**, 162–171.

Wyttenbach, A., von Gunten, H.R., and W. Scherle (1965) "Determination of bromine content and isotopic composition of bromine in stony meteorites by neutron activation", *Geochim. Cosmochim. Acta* **28**, 467.

(Received 5 May 1970)

RUBIDIUM (37)

Gordon G. Goles

Center for Volcanology and Departments of Chemistry and Geology
University of Oregon
Eugene, Oregon 97403

NUMEROUS DETERMINATIONS of meteoritic contents of this trace alkali metal have been made, mostly in conjunction with Rb-Sr geochronology. The majority of these analyses were done by isotope dilution techniques and seem to be of high quality, although some of the earlier results obtained by isotope dilution or distillation and flame photometric techniques are clearly too high. I have excluded from consideration here all results for Rb published before 1957. In addition, I have excluded the results of Von Michaelis, Ahrens and Willis (1969), which were determined by an x-ray fluorescence technique which appears to be near or at the limit of its sensitivity as applied to meteorites. Some of these latter values may well be useful, but they add little to the information available from other authors.

Unlike the case of K, there is no simple way to assess the effect of sampling errors on Rb determinations. Agreement (or lack of it) among Rb contents determined on different samples from a given meteorite is a helpful indication of the magnitude of sampling errors, but this approach is complicated by possible interlaboratory systematic errors. Data for one C chondrite *(Orgueil)*, three H's and two L's, one LL *(St. Séverin)* and the E chondrite *Abee*, all of which have had at least four independent Rb determinations made on them, are listed in Table 1. (Chondrites are classified according to Van Schmus and Wood (1967).) The H5 chondrite *Beardsley* is a special case, with striking differences between Rb contents of fresh and weathered specimens; see discussion in chapter on K. Gast (1962) found that a large fraction of the Rb in a fresh specimen of *Beardsley* was readily leachable. Note the anomalously low K/Rb ratios determined for this meteorite.

There is appreciable variation in the dispersion among independent deter-

TABLE 1 Rubidium in carbonaceous, some H and L-group chondrites, LL and enstatite chondrites

Meteorite (Class)		Rb content (10^{-6} g/g)	Reference	K/Rb (by mass)
Ivuna (C1)		2.27 ± 0.04	(1)	
		2.58	(2)	366
Orgueil (C1)	"1"	2.38	(2)	237
	"2"	2.51		226
	"3"	2.05		286
	"IA, B, C"	1.76 ± 0.02	(3)	
	"II"	2.08		
Cold Bokkeveld (C2)	"A"	1.62	(2)	249
	"B"	1.58		255
Erakot (C2)		2.02	(2)	179
Essebi (C2)		1.39	(2)	260
Haripura (C2)	"A"	1.20	(2)	229
	"B"	1.23		233
Mighei (C2)		1.66 ± 0.02	(1)	
	"A"	1.75	(2)	242
	"B"	1.79		241
Murray (C2)		1.57	(3)	
Santa Cruz (C2)		1.69	(2)	231
Felix (C3)		1.36	(1)	
Grosnaja (C3)		1.68	(2)	267
Lancé (C3)		1.42	(3)	
Mokoia (C3)		1.15	(2)	257
		1.21 ± 0.01	(3)	
Pueblito de Allende (C3)		0.99	(2)	232
Beardsley (H5)		4.90 ± 0.06 (weathered)	(4)	186
	"I"	4.76 ± 0.08 (weathered)	(5)	
	"II"	14.26 (fresh)		
		10.7 ± 0.3 (weathered)	(6)	
	"A"	13.94 (fresh)	(7)	88
	"B"	13.70 (fresh)		87
Forest City (H5)		2.75 ± 0.04	(4)	298
		2.71	(5)	
		3.3 ± 0.2	(6)	
		3.04	(8)	
		2.91 ± 0.11	(9)	
		2.9	(10)	

TABLE 1 (cont.)

Meteorite (Class)		Rb content (10^{-6} g/g)	Reference	K/Rb (by mass)
Forest City (H5)	"I"	2.87	(11)	286
	"II"	2.68		286
	"III"	3.16		286
		2.81	(13)	280
		2.79		282
Ochansk (H4)	"A"	1.64	(1)	
	"B"	1.62		
		2.34	(7)	313
		2.1	(10)	
Bruderheim (L6)	"I"	2.71 ± 0.02	(12)	
	"II"	2.65		
	"III"	2.83		
		2.62	(13)	337
		2.81 ± 0.15	(14), (15)	~320
	"I"	2.73	(16)	
	"II"	2.72		
	"III"	2.79		
	"I"	2.75	(17)	
	"II"	2.75		
Homestead (L5)		3.2 ± 0.2	(6)	
		3.15	(8)	
	"I"	2.88	(16)	
	"II"	3.00		
Chainpur (LL3)		3.053	(18)	285
Cherokee Springs (LL6)		1.938	(18)	450
Chico (LL) (Find)		0.994	(18)	615
Dhurmsala (LL6)		2.36	(1)	
		2.151	(18)	450
Ensisheim (LL6)		0.405	(18)	595
Krähenberg (LL5)	"Light"	1.94	(17)	428
	"Dark"	50.8		236
Lake Labyrinth (LL6) (Find)		1.321	(18)	575
Mangwendi (LL6)		2.36 ± 0.03	(1)	
Ngawi (LL3)		2.183	(18)	330
Oberlin (LL5) (Find)		2.208	(18)	435
Olivenza (LL5)		2.19	(1)	
		3.081	(18)	295
		2.6 ± 0.2	(19)	
Ottawa (LL6)		0.561	(18)	740
Parnallee (LL3)		3.010	(18)	290
St. Mesmin (LL6)		3.163	(18)	300

TABLE 1 (cont.)

Meteorite (Class)		Rb content (10^{-6} g/g)	Reference	K/Rb (by mass)
St. Severin (LL)	"A II"	0.568	(18)	1585
	"B I"	0.774		1340
	"C I"	0.607		1510
	"D I"	0.600		1520
Soko Banja (LL4)	"I"	0.580	(18)	500
	"II"	4.880		350
	"III"	0.515		620
Abee (E4)		4.81	(14)	
		5.05		
		4.7		
	"I"	3.461	(20)	251
	"II"	3.351		251
Adhi Kot (E3)	"I"	2.993	(20)	270
	"II"	3.913		259
Indarch (E4)	"I"	3.813	(20)	244
	"III"	3.749		244
St. Mark's (E5)	"I"	0.863	(20)	863
	"II"	0.824		886
Daniel's Kuil (E6)		1.370	(20)	490
Hvittis (E6)		3.234	(20)	246
Jajh deh Kot Lalu (E6)		3.497	(20)	250
Khairpur (E6)		2.148	(20)	330

(1) Smales *et al.* (1964)
(2) Kaushal and Wetherill (1970)
(3) Murthy and Compston (1965)
(4) Gast (1960)
(5) Gast (1962)
(6) Pinson *et al.* (1965)
(7) Kaushal and Wetherill (1969)
(8) Cabell and Smales (1957)
(9) Webster, Morgan and Smales (1957)
(10) Webster, Morgan and Smales (1958)
(11) Bogard *et al.* (1967)
(12) Burnett and Wasserburg (1967)
(13) Tera *et al.* (1970)
(14) Shima and Honda (1967a)
(15) Shima and Honda (1967b)
(16) Gopalan and Wetherill (1968)
(17) Kempe and Müller (1969)
(18) Gopalan and Wetherill (1969)
(19) Papanastassiou, Wasserburg and Burnett (1970)
(20) Gopalan and Wetherill (1970)

minations. The difference between extreme values for *Bruderheim* (eleven determinations) is only about 8% of the mean value, and that for *Homestead* (four determinations) is only about 10% of the mean. On the other hand, the difference between extreme values reported for *Ochansk* (four analyses) is about 37% of the mean value and that for *Abee* (5 analyses) is about 40% of the mean. The data suggest, however, that sampling errors greater than $\pm 20\%$ of the mean values are unlikely for samples of about a gram in mass, provided that one is not dealing with a meteorite like *Beardsley*.

An average content of about 2.2×10^{-6} g Rb/g seems to be appropriate for C1 chondrites, equivalent to an atomic abundance of about 7.0 atoms Rb/10^6 atoms Si. Corresponding values for the C2 class are 1.6×10^{-6} g Rb/g and 4.0 atoms Rb/10^6 atoms Si, and for the C3 class they are 1.3×10^{-6} g Rb/g and 2.8 atoms Rb/10^6 atoms Si. The significance of these estimates is of course clouded by limited data and variablity within classes. About all that is certain is that Rb tends to decrease, both absolutely and relative to Si, in the sequence C1, 2, 3.

Data of apparently acceptable quality have been reported for 24 H chondrites other than those in Table 1. (These data, and those discussed below for L chondrites, may be found in the references for Table 1 and in Shields, Pinson and Hurley (1966) and Wasserburg, Papanastassiou and Sanz (1969).) If one excludes from consideration the four finds and *Beardsley*, 45 independent determinations of Rb in samples from these 22 H chondrites remain. Treating all 45 determinations as of equal weight, the average Rb content is $(2.5 \pm 0.6) \times 10^{-6}$ g/g. The population approximates a Gaussian distribution, so that it seems valid to estimate the standard deviation in the conventional way. This average content is equivalent to about 4.8 atoms Rb/10^6 atoms Si.

Acceptable data have been reported for 21 L chondrites other than the two in Table 1, of which five are finds. There are 42 independent determinations of Rb in these 18 observed falls, but two of them, both for *Bath Furnace*, are extraordinarily low (0.34 and 0.35×10^{-6} g Rb/g; Gopalan and Wetherill, 1968). Points representing data for *Bath Furnace* lie on a reasonable Rb-Sr isochron in company with other L chondrites studied by Gopalan and Wetherill, so that its anomalously low Rb contents cannot reflect terrestrial (or any other recent) alteration. Nevertheless, it seems preferable to exclude these values from the average for L chondrites, which then becomes $(2.7 \pm 0.5) \times 10^{-6}$ g Rb/g. This population of 40 determinations closely approximates a Gaussian distribution. The average is equivalent to about 4.8 atoms Rb/10^6 atoms Si, just as for the H group.

All data on Rb contents of LL and E chondrites which were deemed acceptable, except for those on mineral separates, are given in Table 1. Both classes (especially the former) have highly variable Rb contents. Omitting data on *Krähenberg* but including those on the three finds, the LL's average about 1.8×10^{-6} g Rb/g; this value clearly has very little significance. The enstatite chondrites average about 3.2×10^{-6} g Rb/g, with *St. Mark's* and *Daniel's Kuil* being conspicuously poor in Rb. There is no apparent correlation between their mineralogical and chemical characteristics as described by Keil (1968) and their Rb contents. It would be interesting to analyse *Saint Sauveur*, the remaining example of Keil's intermediate type, to see if it is Rb-poor like its cousin *St. Mark's.*

Fig. 1 depicts relationships between Rb contents and K/Rb rations for chondrites. There is a very strong and consistent anti-correlation between these parameters for the H, L, LL and E classes. Similar anti-correlations are evident for the three points representing *Orgueil* and for the C2 chondrites exclusive of *Haripura.* Contents of Rb and K are neither independent of one another, nor are they observed in nearly constant ratio. Wasserburg, Sanz and Bence (1968) cautioned that "K/Rb ratios are controlled by mineral partitioning, and ... care must be taken in use of this value as an index of differentiation when sampling is not adequate." I would now go further, and suggest that the concept of a chondritic or "primitive" K/Rb ratio is meaningless until we have a better understanding of the processes which gave rise to the trends depicted in Fig. 1. Note especially that within a given meteorite there is a tendency for K/Rb ratios for individual samples to be relatively constant. Examples in Table 1 are *Forest City*, where the relative range in Rb contents of samples in which K was also measured is about twice that of associated K/Rb ratios, *Krähenberg*, with a factor of about 26 difference in Rb contents but less than a factor of 2 difference in K/Rb between light and dark clasts, and *Soko Banja*, where a variation of as much as a factor of 9.5 in Rb is damped to less than a factor 2 in K/Rb. These observations imply that the effects of sampling errors (i.e., of the *present* distribution of K and Rb among minerals of these meteorites) are minor in establishing the trends of Fig. 1. Sampling errors of course cannot be neglected, but it seems likely that the major features of K-Rb geochemistry in chondrites were determined *before* these objects attained their present mineralogy in which K and Rb are nearly coherent.

Table 2 embodies data on Rb contents of achondrites (identified by the code introduced by Keil, 1969), and of silicates of the *Estherville* mesosiderite. The value reported by Smales et al. (1964; reference 1) for Rb in the latter

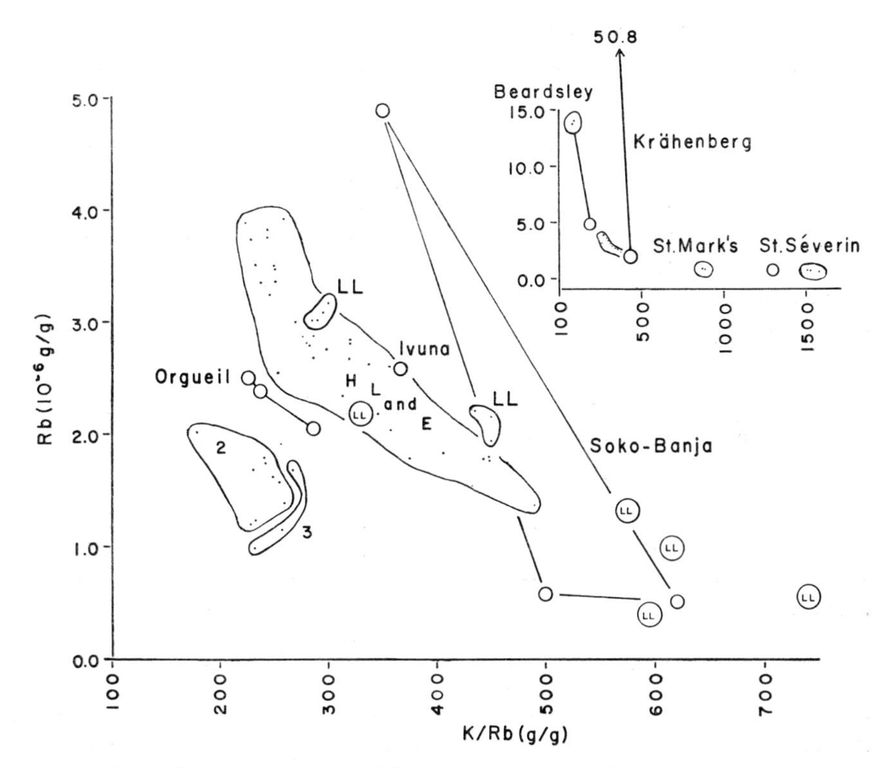

FIGURE 1 Rb contents vs. K/Rb ratios. An anti-correlation between these parameters is seen for H, L and E chondrites, plotted as points within a field across the center of the diagram. Most of the points representing C2 chondrites define a similar anti-correlation displaced toward the origin (see field marked "2"), as do the three separate analyses of the C1 stone *Orgueil*. LL chondrites scatter widely on this diagram, so much so that *St. Séverin* is represented on the smaller scale insert in the upper right (along with *Beardsley*, *St. Mark's*, and the point for the light portion of *Krähenberg*). C3 chondrites plot in the field marked "3".

is much too high, judging from close similarities noted by many previous authors between silicates in mesosiderites and howardites. No Rb determinations on howardites are available, but they should lie in the range of those for the eucrites or somewhat below, according to the mixing model of Jérome (1970). The value reported by Pinson et al. (1965; reference *6*) for silicates from *Estherville* seems to be quite reasonable. Other apparent disagreements in data for a given meteorite in Table 2 are likely to reflect real heterogeneity of the specimens.

TABLE 2 Rubidium in achondrites

Meteorite (Class)		Rb content (10^{-6} g/g)	Reference	K/Rb (by mass)
Bishopville (Ae)	"1–A"	1.77 ± 0.08	(21)	
	"1–B"	1.65		
	"2–A"	2.00 ± 0.04		565
	"2–B"	1.88		600
Khor Temiki (Ae)		1.74	(1)	
Norton Co. (Ae)	"523.3x"		(11)	
	Breccia "I"	0.220		
	"II"	0.221		305
	"IV"	0.157		340
	Enstatite separate "I"	0.160		
	"II"	0.302		
	"III"	0.167		
	"4965E" Breccia	0.354		300
	Enstatite separate	0.306		140
Johnstown (Ab)		0.139	(1)	
		0.04	(8)	
	(isotope dilution)	0.105	(9)	
	(activation analysis)	0.04		
Angra dos Reis (Ae)		0.0311 ± 0.0022	(13)	415
		0.038	(19)	
Nakhla (Ado)		2.8 ± 0.2	(6)	
Bereba (Ap)		0.178	(13)	1450
Jonzac (Ap)		0.405	(13)	807
		0.444	(22)	
Juvinas (Ap)		0.167	(13)	1930
		0.173	(22)	
Moore Co. (Ap)		0.16 ± 0.03	(4)	1200
		0.16	(5)	
		0.0497 ± 0.0014	(13)	3260
		0.051	(19)	
		0.13	(21)	4050 (?)
		0.0524	(22)	
Nuevo Laredo (Ap)		0.37 ± 0.02	(4)	990
		0.38 ± 0.02	(5)	
		0.37	(12)	
		0.324	(13)	1280
		0.325	(22)	
		0.437	(23)	991

TABLE 2 (cont.)

Meteorite (Class)		Rb content (10^{-6} g/g)	Reference	K/Rb (by mass)
Pasamonte (Ap)	"white"	0.21 ± 0.03	(4)	2050
		0.22	(5)	
	"197g"	0.268 ± 0.001	(13)	1220
	"297y"	0.192 ± 0.003		1430
		0.25	(16)	
		0.207	(22)	
Sioux Co. (Ap)		0.28 ± 0.03	(4)	1250
		0.25 ± 0.07	(5)	
		0.19	(12)	
		0.204 ± 0.03	(13)	1480
		0.195	(22)	
Stannern (Ap)		0.696	(13)	944
		0.707	(22)	
Estherville	(silicates)	1.87	(1)	
		0.17 ± 0.03	(6)	

(21) Compston, Lovering and Vernon (1965)
(22) Papanastassiou and Wasserburg (1969)
(23) Murthy, Schmitt and Rey (1970); others are the same as in Table 1.

An interesting feature of these data is the wide range in Rb contents of aubrites, similar to that observed for K (see chapter on K). Rubidium and potassium are at least crudely coherent in aubrites, and it seems that K/Rb ratios increase with increasing Rb, unlike the case for chondrites. Perhaps these observations may be understood in terms of a model whereby in the course of brecciation, K- and Rb-bearing phases such as the oligoclase from *Bishopville* studied by Compston, Vernon and Lovering (1965) are added to the enstatite which makes up the bulk of these meteorites. Of course, there is a wide variety of plausible models which rely on magmatic differentiation and which could be adduced to explain these observations. Magmatic differentiation is a likely cause for the variations in Rb contents and K/Rb ratios observed among eucrites.

Kempe and Müller (1969) studied Rb contents of separated fractions from the *Krähenberg* LL chondrite. The mineralogical character of the material they worked with is not well-defined, but they report 97.6×10^{-6} g Rb/g in one fraction. Their work shows that K and Rb are crudely coherent in this meteorite. Gopalan and Wetherill (1970) have studied fractions of the *Indarch* enstatite chondrite in which Rb contents range from 0.564×10^{-6} g/g (K/Rb = 160) to 28.52×10^{-6} g/g (K/Rb = 280). Additional careful work should be done on the distribution of alkali metals among meteoritic minerals, if possible with well-defined separated fractions.

Smales et al. (1964) report Rb contents for two olivine separates from pallasites, one of which (0.033×10^{-6} g Rb/g for olivine from *Imilac*) might not be too far above the true content in this mineral. Most likely even in that case, the observed Rb is in part terrestrial contamination and in part resided in inclusions or along grain boundaries in the olivine.

I shall not review here Rb contents in samples from iron meteorites, although a substantial literature exists on this topic. Such observations have great geochronological significance but are of little interest in defining distribution of Rb in meteorites in general.

Acknowledgement

The preparation of this review was supported in part by NASA grant NGL 38-003-010.

References

Bogard, D.D., Burnett, D.S., Eberhardt, P. and Wasserburg, G.J. (1967) "^{87}Rb-^{87}Sr isochron and ^{40}K-^{40}Ar ages of the Norton County achondrite." *Earth Planet. Sci. Letters* **3**, 179–189.

Burnett, D.S. and Wasserburg, G.J. (1967) "^{87}Rb-^{87}Sr ages of silicate inclusions in iron meteorites." *Earth and Planet. Sci. Letters* **2**, 397–408.

Cabell, M.J. and Smales, A.A. (1957) "The determination of rubidium and caesium in rocks, minerals and meteorites by neutron-activation analysis." *Analyst* **82**, 390–406.

Compston, W., Lovering, J.F. and Vernon, M.J. (1965) "The rubidium-strontium age of the Bishopville aubrite and its component enstatite and feldspar." *Geochim. Cosmochim. Acta* **29**, 1085–1099.

Gast, P.W. (1960) "Alkali metals in stone meteorites." *Geochim. Cosmochim. Acta* **19**, 1–4.

Gast, P.W. (1962) "The isotopic composition of strontium and the age of stone meteorites—I." *Geochim. Cosmochim. Acta* **26**, 927–943.

Gopalan, K. and Wetherill, G.W. (1968) "Rubidium-strontium age of hypersthene (L) chondrites." *J. Geophys. Res.* **73**, 7133–7136.

Gopalan, K. and Wetherill, G.W. (1969) "Rubidium-strontium age of amphoterite (LL) chondrites." *J. Geophys. Res.* **74**, 4349–4358.

Gopalan, K. and Wetherill, G. W. (1970) "Rubidium-strontium studies on enstatite chondrites: whole meteorite and mineral isochrons." *J. Geophys. Res.* **75**, 3457–3467.

Jérome, D.Y. (1970) *Composition and origin of some achondritic meteorites.* (Ph. D. dissertation, University of Oregon, Eugene).

Kaushal, S. K. and Wetherill, G.W. (1969) "Rb^{87}-Sr^{87} age of bronzite (H group) chondrites." *J. Geophys. Res.* **74**, 2717–2726.

Kaushal, S.K. and Wetherill, G.W. (1970) "Rubidium 87-strontium 87 age of carbonaceous chondrites." *J. Geophys. Res.* **75**, 463–468.

Keil, K. (1968) "Mineralogical and chemical relationships among enstatite chondrites." *J. Geophys. Res.* **73**, 6945–6976.

Keil, K. (1969) "Meteorite composition". chapter 4 in *Handbook of Geochemistry*, edited by K.H.Wedepohl (Springer-Verlag, Berlin) 78–115.

Kempe, W. and Müller, O. (1969) "The stony meteorite Krähenberg." In *Meteorite Research* (editor P.M.Millman, pub. by D.Reidel, Dordrecht, Holland), 418–428.

Murthy, V.R. and Compston, W. (1965) "Rb-Sr ages of chondrules and carbonaceous chondrites." *J. Geophys. Res.* **70**, 5297–5307.

Murthy, V.R., Schmitt, R.A. and Rey, P. (1970) "Rubidium-strontium age and elemental and isotopic abundances of some trace elements in lunar samples." *Science* **167**, 476–479.

Papanastassiou, D.A. and Wasserburg, G.J. (1969) "Initial strontium isotopic abundances and the resolution of small time differences in the formation of planetary objects." *Earth Planet. Sci. Letters* **5**, 361–376.

Papanastassiou, D.A., Wasserburg, G.J. and Burnett, D.S. (1970) "Rb-Sr ages of lunar rocks from the Sea of Tranquillity." *Earth Planet. Sci. Letters* **8**, 1–19.

Pinson, W.H. Jr., Schnetzler, C.C., Beiser, E., Fairbairn, H.W. and Hurley, P.M. (1965) "Rb-Sr age of stony meteorites." *Geochim. Cosmochim. Acta* **29**, 455–466.

Shields, R.M., Pinson, W.H. Jr. and Hurley, P.M. (1966) "Rubidium-strontium analyses of the Bjurböle chondrite." *J. Geophys. Res.* **71**, 2163–2167.

Shima, M. and Honda, M. (1967a) "Determination of rubiaium-strontium age of chondrites using their separated components." *Earth Planet. Sci. Letters* **2**, 337–343.

Shima, M. and Honda, M. (1967b) "Distributions of alkali, alkaline earth and rare earth elements in component minerals of chondrites." *Geochim. Cosmochim. Acta* **31**, 1995–2006.

Smales, A.A., Hughes, T.C., Mapper, D., McInnes, C.A.J. and Webster, R.K. (1964) "The determination of rubidium and caesium in stony meteorites by neutron activation analysis and by mass spectrometry." *Geochim. Cosmochim. Acta* **28**, 209–233.

Tera, F., Eugster, O., Burnett, D.S. and Wasserburg, G.J. (1970) "Comparative study of Li, Na, K, Rb, Cs, Ca, Sr and Ba abundances in a chondrites and in Apollo 11 lunar samples". *Geochim. Cosmochim. Acta, Suppl. I*, 1637–1657.

Van Schmus, W.R. and Wood, J.A. (1967) "A chemical-petrological classification for the chondritic meteorites." *Geochim. Cosmochim. Acta* **31**, 747–765.

Von Michaelis, H., Ahrens, L.H. and Willis, J.P. (1969) "The composition of stony meteorites. (II) The analytical data and an assessment of their quality." *Earth Planet. Sci. Letters* **5**, 387–394.

Wasserburg, G.J., Papanastassiou, D.A. and Sanz, H.G. (1969) "Initial strontium for a chondrite and the determination of a metamorphism or formation interval." *Earth Planet. Sci. Letters* **7**, 33–43.

Wasserburg, G.J., Sanz, H.G. and Bence, A.E. (1968) "Potassium-feldspar phenocrysts in the surface of Colomera, an iron meteorite." *Science* **161**, 684–687.

Webster, R.K., Morgan, J.W. and Smales, A.A. (1957) "Some recent Harwell work on geochronology." *Trans. Amer. Geophys. Union* **38**, 543–545.

Webster, R.K., Morgan, J.W. and Smales, A.A. (1958) "Caesium in chondrites." *Geochim. Cosmochim. Acta* **15**, 150–152.

(Received 2 September 1970)

STRONTIUM (38)

K. Gopalan and **G. W. Wetherill**

Institute of Geophysics and Planetary Physics
University of California at Los Angeles,
Los Angeles 90024.

STRONTIUM IS a trace element in stony meteorites. Its concentration can be accurately measured by modern analytical techniques in all the groups of stone meteorites with the possible exception of pallasites. Strontium is also present in comparable amounts in the silicate inclusions of iron meteorites.

Rankama and Sahama (1950) quoted a value of 26 ppm by weight for the average Sr concentration of silicate meteorites. More recent spectrographic determinations by Pinson, Ahrens and Franck (1953), Greenland and Lovering (1956), and Erlank and Willis (1964) yield a much lower mean of about 11 ppm Sr. While these determinations have served to fix an approximate value for the abundance of Sr, and as such have their significance for comparison with cosmic abundances, the spectrographic techniques employed are inherently unsuitable, at the levels of Sr present in meteorites, for studying any subtle fractionation of this element among the meteorite classes. For this reason, spectrographic and earlier x-ray fluorescence analyses have not been included in this review, though some of these figures are in good agreement with recent isotope dilution measurements.

A large number of accurate and precise strontium determinations for each group of meteorites have been made in the last few years by mass spectrometric isotope dilution techniques. This is a result of the recent work on the rubidium-strontium chronology of stone meteorites and silicate inclusions of iron meteorites (Gast, 1962; Pinson et al., 1965; Murthy and Compston, 1965; Wasserburg and Burnett, 1965; Bogard et al., 1967; Burnett and Wasserburg, 1967; Gopalan and Wetherill, 1968; Kaushal and Wetherill, 1969; and Papanastassiou and Wasserburg, 1969). In these studies, both the Sr^{87}/Sr^{86} ratio and the nonradiogenic strontium concentration for each

TABLE 1

	No. of determinations	Sr (10^{-6} g/g) Range	Mean	Atoms/ 10^6 Si	References
Chondrites:					
Carbonaceous, Type I	2	7.9–9.3	8.6	27	(1)
Carbonaceous, Type II	8	9.6–22.4	12.1	30	(1)
	4	9–15	11.0	27	(2)
Carbonaceous, Type III	2	14.4–14.6	14.5	30	(1)
	3	12–17	14	29	(2)
Bronzite (H group)	15	9.3–11.1	10.0	19	(3)
	12	8–12	9.1	17	(2)
Hypersthene (L group)	15	10.1–11.9	11.1	19	(4)
	20	9–12	10.5	18	(2)
Amphoterite (LL group)	12	10.5–11.9	11.1	19	(5)
Enstatite, Type I	3	6.5–7.6	7.2	14	(6)
	2	7–8	7.5	14	(2)
Enstatite, Type II	5	7.5–8.5	8.2	14	(7)
	3	7	7	12	(2)
Calcium-poor achondrites:					
Enstatite: Bishopville			12.3	14	(8)
Norton County			1.4	2	(9)
Hypersthene–Johnstown			2.1	3	(10)
			1.7		(2)
Hypersthene–Shalka			0	0	(2)
Calcium-rich achondrites					
Eucrites	8	58–87	76	106	(11)
Nakhlite–Nakhla	1	–	60	83	(12)
Stony-irons					
Mesosiderites–Estherville	1		22		(12)
Pallasites–Salta	1		<0.02		(13)
Silicate Inclusions of iron meteorites					
Colomera	6	38–70	53		(13)
Four Corners	6	14–34	21		(13)
Weekeroo Station	7	37–70	44		(14)
Toluca	2	7.8–42	26		(13)
Odessa	1		14		(13)
Linwood	1		12		(13)
Pine River	2	6.2–6.8	6.5		(13)
Kodaikanal	7	33–78	55		(15)

meteorite or silicate nodule are determined mass spectrometrically, the latter usually to better than $\pm 2\%$. Because of the high precision and accuracy, major emphasis in this review has been placed on the isotope dilution measurements of different workers. The very recent and careful measurements of Von Michaelis et al. (1969) by x-ray fluorescence methods have given consistent and reliable data on the different classes of meteorites. However, it was decided not to lump the results of the two methods. Table 1 therefore lists the results of Von Michaelis et al. separately, and Fig. 1 represents only the data of mass spectrometric measurements.

The isotope dilution analyses have usually been carried out on splits of 1–2 gram powdered samples, and are admittedly not representative of the meteorite as a whole. But the mean of so many measurements, often in replicate, is expected to minimize this sampling problem. The following procedure has been adopted in presenting the data of Table 1. When a meteorite has been measured more than once or by more than one set of workers, the mean has been found and taken as a single determination. Data on the same meteorite by different workers are often seen to differ by more than the precision of measurement. This is believed to reflect sampling difficulties. The means so determined have not been given any added weight over single determinations in calculating the mean for a group. Also, in most cases, the non-radiogenic portion of the strontium present has been calculated using 0.699 for the primordial Sr^{87}/Sr^{86} ratio. The few exceptions to this

(1) Murthy and Compston (1965); Kaushal and Wetherill (1970)
(2) Von Michaelis et al. (1969)
(3) Kaushal and Wetherill (1969); Gast (1962; Pinson et al. (1965); Schumacher (1956); Webster et al. (1957); Bogard et al. (1967) and Herzog and Pinson (1956)
(4) Gopalan and Wetherill (1969); Pinson et al. (1965); Gast (1962); Shields et al. (1966); Murthy and Compston (1965); Kempe and Müller (1968); Shima and Honda (1967); Burnett and Wasserburg (1967b) and Herzog and Pinson (1956)
(5) Gopalan and Wetherill (1969); Kempe and Müller (1968)
(6) Gopalan and Wetherill (1970); Shima and Honda (1967)
(7) Gopalan and Wetherill (1970)
(8) Compston et al. (1965)
(9) Bogard et al. (1967)
(10) Webster et al. (1956)
(11) Papanastassiou and Wasserburg (1969); Gast (1962); Burnett and Wasserburg (1967a); Burnett and Wasserburg (1967b); Bogard et al. (1967); Compston et al. (1965); Gopalan and Wetherill (1968) and Schumacher (1956)
(12) Pinson et al. (1965)
(13) Burnett and Wasserburg (1967b)
(14) Burnett and Wasserburg (1967b) and Wasserburg et al. (1965)
(15) Burnett and Wasserburg (1967a)

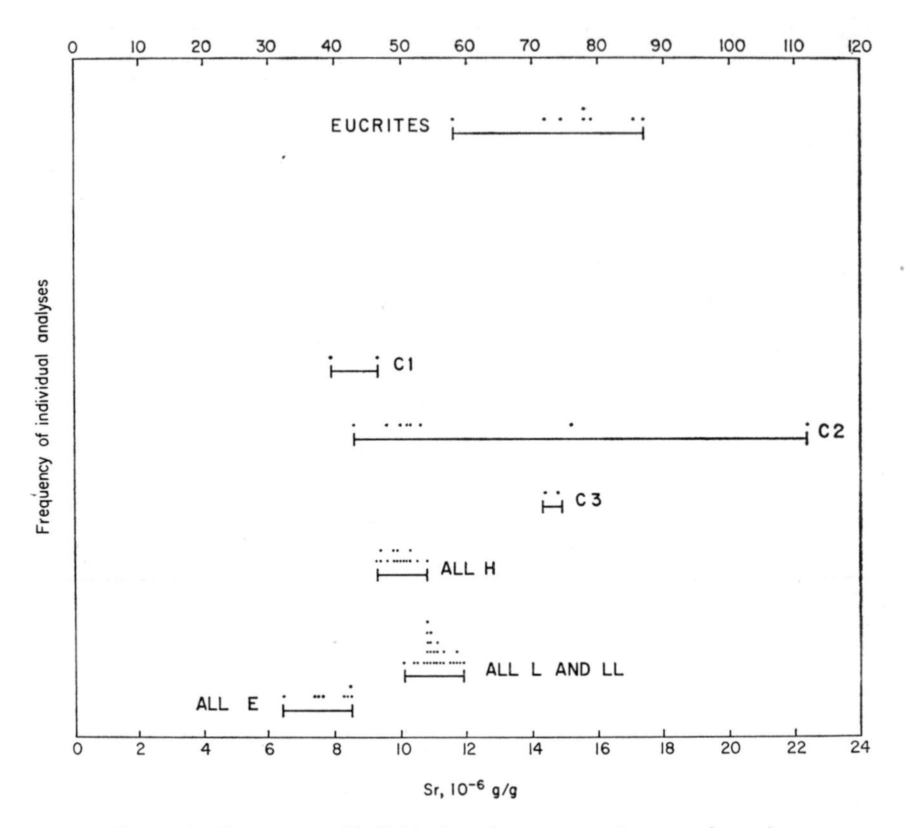

FIGURE 1 Frequency of individual analyses versus the strontium abundance for the various chondritic groups and euctries. The upper scale refers only to the eucrites.

are not serious, as the radiogenic contribution is but a few per cent in most cases. Strontium concentrations of meteorite finds have been omitted from consideration, as there is considerable evidence (Gopalan and Wetherill, 1968; Kaushal and Wetherill, 1969) that the strontium even in fresh samples of chondritic finds is either leached or contaminated or both.

Table 1 shows a fairly uniform level of strontium at 27–30 atoms/10^6 Si for the carbonaceous chondrites, though the statistics are rather poor for the type I and III meteorites. Though there appears to be a slight difference in absolute concentration between bronzites and hypersthenes or amphoterites, the concentrations relative to Si are remarkably uniform (19 atoms/10^6 Si) in these ordinary chondrite groups. The Sr/Si ratio in these groups is less than that in carbonaceous chondrites. The dark portion of the amphoterite chondrite, Krähenberg (Kempe and Müller, 1969), however, shows only 5.5 ppm of Sr as against 11.6 ppm in the light portion. Enstatite chondrites of both types have about the same absolute and relative concentrations of strontium, but they are markedly depleted both absolutely and relative to silicon as compared with carbonaceous and ordinary chondrites. One of the two samples of Adhi Kot measured (Gopalan and Wetherill, in preparation) shows a much smaller Sr content, 2.6 ppm, as compared with the other sample and other enstatite chondrites.

Only 3 calcium-poor achondrites have so far been measured for Sr by isotope dilution. Of the two enstatite achondrites, Bishopville and Norton County, the former has an order of magnitude higher strontium than the latter. The hypersthene achondrite, Johnstown, has about the same amount as Norton County. Norton County and Johnstown meteorites are depleted in Sr in both respects as compared with all the chondritic classes. The calcium-rich achondrites, eucrites in particular, with about 75 ppm, show a considerable enrichment in strontium in both absolute and relative terms. Among the stony irons, the pallasites (only one analysis) are very deficient in strontium; the single mesosiderite reported, Estherville, is enriched absolutely in this element over the chondritic meteorites.

Table 1 also gives the strontium abundances in the silicate inclusions of iron meteorites individually. There is a considerable variation in the concentration of strontium among the silicate nodules within the same meteorite. In many cases, the data indicate an absolute enrichment over the concentrations found in carbonaceous chondrites.

Müller (1968) lists the solar abundance of Sr to be $\log N_{Sr} = 3.02$ ($\log N_H = 12.00$). Using a Si abundance of $\log N_{Si} = 7.24$ (which Müller suggests is a questionable value) the solar abundance of Sr relative to $Si = 10^6$ is 60. If the value of $\log N_{Si} = 7.70$ (Müller, 1968) is used, the strontium abundance becomes 21. The abundance of 27–30 relative to Si in C1 chondrites falls between the above two solar abundance figures.

Strontium is present in several phases in a single meteorite. No mineral in which Sr is an essential constituent is known to occur in meteorites. It is clearly dispersed among the various major and accessory minerals of stone meteorites, substituting mainly for calcium in them. No detailed determination of the distribution of Sr among the likely mineral phases has so far been made. However, the few semiquantitative mineral separations so far made in order to define internal isochrons for individual meteorites and leaching experiments (Gast, 1962; Compston et al., 1965, Bogard et al., 1967 and Honda and Shima, 1967) indicate that a major amount is contained in the feldspar and possibly in the pyroxenes. It is also known to be enriched in the accessory calcium minerals, apatite and whitlockite, which are present in most common chondrites, and oldhamite, which is present in significant amounts only in the enstatite chondrites. 99% pure whitlockite from the St. Severin amphoterite has been found to contain 30 ppm of Sr (Papanastassiou and Wasserburg, 1969). The distribution between plagioclase and pyroxene appears to be variable and to depend on the degree of metamorphism exhibited by individual meteorites.

References

Bogard, D. D., Burnett, D.S., Eberhardt, P. and Wasserburg, G. J. (1967) *Earth Planet. Sci. Letters*, **3**, 179.
Burnett, D.S. and Wasserburg, G.J. (1967a) *Earth Planet. Sci. Letters*, **2**, 137.
Burnett, D.S. and Wasserburg, G.J. (1967b) *Earth Planet. Sci. Letters*, **2**, 397.
Compston, W., Lovering, J.F. and Vernon, M.J. (1965) *Geochim. Cosmochim. Acta*, **29**, 1085.
Erlank, A.J. and Willis, J.P. (1964) *Geochim. Cosmochim. Acta* **28**, 1715.
Gast, P.W. (1962) *Geochim. Cosmochim. Acta* **26**, 927.
Gopalan, K. and Wetherill, G.W. (1968) *J. Geophys. Res.* **73**, 7133.
Gopalan, K. and Wetherill, G.W. (1969) *J. Geophys. Res.* **74**, 4349.
Gopalan, K. and Wetherill, G.W. (1970) *J. Geophys. Res.* **75**, 3457.
Greenland, L. and Lovering, J.F. (1965) *Geochim. Cosmochim. Acta* **29**, 821.
Herzog, L.F. and Pinson, W.H. (1956) *Am. J. Sci.*, **254**, 555.
Kaushal, S.K. and Wetherill, G.W. (1969) *J. Geophys. Res.*, **74**, 2717.
Kaushal, S.K. and Wetherill, G.W. (1970) *J. Geophys. Res.* **75**, 463.
Kempe, W. and Müller, O. (1969) *Meteorite Research* (ed. P.M. Millman), 418. Reidel.
Murthy, V.R. and Compston, W. (1965) *J. Geophys. Res.*, **70**, 5297.
Papanastassiou, D.A. and Wasserburg, G.J. (1969) *Earth Planet. Sci. Letters*, **5**, 361.
Pinson Jr., W.H., Schnetzler, C.C., Beiser, F., Fairbairn, H.W. and Hurley, P.M. (1965) *Geochim. Cosmochim. Acta*, **29**, 455.
Pinson, W.H., Ahrens, L.H. and Franck, M.L. (1953) *Geochim. Acta* **4**, 251.
Rankama, K. and Sahama, T.G. (1950) *Geochemistry*, University of Chicago Press.
Schumacher, E. (1956) *Z. Naturf.*, **11a**, 206.
Shieds, R.M., Pinson Jr., W.H. and Hurley, P.M. (1966) *J. Geophys. Res.*, **71**, 2163.
Shima, M. and Honda, M. (1967) *Earth Planet. Sci. Letters*, **2**, 337.
Von Michaelis, H., Ahrens, L.H. and Willis, J.P. (1969) *Earth Planet. Sci. Letters*, **5**, 387.
Wasserburg, G.J., Burnett, D.S. and Frondel, C. (1965) *Science* **150**, 1814.
Webster, R.K., Morgan, J.W. and Smales, A.A. (1957) *Trans. Am. Geophys. Union*, **38**, 543.

(Received 15 July 1969)

YTTRIUM (39)

Brian Mason

Smithsonian Institution, Washington, D.C.

THE DATA ON THE ABUNDANCE of yttrium in meteoritic matter are comparatively sparse. However, Haskin et al. (1966) have provided figures for meteorites representing all the chondrite classes, and for the principal classes of achondrites and stony-irons, and their data are summarized in Table 1. A few additional determinations, all of them consistent with these data, have been found in the literature:

	10^{-6} g/g	
Holbrook (hypersthene chondrite)	2.3	Mason and Wiik (1961)
Ramsdorf (hypersthene chondrite)	2.60	Haskin et al. (1968)
St. Marks (enstatite chondrite)	1.74	Haskin et al. (1968)
Composite of 9 chondrites	1.96	Haskin et al. (1968)
Average for 5 eucrites	26	Duke and Silver (1967)
Angra dos Reis (angrite)	30, 40	Smales et al. (1970)

The figures show that yttrium is relatively unfractionated between the different classes of chondrites. In terms of the Y/Si atomic ratio, there is a small but systematic decrease from the carbonaceous chondrites through the ordinary (bronzite, hypersthene, amphoterite) chondrites to the enstatite chondrites.

Among the calcium-poor achondrites, the single enstatite achondrite analysed (Norton County) is comparable in Y/Si ratio to the enstatite chondrites, and the two hypersthene achondrites give rather divergent figures but appear significantly lower in yttrium than the chondrites. The calcium-rich achondrites are enriched in this element, moderately in the case of the nakhlites, and quite markedly for the eucrites. Angra dos Reis, a

TABLE 1 Yttrium in meteorites (Haskin et al., 1966)

	No. of detns.	Y, 10^{-6} g/g Range	Mean	Atoms/ 10^6 Si
Chondrites				
Carbonaceous, Type I	2	1.4, 1.7	1.6	4.8
Carbonaceous, Type II	2	1.8, 2.1	2.0	4.6
Carbonaceous, Type III	3	2.4, 2.4, 2.4	2.4	4.9
Bronzite	2	2.1, 2.2	2.2	3.9
Hypersthene	2	2.0, 2.1	2.1	3.4
Amphoterite	2	1.9, 2.0	2.0	3.2
Enstatite	2	1.0, 1.5	1.3	2.1
Calcium-poor achondrites				
Enstatite	1	2.09	2.09	2.6
Hypersthene	2	0.22, 1.22	0.7	0.9
Calcium-rich achondrites				
Nakhlites	2	3.2, 4.4	3.8	5.2
Eucrites	2	17, 28	23	31
Stony-irons				
Pallasites (olivine)	1	1.6	1.6	2.6
Mesosiderites (silicate)	2	2.5, 5.6	4.1	5.1

unique calcium-rich achondrite, has the highest Y/Si ratio so far found for meteorites, and also has the highest calcium content of any meteorite.

For the stony-irons the data are especially sparse. The figure given by Haskin et al. for the pallasite olivine is for the Brenham meteorite; Masuda (1968) has found that their figures for the other lanthanides in this meteorite are erroneously high, probably because of contamination introduced by weathering, and crystal chemistry suggests that the yttrium content of olivine should be much lower than the figure given. The silicates in the two mesosiderites show a modest enrichment of yttrium relative to the chondrites.

Yttrium is essentially lithiophilic in meteorites. Mason and Graham (1970) analysed mineral separates of two chondrites (Modoc and St. Severin) and found yttrium highly enriched in the calcium phosphate minerals, being present in them at concentrations around 200×10^{-6} g/g; it was not detected in metal or olivine, and was present at about 1×10^{-6} g/g in plagioclase, pyroxene, and troilite. The only mineral besides the phosphates enriched in yttrium was calcic clinopyroxene (diopside and pigeonite), which contained about 20×10^{-6} g/g. In ionic radius yttrium is very similar to calcium, and this evidently conditions its tendency to concentrate in calcium-rich minerals.

References

Duke, M.B. and Silver, L.T. (1967) *Geochim. Cosmochim. Acta* **31**, 1637.
Haskin, L.A., Frey, F.A., Schmitt, R.A. and Smith, R.H. (1966) *Phys. Chem. Earth* **7**, 167.
Haskin, L.A., Haskin, M.A., Frey, F.A. and Wildeman, T.R. (1968) *Origin and Distribution of the Elements* (ed. L.H. Ahrens), 889. Pergamon.
Mason, B. and Graham, A.L. (1970) *Smithsonian Contr. Earth Sci.* **3**.
Mason, B. and Wiik, H.B. (1961) *Geochim. Cosmochim. Acta* **21**, 276.
Masuda, A. (1968) *Earth Planet. Sci. Letters* **5**, 59.
Smales, A.A., Mapper, D., Webb, M.S.W., Webster, R.K. and Wilson, J.D. (1970) *Science* **167**, 509.

(Received 25 February 1970)

ZIRCONIUM (40) AND HAFNIUM (72)

William D. Ehmann and **Teofila V. Rebagay**

Department of Chemistry, University of Kentucky, Lexington, Kentucky and the Department of Chemistry, Arizona State University, Tempe, Arizona

ZIRCONIUM AND HAFNIUM have almost identical chemical properties and exhibit a very close geochemical coherence. The pair is regarded by Goldschmidt (1958) to be essentially lithophilic in nature. Hafnium is not known to form minerals of its own, but is found in zirconium-bearing minerals. Numerous investigators have observed that the Zr/Hf weight ratios in terrestrial igneous rocks decrease with evolution from basic to acidic rocks. Kosterin et al. (1958) found the ratios to be 71 for gabbros, 45 for granites, and 29 for hydrothermal veins in rocks from Northern Kirgizia. The variation of the Zr/Hf ratio in three calc-alkali granite masses and a discussion of possible reasons for the observed decrease in the ratio during the crystallization of rocks have been reported recently by Esson et al. (1968).

Zirconium consists of five stable isotopes in nature: ^{90}Zr (51.46%), ^{91}Zr (11.23%), ^{92}Zr (17.11%), ^{94}Zr (17.40%), and ^{96}Zr (2.80%). Hafnium consists of six stable isotopes in nature: ^{174}Hf (0.18%), ^{176}Hf (5.20%), ^{177}Hf (18.50%), ^{178}Hf (27.14%), ^{179}Hf (13.75%), and ^{180}Hf (35.24%). Due to their close chemical similarity, early analytical results for zirconium in rocks were most often reporting the sum of the two abundances, rather than that of zirconium alone. Only recently through the development of techniques of solvent extraction and ion exchange have relatively simple laboratory separation schemes for these two elements been devised.

Early studies of the abundance of zirconium in meteorites were conducted by Hevesy and Wurstlin (1934) using X-ray fluorescence techniques. They reported an average of 30 ppm (μg/g) for two carbonaceous chondrites and values ranging from 40 to 160 ppm for four ordinary chondrites. Thirty years later, Erlank and Willis (1964) using essentially the same technique found the zirconium abundances in eight chondrites to be indistinguishable from standards containing 10 ppm Zr. The latter group utilized the Zr K_α line in the first order and made corrections for the interference from the Sr K_α line.

Pinson et al. (1953) determined zirconium in 21 ordinary chondrites, one calcium-poor achondrite, and one carbonaceous chondrite employing emission spectrometry. For these three classes they obtained abundances of 33, 30, and 1 ppm, respectively. Eleven years later Schmitt et al. (1964), using essentially the same technique, obtained abundances of 9 ppm for 17 ordinary chondrites and 10 ppm for eight carbonaceous chondrites. Schmitt et al. point out that Pinson et al. may have based their determination on the Zr–3391.98 line, so that the adjacent Fe–3392.01 line would provide an interference. Aware of this interference, Schmitt et al. used the Zr–3273.05 line in their determinations.

Merz (1962) and Setser and Ehmann (1964) employed neutron activation analysis for the determination of zirconium. Both methods employed radiochemical separations for zirconium, and in the latter work also for hafnium. Both groups obtained values of approximately 33 ppm for the ordinary chondrites, using ^{95}Zr as the indicator radionuclide. About the same time Schmitt (1962) analyzed four chondrites by neutron activation analysis, using ^{97}Zr as the indicator radionuclide. He isolated ^{97}Zr radiochemically and followed the ^{97}Zr–^{97}Nb equilibrium decay. Schmitt obtained an average zirconium abundance of 13 ppm and suggested that this result might still be too high, due to possible incomplete decontamination from ^{24}Na. Merz and Schrage (1964) then analyzed three ordinary chondrites by activation analysis, using a different radiochemical separation scheme than originally employed by Merz. They again used ^{95}Zr as the indicator radionuclide and obtained a zirconium abundance of 12 ppm. They concluded the higher values obtained earlier by Merz were in error, due to fractionation of zirconium and hafnium during the chemical separations and problems associated with the preparation and stability of the standards used.

The fact that Setser and Ehmann (1964) obtained zirconium abundances in a number of standard rocks and tektites that have since been shown to be correct, suggest that the error in their determinations was not due to incorrect standards. Gordon et al. (1968) have shown that europium may provide a serious spectral interference in the determination of zirconium by activation analysis. ^{152}Eu has a gamma-ray at 0.779 MeV and the 0.723 and 0.757 MeV gamma-rays of ^{154}Eu almost exactly overlap the 0.724 and 0.756 MeV gamma-rays of ^{95}Zr. Both of these radionuclides are produced by thermal neutron capture reactions on europium. Primary interference reactions such as (n, α) reactions on ^{98}Mo and ^{100}Mo to produce ^{95}Zr and ^{97}Zr have been discussed by Choy et al. (1965). Due to the low abundance of molybdenum in chondrites and the highly thermal neutron fluxes used, errors due to primary interference reactions may be regarded as negligible. It now appears that the high results of Setser and Ehmann for chondrites were due to a combination of spectral interferences, most likely including the europium activities.

Recently Ehmann and Rebagay (1970) have redetermined the zirconium abundances in a large number of stony meteorites using a modified neutron activation analysis procedure. Following thermal neutron irradiation, zirconium and hafnium are separated as a group from the bulk material of the chondrite by means of an anion exchange procedure. The chemical yield of the separation procedure for both zirconium and hafnium is based on the amount of zirconium carrier added prior to fusion dissolution of the sample. Automated fast neutron activation analysis is employed in the chemical yield determination. The determination of zirconium is based on high resolution Ge (Li) gamma-ray spectrometry, using the 0.724 MeV gamma-ray of 65 day half-life ^{95}Zr. Half-life determinations, chemical recycling, precise energy calibrations, and photopeak-area ratios have been used to firmly establish the identity of the analytical photopeak. The results of Rebagay and Ehmann clearly confirm the lower abundance level for zirconium in chondrites.

A summary of the recent activation analysis results of Schmitt (1962), Merz and Schrage (1964), and Ehmann and Rebagay (1970) together with the emission spectrographic results of Schmitt et al. (1964) are given in Table 1. In most cases the results of Schmitt et al. and Ehmann and Rebagay are in reasonably good agreement. Exceptions are noted for the enstatite chondrites and the Ca-poor achondrites, where the data of Ehmann and Rebagay are considerably lower. In these two cases, the number of meteorites analyzed is quite small and additional data would be desirable to resolve the discrepancy. However, Schmitt et al. point out that for some of their early determinations by emission spectrometry zirconium abundances of less than 10 ppm were considered to be borderline plate readings and the true abundances could be somewhat lower. For this reason, we feel that the higher sensitivity possible by neutron activation analysis recommends selection of the data of Ehmann and Rebagay.

Hafnium abundances in meteorites have been reported by Merz (1962), Merz and Schrage (1964), Setser and Ehmann (1964), and Ehmann and

TABLE 1 Recent determinations of zirconium in stony meteorites

Classification*	Number of meteorites analyzed	Total number of analyses	Mean Zr abundance in ppm	Atomic abundance Zr (Si = 10^6)†	Ref.
C1	1	1	11	32	(3)
	2	3	9	26	(4)
C2	1	1	13	30	(1)
	5	12	10	23	(3)
	3	5	5.8	13	(4)
C3	2	4	9	18	(3)
	1	2	12	24	(4)
C4	1	3	9.5	18	(4)
All C	1	1	13	30	(1)
	8	17	10	23	(3)
	7	13	8.1	19	(4)
E4	1	2	14	25	(3)
	1	2	3.8	7	(4)
E5	1	2	4.1	7	(4)
E6	1	1	7.7	13	(4)
All E	1	2	14	25	(3)
	3	5	5.2	9	(4)
H4	1	2	5	9	(3)
	1	2	5.7	10	(4)
H5	1	1	14	26	(1)
	1	1	12	22	(2)
	6	13	8	15	(3)
	3	7	6.2	11	(4)
H6	1	1	15	28	(2)
	1	1	7.2	13	(4)
All H	1	1	14	26	(1)
	2	2	13	25	(2)
	9	19	7	14	(3)
	5	10	6.4	11	(4)
L3	1	2	6.3	11	(4)
L4	1	2	6.9	12	(4)
L5	1	4	10	17	(3)
	1	1	6.8	11	(4)
L6	2	2	13	22	(1)
	1	1	9.2	16	(2)
	7	17	11	19	(3)
	5	9	5.1	9	(4)

Classification*	Number of meteorites analyzed	Total number of analyses	Mean Zr abundance in ppm	Atomic abundance Zr (Si = 10^6)†	Ref.
All L	2	2	13	22	(1)
	1	1	9.2	16	(2)
	8	21	11	19	(3)
	9	18	5.9	10	(4)
LL3	1	1	7	11	(4)
LL5	1	2	9.2	15	(4)
LL6	1	3	6	10	(3)
	1	2	6.9	11	(4)
All LL	2	5	9	16	(3)
	3	5	7.7	12	(4)
Achondrites, Ca-poor	2	4	7	8	(3)
	4	6	1.2	1.4	(4)
Achondrites, Ca-rich	5	13	51	68	(3)
	3	4	46	61	(4)

Note: The work of Erlank and Willis (1964) did not assign abundances to individual chondrites, but their average of approximately 10 ppm is in good agreement with the data presented here.

* Classification according to Van Schmus and Wood (1967).

† Using silicon data of Ehmann and Durbin (1968).

(1) Schmitt (1962)
(2) Merz and Schrage (1964)
(3) Schmitt et al. (1964)
(4) Ehmann and Rebagay (1970)

Rebagay (1970). All these groups used neutron activation analysis and determined hafnium via the ^{181}Hf (43 day half-life) indicator radionuclide. The principal gamma-rays of ^{181}Hf which may be used analytically are at 0.13 and 0.48 MeV. Potential spectral interferences include ^{131}Ba, ^{152}Eu, ^{233}Pa, ^{140}La, and ^{169}Yb. Hence, radiochemical separations are to be preferred over strictly instrumental activation procedures for the determination of hafnium in the stony meteorites. Merz in his original work reported hafnium abundances of 0.39 to 2.3 ppm in the ordinary chondrites. In the later work of Merz and Schrage they point out that the hafnium data of Merz are too high, due to problems of zirconium-hafnium fractionation in the chemical separation scheme and instability of the standard hafnium solutions used. They obtained hafnium abundances of 0.30 to 0.46 ppm for three ordinary chondrites. Ehmann and Rebagay employed a high resolution Ge(Li) detector to measure the ^{181}Hf, and hence minimize problems of spectral interferences. They report an average of 0.19 ppm Hf in a wide variety of chondrites, which is in good agreement with the average of 0.20 ppm Hf reported originally by Setser and Ehmann.

The results of hafnium analyses of stony meteorites, excluding the high data of Merz (1962), are presented in Table 2. It is felt that the three analyses of Merz and Schrage (1964) may reflect spectral interferences that would not be resolved with the poorer resolution NaI (Tl) detector used in their studies.

Based on the more extensive sets of data by Schmitt et al. (1964) and Ehmann and Rebagay (1970), it is seen that the absolute abundances zirconium and hafnium are relatively uniform among the various major classes of chondrites. The majority of the analyses fall in the range of 7 ± 3 ppm Zr and 0.19 ± 0.07 ppm Hf. As also noted by Schmitt et al. for zirconium, these elements appear to be slightly enriched in the carbonaceous chondrites relative to the ordinary chondrites. The data of Ehmann and Rebagay do not show enrichment of these elements in the enstatite chondrites relative to the ordinary chondrites, as suggested by the data of Schmitt et al. It is also of interest to note that the data of Ehmann and Rebagay indicate that the C2 group chondrites are depleted in zirconium and hafnium as compared to both the C1 and the C3 groups. Anders (1964) has suggested that the C1, C2, and C3 chondrites differ chemically because they consist of different proportions of a component that condensed at a high temperature (presumably the chondrules) and a second component that condensed at a low temperature (presumably the matrix material). Lord (1965) has indicated that Al_2O_3, W, and ZrO_2 would be among the first substances to condense from solar nebula material at high temperatures and a hydrogen pressure of 0.5 atmosphere. It would be reasonable to expect, therefore, that zirconium and hafnium might well be concentrated in the high temperature minerals of the chondrules. Larimer and Anders (1967) have suggested that the C3 chondrites contain approximately 68% of the high temperature material, while the C2 chondrites contain only approximately 45%. The C1 chondrites are regarded as objects that condensed at low temperatures and essentially retain the original "cosmic" abundances of the elements.

The higher zirconium and hafnium abundances in the C3 chondrites with respect to the C2 chondrites may then reflect an enrichment of these elements

TABLE 2 Hafnium abundances in various classes of stony meteorites

Classification*	Number of meteorites analyzed	Total number of analyses	Mean Hf abundance in ppm	Atomic abundance Hf (Si = 10^6)†	Ref.
C1	2	3	0.32	0.47	(3)
C2	2	2	0.26	0.30	(2)
	3	5	0.19	0.22	(3)
C3	1	2	0.25	0.25	(3)
C4	1	2	0.29	0.28	(3)
All C	2	2	0.26	0.30	(2)
	7	12	0.25	0.30	(3)
E4	1	1	0.17	0.15	(2)
	1	2	0.10	0.09	(3)
E5	1	1	0.15	0.14	(2)
	1	2	0.12	0.11	(3)
E6	1	1	0.21	0.18	(3)
All E	2	2	0.16	0.14	(2)
	3	5	0.14	0.13	(3)
H4	1	2	0.10	0.09	(3)
H5	1	1	0.29	0.27	(1)
	3	6	0.20	0.19	(2)
	3	5	0.22	0.21	(3)
H6	1	1	0.46	0.44	(1)
	1	1	0.14	0.13	(3)
All H	2	2	0.37	0.35	(1)
	3	6	0.20	0.19	(2)
	5	8	0.18	0.17	(3)
L3	1	2	0.15	0.13	(3)
L4	1	2	0.18	0.15	(3)
L5	1	1	0.24	0.21	(3)
L6	1	1	0.30	0.26	(1)
	3	3	0.19	0.17	(2)
	4	8	0.14	0.12	(3)
All L	1	1	0.30	0.26	(1)
	4	4	0.19	0.17	(2)
	8	15	0.17	0.15	(3)
LL3	1	1	0.16	0.13	(3)
LL5	1	1	0.21	0.17	(3)
LL6	1	1	0.19	0.16	(2)
	1	2	0.14	0.12	(3)
All LL	1	1	0.19	0.16	(2)
	3	4	0.17	0.14	(3)

Classification*	Number of meteorites analyzed	Total number of analyses	Mean Hf abundance in ppm	Atomic abundance Hf ($Si = 10^6$)†	Ref.
Achondrites, Ca-poor	3	3	0.13	0.08	(2)
	4	6	0.02	0.01	(3)
Achondrites, Ca-rich	1	1	0.21	0.14	(2)
	3	4	0.78	0.53	(3)

* Classification according to Van Schmus and Wood (1967).
† Using silicon data of Ehmann and Durbin (1968).

(1) Merz and Schrage (1964)
(2) Setser and Ehmann (1964)
(3) Ehmann and Rebagay (1970)

in the chondrules relative to the low temperature condensate represented by the matrix of these two classes. Since the C1 chondrites also have high zirconium and hafnium abundances, this model would require that the C1 chondrites did not condense under conditions identical to those for the matrix component of the C2 and C3 chondrites. It must be pointed out, however, that with the relatively few data available for zirconium and hafnium among the various classes of carbonaceous chondrites sampling errors could account for the observed differences in the abundances of these elements among the various classes of carbonaceous chondrites.

It is interesting to note that the calcium-rich achondrites are enriched and the calcium-poor achondrites are depleted in zirconium and hafnium with respect to all other classes of stony meteorites. Mason (1962) notes that the mineralogy and chemical composition of achondrites closely resemble those of certain igneous rocks. Crystallochemical factors must have played an important role in the observed isomorphism between zirconium or hafnium and calcium. In an environment rich in calcium, zirconium and hafnium with their smaller radii and higher lattice energy coefficients could readily replace calcium isomorphously. Wager and Mitchell (1951) have shown that zirconium enters the early pyroxenes by replacing calcium. In calcium-poor environments, elements with more desirable crystallochemical properties may provide greater competition with zirconium and hafnium for the available calcium sites.

The Zr/Hf ratios for various classes of stony meteorites as given by Ehmann and Rebagay (1970) are presented in Table 3. The weight ratios for the C, E, H, L, and LL groups are 33, 36, 42, 36, and 46. The average ratio

TABLE 3 Zirconium/Hafnium ratios in stony meteorites (Ehmann and Rebagay, 1970)

Classification*	Zr, ppm/Hf, ppm	Zr, atoms/Hf, atoms
C1	28	55
C2	32	63
C3	48	94
C4	33	65
All C	33	65
E4	38	75
E5	34	67
E6	37	73
All E	36	71
H4	57	110
H5	34	67
H6	51	100
All H	42	82
L3	42	82
L4	38	75
L5	28	55
L6	37	73
All L	36	71
LL3	44	86
LL5	44	86
LL6	49	96
All LL	46	90
Ca-poor Achondrites	66	130
Ca-rich Achondrites	57	110

* Classification according to Van Schmus and Wood (1967).

for all stony meteorites analyzed is approximately 43. The data of Taylor (1964) for elemental abundances in the continental crust yield a Zr/Hf ratio of 55. Vlasov (1966) gives a value of 50 for the earth's crust, and Horn and Adams (1966) give a value of 41 for this ratio in igneous rocks. The new data suggest, therefore, that there is no large-scale fractionation of these elements in the earth relative to the chondrites, as was earlier suggested by the data of Setser and Ehmann (1964). Trends in the Zr/Hf ratio with petrologic type within the principal chemical groups of chondrites (Van Schmus and Wood,

1967) are not observed for the ordinary and enstatite chondrites, based on the data of Table 3. A suggestion of an increase in the Zr/Hf ratio in the order C1–C2–C3 is noted among the carbonaceous chondrites. However, the number of carbonaceous chondrites analyzed is small and the estimated uncertainties in the Zr and Hf data are of the order of $\pm 10\%$. Based on the data available, it would be unwise to attach any major significance to the small variations in the Zr/Hf ratio among either the chemical or petrologic classes of chondrites. In spite of the great difference in the absolute abundances of zirconium and hafnium in the calcium-rich and calcium-poor achondrites, the Zr/Hf ratios of 57 and 66, respectively, reflect little fractionation of these elements in the formation of these meteorites.

Ahrens (1967) has calculated the Si/Al, Si/Mg, and Si/Ca ratios in the carbonaceous, enstatite, and ordinary chondrites. He noted that the enstatite chondrites exhibited the greatest depletion in the major abundance nonvolatile elements Al, Ca, and Mg relative to Si with the carbonaceous chondrites showing the least fractionation. Reference to the relative atomic abundances of zirconium and hafnium presented in Tables 1 and 2, shows a similar trend for these nonvolatile trace elements.

Little data obtained by the improved analytical methods are available for separated meteoritic phases. Ehmann and Rebagay (1970) obtained 2.8 ppm Zr and 0.03 ppm Hf in olivine derived from the Brenham pallasite. The Zr value is in good agreement with the value of 3.0 ppm obtained by Schmitt et al. (1964). Ehmann and Rebagay report values of 1.4 ppm Zr and 0.01 ppm Hf in troilite extracted from the Canyon Diablo siderite, confirming the small chalcophilic tendencies of these elements. Since these values approach the detection limit of the technique as it was ordinarily conducted and the freedom of the troilite from micro-inclusions of silicates was not tested, these values are best regarded as upper limits for the abundances in the troilite phase.

The new zirconium and hafnium data summarized in this paper yield "cosmic" atomic abundances relative to silicon equal to 10^6 atoms that are in reasonable agreement with solar abundances and abundances calculated from theories of nucleosynthesis. Aller (1965) reports a solar atomic abundance for zirconium of 14 (Si $= 10^6$). Clayton and Fowler (1961) report calculated relative atomic abundances for Zr and Hf of 22 and 0.18, respectively. It would appear that the "Zr–Hf dilemma" (Erlank and Willis, 1964) has now been satisfactorily resolved.

Addendum

After submittal of this manuscript additional data on zirconium abundances in stony meteorites have been published by Von Michaelis et al. (1969). In general their data are in good agreement with the more recent data presented in Table 1. Zirconium abundances for seven enstatite chondrites ranged from 6 to 10 ppm, for seven C2 and C3 carbonaceous chondrites from 9 to 13 ppm, for twelve H-group chondrites from 6 to 9 ppm, and for twenty L-group chondrites from 7 to 11 ppm. Three Ca-poor achondrites were reported to have zirconium abundances of 3 to 4 ppm and three Ca-rich achondrites contained 42 to 58 ppm. The technique used was x-ray fluorescence spectrometry.

Von Michaelis, H.; Ahrens, L.H.; and Willis, J.P. (1969) *Earth Planet Sci. Letters* 5, 387–394.

References

Ahrens, L.H. (1967) *Geochim. Cosmochim. Acta* **31**, 861–868.
Aller, L.H. (1965) *Advances in Astronomy and Astrophysics*, Vol. 3, pp. 1–25, Academic Press, New York.
Anders, E. (1964) *Space Sci. Rev.* **3**, 583–714.
Choy, T.K., Lukens, H.R. and Anderson, G.H. (1965) *Nucl. Appl.* **1**, 179–183.
Clayton, D.D. and Fowler, W.A. (1961) *Ann. Phys.* **16**, 51–68.
Ehmann, W.D. and Durbin, D.R. (1968) *Geochim. Cosmochim. Acta* **32**, 461–464.
Ehmann, W.D. and Rebagay, T.V. (1970) *Geochim. Cosmochim. Acta* **34**, 649–658.
Erlank, A.J. and Willis, J.P. (1964) *Geochim. Cosmochim. Acta* **28**, 1715–1728.
Esson, J., Hahn-Weinheimer, P. and Johanning, H. (1968) *Talanta* **15**, 1111–1118.
Goldschmidt, V.M. (1958) *Geochemistry*, Oxford University Press, London.
Gordon, G.E., Randle, K., Goles, G., Corliss, J.B., Beeson, M. and Oxley, S. (1968) *Geochim. Cosmochim. Acta* **32**, 369–396.
Hevesy, G.V. and Wurstlin, K. (1934) *Z. anorg. Allgem. Chem.* **216**, 305–311.
Horn, M.K. and Adams, J.A.S. (1966) *Geochim. Cosmochim. Acta* **30**, 279–297.
Kosterin, A.V., Zuev, V.N. and Shevaleeskii, I.D. (1958) *Geochemistry* 1958, No. **1**, 116–119.
Larimer, J. and Anders, E. (1967) *Geochim. Cosmochim. Acta* **31**, 1239–1270.
Lord, H.C. (1965) *Icarus* **4**, 279–288.
Mason, B. (1962) *Meteorites*, John Wiley & Sons, Inc., New York.
Merz, E. (1962) *Geochim. Cosmochim. Acta* **26**, 347–350.
Merz, E. and Schrage, E. (1964) *Geochim. Cosmochim. Acta* **28**, 1873–1877.
Pinson, W.H., Ahrens, L.H. and Franck, M.L. (1953) *Geochim. Cosmochim. Acta* **4**, 251–260.
Schmitt, R.A. (1962) *General Atomic Report GA-3687*.

Schmitt, R.A., Bingham, E. and Chodos, A.A. (1964) *Geochim. Cosmochim. Acta* **28**, 1961–1979.

Setser, J.L. and Ehman, W.D. (1964) *Geochim. Cosmochim. Acta* **28**, 769–782.

Taylor, S.R. (1964) *Geochim. Cosmochim. Acta* **28**, 1273–1285.

Van Schmus, W.R. and Wood, J.A. (1967) *Geochim. Cosmochim. Acta* **31**, 747–765.

Vlasov, K.A. (1966) *Geochemistry of Rare Elements*, Vol. 1. Israel Program for Scientific Translation.

Wager, L.R. and Mitchell, R.L. (1951) *Geochim. Cosmochim. Acta* **1**, 129–208.

(Received 1 May 1969)

NIOBIUM (41)*

Peter R. Buseck

Departments of Geology and Chemistry
Arizona State University
Tempe, Arizona 85281

VERY LITTLE INFORMATION is available regarding the distribution of niobium in meteorites—it is probably one of the least studied elements. The only study that is even close to comprehensive is that of Rankama (1948), who analyzed ten meteorites. The results of his investigations are given in Table 1.

The analytical procedure used by Rankama was to first dissolve the meteorite. This was followed by precipitation by phenylarsonic acid and then measurement by spectrochemical analysis in the carbon arc cathode layer using a logarithmic wedge sector. Error estimates for the measurements were not provided, although Rankama does state that the indicated values are upper concentration limits. Lovering (1957) also spectrographically measured the niobium content of the olivine of some pallasites and determined upper concentration limits. These figures are included in Table 1.

Of all the measured meteorites, only two show niobium contents above the detectability limits. The Lake Labyrinth chondrite (group LL6) contains 1 ppm and the St. Michel chondrite (group L6) contains 0.2 ppm niobium.

As so many of the meteorites have niobium contents below the detectability limits, it is difficult to arrive at a reliable mean value. If we accept the data of Rankama (1948), the chondrites (3 falls and 2 finds) contain a mean of <0.2 ppm and the irons <0.02 ppm niobium. Rankama also indicates that the Noddacks (1934) found an average value of 0.02 ppm for the meteorites.

* Contribution No. 49 from Arizona State University Center for Meteorite Studies

TABLE 1 Niobium content of meteorites

Type	Meteorite	Concentration, ppm (Reference)	Atoms/10^6 Si Atoms
Chondrites*			
E6	Hvittis (Huittinen)	<0.01 (1)	<0.02
H5	Plainview	<0.002 (1)	<0.004
L4	Bjurböle	<0.02 (1)	<0.03
L6	St. Michel (Mikkeli)	0.2 (1)	0.3
LL6	Lake Labyrinth	1 (1)	1.6
Irons†			
Om	Casas Grandes	<0.03 (1)	–
	Henbury (troilite)	<20 (2)	–
	Kyancutta	<0.003 (1)	–
	Toluca (troilite)	<20 (2)	–
Og	Coolac	<0.02 (1)	–
H	Rio Loa (troilite)	<20 (2)	–
Pallasites†			
	Admire (olivine)	<20 (2)	–
	Albin (olivine)	<20 (2)	–
	Brenham (olivine)	<20 (2)	–
	(troilite)	<20 (2)	–
	Marjalahti (olivine)	<0.07 (1)	–
	(metal	<0.3 (1)	–
	Springwater (olivine)	<20 (2)	–
	(troilite)	<20 (2)	–

* Classification according to Van Schmus and Wood (1967).
† Classification according to Hey (1966).

(1) Rankama, 1948
(2) Lovering, 1957

The meteoritic data hardly permit an accurate evaluation of the geochemical character of niobium. Based on its terrestrial occurrence in oxide and silicate minerals, especially those produced during late-stage magmatic crystallization, niobium is a strongly lithophilic element. Its ionic radius is similar to Ti^{4+} and thus it probably is concentrated in titanium-rich minerals. Geochemically, niobium is also very similar to tantalum; analyes for both elements are available only for the Plainview and Bjurböle chondrites (Ehmann, 1965; Atkins and Smales, 1960). The ratio of Ta/Nb are <11 and <1, respectively. Additional work to look for chemical fractionations between these geochemically similar elements would be of interest.

The cosmic abundance of niobium, based on abundances in the chondrites, was estimated as 1 atom niobium per 10^6 atoms silicon (Suess and Urey, 1956). This was revised to 1.15 atoms by Cameron (1968). The nucleosynthetic calculations of Clayton and Fowler (1961) indicate an abundance of 3.5 atoms, and the astrophysical data of solar photospheric abundances of Müller (1968) indicate 6.3 atoms. Surely further work on niobium abundances is in order to more fully understand the geochemistry, cosmic abundance, and distribution between phases of this element.

Acknowledgements

This report was supported in part by Grant GA–1200 from the National Science Foundation. Helpful assistance and comments were provided by Mr. Jerrald Durtsche and Dr. J.W. Larimer.

References

Atkins, D.H.F. and Smales, A.A. (1960) *Anal. Chim. Acta* **22**, 462.
Cameron, A.G.W. (1968) in *Origin and Distribution of the Elements* (L.H. Ahrens, ed.). Pergamon.
Clayton, D.D. and Fowler, W.A. (1961) *Ann. Phys.* **16**, 51.
Ehmann, W.D. (1965) *Geochim. Cosmochim. Acta* **29**, 43.
Hey, M.H. (1966) *Catalogue of Meteorites* (3rd edition). British Museum.
Lovering, J.F. (1957) *Geochim. Cosmochim. Acta* **12**, 253.
Müller, E.A. (1968) in *Origin and Distribution of the Elements* (L.H. Ahrens, ed.). Pergamon.
Noddack, I. and Noddack, W. (1934) *Svensk. Kem. Tidskr.* **46**, 173.
Rankama, K. (1948) *Ann. Acad. Sci. Fennicae* **AIII, 13.**
Suess, H.E. and Urey, H.C. (1956) *Rev. Mod. Phys.* **28**, 53.
Van Schmus, W.R. and Wood, J.A. (1967) *Geochim. Cosmochim. Acta* **31**, 747.

(Received 7 April 1970)

respectively. Additional work to look for chemical incompatibilities between

Acknowledgements

References

MOLYBDENUM (42)

Michael E. Lipschutz

Departments of Chemistry and Geosciences
Purdue University
Lafayette, Indiana 47907

MOLYBDENUM IS PRESENT as a trace element ($0.5–30 \times 10^{-6}$ g/g) in meteorites and has been determined by a variety of chemical techniques: spectrophotometry (Kuroda and Sandell, 1954) in 14 chondrites and 2 irons; isotope dilution (Murthy, 1962, 1963; Wetherill, 1964) in 1 pallasite and 14 irons; and neutron activation in 25 chondrites (Case et al., 1971), 1 pallasite and 66 irons (Smales et al., 1967) and 3 chondrites (Kiesl and Hecht, 1969). In 16 instances, different samples of the same stony or iron meteorite have been studied and except for two irons (Canyon Diablo and Henbury) where the spectrophotometric results are higher than those obtained by other techniques, and Grosnaja, where the datum of Kiesl and Hecht (1969) is suspiciously low, the extreme results differ by less than 30% (Tables 1 and 2). Except for the achondrites and enstatite chondrites, for which there exist no molybdenum analyses, coverage of the principal meteoritic types may be considered as reasonably adequate.

In meteorites, molybdenum is a dispersed element and is primarily siderophile in geochemical behavior although it also has a strong chalcophile tendency as well. Kuroda and Sandell (1954) have determined the molybdenum contents in separated metal, sulfide and silicate portions of two composites (each consisting of H- and L-group finds) to be, respectively, 7.8, 5.9, 0.7 and 8.2, 5.4, 0.5×10^{-6} g/g. Kiesl and Hecht (1969) report a range of $4.2–6.7 \times 10^{-6}$ g/g in troilite from a hexahedrite and two octahedrites and an upper limit of $<1 \times 10^{-6}$ g/g in a troilite-metal grain from the H-group chondrite Mocs. They also report the molybdenum contents in separated metal and stony (doubtless containing sulfide) portions of the siderophyre Steinbach as 0.53 and 0.10×10^{-6} g/g, respectively. The primary siderophile

TABLE 1 Molybdenum in chondrites

Type	Number analyzed	No. of detns.	Mo contents range (ref.) (10^{-6} g/g)	mean (10^{-6} g/g)	(atoms/10^6 Si atoms)
C1	2	4	1.2–1.6 (1)	1.4	4.0
C2	3	4	1.2–1.8 (1)	1.5	
	1	1	1.5 (3)	1.5	
	3	5	1.2–1.8 (All)	1.5	3.4
C3	6	7	1.7–2.3 (1)	2.0	
	2	2	0.5–2.1 (3)	1.3	
	6	8	1.7–2.3 (All*)	2.0	3.8
H	6	7	1.3–2.0 (1)	1.7	
	7	7	1.4–2.0 (2)	1.6	
	12	14	1.3–2.0 (All)	1.7	2.9
L	4	4	1.0–1.6 (1)	1.2	
	7	7	1.2–1.7 (2)	1.5	
	11	11	1.0–1.7 (All)	1.4	2.2
LL	4	4	0.82–1.4 (1)	1.2	1.9

(1) Case et al. (1971)
(2) Kuroda and Sandell (1954)
(3) Kiesl and Hecht (1969)
* Omitting the Grosnaja datum of Kiesl and Hecht (1969).

geochemical behavior of molybdenum is also demonstrated by the analyses of whole-rock samples and separated metal and/or silicate phases of 4 individual chondritic finds (Kuroda and Sandell, 1954) and the relatively high concentrations in iron meteorites. Previous reports of molybdenum isotopic anomalies in iron meteorites (Murthy, 1962, 1963) were not confirmed by Wetherill (1964), who showed that the Mo^{92}/Mo^{100} ratios in 7 iron meteorites and the metallic phase of a pallasite are the same as that of terrestrial molybdenum to within 1%. The isotopic composition in one L6 chondrite fall (Forksville) also appears to be the same as that of terrestrial molybdenum (Murthy, 1963).

The data for 21 carbonaceous and ordinary chondrite falls (Case et al., 1971; Kiesl and Hecht, 1969) and 17 ordinary chondrite finds (Kuroda and Sandell, 1954; Case et al., 1971) are illustrated in Figure 1 and summarized in Table 1. These data indicate no systematic difference in the molybdenum analyses for falls and finds. The concentrations within each chondritic group and, indeed, within all chondrites appear to be rather similar (Figure 1) although the abundances (Table 1) are higher in the carbonaceous than in

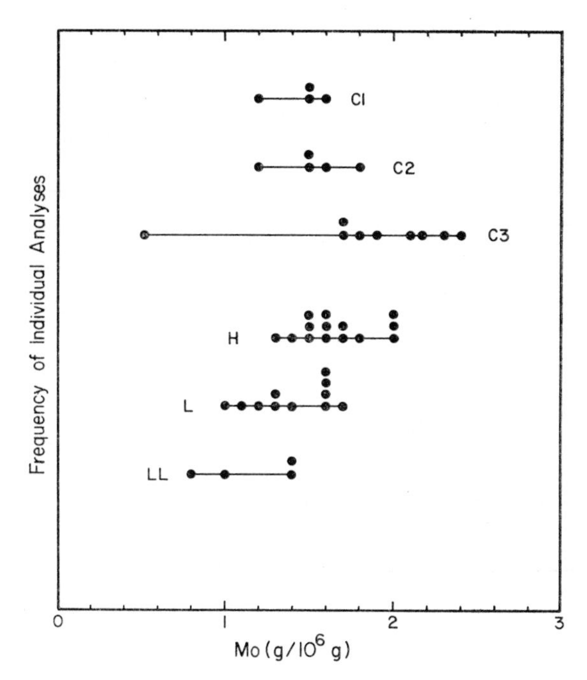

FIGURE 1 Molybdenum concentrations in carbonaceous and ordinary chondrites.

the ordinary chondrites (Case et al., 1971). While it seems that the molybdenum concentrations in ordinary chondrites decrease in the order H–L–LL (Figure 1, Table 1) as might be expected from its geochemical behavior, the ranges in each group overlap to such an extent that this tendency does not appear to be statistically significant (Case et al., 1971). The molybdenum concentrations within each chondritic group seem independent of chemical-petrologic type and of the degree of equilibration of the unequilibrated ordinary chondrites (Case et al., 1971). The similarities in abundance in the various types of carbonaceous chondrites (Table 1) lead to mean "cosmic" abundance values for molybdenum of 3.7 atoms/10^6 Si atoms based on all carbonaceous chondrites or 4.0 atoms/10^6 Si atoms based solely on Type C1 (Case et al., 1970). These values are in excellent agreement with the solar photospheric value of 4.0 atoms/10^6 Si atoms obtained from the data of Müller (1968) using her preferred value for the relative abundance of silicon. These values are somewhat higher than Aller's (1961) solar photospheric value of 2.5 atoms/10^6 Si atoms which is identical to the mean molybdenum abundance in ordinary chondrites (Table 1).

The molybdenum concentrations in iron meteorites are illustrated in Figure 2 and summarized in Table 2. These data indicate that the mean molybdenum concentrations generally are constant in iron meteorites. A significant exception to this is the nickel-rich ataxites which, on the average,

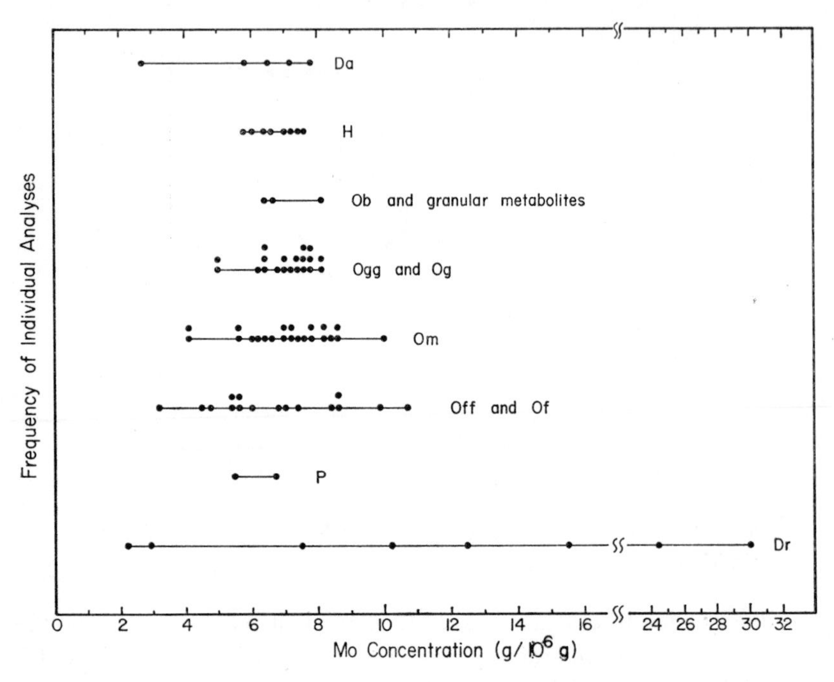

FIGURE 2 Molybdenum concentrations in pallasites and iron meteorites. For convenience, Ogg and Og have been grouped as have Of and Off.

contain about twice as much molybdenum as do the other iron meteorites. In Smales et al.'s study (1967), a number of other elements (Ag, As, Cr, Cu, Ga, Ge, In, Pd and Zn) were measured in the same samples and molybdenum was not found to be correlated with any of these. Based on the nickel values listed by Hey (1966) there also appears to be no direct correlation between molybdenum and nickel. However it is interesting to note that there seems to be a tendency for the molybdenum concentrations in iron meteorite groups to be more variable as the mean nickel concentration increases (Table 2, Figure 2). This trend is not without exception although it is paralleled by a similar variation in the cooling rate ranges for all meteorites studied by Goldstein and Short (1967) or just for those in which molybdenum has been determined (Table 2).

TABLE 2 Molybdenum in pallasites and iron meteorites

Structural type	Nickel content*(%)	Cooling rate range‡ (°C/m.y.)	Number analyzed	No. of detns.	Mo concentration (10^{-6} g/g) range (ref.)	mean
Da	5.5 (4)		5[b]	5	2.7–7.8 (1)	6.0
H	5.6 (6)		7	8	5.9–7.5 (1), (2)	6.8
Granular metabolite†	6.3 (1)		2	2	6.3–8.1 (1)	7.2
Ob	8.0 (1)		1	1	6.6 (1)	6.6
Ogg	6.6 (6)	>2 (1)	6	7	5.9–8.1 (1), (2), (3)	6.9
Og	7.2 (9)	2–3 (6)	10	13[a]	4.9–8.1 (1), (2), (3)	6.9
Om	8.5 (19)	1–12 (16)	20	22	4.1–10 (1), (2), (4)	7.0
Of	9.1 (11)	1.5–100 (10)	11	12	3.2–11 (1), (2)	6.7
Off	10 (4)	50–200 (3)	4	4	4.8–8.5 (1)	6.9
P	10 (2)	0.4 (2)	2	2	5.5–6.7 (1), (3)	6.1
Dr	19 (8)	9–400 (4)	8	8	2.2–30 (1), (3)	13

(1) Smales et al. (1967)
(2) Murthy (1963)
(3) Wetherill (1964)
(4) Kuroda and Sandell (1954)

[a] A suspiciously high value of 17×10^{-6} g/g for Canyon Diablo (Kuroda and Sandell, 1954) which differs markedly from other results (Murthy, 1963; Wetherill, 1964; Smales et al., 1967) has been omitted.

[b] One of these meteorites (Santa Rosa) is classified by Hey (1966) as a possible metabolite.

* The values listed are the means of the nickel determinations reported (Hey, 1966) for those meteorites in which molybdenum has been determined. The numbers in parentheses are the number of different meteorites whose nickel analyses were used in calculating the means.

† It has been demonstrated (Jain and Lipschutz, 1968) that granular metabolites can be produced from shock-loaded octahedrites by extended annealing below the α–γ transformation temperature. These samples may well be heat-altered, shock-loaded octahedrites.

‡ Data from Goldstein and Short (1967). The numbers in parentheses are the numbers of meteorites whose cooling rates are included in the respective ranges.

Acknowledgements

This research was supported in part by the U.S. National Science Foundation, grant GA–1474.

References

Aller, L.H. (1961) *The Abundance of the Elements*, Interscience.
Case, D.R., Laul, J.C., Pelly, I.Z., Wechter, M.A., Schmidt-Bleek, F. and Lipschutz, M.E. (1971) In preparation.
Goldstein, J.I. and Short, J.M. (1967) *Geochim. Cosmochim. Acta* **31**, 1733.
Hey, M.H. (1966) *Catalogue of Meteorites*. Third edition, British Museum.
Jain, A.V. and Lipschutz, M.E. (1968) *Nature* **220**, 139.
Kiesl, W. and Hecht, F. (1969) *Meteorite Research* (P.M. Millman, ed.), 67–74, Reidel.
Kuroda, P.K. and Sandell, E.B. (1954) *Geochim. Cosmochim. Acta* **6**, 35.
Müller, E. (1968) *Origin and Distribution of the Elements* (L.H. Ahrens, ed.) 155–176, Pergamon.
Murthy, V.R. (1962) *J. Geophys. Res.* **67**, 905.
Murthy, V.R. (1963) *Geochim. Cosmochim. Acta* **27**, 1171.
Smales, A.A., Mapper, D. and Fouché, K.F. (1967) *Geochim. Cosmochim. Acta* **31**, 673.
Wetherill, G.W. (1964) *J. Geophys. Res.* **69**, 4403.

(Received 12 September 1969; revised 1 June 1970)

RUTHENIUM (44)

Walter Nichiporuk

Center for Meteorite Studies
Arizona State University
Tempe, Arizona

RUTHENIUM, BEING the first member of the platinum group of metals, occurs in considerable amounts in meteorites. In most meteorites, ruthenium is mainly siderophile, but also has some chalcophile character; in bronzite and hypersthene chondrites, it is roughly as chalcophile as it is siderophile. Although ruthenium is regarded as having little or no lithophile character, it is occasionally looked for in the analyses of the achondritic meteorites. The ruthenium content of the meteorites has been calculated from many stony and iron meteorite analyses, and the distribution of ruthenium between the various mineral phases of the meteorites has been investigated. Goldschmidt (1954) summarized early work on ruthenium and other platinum group metals and reported average ruthenium abundance of 10×10^{-6} g/g in meteoritic nickel-iron, 9×10^{-6} g/g in troilite, and 2.23×10^{-6} g/g in mean meteoritic matter assumed to consist of 10 parts silicate with a negligible ruthenium content, 2 parts nickel-iron and 1 part troilite. The computed atomic abundance for ruthenium in relation to silicon taken as 10^6 atoms was 3.6, compared to 2.3 for the solar atmosphere (Russell, 1929). Goldschmidt also pointed out special difficulties in the determination of ruthenium, including volatility of its tetroxide. This, coupled with the fact that platinum metals were among the most common contaminants in the analytical reagents used in early ruthenium analyses, creates some doubts as to the accuracy of the ruthenium data summarized by Goldschmidt.

Ruthenium in meteorites has been studied in recent years by Bate and Huizenga (1963), Crocket et al. (1967) and Herr et al. (1958) using neutron activation analysis; by Hara and Sandell (1960), Sen Gupta and Beamish (1963) and Sen Gupta (1968a and 1968b) using spectrophotometric methods;

TABLE 1 Ruthenium in stony meteorites

Name	Type*	Number of meteorites analyzed	Total number of determinations	Range, 10^{-6} g/g†	Mean Ru, 10^{-6} g/g	Atomic Ru abundance, (Si = 10^6 atoms)[a]
Chondrites						
Carbonaceous	C1	3	7	0.58–0.78	0.69	1.9
Carbonaceous	C2	3	7	0.69–0.88	0.83	1.8
Carbonaceous	C3	1	4	1.0–1.1	1.0	1.8
Bronzite	H, H4–H6	14	21	0.82–1.4	1.1	1.8
Hypersthene	L, L4–L6	11	16	0.60–0.82	0.75	1.1
Amphoterite	LL6	1	1	0.50	0.50	0.74
Enstatite	E4	2	6	0.92–1.1	1.0	1.6
Achondrites						
Hypersthene		1	1	0.0029	0.0029	

* Classification of Van Schmus and Wood (1967).

† From the work of Crocket et al. (1967), Bate and Huizenga (1963), Hara and Sandell 1960) and Sen Gupta (1968a and 1968b).

[a] Using standard average silicon values for individual chondrite groups supplied for the purposes of these reviews by B. Mason.

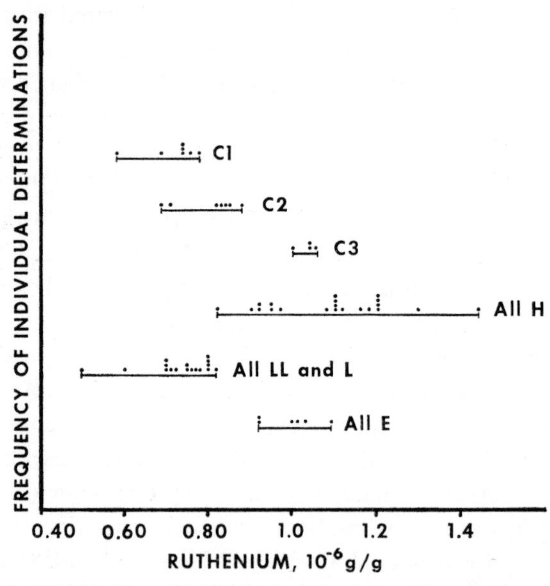

FIGURE 1 Distribution of individual ruthenium determinations for the major classes of chondrites.

and by Nichiporuk and Brown (1965) and Yavnel' (1950) using emission spectrography. The data on ruthenium in the various classes of stony meteorites are summarized in Table 1. Atomic abundances of ruthenium in the chondrites relative to Si = 10^6 atoms are also listed in Table 1. The distribution of the individual data points including replicate determinations on individual meteorites is indicated in Figure 1. The data on ruthenium in different classes of the iron meteorites are collected in Table 2 and are shown graphically in relation to the nickel contents in Figure 2.

TABLE 2 Ruthenium in iron meteorites

Class	Number of meteorites analyzed	Total number of determinations	Range* 10^{-6} g/g	Mean Ru 10^{-6} g/g
Normal hexahedrites	4	5	9.1–23.7	17.3
Granular hexahedrites	3	5	5.0–21.0	11.3
Coarse octahedrites†	7	10	3.8–16.6	8.6
Medium octahedrites	13	22	2.5–13.4 (52.4)[a]	6.0
Fine octahedrites	6	9	1.9–12.0	5.1
Finest octahedrites	2	4	0.5–0.8	0.6
Ni-rich ataxites	2	3	13.3–34.8	23.8

* Based on data of Hara and Sandell (1960), Nichiporuk and Brown (1965), Sen Gupta and Beamish (1963), Sen Gupta (1968a and 1968b), Herr et al. (1958), Yavnel' (1950).

† Including two coarsest octahedrites, Ainsworth and Arispe.

[a] Meteorite Carbo; excluded from calculation of the mean.

The problem in the determination of ruthenium in meteorites by spectrographic and spectrophotometric methods is principally one of the amount of the meteorite material used. While these techniques give accurate results, the amounts of the sparse material that are actually used are often very large, 20 g or more. In the activation analysis, samples as small as 50–100 mg are irradiated in a reactor to produce 215-keV, 2.9-day ^{97}Ru and 500-keV, 40-day ^{103}Ru, both of which are used in ruthenium determinations. However, ^{103}Ru is also produced in the fission of ^{235}U. Therefore, a correction must be made for this production if the uranium content of the sample is comparable to its ruthenium content, as is indeed the case with the achondritic meteorites. After subtracting the fission-produced ^{103}Ru gamma contribution to 500-keV gamma photopeak, Bate and Huizenga (1963) attributed the remaining 500-keV activity to natural ruthenium in their analysis of the

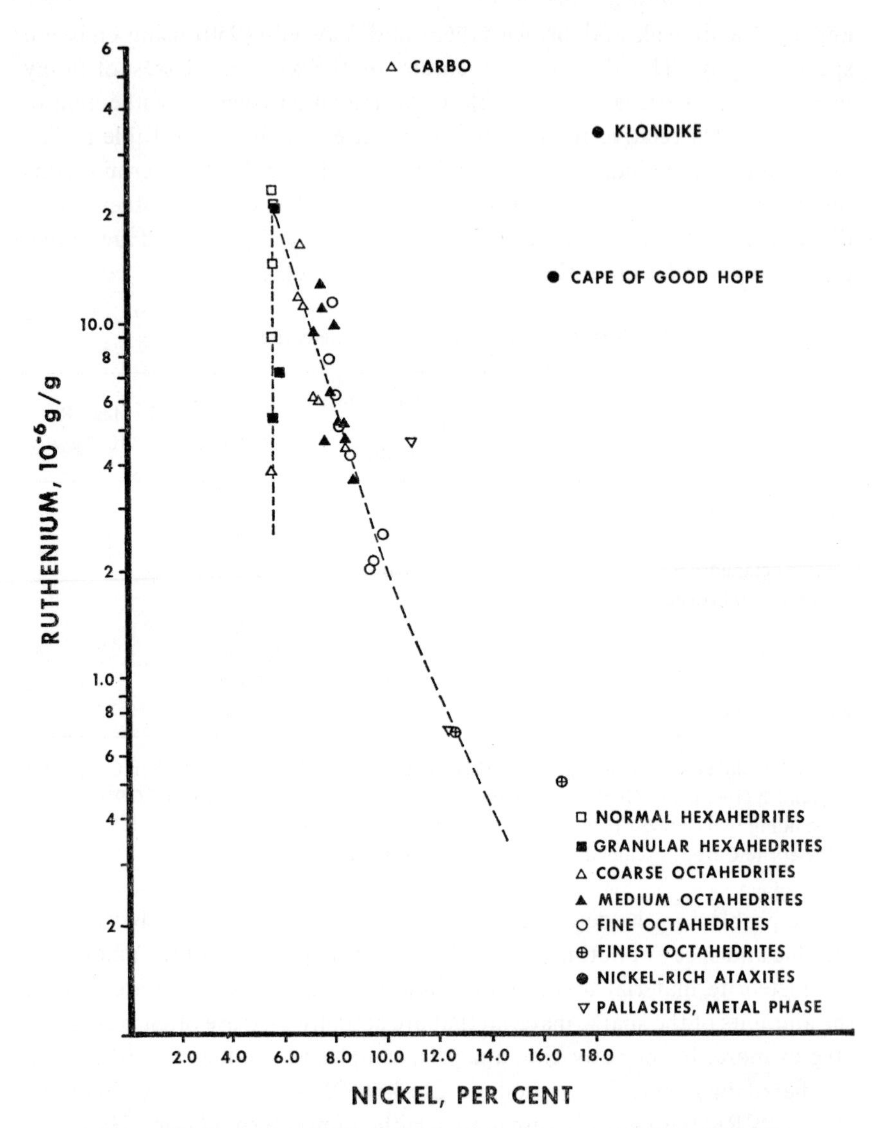

FIGURE 2 Variation of ruthenium contents of iron meteorites and of the metal phase of pallasites with nickel content. Sources of nickel data: Goldberg et al. (1951), Yavnel' (1954), Lovering et al. (1957), Nichiporuk (1958), Wasson and Kimberlin (1967), Sen Gupta (1968a), and Moore et al. (1969)

hypersthene achondrite Johnstown and in this way calculated its ruthenium content.

Table 1 shows a uniform ruthenium level of about 1.8 atoms/10^6 Si atoms in the C1, C2, and C3 carbonaceous chondrites, even though the experimental data points for these meteorites are relatively few and the individual determinations do exhibit some scatter (Figure 1). Based on these studies, it is interesting to note that ruthenium actually departs from the trend observed for several other siderophile elements (Larimer and Anders, 1967). Larimer (1967) does not include ruthenium in his listing of depleted elements and Crocket et al. (1967) calculated a condensation temperature of 1875°K at 1 atm. for ruthenium, suggesting that it was among the very first elements to be collected into a condensed phase from a cooling solar nebula. The ruthenium atomic abundance in all three types of carbonaceous chondrites based on this summary is in a fairly satisfactory agreement with the abundance of 1.49 Ru atoms/10^6 Si atoms selected by Suess and Urey (1956) for the composition of the nonvolatile part of solar system material, but is about a factor of 2 larger than the following atomic abundances: 0.87 selected by Cameron (1959), 0.83 calculated by Clayton and Fowler (1961) from theories of nucleosynthesis, and the value of 0.85 for the solar atmosphere according to Aller (1961).

In ordinary chondrites the ruthenium abundances are comparable to those in the carbonaceous chondrites, with the exception of the hypersthene chondrites, which show a moderate depletion. The unique meteorite Johnstown shows extreme depletion; this meteorite is composed almost entirely of an oxidized magnesium-iron metasilicate phase which does not accommodate ruthenium.

Hara and Sandell (1960) report 4.3×10^{-6} g Ru/g and 5.3×10^{-6} g Ru/g, respectively, for the separated metal phases of the two composites of six and nine different bronzite and hypersthene chondrites, and 6.3×10^{-6} g Ru/g and 5.2×10^{-6} g Ru/g, respectively, for the separated troilite phases of these composites. Although very similar ruthenium contents of 5.4×10^{-6} g/g to 6.5×10^{-6} g/g have been reported by Nichiporuk and Brown (1965) for the metal phases of five different bronzite chondrites, smaller ruthenium values of 2.3×10^{-6} g/g to 3.2×10^{-6} g/g have been reported by Bate and Huizenga (1963) for the metal phases of the bronzite chondrites Forest City and Ochansk. These investigators also report a large and variable ruthenium content of 0.2×10^{-6} g/g to 0.7×10^{-6} g/g for the silicate phases of these two chondrites, and a very small ruthenium content of some 0.03×10^{-6} g/g for the Canyon Diablo troilite. In the metal phases of the Admire and Brenham pallasites, Hara and Sandell have found 0.7×10^{-6} g Ru/g and 4.6×10^{-6} g Ru/g, respectively. While all of these data point to a principally siderophile character of ruthenium, it is clear that in order to establish with certainty the degree of its chalcophile and lithophile character, further work on ruthenium abundance distributions in the separated phases of meteoritic materials is highly desirable.

Ruthenium in the large groups of iron meteorites has been studied by Hara and Sandell (1960) and by Nichiporuk and Brown (1965). Additional data on the various iron meteorites have been reported in publications dealing with smaller numbers of meteorites (e.g. Sen Gupta and Beamish, 1963) or with single meteorites (Yavnel', 1950; Herr et al., 1958). Data of different investigators for the overlapping meteorites show only a moderate scatter, indicating that individual meteorites are probably very nearly uniform in composition. The ranges of the ruthenium content of the iron meteorites based on the analyses summarized in Table 2 are only for the structural classes as indicated in the work of Hara and Sandell and of Nichiporuk and Brown. Below about 8% nickel (Figure 2) and in the region where the octahedrites gradually pass into structureless normal hexahedrites, the ruthenium content rises sharply with decreasing nickel concentration (note, however, prominent exceptions provided by Ni-rich ataxites Cape of Good Hope and Klondike, and medium octahedrite Carbo). It appears that, in this region, the content of ruthenium depends on the structure of the iron as well as on the nickel content. Nichiporuk and Brown noted also that there appears to be a relatively strong positive correlation of ruthenium content with platinum content in the iron meteorites.

References

Aller, L.H. (1961) *The abundance of the elements*. Interscience.
Bate, G.L. and Huizenga, J.R. (1963) *Geochim. Cosmochim. Acta* **27**, 345.
Cameron, A.G.W. (1959) *Astrophys. J.* **129**, 676.
Clayton, D.D. and Fowler, W.A. (1961) *Ann. Phys.(N.Y.)* **16**, 51.
Crocket, J.H., Keays, R.R. and Hsieh, S. (1967) *Geochim. Cosmochim. Acta* **31**, 1615.
Goldberg, E., Uchiyama, A. and Brown, H. (1951) *Geochim. Cosmochim. Acta* **2**, 1.
Goldschmidt, V.M. (1954) *Geochemistry*. Edited by A. Muir. Clarendon Press.
Hara, T. and Sandell, E.B. (1960) *Geochim. Cosmochim. Acta* **21**, 145.
Herr, W., Merz, E., Eberhardt, P., Geiss, J., Land, C. and Signer, P. (1958) *Geochim. Cosmochim. Acta* **14**, 158.
Larimer, J.W. (1967) *Geochim. Cosmochim. Acta* **31**, 1215.
Larimer, J.W. and Anders, E. (1967) *Geochim. Cosmochim. Acta* **31**, 1239.

Lovering, J., Nichiporuk, W., Chodos, A. and Brown, H. (1957) *Geochim. Cosmochim. Acta* **11**, 263.
Moore, C.B., Lewis, C.F. and Nava, D. (1969) *Meteorite Research* (ed. P. Millman), 738. Reidel.
Nichiporuk, W. (1958) *Geochim. Cosmochim. Acta* **13**, 233.
Nichiporuk, W. and Brown, H. (1965) *J. Geophys. Res.* **70**, 459.
Russell, H.N. (1929) *Astrophys. J.* **70**, 11.
Sen Gupta, J.G. (1968a) *Anal. Chim. Acta* **42**, 481.
Sen Gupta, J.G. (1968b) *Chem. Geol.* **3**, 293.
Sen Gupta, J.G. and Beamish, F.E. (1963) *Am. Mineral.* **48**, 379.
Suess, H.E. and Urey, H.C. (1956) *Revs. Mod. Phys.* **28**, 53.
Van Schmus, W.R. and Wood, J.A. (1967) *Geochim. Cosmochim. Acta* **31**, 747.
Wasson, J.T. and Kimberlin, J. (1967) *Geochim. Cosmochim. Acta* **31**, 2065.
Yavnel', A.A. (1950) *Meteoritika* **8**, 134.
Yavnel', A.A. (1954) *Meteoritika* **11**, 107.

(Received 30 July 1969)

RHODIUM (45)

Walter Nichiporuk

Center for Meteorite Studies
Arizona State University
Tempe, Arizona

RHODIUM, EVEN MORE than ruthenium, is a typical siderophile element, and in the analyses of meteorites is usually determined along with ruthenium. Analyses for rhodium were published by the Noddacks (1930, 1931) and Goldschmidt and Peters (1932), and were summarized by Goldschmidt (1954). He reported an average content of 5×10^{-6} g Rh/g for meteoritic nickel-iron and a content of 0.4×10^{-6} g Rh/g for meteoritic troilite. From the average composition of the meteorites which he estimated at 10 parts silicate, 2 parts nickel-iron, and 1 part troilite, Goldschmidt obtained 0.80×10^{-6} g Rh/g for mean meteoritic matter. The calculated atomic abundance relative to silicon taken at 10^6 was 1.3, compared to 0.14 for the solar atmosphere according to Russell (1929).

In more recent years data on rhodium in meteorites were obtained by Schindewolf and Wahlgren (1960) using modern techniques of neutron activation analysis, by Sen Gupta and Beamish (1963) and Sen Gupta (1968a and 1968b) using spectrophotometric methods, and by Yavnel' (1950) and Nichiporuk and Brown (1965) using emission spectrography. Since the number of analyses of chondritic meteorites is small, all available data are listed individually by meteorites in Table 1. Atomic abundances for the different chondrite groups relative to Si $= 10^6$ are also included in Table 1. The data on rhodium in the various classes of iron meteorites are summarized in Table 2.

The rhodium content of the chondrites shows a considerable spread, so that the arithmetic means are uncertain to about 25 per cent. Note the unique chondrite Benton; it has the largest concentration of rhodium and the smallest content of metallic nickel-iron, 0.9 per cent by weight, of any of the

TABLE 1 Rhodium in chondrites

Name	Type*	Rh content 10^{-6} g/g	Reference	Mean Rh 10^{-6} g/g	Atomic abundance (Si = 10^6 atoms)†
Beardsley	H5	0.21	(1)		
Forest City	H5	0.21	(1)	0.25	0.40
Hessle	H5	0.20	(1)		
Belly River	H	0.36, 0.40	(2)[a]		
Bruderheim	L6	0.25	(2)[a]		
Holbrook	L6	0.15	(1)	0.22	0.31
Modoc	L6	0.16	(1)		
Peace River	L6	0.30, 0.31	(2)[a]		
Benton	LL6	0.48	(2)[a]		0.70
Abee	E4	0.25, 0.25	(2)[a]		0.36

* Classification according to Van Schmus and Wood (1967). References: (1) Schindewolf and Wahlgren (1960); (2) Sen Gupta (1968a and 1968b).

† Using standard average silicon values for individual chondrite groups supplied for the purposes of these reviews by B. Mason.

[a] It may be noted that the values reported by Sen Gupta indicate systematically a greater rhodium content for a given type of chondrite than the values reported by Schwindewolf and Wahlgren.

TABLE 2 Rhodium in iron meteorites

Class	Number of meteorites analyzed	Total number of determinations	Range* 10^{-6} g/g	Mean Rh 10^{-6} g/g
Normal hexahedrites	3	3	2.3–2.5	2.4
Granular hexahedrites	3	4	0.9–2.1	1.6
Coarse octahedrites†	5	6	1.4–3.9	2.0
Medium octahedrites	11	14	1.1–4.1	2.0
Fine octahedrites	5	6	0.7–4.2	1.8
Finest octahedrites	2	2	0.14–0.3	0.22
Ni-rich ataxites	2	3	1.7–5.5	3.5

* Based on data of Sen Gupta and Beamish (1963), Nichiporuk and Brown (1965), Sen Gupta (1968a and 1968b) and Yavnel' (1950).

† Including one coarsest octahedrite.

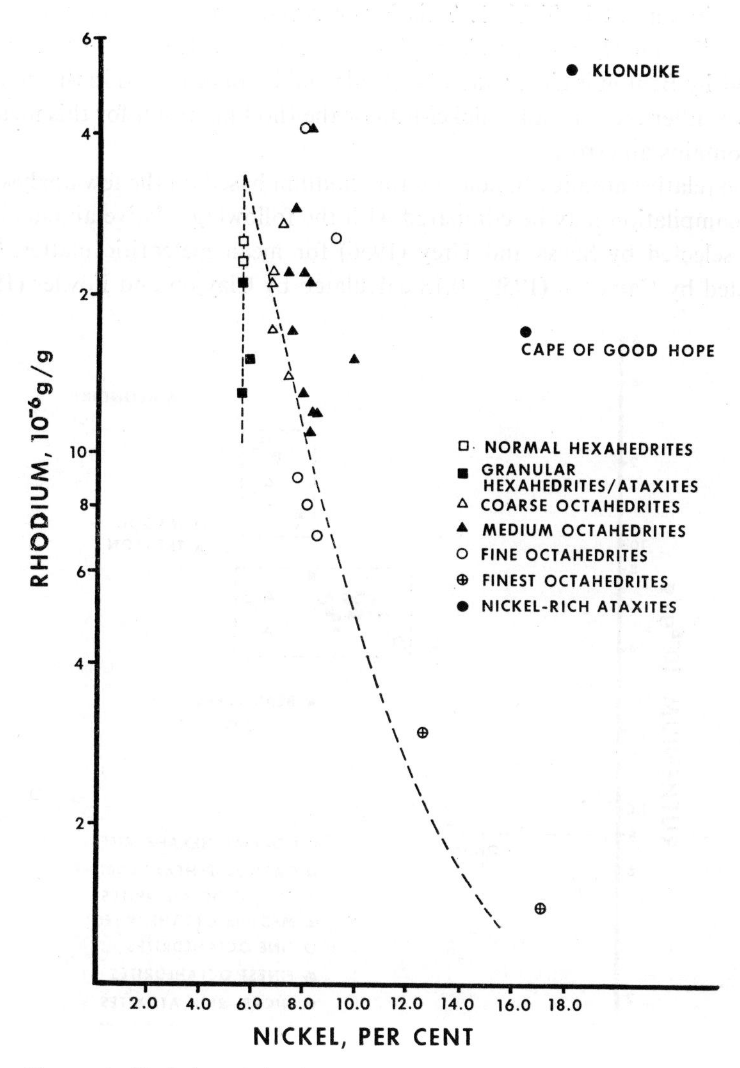

FIGURE 1 Variation of rhodium contents of iron meteorites with nickel content. The figure is from Nichiporuk and Brown (1965), supplemented by the data on rhodium from Sen Gupta and Beamish (1963), Sen Gupta (1968a and 1968b), and Yavnel' (1950). Sources of the nickel data: Goldberg et al. (1951), Yavnel' (1954), Lovering et al. (1957), Nichiporuk (1958), Wasson and Kimberlin (1967), Sen Gupta (1968a), and Moore et al. (1969).

chondrites listed in Table 1. This is the reverse of a relation one would expect for an element as siderophile as rhodium. Thus rhodium in the chondrite Benton is either not siderophile and is concentrated in the mineral phases other than metallic nickel-iron or the rhodium result for this meteorite contains an error.

The relative atomic abundances for rhodium based on the few analyses in this compilation may be compared with the following relative abundances: 0.21 selected by Suess and Urey (1956) for mean meteoritic matter, 0.15 selected by Cameron (1959), 0.13 calculated by Clayton and Fowler (1961)

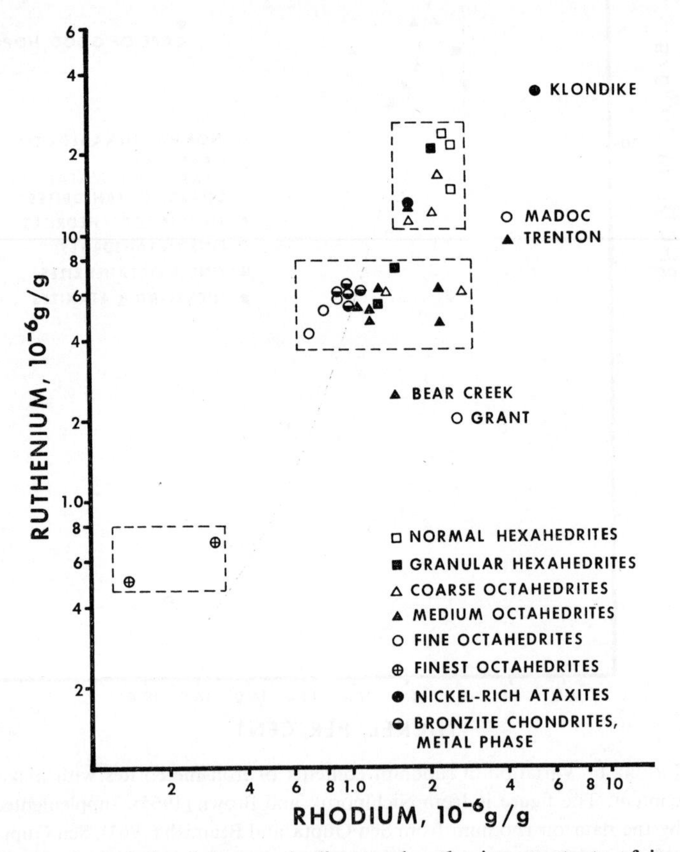

FIGURE 2 Relation between rhodium and ruthenium contents of iron meteorites and of the metallic nickel-iron phases of chondrites. The figure is from Nichiporuk and Brown (1965); the data on rhodium as in the caption to Figure 1. Sources of the data on ruthenium: Hara and Sandell (1960), Nichiporuk and Brown (1965), Sen Gupta and Beamish (1963), Sen Gupta (1968a and 1968b), Yavnel' (1950), and Herr et al. (1958).

from theories of the origin of the elements, and 0.19 given by Aller (1961) for the solar atmosphere.

The concentration of rhodium in the metallic nickel-iron phase of chondrites has been determined by Nichiporuk and Brown (1965). Studies of the separated metal phases of the chondrites Alamogordo, Gilgoin Station, Gladstone, Ochansk, and Plainview show that these phases exhibit a very small range of values for rhodium from 0.9×10^{-6} g/g to 1.1×10^{-6} g/g. These investigations showed that the concentrations of rhodium in the metallic nickel-iron phase of the chondrites are in general compatible with the total rhodium contents of these meteorites.

The results on iron meteorites represent the majority of published results. In iron meteorites almost all rhodium is present in solution in the metallic nickel-iron, usually in the range of 0.7×10^{-6} g/g to 2.5×10^{-6} g/g, but as much as 4.1×10^{-6} g/g was recorded in the medium octahedrite Trenton (Sen Gupta and Beamish) and up to 5.5×10^{-6} g/g in the nickel-rich ataxite Klondike (Sen Gupta). An interesting variation of rhodium with nickel, shown in Figure 1, and of rhodium with ruthenium, shown in Figure 2, has been reported in the literature for iron meteorites. Possibly other elements showing variations, such as gallium, germanium, osmium, rhenium, and iridium (Wasson and Kimberlin, 1967; Lovering et al., 1957) are related to these.

References

Aller, L.H. (1961) *The abundance of the elements*. Interscience.

Cameron, A.G.W., (1959) *Astrophys. J.* **129**, 676.

Clayton, D.D. and Fowler, W.A. (1961) *Ann. Phys. (N.Y.)* **16**, 51.

Goldberg, E., Uchiyama, A., and Brown, H. (1951) *Geochim. Cosmochim. Acta* **2**, 1.

Goldschmidt, V.M. (1954) *Geochemistry*. Edited by A.Muir. Clarendon Press.

Goldschmidt, V.M. and Peters, Cl. (1932) *Nachr. Ges. Wiss. Göttingen, math.-phys. Kl.* **4**, 377.

Hara, T. and Sandell, E.B. (1960) *Geochim. Cosmochim. Acta* **21**, 145.

Herr, W., Merz, E., Eberhardt, P., Geiss, J., Land, C. and Signer, P. (1958) *Geochim. Cosmochim. Acta* **14**, 158.

Lovering, J., Nichiporuk, W., Chodos, A., and Brown, H. (1957) *Geochim. Cosmochim. Acta* **11**, 263.

Moore, C.B., Lewis, C.F. and Nava, D. (1969) *Meteorite Research* (ed. P.M.Millman), 738, Reidel.

Nichiporuk, W. (1958) *Geochim. Cosmochim. Acta* **13**, 233.

Nichiporuk, W. and Brown, H. (1965) *J. Geophys. Res.* **70**, 459.

Noddack, I. and Noddack, W. (1930) *Naturwiss.* **18**, 757.

Noddack, I. and Noddack, W. (1931) *Z. phys. Chem., Bodenstein-Festband*, **890**.
Russell, H.N. (1929) *Astrophys. J.* **70**, 11.
Schindewolf, U. and Wahlgren, M. (1960) *Geochim. Cosmochim. Acta* **18**, 36.
Sen Gupta, J.G. (1968a) *Anal. Chim. Acta* **42**, 481.
Sen Gupta, J.G. (1968b) *Chem. Geol.* **3**, 293.
Sen Gupta, J.G. and Beamish, F.E. (1963) *Am. Mineral.* **48**, 379.
Suess, H.E. and Urey, H.C. (1956) *Revs. Mod. Phys.* **28**, 53.
Van Schmus, W.R. and Wood, J.A. (1967) *Geochim. Cosmochim. Acta* **31**, 747.
Wasson, J.T. and Kimberlin, J. (1967) *Geochim. and Cosmochim. Acta* **31**, 2065.
Yavnel', A.A. (1950) *Meteoritika* **8**, 134.
Yavnel', A.A. (1954) *Meteoritika* **11**, 107.

(Received 10 August 1969)

PALLADIUM (46)

Walter Nichiporuk

Center for Meteorite Studies
Arizona State University
Tempe, Arizona

OF THE THREE platinum group metals, ruthenium, rhodium and palladium, palladium is the most frequently determined metal in the meteorites. Goldschmidt (1958), who reviewed early X-ray spectroscopic (Noddack and Noddack, 1930 and 1931) and emission spectroscopic (Goldschmidt and Peters, 1932) analyses for palladium, reported an average of 9×10^{-6} g/g for the meteoritic nickel-iron phase and an average of 2×10^{-6} g/g for the meteoritic troilite phase. In 1951 E. Goldberg, A. Uchiyama, and H. Brown first used modern techniques of neutron activation analysis for the determination of palladium in a wide range of iron meteorites. They reported a range from 1.4×10^{-6} g Pd/g to 9.9×10^{-6} g Pd/g with a mean of 3.7×10^{-6} g Pd/g. Goldschmidt calculated an average for palladium in meteorites, but in order to get a fairly representative estimate he calculated the palladium in a mixture of 10 parts silicate which he assumed to contain no palladium, 2 parts nickel-iron, and 1 part troilite, using the averages of the older data. This gave 1.5×10^{-6} g Pd/g. Using the more recent iron meteorite value of Goldberg et al. (1951) this average would be approximately 0.6×10^{-6} g Pd/g. For comparison with the palladium abundance of 0.53 atoms/10^6 atoms Si in the solar atmosphere, based on Russell's data (1929), Goldschmidt calculated the palladium abundance of 2.5 atoms/10^6 atoms Si for mean meteoritic matter.

Since the work of Goldberg et al. was done, many additional data on palladium in meteorites have become available. Table 1 gives a list of the recent determinations of palladium in chondrites and achondrites. These determinations have been made by neutron activation analysis, except the data of Sen Gupta (1967), which are spectrophotometric. The palladium data

TABLE 1 Palladium in chondrites and achondrites

Name	Classification*	Pd, 10^{-6} g/g (Reference)
Alais	C1	0.33 (3)
Ivuna	C1	1.8 (1); 0.53, 0.55 (2); 0.61, 0.54 (3)
Orgueil	C1	1.6 (1); 0.59, 0.54 (2); 0.52, 0.62, 0.62 (3)
Cold Bokkeveld	C2	0.60, 0.56 (2); 0.57 (3)
Mighei	C2	1.3 (1); 0.59, 0.59 (2); 0.62 (3)
Murray	C2	0.94 (1); 0.79 (4)
Nogoya	C2	0.75 (3)
Felix	C3	2.2 (1)
Lancé	C3	2.0 (1); 0.68, 0.66 (2); 0.74, 0.88 (3); 1.0 (4)
Mokoia	C3	1.3 (1)
Ornans	C3	0.68, 0.72 (2)
Vigarano	C 3	1.4 (1)
Warrenton	C3	1.9 (1)
Dimmitt	H3–4	0.91 (4)
Beaver Creek	H4	1.1 (4)
Ochansk	H4	1.1 (4)
Allegan	H5	1.5 (1)
Forest City	H5	1.4, 0.76, 0.95 (6)
Pantar	H5	1.3 (1); 0.52, 0.53 (5)
Belly River	H	0.77, 0.80 (7)
Ehole	H	1.5 (1)
Bruderheim	L6	1.0 (1); 0.50 (7)
Mocs	L6	0.84 (1)
Modoc	L6	0.82, 0.63, 0.73 (6)
Peace River	L6	0.89 (1); 0.41, 0.44 (7)
Walters	L6	1.0 (1)
Chainpur	LL3	0.83 (1)
Benton	LL6	0.60 (7)
Ensisheim	LL 6	~0.2, 0.13 (5)
Abee	E4	1.3 (1); 0.79 (3); 0.46, 0.44 (7)
Indarch	E4	2.1 (1); 0.80 (3)
Daniel's Kuil	E6	2.2 (1)
Hvittis	E6	1.1 (1)
Khairpur	E6	1.4 (1)
Pillistfer	E6	1.9 (1)
Nuevo Laredo	Eu	not detected (6)

* Classification according to Van Schmus and Wood (1967).

(1) Greenland (1967)
(2) Fouché and Smales (1967)
(3) Crocket et al. (1967)
(4) Rieder and Wänke (1969)
(5) Reed (1963)
(6) Hamaguchi et al. (1961)
(7) Sen Gupta (1967)

are somewhat variable in quality and the replicate analyses, particularly of the carbonaceous chondrites, show considerable spread, so it is difficult to determine whether the spread is due to errors of analyses or to sample errors, since many small samples were used in the analyses by neutron activation. But neither of these errors seems to apply, for example, to ruthenium, for which also many small samples were used (see an earlier chapter on ruthenium). Clearly, further determinations of palladium on different specimens of the same chondrite are needed.

TABLE 2 Palladium abundances in chondrite groups

Name	Type	Number of meteorites analyzed	Mean Pd* 10^{-6} g/g	Atomic Pd abundance Si = 10^6 atoms†
Chondrites				
Carbonaceous	C1	3	0.74 (0.49)[a]	1.9 (1.3)[a]
Carbonaceous	C2	4	0.76 (0.68)[a]	1.5 (1.3)[a]
Carbonaceous	C3	6	1.4 (0.77)[a]	2.4 (1.3)[a]
Bronzite	H, H3–H5	8	1.1 (0.91)[a]	1.7 (1.4)[a]
Hypersthene	L6	5	0.80 (0.55)[a]	1.1 (0.78)[a]
Amphoterite	LL3, LL6	3	0.54 (0.2)[b]	0.76 (0.28)[b]
Enstatite	E4	2	1.2 (0.83)[c]	1.9 (1.3)[c]
Enstatite	E6	4	1.7	2.3

* Based on data of Table 1.

† Using standard silicon values for chondrite groups supplied for the purposes of these compilations by B. Mason.

[a] Omitting analyses by Greenland (1967).

[b] Omitting analyses by Greenland (1967) and by Sen Gupta (1967).

[c] Omitting analysis of Indarch by Greenland (1967).

The averages for the chondrite groups derived from the data of Table 1 and the atomic abundances calculated from them are given in Table 2. Whenever the average value is in parentheses, it means that it has been calculated with the omission of the high values in Table 1 and is probably to be preferred. Based on these preferred averages, it is interesting to note that the palladium level in the carbonaceous chondrites is constant at about 1.3 atoms/10^6 atoms Si, regardless of the meteorite type. The palladium level is also constant at 1.3 in the bronzite and enstatite E4 chondrites but decreases steadily from the bronzite to the hypersthene to the amphoteric chondrites.

The atomic abundances in the C1 chondrites are often regarded as being representative of the initial abundances in undifferentiated solar system material. The C1 atomic abundance of 1.3 in this compilation is about a factor of two greater than the value of 0.675 selected by Suess and Urey (1956) and by Cameron (1959) and slightly more than a factor of two greater than the value of 0.601 calculated by Clayton and Fowler (1961) from theories of nucleosynthesis. The abundance of palladium at 0.51 in the solar atmosphere, based on Aller's (1961) data, is about 2/5 of the value given in this compilation for the carbonaceous material.

Larimer and Anders (1967) did not list palladium among the elements showing a normal depletion ratio of 1/0.6/0.3 in the groups C1, C2, and C3. From the data of Table 2, the abundance ratio for palladium relative to C1 group is 1.0/1.0/1.0. Although the analytical data are relatively few for these meteorites and have been selected in this compilation, it appears that within these stated restrictions palladium is a typical unfractionated platinum metal in the carbonaceous chondrites.

Fouché and Smales (1967) have studied palladium in the separated metallic and silicate phases of seven bronzite, nine hypersthene and four enstatite chondrites. They obtained an average metal phase palladium level of 4.4×10^{-6} g/g for the bronzite chondrites, 8.0×10^{-6} g/g for the hypersthene chondrites and 3.5×10^{-6} g/g for the enstatite chondrites. The observed palladium levels in the separated silicate phases varied widely: 0.023×10^{-6} g/g – 0.14×10^{-6} g/g in the bronzite chondrites, 0.015×10^{-6} g/g – 0.18×10^{-6} g/g in the hypersthene chondrites, and 0.016×10^{-6} g/g – 0.038×10^{-6} g/g in the enstatite chondrites. Palladium has been also studied in the separated metal phases of the bronzite chondrites by Nichiporuk and Brown (1965), and their average value of 4.1×10^{-6} g/g agrees very closely with the corresponding value of Fouché and Smales. Whereas the concentration of palladium in the metal phase of the pallasite Admire was 4.9×10^{-6} g/g, the concentration of palladium in the olivine phase of that pallasite was less than 0.04×10^{-6} g/g, according to a report by Hamaguchi et al. (1961). Obviously palladium favors the metal phase over the silicate phase with a concentration factor of the order of 100 in the bronzite and enstatite chondrites and with a concentration factor of the order of 200 in the hypersthene chondrites.

Some rather interesting results have been reported by Reed (1963) and by Rieder and Wänke (1969), from their investigations of separated light and dark fractions of gas-rich bronzite chondrites. Within experimental error palladium was found in identical concentrations in the light and dark fractions of the chondrites Pantar, Leighton, and Fayetteville, although the results of Rieder and Wänke indicated about twice as much palladium in the chondrite Pantar as did the results of Reed. According to Reed, the amphoteric chondrite Ensisheim which he investigated as a possible example of the Pantar type meteorite also contained nearly equal amounts of palladium in its light and dark fractions.

Table 3 gives a summary of the recent determinations of the amounts of palladium in the iron meteorites. Smales, Mapper, and Fouché (1967) have published the most recent analyses for palladium in a number of iron meteorites. Only those meteorites which fell into five distinct structural and trace

TABLE 3 Palladium in iron meteorites

Class	Number of meteorites analyzed	Total number of determinations	Range* 10^{-6} g/g	Mean Pd 10^{-6} g/g
Hexahedrites	17	22	1.44–2.7 (6.9)†	2.0
Nickel-poor ataxites	4	5	1.5–2.4	1.9
Coarsest octahedrites	7	9	2.4–3.5	3.2
Coarse octahedrites	14	30	2.0–5.1	3.4
Medium octahedrites	42	62	1.2–7.1	3.8
Fine octahedrites	16	27	1.7–6.7	4.3
Finest octahedrites	5	7	3.1–7.7	4.8
Nickel-rich ataxites	10	14	4.1–19.7	9.7

* From the work of Smales et al. (1967), Goldberg et al. (1951), Nichiporuk and Brown (1965), Chakraburtty et al. (1964), Hamaguchi et al. (1961), Sen Gupta and Beamish (1963), Sen Gupta (1967), Yavnel' (1950).

† Omitted from the mean.

siderophile element groups were tabulated separately by those authors. Many other meteorites, while apparently similar structurally to the classified meteorites, differed from them in the concentration of a wide range of siderophile elements, and were not included in the observed groupings. Although Smales et al. feel that their palladium data are consistent with the proposed five groups, it appears that in view of such limited memberships of the groups it would actually be a very difficult task in this computation to attempt to classify on the basis of the proposed system some 115 iron meteorites for which palladium contents are listed in the literature. Consequently the entries in Table 3 are only for the conventional structural classes (Hey,

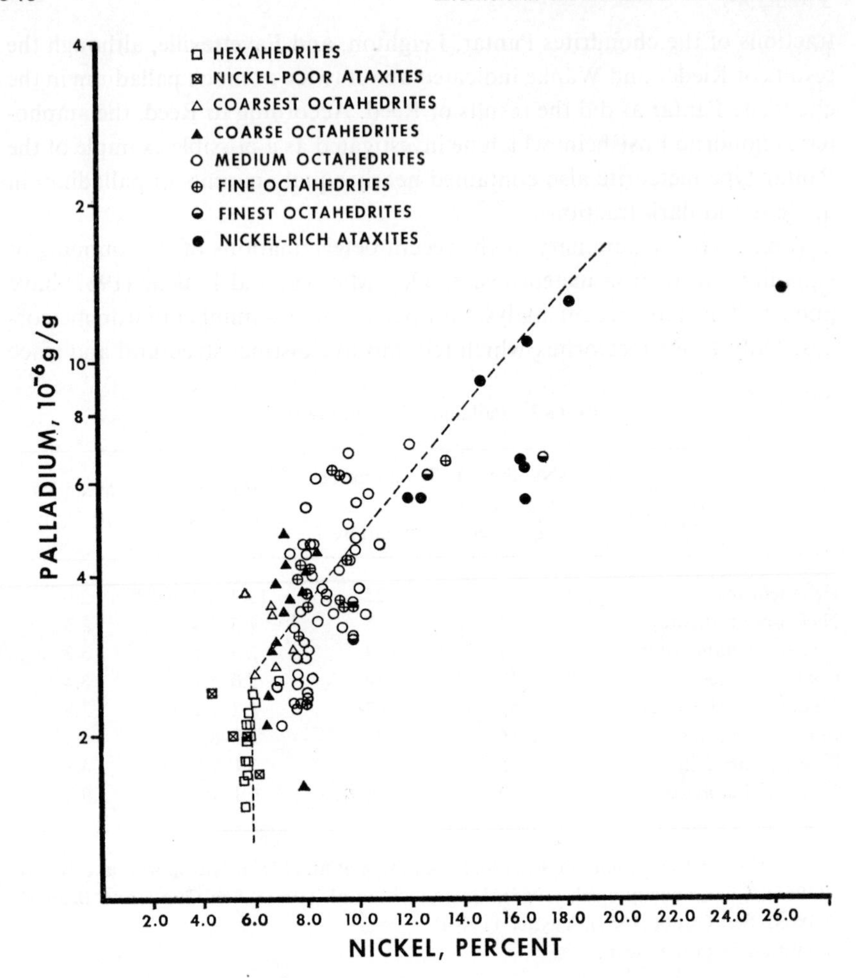

FIGURE 1 Plot of palladium contents of iron meteorites against nickel content. The figure is from Goldberg et al. (1951) and from Nichiporuk and Brown (1965), supplemented by data on palladium from Smales et al. (1967), Chakraburtty et al. (1964), Hamaguchi et al. (1961), and Sen Gupta and Beamish (1963). Sources of the nickel data: Goldberg et al. (1951), Lovering et al. (1957), Nichiporuk (1958), Wasson and Kimberlin (1967), Goldstein and Short (1967), Sen Gupta (1968), Moore et al. (1969).

1966) and all meteorites have been classified as belonging to these classes. Analyses of four iron meteorites (Canyon Diablo, Henbury, Toluca, and Grant) for palladium by different workers in each case showed considerable spread of data, so that the averages for these meteorites are uncertain to at least 30 per cent. It can be seen that the palladium content of iron meteorites increases progressively from hexahedrites to nickel-rich ataxites and that the total range of palladium in iron meteorites is from about 1.5×10^{-6} g/g to 20×10^{-6} g/g.

Figure 1 is a composite of similar figures from Goldberg et al. (1951) and from Nichiporuk and Brown (1965), and illustrates the tendency of the palladium concentrations to increase with nickel concentrations. Fouché and Smales (1966) reported a strong positive correlation of palladium with gold ($r = 0.68$) in iron meteorites which was weakened when the log values were taken ($r = 0.52$). Palladium however appears to be negatively correlated with platinum in iron meteorites (Nichiporuk and Brown).

References

Aller, L.H. (1961) *The abundance of the elements*. Interscience.

Cameron, A.G.W. (1959) *Astrophys. J.* **129**, 676.

Chakraburtty, A.K., Stevens, C.M., Rushing, H.C. and Anders, E. (1964) *J. Geophys. Res.* **69**, 505.

Clayton, D.D. and Fowler, W.A. (1961) *Ann. Phys. (N.Y.)* **16**, 51.

Crocket, J.H., Keays, R.R. and Hsieh, S. (1967) *Geochim. Cosmochim. Acta* **31**, 1615.

Fouché, K.F. and Smales, A.A. (1966) *Chem. Geol.* **1**, 329.

Fouché, K.F. and Smales, A.A. (1967) *Chem. Geol.* **2**, 105.

Goldberg, E., Uchiyama, A. and Brown, H. (1951) *Geochim. Comochim. Acta* **1**, 1.

Goldschmidt, V.M. (1954) *Geochemistry*. Edited by A. Muir. Clarendon Press.

Goldschmidt, V.M. and Peters, C. (1932) *Nach. Ges. Wiss. Gottingen, Math-phys. Kl.* **4**, 372.

Goldstein, J.I. and Short, J.M. (1967) *Geochim. Cosmochim. Acta* **31**, 1733.

Greenland, L. (1967) *Geochim. Cosmochim. Acta* **31**, 849.

Hamaguchi, H., Nakai, T. and Kamemoto, Yu. (1961) *Nippon Kagaku Zasshi* **82**, 1489.

Hey, M.H. (1966) *Catalogue of meteorites*. British Museum (Natural History).

Larimer, J.W. and Anders, E. (1967) *Geochim. Cosmochim. Acta* **31**, 1239.

Lovering, J.F., Nichiporuk, W., Chodos, A. and Brown, H. (1957) *Geochim. Cosmochim. Acta* **11**, 263.

Moore, C.B., Lewis, C.F. and Nava, D. (1969) *Meteorite research* (ed. P.M. Millman), 738. Reidel.

Nichiporuk, W. (1958) *Geochim. Cosmochim. Acta* **13**, 233.

Nichiporuk, W. and Brown, H. (1965) *J. Geophys. Res.* **70**, 459.

Noddack, I. and Noddack, W. (1930) *Naturwiss.* **18**, 757.

Noddack, I. and Noddack, W. (1931) *Z. phys. Chem., Bodenstein-Festband,* **890**.

Reed, G.W. (1963) *J. Geophys. Res.* **68**, 3531.

Rieder, R. and Wänke, H. (1969) *Meteorite research* (ed. P.M.Millman), 76. Reidel.

Russell, H.N. (1929) *Astrophys. J.* **70**, 11.

Sen Gupta, J.G. (1967) *Anal. Chem.* **39**, 18.

Sen Gupta, J.G. (1968) *Anal. Chim. Acta* **42**, 481.
Sen Gupta, J.G. and Beamish, F.E. (1963) *Am. Mineral.* **38**, 379.
Smales, A.A., Mapper, D. and Fouché, K.F. (1967) *Geochim. Cosmochim. Acta* **31**, 673.
Suess, H.E. and Urey, H.C. (1956) *Revs. Mod. Phys.* **28**, 53.
Van Schmus, W.R. and Wood, J.A. (1967) *Geochim. Cosmochim. Acta* **31**, 747.
Wasson, J.T. and Kimberlin, J. (1967) *Geochim. Cosmochim. Acta* **31**, 2065.
Yavnel', A.A. (1950) *Meteoritika* **8**, 134.

(Received 20 August 1969)

SILVER (47)*

Peter R. Buseck

Departments of Geology and Chemistry
Arizona State University
Tempe, Arizona

SILVER IS PRESENT in meteorites at low but variable levels. In stony meteorites (26 falls and 1 find) the mean, median, and range of abundances are 0.18 ppm, 0.12 ppm, and 0.031 to 0.57 ppm, respectively; in iron meteorites (1 fall and 71 finds) the mean is 0.03 ppm, the median 0.01 ppm, and the range <0.01 to 0.20 ppm. Isotopic studies are contradictory, but probably indicate no enichment in the Ag^{107}/Ag^{109} ratio, relative to terrestrial silver. Values of Ag^{107}/Ag^{109} for 8 irons (1 fall and 7 finds) have a mean, median, and range of 1.09, 1.092, and 1.02 to 1.15, respectively.

Except for the spectrographic work of Lovering (1957), all of the bulk silver analyses were made by neutron activation analysis, while the isotopic data was performed using mass spectrographic techniques. Although there is the possibility of silver contamination from handling of coins, e.g. the Ehole chondrite, the relatively consistent ratios of selenium/silver (Greenland, 1967) for many meteorites, suggest that contamination is not a serious problem. Schindewolf and Wahlgreen (1960) estimate an accuracy of a factor of two for their analyses.

The silver contents of chondrites have been determined by Greenland (1967), Schindewolf and Wahlgren (1960), and Reed (1963). The results are relatively consistent, as indicated in Table 1 and Figure 1. They show a considerable spread in the concentrations in the carbonaceous chondrites, especially in group C3. There appears to be a decreasing trend going from group C1 to C3, with C1 chondrites having double the silver contents of C2 and C3 chondrites. The H, L and LL groups are slightly more depleted, although there is no apparent trend between them. The enstatite chondrites exhibit

* Contribution No. 46 from Arizona State University Center for Meteorite Studies.

TABLE 1 Silver contents of the stony meteorites

Type	Meteorite*†	Silver concentration ppm (Reference)	Mean, ppm	Mean atoms/10^6 Si atoms
C1	Ivuna	0.36 (1)		
	Orgueil	0.42 (1)	0.39	0.96
C2	Mighei	0.15 (1)		
	Murray	0.19 (1)	0.17	0.34
C3	Felix	0.28 (1)		
	Lancé	0.57 (1)		
	Mokoia	0.12 (1)	0.25	0.41
	Vigarano	0.12 (1)		
	Warrenton	0.16 (1)		
All C	–	–	0.26	0.48
E4	Abee	0.50 (1)		
	Indarch	0.41 (1)	0.46	0.70
E6	Hvittis	0.074 (1)		
	Khairpur	0.23 (1)	0.12	0.16
	Pillistfer	0.052 (1)		
All E	–	–	0.25	0.35
H5	Allegan	0.042 (1)		
	Pantar	0.033; 0.031, 0.035 (1) (2)		
	Beardsley	0.12 (3)	0.08	0.09
	Forest City	0.13 (3)		
	Hessle	0.06 (3)		
L6	Bruderheim	0.058 (1)		
	Mocs	0.17 (1)		
	Peace River	0.084 (1)		
	Walters	0.089 (1)	0.09	0.12
	Holbrook	0.04 (3)		
	Modoc	0.12 (3)		
LL3	Chainpur	0.12 (1)	0.12	0.16
LL6	Ensisheim	0.034 (2)	0.034	0.047
All L and LL	–	–	0.089	0.12

* Classification according to Van Schmus and Wood, 1967.

† Lovering, 1957, has found <0.5 ppm silver in olivine of the pallasites Admire, Albin, Brenham and Springwater. He also found <0.5 ppm silver in the troilite phase of Brenham and Springwater.

(1) Greenland, 1967

(2) Reed, 1963

(3) Schindewolf and Wahlgren, 1960

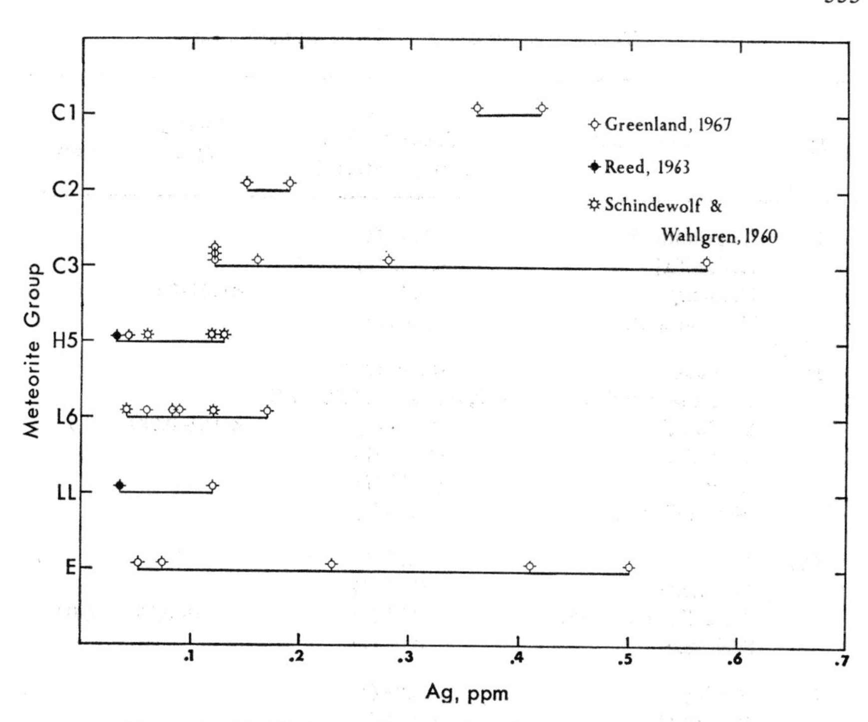

FIGURE 1 Distribution of silver analyses for the several groups of chondrites.

a greater range in silver than any of the other groups. Group E6 is depleted in silver relative to E4.

Fragments from the dark and light portions of the Pantar chondrite and from the light portion of the Ensisheim chondrite were measured by Reed (1963). The silver concentrations—0.035 (Pantar-dark), 0.031 (Pantar-light) and 0.034 ppm (Ensisheim-light)—are relatively constant, but lower that those for many other chondrites. There is an apparent 11 % difference between the silver concentrations in the light and dark fractions of Pantar. The silver is enriched in the dark portions, but considerably less so than is Ar, Xe or Bi.

An extensive study of the silver contents of iron meteorites was made by neutron activation analysis (Smales et al., 1967). Their results are summarized in Table 2 and Figure 2. Sampling was performed by selectively drilling the metal phase. Thus, their values only approximate a total meteorite analysis. It is indeed quite possible that there is an enrichment of silver in the troilite phase. In this event the values of Smales et al. would be systematically low, if interpreted as representative of the whole meteorite.

TABLE 2 Silver contents of the iron meteorites

Type	Meteorite*	Silver Concentration, ppm (Reference)	Range, ppm	Mean, ppm
Da	La Primitiva†	0.03 (1)		
	Nedagolla†	<0.01 (1)		
	Otumpa†	0.02 (1)	<0.01–0.03	0.02
	Tombigbee River†	(0.01 (1)		
H	Coahuila	<0.01 (2)		
	Sandia Mountains	0.032, 0.034, 0.071, 0.087 (2)		
	San Martin	<0.01 (1)	<0.01–0.087	0.02
	Tocopilla	<0.01 (1)		
	Uwet	<0.01 (1)		
	Walker County	0.04 (1)		
Ogg	Arispe	0.05 (1)		
	Gladstone	0.04 (1)		
	Sao Juliao de Moreira	0.01 (1)	0.01–0.05	0.03
	Seelasgen	0.03 (1)		
Og	Bendego	0.03 (1)		
	Bennett County	<0.01 (1)		
	Billings	0.01 (1)		
	Bischtube	0.03 (1)		
	Bohumilitz	0.03 (1)		
	Canyon Diablo	0.04; 0.098, 0.068, 0.072, 0.077, 0.066, 0.072, 0.055, 0.091. 0.087, 0.074, 0.088, 0.090, 0.079; 0.0408, 0.0588 0.0927 (1) (2) (3)	<0.01–0.177	0.05
	Cranbourne	0.03 (1)		
	Odessa	0.116, 0.177, 0.085, 0.112 (2)		
	Pan de Azucar	0.07 (1)		
	Wichita County	0.05 (1)		
	Youndegin	0.05 (1)		
Om	Breece	0.05 (1)		
	Campbellsville	0.02 (1)		
	Canton	<0.01 (1)		
	Caperr	<0.01 (1)		
	Cape York	<0.01 (1)		
	Chinautla	<0.01 (1)		
	Clark County	<0.01 (1)		
	Colfax	0.07 (1)	<0.01–0.132	0.02

TABLE 2 (cont.)

Type	Meteorite*	Silver concentration, ppm (Reference)	Range, ppm	Mean, ppm
Om	Cumpas	0.03 (1)		
	Descubridora	<0.01 (1)		
	Gun Creek	0.03 (1)		
	Henbury	0.10‡ (1)		
	Kingston	0.02 (1)		
	La Caille	<0.01 (1)		
	Mount Edith	<0.01 (1)		
	Puquios	0.02 (1)		
	Rhine Villa	<0.01 (1)		
	Sams Valley	0.02 (1)		
	Toluca	0.01; 0.130, 0.132, 0.103, 0.063 (1) (2)		
Of	Boogaldi	<0.01 (1)		
	Bristol	0.034, 0.022, 0.197, 0.165 (2)		
	Carlton	0.01 (1)		
	Gibeon	<0.01 (1)		
	Grant	0.050, 0.047, 0.079, 0.084 (2)		
	Kamkas	<0.01 (1)		
	Grand Rapids	<0.01 (1)	<0.01–0.197	0.03
	Huizopa	<0.01 (1)		
	Lockport	0.01 (1)		
	Moonbi	0.05 (1)		
	Obernkirchen	<0.01 (1)		
	Otchinjau	0.01 (1)		
Off	Bacubirito	<0.01 (1)		
	Ballinoo	0.04 (1)	<0.01–0.04	0.02
	Mount Magnet	0.01 (1)		
	Salt River	<0.01 (1)		
All O	–	–	<0.01–0.197	0.03
Ob and Gm	Barranca Blanca	0.04 (1)		
	Kopjes Vlei	0.05 (1)		
	Murnpeowie	0.02 (1)	<0.01–0.05	0.03
	Santa Rosa	0.01 (1)		
	Weekeroo Station	<0.01 (1)		
Dr	Babb's Mill†	<0.01 (1)		
	Hoba†	<0.01 (1)		
	Illinois Gulch†	<0.01 (1)		

TABLE 2 (cont.)

Type	Meteorite*	Silver Concentration, ppm (Reference)	Range, ppm	Mean, ppm
Dr	Pinon†	0.010, 0.014, 0.025, 0.011 (2)	<0.01–0.11	0.03
	San Cristobal†	0.11 (1)		
	Santa Catharina†	0.03 (1)		
	Smithland†	<0.01 (1)		

* Classification according to Reference unless otherwise noted.

† Classification according to Mason, 1963.

‡ Lovering, 1957, found >0.5 ppm in troilite phase. He also found <0.5 ppm in troilite phase of Rio Loa (Type H).

(1) Smales et al., 1967

(2) Charakraburtty et al., 1964

(3) Dews and Newbury, 1966

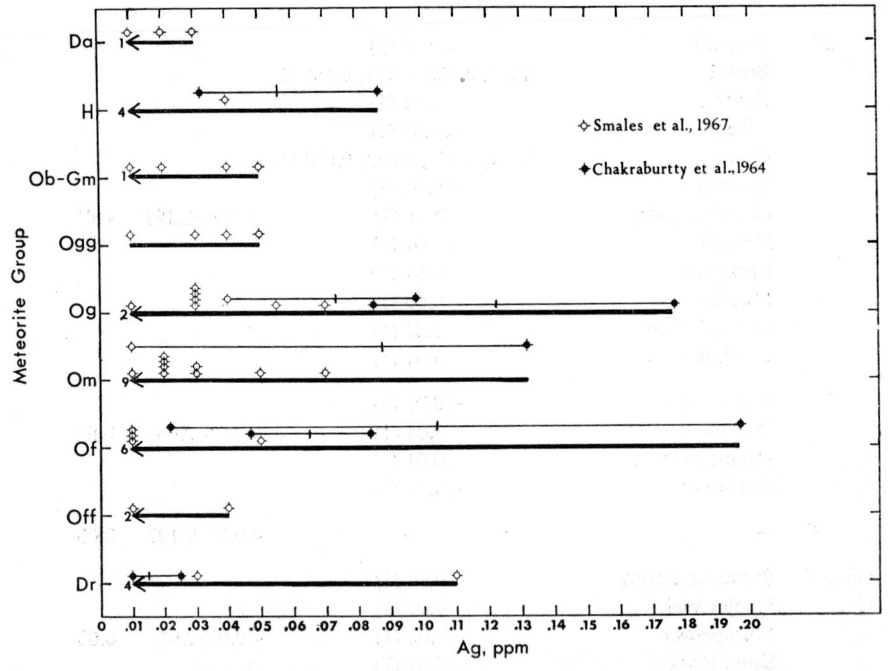

FIGURE 2 Distribution of silver analyses for the several groups of iron meteorites. For meteorites having replicate analyses, the range is represented by a solid line connecting the extreme values, and the mean is indicated by a vertical slash. The number of specimens with silver abundances below the analytical detectability limit is indicated to the left of the arrow showing the total range for each meteorite group.

For the iron meteorites the reported silver detectability limit is 0.01 ppm and some 27 meteorites (40% of the total measured) had values below this. The highest measured silver content is 0.11 ppm for the Ni-rich ataxite San Cristobal and only 2 others contain more than 0.05 ppm. Thus, 37 iron meteorites (over 50% of the total measured) have silver contents of 0.01 to 0.05 ppm. The coarse and coarsest octahedrites have distinctly higher silver concentrations than the hexahedrites and the medium and fine octahedrites.

In the process of sudying the isotopic composition of silver in iron meteorites, several authors reported total silver values. In the case of Canyon Diablo, the values of Smales et al. (1967) are within a factor of 2½ of the results of Dews and Newbury (1966), Chakraburtty et al. (1964) and Hess et al. (1957). There is greater discrepancy for the Toluca octahedrite. Chakraburtty et al. report values an order of magnitude greater than those of Smales et al. Because of the large number of analyses and the internal consistency of the results of Smales et al., these are preferred. As shown in Figure 2 they indicate minimal spreads in the silver values of the iron meteorites.

The silver content in troilite from 3 irons (Rio Loa, Henbury and Toluca) and 2 pallasites (Brenham and Springwater), and in the olivine from 4 pallasites (Admire, Albin, Brenham and Springwater) is below the reported detectability limit of 0.5 ppm (Lovering, 1957).

Considerable work has been done on the isotopic composition of silver in the iron meteorites in an attempt to locate the stable decay products of the extinct radioactive nuclide 6.8 m.y. Pd^{107}. An excess in Ag^{107}, relative to terrestrial silver, would indicate the former presence of Pd^{107} (Hess et al., 1957). The early results of Murthy (1960, 1962) indicate a 2% excess of Ag^{107} in Toluca troilite, a 3% excess in Sikhote Alin iron, and a 4% excess in Canyon Diablo iron. The later results of Dews and Newbury (1966) and Chakraburtty et al. (1964) do not support these results. Both sets of authors report no detectable enrichment of Ag^{107} in their measurements of the Canyon Diablo, Bristol, Grant, Odessa, Piñon, Sandia Mountains and Toluca iron meteorites. These data are summarized in Table 3.

The overall abundance of silver is greater in the stony meteorites than it is in the irons. Smales et al. (1967, p. 705) indicated that, based on two sulphur-rich meteorites, the distribution of silver does not appear to be influenced by the possibility of troilite inclusions. Nonetheless, silver does not appreciably enter the silicate lattice and thus a partial explanation for the difference in measured silver concentrations between stony and iron meteorites may well the be selective avoidance of troilite in sampling the iron meteorites. If

TABLE 3 Isotopic silver contents of the iron meteorites

Type	Meteorite	$Silver^{107}/Silver^{109}$ (Reference)	Range	Mean
Dr	Piñon*	1.110, 1.083, 1.15 (1)	1.083–1.15	1.11
H	Sandia Mtns.†	1.093 (1)		
	Sikhote-Alin‡	1.091 (2)	1.091–1.093	1.092
Og	Canyon Diablo†	1.110, 1.102, 1.097, 1.089, 1.092, 1.096, 1.090; 1.107; 1.020, 1.050, 1.081, 1.080, 1.079, 1.070, 1.086, 1.077 (1) (2) (3)	1.020–1.110	1.087
	Odessa†	1.090, 1.091 (1)		
Om	Toluca† (troilite)	1.106, 1.106, 1.094 (1) 1.087; 1.097 (2) (4)	1.087–1.106	1.098
Of	Grant†	1.107, 1.090 (1)		
	Bristol†	1.085, 1.094 (1)	1.085–1.107	1.094
All O	–	–	1.020–1.110	1.093

* Classification according to Mason, 1962.
† Classification according to Reference.
‡ Classification according to Hey, 1966.
(1) Chakraburtty et al., 1964
(2) Murthy, 1962
(3) Dews and Newbury, 1966
(4) Murthy, 1960

this is so, these analyses would confirm the chalcophilic character of silver. Terrestrially, silver commonly occurs as a sulfide. No meteoritic minerals of silver have been recognized.

The mean silver abundances of the stony meteorites, relative to 10^6 Si atoms, are given in Table 1. All of the values are relatively similar; they are within a factor of 4 of the cosmic value that Suess and Urey (1956) determined from the chondrites. Müller (1968) gives a value of 0.18 determined from astrophysical data of solar photospheric elemental abundances. Clayton and Fowler (1961) derived a similarly low value of 0.17 from nucleosynthetic calculations. Silver belongs to the strongly depleted group of elements of Larimer and Anders (1967). They give a value of 0.95 for Type I carbonaceous chondrites, essentially identical to the value in this paper, but considerably higher than the cosmic abundance values based on nucleosynthetic and solar abundance data.

Acknowledgements

This report was supported in part by Grant GA-1200 from the National Science Foundation. Helpful assistance and comments were provided by Mr. Jerrald Durtsche and Dr. J.W. Larimer.

References

Chakraburtty, A.K., Stevens, C.M., Rushing, H.C. and Anders, E. (1964) *J. Geophys. Res.* **69**, 505.
Clayton, D.D. and Fowler, W.A. (1961) *Ann. Phys.* **16**, 51.
Dews, J.R. and Newbury, R.S. (1966) *J. Geophys. Res.* **7**, 3069.
Greenland, L. (1967) *Geochim. Cosmochim. Acta* **31**, 849.
Hess, D.C., Marshall, R.R. and Urey, H.C. (1957) *Sci.* **126**, 1291.
Hey, M.H. (1966) *Catalogue of Meteorites* (3rd. edition). British Museum.
Larimer, J.W. and Anders, E. (1967) *Geochim. Cosmochim. Acta* **31**, 1239.
Lovering, J.F. (1957) *Geochim. Cosmochim. Acta* **12**, 253.
Mason, B. (1962) *Meteorites*, Wiley and Sons.
Müller, E.A. (1968) *Origin and Distribution of the Elements*, (L.H. Ahrens. ed.). Pergamon.
Murthy, V.R. (1960) *Phys. Rev. Letters* **5**, 539.
Murthy, V.R. (1962) *Geochim. Cosmochim. Acta* **26**, 481.
Reed, G.W. (1963) *J. Geophys. Res.* **68**, 3531.
Schindewolf, U. and Wahlgreen, M. (1960) *Geochim. Cosmochim. Acta* **18**, 36.
Smales, A.A., Mapper, D. and Fouché, K.F. (1967) *Geochim. Cosmochim. Acta* **31**, 673.
Suess, H.E. and Urey, H.C. (1956) *Rev. Mod. Phys.* **28**, 53.
Van Schmus, W.R. and Wood, J.A. (1967) *Geochim. Cosmochim. Acta* **31**, 747.

(Received 7 April 1970)

CADMIUM (48)*

Peter R. Buseck

Departments of Geology and Chemistry
Arizona State University
Tempe, Arizona 85281

THERE ARE RELATIVELY FEW recent measurements of cadmium abundances in meteorites. For some meteorites the results of replicate analyses are in good agreement, whereas for others there is up to a factor of 100 difference between duplicates. Clearly, considerably more work will have to be done before the occurrence and distribution of cadmium in meteorites is fully understood.

The cadmium concentrations in 29 chondrites (28 falls) range from 0.007 to 3.3 ppm and have a mean of 0.34 ppm and a median of 0.087 ppm. For six achondrites (5 falls), cadmium ranges from less than 0.010 to 1.79 ppm, with a mean of 0.4 ppm and a median of 0.065 ppm. Seven iron meteorites (1 fall) range from 0.0085 to 0.056 ppm, with a mean of 0.03 ppm and a median of 0.022 ppm. Although cadmium has eight stable isotopes no measurements of their relative abundances are available.

Recent measurements of the abundance of cadmium in meteorites have been made by Greenland (1967) and by Schmitt and co-workers (1962, 1963a, 1963b). Older determinations were made by Noddack and Noddack (1930, 1931, 1934) and Goldschmidt and Peters (1933). The pre-1940 measurements were made using spectrographic techniques; according to Schmidt et al. (1963) the reliability of all such cadmium measurements is highly questionable. Indeed, there is generally poor agreement between these results and more recent ones. Only the latter are reviewed here.

Greenland (1967) and Schmitt et al. (1963) both made their measurements by neutron activation analysis. Error estimates are not provided by Greenland. Tolerances for the cadmium values of Schmitt et al. (1963b) indicate

* Contribution No. 47 from the Arizona State University Center for Meteorite Studies.

TABLE 1 Cadmium contents of the chondrites

Type	Meteorite*	Concentration, ppm (Reference)	Mean, ppm	Mean Atoms/10^6 Silicon Atoms
C1	Ivuna	0.91: 1.7 (1) (2)	1.0	2.4
	Orgueil†	1.11; 0.40 (1) (2)		
C2	Boriskino	0.46 (3)		
	Mighei	1.16; 0.011 (1) (2)	0.44	0.80
	Mokoia	0.52; 0.012 (1) (2)		
	Murray	0.59; 0.33 (1) (2)		
C3	Felix	0.34 (2)		
	Grosnaja	0.38 (3)		
	Lancé	0.17 (2)	0.20	0.32
	Vigarano	0.039 (2)		
	Warrenton	0.052 (2)		
All C		–	0.44	0.77
E4	Abee	3.3 (1)	2.1	3.1
	Indarch	1.7; <0.01 (1) (2)		
E5	St. Marks	0.042 (1)	0.042	0.061
E6	Hvittis	0.11 (2)		
	Khairpur	0.12 (2)	0.10	0.13
	Pillistfer	0.082 (2)		
All E		–	0.75	1.04
H_	Ehole	0.015 (2)	0.015	0.022
H5	Allegan	0.015; 0.037 (1) (2)		
	Miller	0.12 (1)	0.047	0.069
	Pantar	0.023 (2)		
	Richardton	0.020 (1)		
All H		–	0.041	0.060
L6	Holbrook	0.047 (1)		
	Kyushu	0.087 (1)	0.072	0.069
	Peace River	0.073 (2)		
	Walters	0.080 (2)		
LL_	Vavilovka	0.007 (3)	0.007	0.004
LL3	Chainpur	0.039 (2)	0.039	0.051
LL6	Manbhoom	0.053 (1)	0.053	0.033
All L and LL		–	0.05	0.07

* Classification according to Van Schmus and Wood, 1967.

† Goldschmidt and Peters (1933) found about 10 ppm.

(1) Schmitt et al., 1963a

(2) Greenland, 1967

(3) Schmitt et al., 1963b

one standard deviation due to counting statistics. Surely these are modest estimates of precision. A limited estimate of the accuracy can be made by considering the analyses of olivine from the Thiel Mountain pallasite (Table 3). Although these measurements are from two separate crystals, there is evidence that the pallasites have achieved a high degree of internal equilibrium via their slow cooling rates (Buseck and Goldstein, 1969). Thus, it is highly likely that these crystals have the same cadmium contents. As the reported values are 0.17 ± 0.02 and 0.33 ± 0.01 ppm, it appears probable that the true error on these cadmium measurements is at least 0.08 ppm.

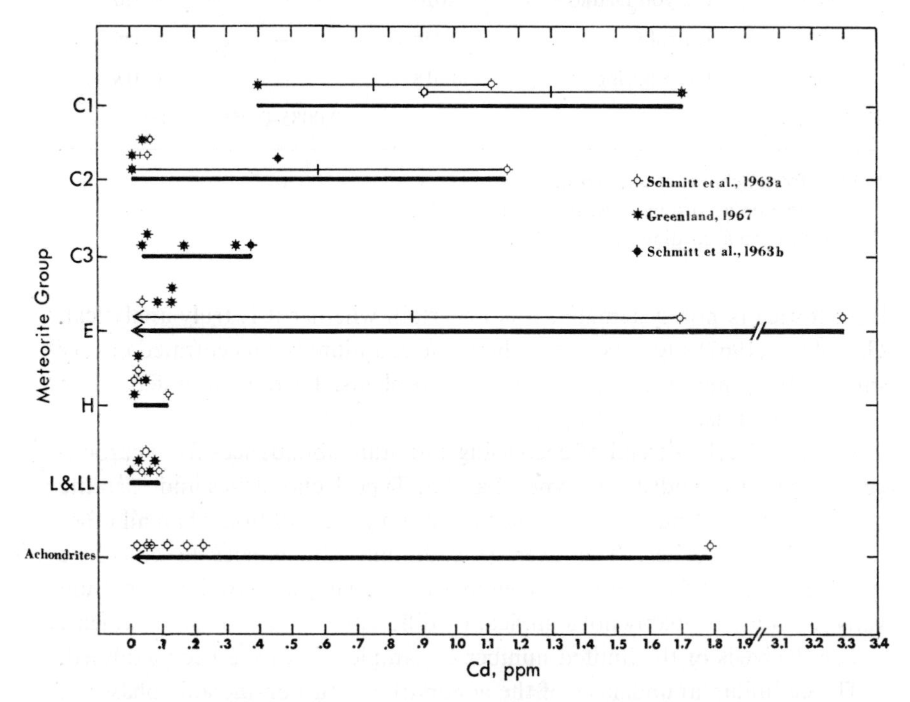

FIGURE 1 Distribution of cadmium analyses for the achondrites and chondrites. For meteorites having replicate analyses, the range is represented by a solid line connecting the extreme values, and the mean is indicacted by a vertical slash.

The cadmium contents of stony meteorites are summarized in Table 1 and Figure 1. The first feature that is apparent is the large difference between certain duplicate analyses performed by different laboratories. For some meteorites the agreement is good, but for Mighei and Indarch the difference is a factor of 100. For Mokoia it is a factor of 50. As the difference between

TABLE 2 Cadmium contents of the iron meteorites*

Type	Meteorite	Concentration, ppm	Range, ppm	Mean, ppm
H	Sandia Mountains†	0.0085		
	Sikhote Alin†	0.022	0.0085–0.022	0.015
Ogg	Arispe‡	0.022	–	0.022
Og	Odessa‡	≤0.04	–	≤0.04
Om	Canyon Diablo‡	0.020	–	0.020
Of	Bristol‡	0.056	–	0.056
Dr	Deep Springs‡	0.018	–	0.018
All Irons		–	0.0085–0.056	0.03

* Data from Schmitt et al., 1963a.
† Classification according to Schmitt et al., 1963a.
‡ Classification according to Hey, 1966.

laboratories is not systematic, it is uncertain whether it is truly analytical. Greenland (1967) suggests that perhaps the cadmium is concentrated in very small grains, thus resulting in sampling problems. There is no independent evidence for this.

There is clearly a trend of decreasing cadmium abundances from carbonaceous type 1 chondrites to types 2 and 3. Type 1 chondrites and enstatite chondrite type E4 have much higher cadmium concentrations than all other chondritic meteorites. The average cadmium abundances in chondrite types H, L, LL, E5 and E6 are given in Table 1. They all have low cadmium values, but are probably not significantly different from one another, at least not on the basis of the limited number of samples that have been analysed.

The cadmium abundances of the achondrites and non-metallic phases of stony-iron meteorites are reviewed in Table 3. Nakhla, with a concentration of 1.79 ppm, contains more cadmium than any chondrite other than Abee. It contains more than five times as much cadmium as the next richest achondrite, Johnstown.

Only seven iron meteorites have been analyzed for cadmium (Schmitt et al., 1963a). The values are given in Table 2. Based on these few analyses there are no striking variations between iron meteorites of differing types. The mean of the cadmium in iron meteorites is lower than that of any of the classes of chondritic or achondritic meteorites.

TABLE 3 Cadmium contents of the achondrites and non-metallic phases of the stony iron meteorites

Type	Meteorite†	Concentration, ppm†	Range, ppm	Mean, ppm	Mean Cadmium/10^6 Silicon Atoms
Achondrites					
Calcium-rich	Juvinas	0.065			
	Stannern	0.022	0.022–0.065	0.044	0.047
Nakhlitic	Nakhla	1.79			
	Lafayette	0.18	0.18–1.79	0.98	1.08
Calcium-poor	Johnstown	0.33			
	Norton County	≤0.010	≤0.010–0.33	0.17	0.17
All Achondrites		–	<0.010 1.79	0.40	0.43
Stony irons					
Mesosiderites (non-metal phase)	Estherville	<0.004			
	Veramin	0.056	<0.004–0.056	0.03	
Pallasites* (olivine)	Brenham	0.13	0.13–0.33	0.19	
	Thiel Mountains	0.33, 0.17			
All Stony Irons		–	<0.004–0.33	0.11	

* Lovering, 1957, has found <20 ppm cadmium in olivine of the pallasites Admire, Albin, Brenham, and Springwater, and in the troilite phases of Brenham, Springwater, and the irons Rio Loa (H), Henbury (Om), and Toluca (Om).
† Classification according to Schmitt et al., 1963a.
‡ Data from Schmitt et al., 1963a.

Cadmium abundances in separated phases of meteorites have not been reported in recent analyses. However, I. and W. Noddack (1930) reported high cadmium concentrations in troilite. Based on the high abundances of cadmium in the stony meteorites, relative to the irons, it is reasonable to conclude that cadmium is not siderophilic and if it is truly localized in troilite, this would indicate it is chalcophilic. This is accepted by Schmitt et al (1963b) who note that in those meteorites where cadmium is high, the other chalcophilic elements are also abundant. The only common terrestrial mineral of cadmium is the sulfide greenockite.

Geochemically, cadmium and zinc are very similar. Terrestrially, much cadmium occurs in zinc minerals, where it fits into the zinc lattice sites.

TABLE 4 Cadmium/zinc values for the meteorites

Type	Meteorite	Mean cadmium, ppm	Zinc, ppm (Reference)	Mean zinc, ppm	Mean cadmium / zinc ($\times 10^4$)	Group mean cadmium / zinc ($\times 10^4$)
Chondrites*						
C1	Ivuna	1.3	320; 570 (1) (5)	445	29	
	Orgueil	0.76	340; 228; 550, 390 (1) (3) (5)	377	20	25
C2	Mighei	0.59	150; 230, 200, 200 (1) (5)	195	30	
	Mokoia	0.27	80; 64.2; 130, 210 (1) (3) (5)	121	22	27
	Murray	0.46	200; 64, 210, 180 (1) (5)	166	28	
C3	Felix	0.34	110; 130, 110 (1) (5)	117	29	
	Lancé	0.17	110; 57.4; 160 (1) (3) (5)	109	16	13
	Vigarano	0.039	85; 170 (1) (5)	128	3.0	
	Warrenton	0.052	98; 150 (1) (5)	129	4.0	
All C		0.44	–	198.5		20
E4	Abee	3.3	720; 730; 112; 210; 350 (1) (2) (3) (5)	424	78	53
	Indarch	0.86	315 (3)	315	27	
E5	St. Marks	0.042	85 (4)	85	4.9	4.9
E6	Hvittis	0.11	20; 16; <5; 25; 29 (1), (2), (3), (4), (5)	19	58	
	Khairpur	0.12	22; 36.1; 30, 22 (1), (3), (5)	28	43	53
	Pillistfer	0.072	7.8; 24.4; 10 (1), (3), (5)	14	59	
All E		0.75		147.5		45
H_	Ehole	0.015	63; 62 (1), (5)	62	2.4	
H5	Allegan	0.026	43; 42.9; 48 (1), (3), (5)	45	5.8	
	Pantar	0.023	52 (1)	52	4.4	4.4
	Richardton	0.020	65 (2)	65	3.1	
All H		0.021		56		3.9
L6	Holbrook	0.047	39 (2)	39	12	
	Kyushu	0.087	56 (2)	56	16	13

TABLE 4 (cont.)

Type	Meteorite	Mean cadmium, ppm	Zinc, ppm (Reference)	Mean zinc, ppm	Mean cadmium / zinc ($\times 10^4$)	Group mean cadmium / zinc ($\times 10^4$)
L6	Peace River	0.073	52; 56 (1), (5)	54	14	13
	Walters	0.080	63 (1)	63	13	
LL3	Chainpur	0.039	58; 190 (1), (5)	124	3.1	3.1
All L and LL		0.065		67		11
Irons†						
Og	Canyon Diablo	0.020	37; 52.0 (2), (6)	44.5	4.5	4.5
Of	Bristol	0.056	13.6 (6)	13.6	41	41
All O		0.038		29.1		23
Dr	Deep Springs	0.018	13.2 (6)	13.2	14	14
Achondrite‡						
Calcium-poor	Johnstown	0.33	3 (2)	3	1100	1100

* Mean cadmium values calculated from data in Table 1.
† Mean cadmium values taken from Table 2.
‡ Mean cadmium value taken from Table 3.

(1) Greenland, 1967
(2) Nishimura and Sandell, 1964
(3) Greenland and Lovering, 1965
(4) Mason, 1966
(5) Greenland and Goles, 1965
(6) Nava, 1968

Because of this similarity it was thought of interest to compare abundances. The ratios of cadmium/zinc are summarized in Table 4. Mean values of both zinc and cadmium were used. Clearly zinc is far more abundant than cadmium. The cadmium/zinc values are highest for the enstatite chondrites, followed by carbonaceous chondrites types 1 and 2. The H group chondrites have considerably higher ratios than L group chondrites. There is the appearance of the ratio being higher in the more primitive meteorite types.

The cosmic abundances of cadmium, based on type 1 carbonaceous chondrites, is 2.4 atoms per 10^6 atoms silicon. This compares to 2.1 given by Larimer and Anders (1967), placing it into their "strongly depleted" group. Müller (1968) gives 1.1 atoms relative to 10^6 silicon atoms based on astrophysical data of solar photospheric abundances. Clayton and Fowler (1961), using nucleosynthetic calculations, derive a value of 0.80, while Suess and Urey (1956) record a value of 0.89, based on chondritic data.

Acknowledgements

This report was supported in part by Grant GA-1200 from the National Science Foundation. Helpful assistance was provided by Mr. Jerrald Durtsche and Dr. J. W. Larimer.

References

Buseck, P.R. and Goldstein, J.I. (1969) *Geol. Soc. Am. Bull.* **80**, 2141.
Clayton, D.D. and Fowler, W.A. (1961) *Ann. Phys.* **16**, 51.
Goldschmidt, V.M. and Peters, C. (1933) *Nachr. Ges. Wiss. Göttingen, math.-phys. Kl.* **278**.
Greenland, L. (1967) *Geochem. Cosmochim. Acta* **31**, 849.
Greenland, L. and Goles, G.G. (1965) *Geochim. Cosmochim. Acta* **29**, 1285.
Greenland, L. and Lovering, J.F. (1965) *Geochim. Cosmochim. Acta* **29**, 821.
Hey, M.H. (1966) *Catalogue of Meteorites* (3rd. edition). British Museum.
Larimer, J.W. and Anders, E. (1967) *Geochim. Cosmochim. Acta* **31**, 1239.
Lovering, J.F. (1957) *Geochim. Cosmochim. Acta* **12**, 253.
Mason, B. (1966) *Geochim. Cosmochim. Acta* **30**, 23.
Müller, E.A. (1968) in *Origin and Distribution of the Elements* (L.H. Ahrens, Ed.), Pergamon.
Nava, D.F. (1968) Ph. D. Thesis, Arizona State Univ.
Nishimura, M. and Sandell, E.B. (1964) *Geochim. Cosmochim. Acta* **28**, 1055.
Noddack, I. and Noddack, W. (1930) *Naturwiss.* **18**, 757.
Noddack, I. and Noddack, W. (1931) *Z. Phys. Chem.* **154**, 207.
Noddack, I. and Noddack, W. (1934) *Svensk. Kem. Tidskr.* **46**, 173.
Schmitt, R.A., Smith, R.H. and Olehy, D.A. (1963a) *Geochim. Cosmochim. Acta* **27**, 1077.
Schmitt, R.A., Smith, R.H. and Olehy, D.A. (1963b) *Gen. At. Report* GA-4493.
Schmitt, R.A., Smith, R.H., Perry, K.I. and Olehy, D.A. (1962) *Gen. At. Report* Ga-3687.
Suess, H.E. and Urey, H.C. (1956) *Rev. Mod. Phys.* **28**, 53.
Van Schmus, W.R. and Wood, J.A. (1967) *Geochim. Cosmochim. Acta* **31**, 747.

(Received 7 April 1970)

INDIUM (49)

P. A. Baedecker

Institute of Geophysics and Planetary Physics
University of California, Los Angeles

In RECENT YEARS it has become apparent that the relatively high volatility of certain trace elements and their compounds has played an important role in determining how these elements are distributed between and within the various chondrite groups (Anders, 1964; Larimer and Anders, 1967). Indium is one of the most volatile elements, and in the sequence proposed by Larimer (1967) for the condensation of the elements, it is one of the last to condense. An extremely wide range of In concentrations has been observed for the chondrites, varying from approximately 120 ppb for the C1 and E4 chondrites, to as low as 0.03 ppb for some L6 chondrites. (The chondrite classification of Van Schmus and Wood (1967) is used throughout this paper.) Neglecting the unequilibrated ordinary chondrites, it shows the same general distribution pattern between chondrites as all volatile elements: C1 > C2 > C3 > H $\simeq$ L, and E4 > E5 and E6. Recently, Tandon and Wasson (1968) have shown that the In concentration varies from 0.05 ppb to 64 ppb within the L group alone, and that there is a linear one to one correlation between the In content and the concentration of primordial Ar^{36}. It is extremely difficult to propose a mechanism to explain how two elements with such different volatilities and different chemical properties could be coherently distributed within the chondrites. This problem is one of the most challenging enigmas confronting cosmochemists at the present time.

Goldschmidt (1954) describes In as being concentrated in the Ni-Fe and troilite phases of meteorites, based on the early spectrographic work of I. and W. Noddack (1930, 1934) along with his own (with H. Horman) unpublished results. Subsequent activation analysis results by Fouché and Smales (1967), Smales et al. (1967) and Rieder and Wänke (1969) show that In is concentrated less strongly in the metal phases than Goldschmidt

TABLE 1 In in chondrites

Meteorite Type	No. of determinations	No. of meteorites analyzed	Concentration (10^{-9}g/g) Range	Mean	Abundance (Si = 10^6)	Reference
C1	2	1	110–120	115	0.27	Akaiwa
	2	2	82–88	85	0.20	Fouché
	3	2	64–80	75	0.18	Schmitt
C2	3	2	23–54	41	0.077	Akaiwa
	2	2	46–49	48	0.090	Fouché
	1	1		49	0.090	Rieder
	2	2	47–64	56	0.10	Schmitt
C3	2	1	30–32	31	0.049	Akaiwa
	2	2	22–25	24	0.038	Fouché
	3	3	11–26	19	0.030	Rieder
	3	3	23–32	27	0.043	Schmitt
H-group	9	8	0.6–8.2	2.9	0.0042	Fouché
	9	9	0.48–24	5.9	0.0084	Rieder
	3	3	0.8–1.3	1.0	0.0014	Schindewolf
	3	3	0.1–2.1	0.8	0.0011	Schmitt
L-group	9	9	0.5–1.3	0.7	0.0009	Fouché
	2	2	0.3–0.6	0.5	0.0007	Schindewolf
	5	5	0.3–14	5.6	0.0073	Schmitt
	59	26	0.048–64	15.6	0.020	Tandon
E1	2	1	120–150	135	0.19	Akaiwa
	2	2	56–76	66	0.095	Fouché
	3	2	56–130	92	0.13	Schmitt
E2	2	2	4.2–4.6	4.4	0.0055	Fouché
	1	1		6.12	0.0077	Rieder
	3	3	0.22–2.9	1.12	0.0014	Schmitt
LL-group	1	1		74	0.10	Schmitt

had assumed. Fouché and Smales could detect no In in the Ni-Fe phases of six ordinary chondrites, and placed an upper limit of 0.5 ppb on the chondrite metal phase based on the sensitivity of their technique (Table 3). Rieder and Wänke found 20 times more In in the non-magnetic (silicate plus troilite) portion of the Hvittis E6 chondrite than the magnetic portion of the same meteorite (Table 3).

The data which have been obtained for the concentration of In in chondrites are summarized in Table 1 and Figure 1. All data were obtained by neutron-activation analysis. There are large differences in the results obtained by

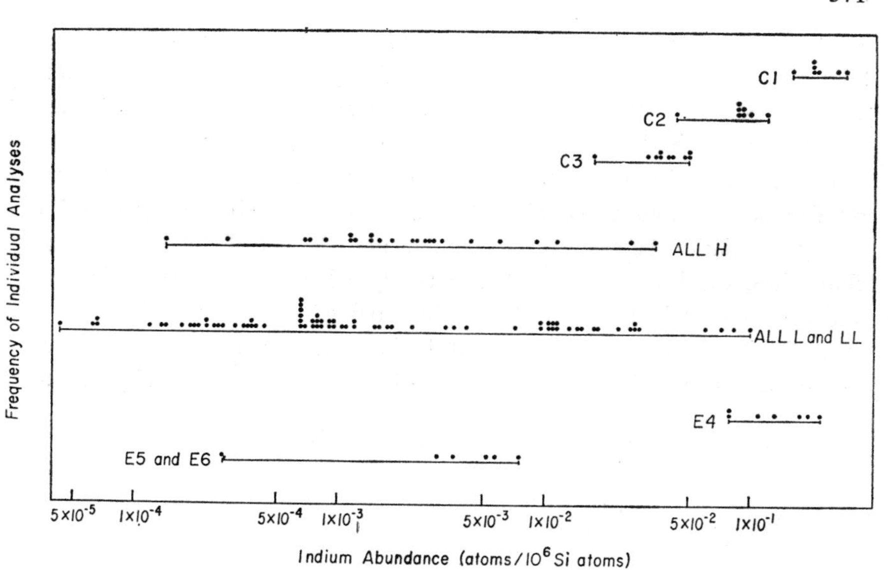

FIGURE 1 Frequency of individual analyses versus indium abundance (atoms/10^6 Si atoms) plotted on a log scale, for various chondrite groups.

different workers, in some cases for the same meteorite. For example, Schmitt and Smith (1968) reported values of 64 and 80 ppb, Akaiwa (1966) 110 and 120 ppb, and Fouché and Smales a value of 82 ppb for the Orgueil meteorite. Schmitt and Smith obtained values of 56 and 130 ppb, Akaiwa

TABLE 2 In in iron meteorites

Ga–Ge group	No. of determinations	No. of meteorites analyzed	Concentration (10^{-9} g/g) Range	Mean*	Reference
I	8	8	4–12	10 (8)	Smales
IIA	7	7	0.5–< 10	1.2 (5)	Smales
IIB	1	1		4.0	Smales
IIC	2	2	1.1–< 10	1.1 (1)	Smales
IID	1	1	< 10	—	Smales
IIIA	6	6	< 10–20	20 (2)	Smales
IIIB	4	4	0.5–< 10	0.6 (2)	Smales
IVA	7	7	0.4–< 10	0.4 (1)	Smales
IVB	1	1	< 10	—	Smales
Anom.	27	27			Smales

* Values listed as upper limits have been ignored in calculating means. Number of analyses is listed in parenthesis.

120 and 150 ppb, and Fouché and Smales 56 ppb for the E4 chondrite Abee. Such variable results could be due to sampling problems, or contamination of the specimens. In some other cases the agreement between different research groups on individual meteorites is quite satisfactory.

Tandon and Wasson (1968) have analyzed a petrologic suite (as defined by Van Schmus and Wood (1967)) of L-group chondrites. They found a strong correlation between the In content and the petrologic grade of the chondrites, the least metamorphosed being rich in In, and those having undergone increasingly more metamorphism being successively depleted in In. One criterion for determining the petrologic grade of a chondrite according to Van Schmus and Wood is the degree of pyroxene inhomogeneity (varia-

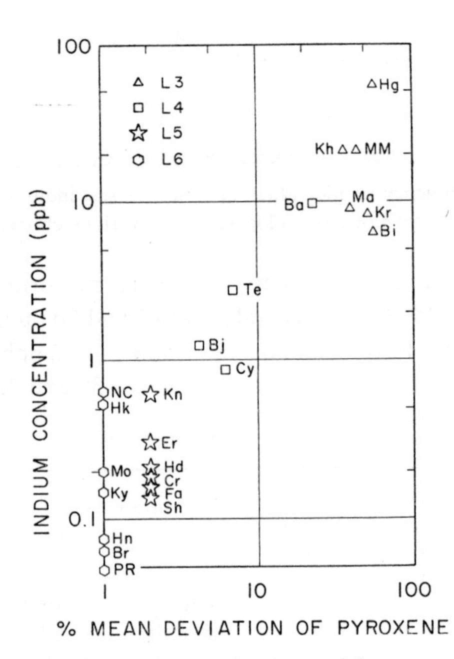

FIGURE 2 A plot of In vs. degree of pyroxene inhomogeneity. The pyroxene data are from Dodd et al. (1967). The per cent mean deviations of the L5 and L6 stones have been arbitrarily set at values of 2 and 1% respectively.

tion in the $FeSiO_3$ content) within the chondrite. Fig. 2 is a plot of In variation versus per cent mean deviation of pyroxene for the L-group chondrites analyzed by Tandon and Wasson.

A similar trend of In content with petrologic grade has been observed by Rieder and Wänke (1969) for the H-group chondrites. Therefore the means reported in Table 1 for the ordinary chondrite groups (and to some extent for the enstatite chondrites) will be highly dependent upon sampling. The data of Rieder and Wänke and Tandon and Wasson most fully reflect the ranges of In concentration that are to be found within the ordinary chondrite groups.

Rieder and Wänke (1969) have determined In in the separated light and dark portions of five brecciated H-group chondrites, and their data are summarized in Table 3. They found that In was enriched in the dark portion relative to the light portion by approximately a factor of 10 in each case. This depletion of In in the light portion is particularly striking, since the authors report that the light portion was contaminated by up to 10% with the dark material. One chondrite, Leighton, has a greater In concentration in the dark portion (151 ppb) than has been observed for the type 1 carbonaceous chondrites.

TABLE 3 In in separated fractions from meteorites

Material	No. of determinations	No. of meteorites analyzed	Concentration (10^{-9} g/g) Range	Mean	Reference
H-chondrites					
magnetic	3	3		<0.5	Fouché
non-magnetic	9	8	0.6–10	3.5	Fouché
L-chondrites					
magnetic	3	3		<0.5	Fouché
non-magnetic	9	9	0.5–1.5	0.7	Fouché
E6 chondrites					
magnetic	1	1		0.43	Rieder
non-magnetic	1	1		8.3	Rieder
H-chondrites					
dark	8	5	9.1–180	71	Rieder
light	5	5	1.1–16	4.3	Rieder
Pallasite					
metal phase	1	1		<10	Smales

The high In concentration observed for the one LL-type chondrite analyzed by Schmitt and Smith (1968) (Table 1) was obtained for the Chainpur chondrite, an unequilibrated chondrite (LL3) in the Van Schmus and Wood classification scheme.

The only data which have appeared concerning the concentration of In in iron meteorites are those of Smales et al. (1967), who studied In in 64 iron meteorites by neutron activation. In about two-thirds of the determinations they were only able to report upper limits. Their data are summarized in Table 2 for the nine chemical groups as defined by Wasson (see chapters on Ga and Ge). Due to the limited amount of data available for irons belonging to the different chemical groups, it is difficult to make any firm conclusions regarding systematic variations in the In contents for the different types of iron meteorites. However, an examination of the data seems to suggest that the type I and IIIA meteorites are enriched in In relative to the other chemical groups.

The only data on the In content of the achondrites are those obtained by Schmitt and Smith (1968) on two eucrites and two Ca-poor achondrites, which are summarized in Table 4.

TABLE 4 In in achondrites

Meteorite Type	No. of deter-minations	No. of meteorites analyzed	Concen-tration Range	(10^{-9} g/g) Mean	Abundance (Si = 10^6)	Reference
Ca-rich						
eucrites	2	2	0.24–1.6	0.92	0.001	Schmitt
Ca-poor						
aubrite	1	1		0.34	0.0004	Schmitt
diogenite	1	1		0.40	0.0004	Schmitt

No determinations of the isotopic composition of In in meteorites have been reported in the literature.

Müller (1968) lists a solar In abundance of 1.62 while Grevesse et al. (1968) list an abundance of 0.87 relative to Si = 10^6. Much of the difference in the two values rests in the abundances which one accepts for the silicon abundance. Nevertheless, both values are significantly above the C1 chondrite values of 0.15–0.28 atoms/10^6 Si atoms.

References

Akaiwa, H. (1966) *J. Geophys. Res.* **71**, 1919–1923.

Anders, E. (1964) *Space Sci. Rev.* **3**, 583–714.

Dodd, R.T., Van Schmus, W.R. and Koffman, D.M. (1967) *Geochim. Cosmochim. Acta* **31**, 921–951.

Fouché, K.F. and Smales, A.A. (1967) *Chem. Geol.* **2**, 5–33.
Goldschmidt, V.M. (1954) *Geochemistry*, Oxford University Press, Oxford, England.
Grevesse, N., Banquet, G. and Boury, A. (1968) *Origin and Distribution of the Elements*, edited by L.H. Ahrens, Pergamon Press, Oxford, England, 177–182.
Larimer, J.W. (1967) *Geochim. Cosmochim. Acta* **31**, 1215–1238.
Larimer, J.W. and Anders, E. (1967) *Geochim. Cosmochim. Acta* **31**, 1239–1270.
Müller, E. (1968) *Origin and Distribution of the Elements*, edited by L. H. Ahrens, Pergamon Press, Oxford, England, 155–176.
Noddack, I. and Noddack, W. (1930) *Naturwiss.* **18**, 757–764.
Noddack, I. and Noddack, W. (1934) *Svensk. Kem. Tid.* **46**, 173–201.
Rieder, R. and Wänke, H. (1969) *Meteorite Research*, edited by P.M. Millman, IAEA, Vienna, 75–86.
Schmitt, R.A. and Smith, R.H. (1968) *Origin and Distribution of the Elements*, edited by L.H. Ahrens, Pergamon Press, Oxford, England, 281–300.
Schindewolf, U. and Wahlgren, M. (1960) *Geochim. Cosmochim. Acta* **18**, 36–41.
Smales, A.A., Mapper, D. and Fouché, K.F. (1967) *Geochim. Cosmochim. Acta* **31**, 673–720.
Tandon, S.N. and Wasson, J.T. (1968) *Geochim. Cosmochim. Acta* **32**, 1087–1109.
Van Schmus, W.R. and Wood, J.A. (1967) *Geochim. Cosmochim. Acta* **31**, 747–765.

References for Tables

Akaiwa (1966)
Fouché and Smales (1967)
Rieder and Wänke (1969)
Schindewolf and Wahlgren (1960)
Schmitt and Smith (1968)
Smales, Mapper and Fouché (1967)
Tandon and Wasson (1968)

(Received 25 July 1969)

TIN (50)*

Peter R. Buseck

Departments of Geology and Chemistry
Arizona State University
Tempe, Arizona

CONSIDERABLE ATTENTION has been given to the distribution of tin in meteorites, but the results are not entirely satisfactory. There are moderately large ranges in the reported tin values both within and between meteorites. It is unclear to what extent these variations are real rather than reflecting analytical difficulties. The lack of reproducibility in duplicate, high-precision analyses from differing laboratories suggests that there are still unresolved analytical problems.

The tin concentrations in 34 chondrites (23 falls and 11 finds) range from less than 0.1 to 2.4 ppm; they have a mean of 0.8 ppm and a median of 0.86 ppm. For 50 irons (4 falls and 46 finds) the concentrations are appreciably higher, from less than 0.1 to 20.6 ppm, with a mean of 2.7 ppm and a median of 1 ppm. The selective concentration of tin in the metallic fraction of chondrites suggests that it is siderophilic. This is confirmed by the relatively high tin concentrations in the iron meteorites. Tin has the greatest number of stable isotopes of any element. De Laeter and Jeffery (1967) found no unexpected isotopic abundance anomalies, nor significant differences between terrestrial and meteoritic abundances.

A variety of analytical techniques have been employed for the measurement of tin. Neutron activation analysis has been used by Hamaguchi et al. (1969) and by Kiesl et al. (1967). Hamaguchi et al. estimate their experimental error as being within 10% of the reported values. A few anomalously high values are believed to be due to pre-laboratory contamination and were therefore rejected by them. Colorimetric or photometric methods were employed by Shima (1964), Onishi and Sandell (1957), and Winchester and

* Contribution No. 45 from the Arizona State University Center for Meteorite Studies.

TABLE 1 Tin contents of the stony meteorites

Type	Meteorite*	Tin concentration, ppm (Reference)	Mean, ppm	Mean Atoms/10^6 Si Atoms
Chondrites				
C1	Orgueil	1.9, 1.5, 1.5 (1)	1.6	3.6
C2	Boriskino	0.79, 0.97 (1)		
	Cold Bokkeveld	0.82 (1)		
	Mighei	0.84, 0.50 (1)	0.85	1.5
	Murray	0.86, 1.1; 0.84, 1.22, 1.09 (1), (2)		
C3	Lancé	0.88 (1)	0.88	1.3
All C	–	–	1.0	1.8
E4	Abee	1.9, 2.4; 0.80, 0.96; 0.7, 0.8 (1), (2), (3)	1.4	1.9
	Indarch	1.5 (1)		
H_	Cocklebiddy	1.1, 1.4 (3)		
	Ehole	1.0, 0.72 (2)		
	Hat Creek	<0.1 (2)	0.6	0.8
	Rochester	0.07 (2)		
H3	Tieschitz	0.39, 1.1 (1)	0.74	1.9
H5	Forest City	0.24; 0.08 (1), (2)		
	Homestead	0.2, 0.3, 0.1, 0.79 (2)		
	Richardton	0.54, 0.38 (2)	0.3	0.4
	Tulia†	0.07 (2)		
H6	Estacado†	1.09; 1 (2), (4)	1	1
All H	–	–	0.6	0.8
L_	Woolgorong	<0.1 (3)	<0.1	<0.1
L5	Bluff	0.3 (2)		
	Knyahinya†	≤1 (5)	0.6	0.8
L6	Bruderheim†	0.85 (2)		
	Holbrook†	<0.07; <2 (2), (5)		
	Ladder Creek	1.01 (2)		
	Lake Brown	1.9 (3)		
	Leedey	0.74, 0.82 (2)		
	Long Island†	≤1 (5)	0.9	1.1
	Mocs†	≤2 (5)		
	Modoc	0.28; 0.1 (1), (2)		
	Ness County	0.08 (2)		

TABLE 1 (cont.)

Type	Meteorite*	Tin Concentration, ppm (Reference)	Mean, ppm	Mean Atoms/10^6 Si Atoms
L6	Potter†	0.5 (4)		
	Waconda†	≤1 (5)		
All L	–	–	0.7	0.9
LL3	Chainpur	0.33 (1)	0.33	0.41
LL4	Soko-Banja†	≤1 (5)	≤1	≤1
Achondrite				
Eu	Stannern‡	≤1 (5)	≤1	≤1

* Classification according to Reference unless otherwise noted.
† Classification according to Van Schmus and Wood, 1967.
‡ Classification according to Mason, 1962.
(1) Hamaguchi et al., 1969
(2) Shima, 1964
(3) De Laeter and Jeffery, 1967
(4) Onishi and Sandell, 1957
(5) Kiesl et al., 1967

Aten (1957). The respective error estimates, as given by the various authors, are ±10% to ±50% (Shima, 1964) and ±0.5 ppm for stones and ±1 ppm for iron meteorites (Onishi and Sandell, 1957). Mass spectrometry was used by de Laeter and Jeffery (1967); they estimate an error of ±5% for their values.

Recent analyses of tin in chondrites are summarized in Table 1 and Figure 1. Earlier analyses were made by Noddack and Noddack (1930, 1931, 1934) and Goldschmidt and Peters (1933). These pre-1940 measurements are, on the average, considerably higher than the more recently determined numbers. Because of this systematic but indeterminable error, these values are excluded from Table 1 and Figure 1. However, these early tin measurements are all conveniently tabulated in Table 1 of Onishi and Sandell (1957).

Perhaps the most striking aspect of the tin analyses of chondrites is their scatter. In the case of the largest spread, Abee (0.7 to 2.4 ppm), this may be partly due to the widely different analytical techniques used by the several investigators. Nonetheless, the replicates of single laboratories also show considerable spreads, e.g. Tieschitz (0.39 to 1.1 ppm) and Homestead (0.1 to

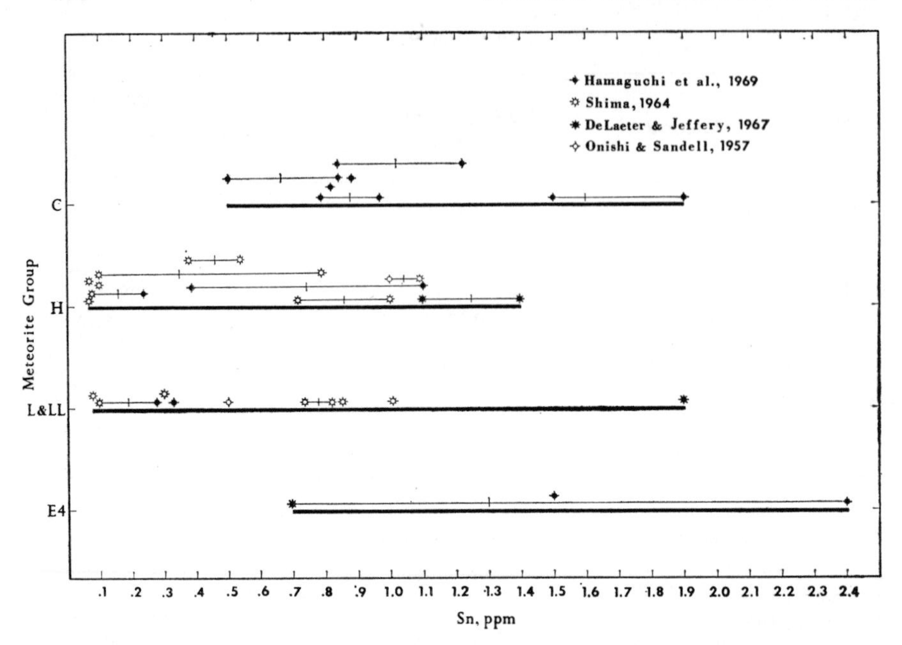

FIGURE 1 Distribution of tin analyses for the several groups of chondrites.

0.79 ppm). It is unclear whether the spread is a reflection of the difficulties of the analyses or whether it indicates a lack of tin homogeneity in these meteorites. Shima (1964) indicates an agreement of 10% for duplicate analyses and therefore believes the meteorites are in fact inhomogeneous in tin. Some of the analyses of Hamaguchi et al. (1969) appear to be systematically higher than those of Shima (1964) by a factor of two or three.

Owing to the scatter in the results, it is difficult to arrive at reliable mean values. Nonetheless, on the basis of her data, Shima (1964) feels that there are no significant differences in tin values between "falls" and "finds". The chondrite data in Table 1 include 23 falls and 11 finds, and they have respective mean tin concentrations of 0.81 and 0.75 ppm. Of the fifty analyzed iron meteorites, only four are falls. Consequently the precise mean values are probably not significant, although again no pronounced difference is apparent between falls and finds. Natural terrestrial contamination does not appear to be significant.

The tin content of iron meteorites has received considerable study; the results are given in Table 2 and Figure 2. The most consistent feature of the iron meteorites is that most contain significantly more tin than the chondrites. Nevertheless, as in the case of the chondrites, there is much scatter between

TABLE 2 Tin contents of the iron and stony iron meteorites

Type	Meteorite	Concentration, ppm (Reference)	Range, ppm	Mean, ppm
Irons				
Da	Locust Grove	13.2, 7.3, 2.70 (1)	2.7–13.2	7.7
Dr	Chinga	0.1 (2)		
	Deep Springs	0.55 (3)		
	Mt. Magnet	1.2 (2)	0.1–13	3
	San Cristobal†	13 (5)		
	Warburton Range	0.2 (2)		
H	Boguslavka	0.2 (2)		
	Coahuila	<0.1, <0.1 (4)		
	Coya Norte	<0.1 (4)		
	Mayodan	<0.2 (3)		
	Negrillos	<0.3 (3)	<0.1–20.6	3
	Sandia Mtns.	<0.1 (4)		
	Sierra Gorda	<0.1 (3)		
	Tocopilla	20.6, 19.9. (1)		
Og	Aroos	5.47, 4.95, 5.26 (3)		
	Canyon Diablo[a]	8.0, 6.1, 6.3; 4.6; 4.57, 4.32, 5.43, 5.50, 5.32, 5.80; 7, 10 (1), (2), (3), (4)		
	Deport	6 (4)		
	Magura†	≤1 (5)		
	Mt. Stirling	5.3 (2)		
	Odessa	6.06, 5.42 (3)	<1–10	5
	(troilite)	<1 (3)		
	Sardis	4, 4.13 (3)		
	Youndegin III	7.6 (2)		
Ogg	Ainsworth	1 (4)		
	Arispe	3 (4)		
	Mount Joy	1 (4)		
	Sao Juliao de Moreira	5.7, 3.9 (1)	0.1–5.7	2
	Sikhote Alin	0.3; 0.1, 0.303 (2), (3)		
Om	Clark County	<0.2 (3)		
	Costilla Peak	<0.1 (4)		
	Goose Lake†	7 (4)		
	Haig	0.1 (2)		
	Henbury	<2.6, <1.0; 0.1, 0.2; 5, 4 (1), (2), (4)		

TABLE 2 (cont.)

Type	Meteorite	Concentration, ppm (Reference)	Range, ppm	Mean, ppm
Om	Kenton County†	≤1 (5)		
	Milly Milly	0.2, 0.1 (2)		
	Mt. Dooling	0.5 (2)		
	Sacramento Mtns.	<0.1 (4)	<0.1–7.8	1
	Spearman	1 (4)		
	Tamarugal	<0.6, <0.8 (1)		
	Toluca[b]	7.8, 7.3; 5.2, 5.4; 6, 7.72 (1), (2), (3)		
	Treysa	<0.5 (3)		
	Williamstown	<0.2 (3)		
	Willow Creek	1 (4)		
Of	Altonah	1 (4)		
	Duchesne†	3 (4)		
	Grant	<0.1 (3)		
	Kumerina	0.7 (2)	<0.1–11.1	3
	Muonionalusta	9.1, 11.1 (1)		
	Wallapai	1.5 (4)		
Off	Edmonton†	5 (4)		
	Perryville	<0.1 (4)	<0.1–5	3
All O	–	–	<0.1–11.1	3
Stony irons‡				
M	Bencubbin I	0.2 (2)		
	(silicate)	0.1 (2)	0.1–0.2	0.15
P	Admire			
	(metal)	2.35 (3)		
	(stone)	0.77, 0.3 (3)		
	(troilite)	<0.1 (3)		
	Brenham		<0.1–2.35	0.7
	(metal)	0.8 (3)		
	(stone)	0.53, 0.6 (3)		
	(troilite)	<0.1 (3)		

* Classification according to Reference unless otherwise noted.
† Classification according to Hey, 1966.
[a] Noddacks (1931) found 50 ppm.
[b] Noddacks (1931) found 60 ppm.
‡ Lovering, 1957, has determined the tin content of olivine in the pallasites Admire (< 10 ppm), Albin (~10 ppm), Brenham (< 10 ppm) and Springwater (< 10 ppm).

(1) Winchester and Aten, 1957
(2) De Laeter and Jeffery, 1967
(3) Shima, 1964
(4) Onishi and Sandell, 1957
(5) Kiesl et al., 1967

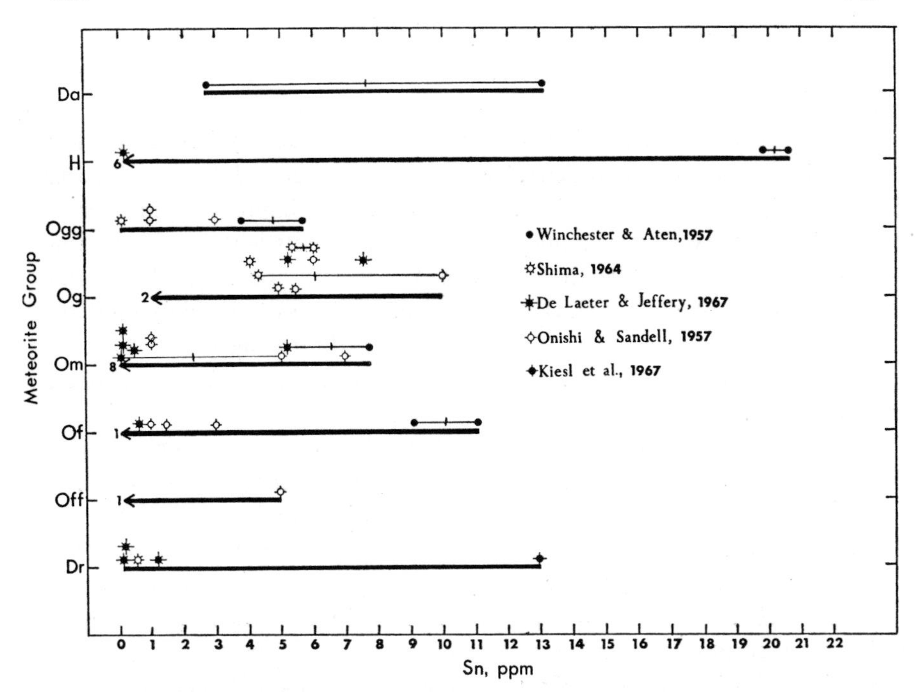

FIGURE 2 Distribution of tin analyses for the several groups of iron meteorites. For meteorites having replicate analyses that deviate >0.25 ppm from the mean, the range is represented by a solid line connecting the extreme values, and the mean is indicated by a vertical slash. The means alone are indicated for those replicate analyses which deviate ≤0.25 ppm from the mean. The number of specimens with values below the analytical detectability limit is indicated by a number to the left of the arrow showing the total range for each meteorite group.

analyses, e.g. Locust Grove ranges from 2.7 to 13.2 ppm, and 12 analyses of Canyon Diablo range from 4.3 to 10 ppm. Other meteorites such as Henbury have differing but consistent values, depending on the analyst. Thus Winchester and Aten (1957) report <0.1 ppm values, whereas Onishi and Sandell (1957) show duplicate values of 4 and 5 ppm, and de Laeter and Jeffery (1967) indicate duplicate values of 0.1 and 0.2 ppm. Nonetheless, both Shima (1964) and de Laeter and Jeffery (1967) indicate that they find tin to be uniform within given meteorites. If this is so it indicates analytical difficulties.

Some iron meteorites have much higher tin contents than the rest of the members of their group. Thus, Tocopilla is 20 ppm higher in tin than the seven other measured hexahedrites, and San Cristobal is 12 ppm higher than the four other Ni-rich ataxites. It is an open question whether these indicate strongly anomalous values, or whether we simply have a very poor and misleading sample of meteorites from these groups.

The mass spectrometric tin values of de Laeter and Jeffery (1967) are lower than those of the other workers. Those of Winchester and Aten (1957) are, in general, the highest tin values. It is probable that improved instrumental techniques have permitted more accurate measurements with lower contamination problems. If the tin values of Winchester and Aten (1957) and the only value of Kiesl et al. (1967) that is above their detectability limit are eliminated, both the range and mean tin values of the iron meteorites decrease significantly, from <0.1–20.6 to <0.1–10 ppm, and from 2.7 to 1.9 ppm, respectively.

The medium, coarse and coarsest octahedrites are the only groups with sufficient analyses to permit even moderately confident generalizations. It does appear that, as a group, the coarse octahedrites are richest in tin. Several authors have pointed out that the medium octahedrites fall into pronounced tin-rich and tin-poor groups; these have respective averages of 6.8 and 0.5 ppm. Henbury spans the gap and was not included in the above averages. The hexahedrites and ataxites exhibit variable tin abundances, but do contain relatively tin-rich members. There is no evident simple but general relation between tin and nickel.

Little information is available for the stony-irons. De Laeter and Jeffery (1967) found roughly equal amounts of tin in the silicate (0.1 ppm) and metal (0.2 ppm) fractions of the Bencubbin mesosiderite. Shima (1964) measured tin in the metal, troilite, and stony (olivine) fractions of the Admire and Brenham pallasites (Table 2). In Brenham almost equal concentrations of tin in the stone and metal fractions were found, although in Admire tin is concentrated in the metal.

Comparisons of the tin contents in the metal phases and the non-metal phases in chondrites were made by Shima (1964) and Onishi and Sandell (1957). The latter prepared two composite samples consisting of seven and eight chondrites, respectively. In both cases the metal phase contained 5 ppm tin, and the non-magnetic fraction contained 0.4 ± 0.2 ppm tin. The metal phase of another composite, consisting of Estacado, Haven and Plainview contained 6 ppm tin. Shima (1964) separated the H-group chondrite Ehole into magnetic and non-magnetic fractions. Using her estimated value of the per cent metal her results provide a corrected value of 3.5 ppm for the metal fraction and 0.12 for the silicate. Thus, it appears clear that tin is selectively concentrated into the metallic fraction of chondrites.

A few measurements have been made of the tin content of troilite. Shima (1964) found tin to be below the detectability limit in troilite from the Admire and Brenham pallasites (<0.1 ppm) and the Odessa octahedrite (<1 ppm). Goldschmidt and Peters (1933) measured the co-existing metal and troilite phases in five iron meteorites. Although their absolute abundances are too high, the ratio of 7 : 1 for tin in the metal versus sulfide phases probably points in the right direction; it is in opposition to the results of Noddack and Noddack (1931).

Based on the above distributions tin appears to be strongly siderophile, weakly lithophile, and without appreciable chalcophilic characteristics. These conclusions are confirmed by the thermodynamic calculations of Shima (1964). Her results indicate that tin is most siderophilic at high temperatures (1500 °K) and lithophilic at low temperatures (298 °K), and without chalcophilic properties at all temperatures. This is somewhat surprising in view of the terrestrial occurrence of tin in sulfide minerals.

The isotopic composition of tin is of considerable interest. It contains more stable isotopes (10) than any other element, being an element with a "magic proton number". As the isotopes were produced by various nucleosynthetic processes, the *s*-, *r*- and *p*-processes, the relative abundances of the isotopes can be used to evaluate the various nucleosynthetic models. The abundances have been studied by de Laeter and Jeffery (1967), but they found no significant anomalies. The data for the meteorites is given in their tables 3, 4 and 5.

Judging by the abundance of tin in the "ordinary" chondrites, tin falls in the "normal" group of elements of Larimer and Anders (1967). For its cosmic abundance Müller (1968) lists values of 1.1 and 3.5 determined on a 10^6 Si atoms scale, determined from astrophysical data of solar photospheric abundances. Based on nucleosynthetic calculations Clayton and Fowler (1961) derived a value of 1.9, and Suess and Urey (1956) gave 1.3, based on chondritic data. The cosmic abundance of Cameron (1968), determined from type 1 carbonaceous chondrites, can be revised downward slightly. Based on the analyses of Orgueil, the revised value should be 3.6. This is in good agreement with the higher value of Müller (1968).

Acknowledgements

This report was supported in part by Grant GA-1200 from the National Science Foundation. Helpful assistance and comments were provided by Mr. Jerrald Durtsche and Dr. J. W. Larimer.

References

Cameron, A.G.W. (1968) in *Origin and Distribution of the Elements* (L.H. Ahrens, ed.). Pergamon.

Clayton, D.D. and Fowler, W.A. (1961) *Ann. Phys.* **16**, 51.

De Laeter, J.R. and Jefferey (1967) *Geochim. Cosmochim. Acta* **31**, 969.

Goldschmidt, V.M. and Peters, C. (1933) *Nachr. Ges. Wiss. Göttingen* **37**, 278.

Hamaguchi, H., Onuma, N., Hirao, Y., Yokoyama, H., Bando, S. and Furukawa, M. (1969) *Geochim. Cosmochim. Acta* **33**, 507.

Hey, M.H. (1966) *Catalogue of Meteorites* (3rd. edition). British Museum.

Kiesl, W., Seitner, H., Kluger, F. and Hecht, F. (1967) *Monatsh. Chem.* **98**, 972.

Larimer, J.W. and Anders, E. (1967) *Geochim. Cosmochim. Acta* **31**, 1239.

Lovering, J.F. (1957) *Geochim. Cosmochim. Acta* **12**, 253.

Mason, B. (1962) *Meteorites*, Wiley and Sons.

Müller, E.A. (1968) in *Origin and Distribution of the Elements* (L.H. Ahrens, ed.). Pergamon.

Noddack, I. and Noddack, W. (1930) *Naturwiss.* **18**, 757.

Noddack, I. and Noddack, W. (1931) *Z. Physik. Chem.* (Leipzig) **A154**, 207.

Noddack, I. and Noddack, W. (1934) *Svensk. Kem. Tidskr.* **46**, 173.

Onishi, H. and Sandell, E.B. (1957) *Geochim. Cosmochim. Acta* **12**, 262.

Shima, M. (1964) *Geochim. Cosmochim. Acta* **28**, 517.

Suess, H.E. and Urey, H.C. (1956) *Rev. Mod. Phys.* **28**, 53.

Van Schmus, W.R. and Wood, J.A. (1967) *Geochim. Cosmochim. Acta* **31**, 747.

Winchester, J.W. and Aten, A.H. (1957) *Geochim. Cosmochim. Acta* **12**, 57.

(Received 7 April 1970)

ANTIMONY (51)

William D. Ehmann

Department of Chemistry, University of Kentucky, Lexington, Kentucky and the Department of Chemistry, Arizona State University, Tempe, Arizona

EARLY DETERMINATIONS of antimony in meteorites were reported by the Noddacks (1930, 1931 and 1934). Their average of 0.64 ppm for the chondrites has been clearly shown by more recent work to be too high. Onishi and Sandell (1955) analyzed two composite samples made up of seven chondrites each and obtained an average chondritic abundance of 0.1 ppm for antimony. They also obtained a chondritic metal phase abundance of approximately 0.5 ppm and reported antimony abundances of 0.4 to 0.9 ppm in two siderites. The analyses were done using photometric techniques.

More recently antimony has been determined in meteorites by numerous investigators using the sensitive technique of neutron activation analysis. Two stable isotopes of antimony exist in nature, ^{121}Sb (57.25%) and ^{123}Sb (42.75%). In the activation analysis procedures for antimony the specimens are irradiated in the thermal flux from a nuclear reactor for periods of several days to several weeks. The radionuclides ^{122}Sb (half-life 2.8 days) and ^{124}Sb (half-life 60.2 days) are produced by thermal capture reactions. Potential interfering reactions on tellurium and iodine are regarded by Tanner and Ehmann (1967) as negligible in meteorite analyses. Radiochemical separations of antimony were used in all the activation analysis procedures. Both beta counting and gamma-ray scintillation spectrometry have been employed in the measurement of the indicator radionuclides. The use of ^{122}Sb is to be preferred when ready access to a reactor is available, due to the higher abundance and capture cross section of ^{121}Sb and the shorter irradiation time required. When transportation of the irradiated sample to a distant laboratory is required, ^{124}Sb is, by necessity, the choice for the indicator radionuclide. Error limits for the activation data are difficult to assign.

TABLE 1 Antimony abundances in various classes of stony meteorites

Classification*	Number of meteorites analyzed	Total number of analyses	Mean Sb abundance in ppm†	Atomic abundance Sb (Si = 10^6)‡
C1	2	4	0.17	0.36
C2	5	7	0.13	0.22
C3	2	2	0.13	0.19
C4	1	1	0.11	0.16
All C	10	14	*0.13*	*0.22*
E4	2	4	0.25	
E5	1	1	0.22	
E6	4	4	0.15	
All E	7	9	*0.19*	*0.25*
H3	1	1	0.09	
H4	2	2	0.11	
H5	6	8	0.09	
H6	2	2	0.08	
All H	12	14	*0.09*	*0.12*
L4	1	1	0.24	
L5	5	6	0.12	
L6	12	16	0.10	
All L	19	24	*0.11*	*0.14*
LL4	1	1	0.05	
LL6	2	2	0.12	
All LL	3	3	*0.10*	*0.12*
Achondrites, Ca-poor	5	5	0.03	
Achondrites, Ca-rich	5	5	(0.01–0.91)	

* Classification according to Van Schmus and Wood (1967).
† Based on data of Hamaguchi et al. (1961), Fouché and Smales (1967), Kiesl et al. (1967), Tanner and Ehmann (1967), Kiesl (1969) and Seitner et al. (1968). Seven analyses yielding exceptionally high values are not included.
‡ Using silicon data of Vogt and Ehmann (1965).

Tanner and Ehmann (1967) estimated error limits of $\pm 10\%$ based on all accountable errors and reported average deviations from the mean for replicate antimony standards of ± 2 to 3%.

A summary of recent activation analysis data for antimony in stony meteorites is given in Table 1. In general, the data of Hamaguchi et al. (1961), Fouché and Smales (1967), and Tanner and Ehmann (1967) are in very good agreement with each other and with the earlier composite averages of Onishi and Sandell (1955). Specific comparisons of the data may be found in Fouché and Smales. Several analyses ranging from 0.47 to 3.30 ppm by Kiesl et al. (1957) and Seitner et al. (1968) for four L-group chondrites are omitted from Table 1. The activation analysis procedures generally require small (0.1 to 0.5 gram) samples. Antimony along with other siderophilic elements would tend to concentrate in the chondritic metal phase, and occasional discordant high values may be expected due to chance inclusion of larger metal phase aggregates. The distribution of individual antimony analyses for the chondrites, omitting the few values above 0.4 ppm, is given in Figure 1.

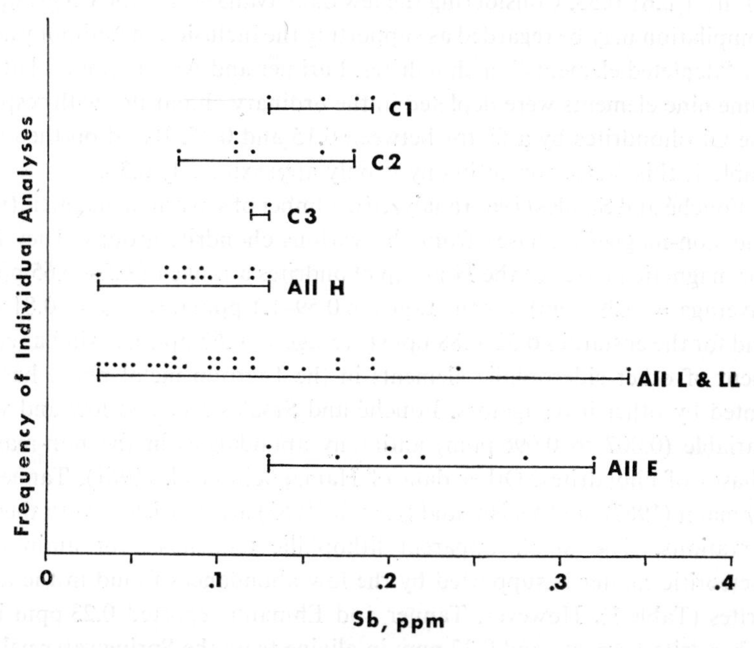

FIGURE 1 Distribution of individual antimony determinations for the principal classes of chondrites. Seven analyses ranging from 0.46 to 3.30 ppm in the C3 and L-groups appear to be discordant and are not plotted.

The abundance of an element in Type I carbonaceous chondrites is often taken as being representative of the abundance in average solar system primordial material. The relative atomic abundance of 0.36 (Si = 10^6) in the C1 chondrites derived from this compilation is somewhat larger than the values selected by Suess and Urey (1956)—0.25, and computed by Clayton and

Fowler (1961)—0.091, and Cameron (1959)—0.15. The recently determined C1 abundances of Pd, Ag, Cd, In, Sn and Te in this region of the abundance curve also lie considerably above the calculated curve of Clayton and Fowler based on theories of nucleosynthesis. If the C1 chondrites are taken as truly representative of non-volatile primordial material, a re-examination of the nuclear processes used in the establishment of the calculated abundance curve in this mass region would appear to be necessary.

Antimony is listed as one of nine elements exhibiting a "normal" depletion pattern of 1/0.6/0.3 in the groups C1/C2/C3 (Larimer and Anders, 1967). Using the data of Table 1, the atomic abundance depletion ratios relative to C1 are 1/0.61/0.53. Considering the few data available for the C3 group, this compilation may be regarded as supporting the inclusion of antimony among the "depleted elements" in chondrites. Larimer and Anders found that these same nine elements were depleted in the ordinary chondrites with respect to the C1 chondrites by a factor between 0.15 and 0.25. Based on the data of Table 1, this factor for antimony is only approximately 0.36.

Fouché and Smales (1967) analyzed a number of separated magnetic (metal) and non-magnetic phases from the various chondrite groups. They found for magnetic phases of the H-group chondrites a range of 0.24–0.65 ppm Sb (average = 0.39 ppm), for the L-group 0.59–1.1 ppm (average = 0.81 ppm), and for the enstatites 0.52–0.88 ppm (average = 0.67 ppm). A similar enrichment of other siderophilic elements in the L-group metal phase has been noted by other investigators. Fouché and Smales reported low and widely variable (0.007 to 0.096 ppm) antimony abundances in the non-magnetic phases of chondrites. Other data of Hamaguchi et al. (1961), Tanner and Ehmann (1967), and Onishi and Sandell (1955) are consistent with these observations. The small apparent lithophilic tendencies for antimony in meteoritic matter is supported by the low abundances found in the achondrites (Table 1). However, Tanner and Ehmann reported 0.23 ppm in the achondrite Kapoeta and 0.37 ppm in olivine from the Springwater pallasite, based on single analyses. Lithophilic tendencies for antimony in meteorites are still uncertain and are deserving of further investigation.

Meteoritic troilite abundances ranging from 0.11 to 0.96 ppm Sb have been reported by Smales et al. (1958) and Tanner and Ehmann (1967). This is well within the range of abundances found for siderites (Table 2). In fact Smales et al. reports an abundance of 0.64 ppm Sb in the interior of a Canyon Diablo troilite nodule, a value that is approximately twice as large as the abundance found in the metal phase of this meteorite. Hence it appears that antimony exhibits appreciable chalcophilic as well as siderophilic properties

TABLE 2 Antimony abundances in some distinct structural groups of siderites (As given in Smales et al., 1967)

Class	Number of meteorites analyzed	Range of Sb abundances, ppm	Mean Sb abundance, ppm
Hexahedrites	7	0.034–0.11	0.059
Coarsest octahedrites	3	0.27–0.33	0.30
Coarse octahedrites	7	0.32–0.48	0.37
Medium octahedrites	12	0.027–0.28	0.11
Fine octahedrites	6	0.003–0.058	0.017

in meteoritic matter. It is not certain to what extent the observation of geochemical affinities in the siderites would apply to the chondrites. No data for antimony in separated troilite phases from chondrites are available.

Data on antimony abundances in siderites have been published by Onishi and Sandell (1955), Hamaguchi et al. (1961), Smales et al. (1958), Smales et al. (1967), Tanner and Ehmann (1967), Kiesl et al. (1967), and Seitner et al. (1968). The largest number of specimens was analyzed by Smales. A summary of Smales' data for some distinct structural groups of siderites is given in Table 2. For the most part, the data of the various investigators are in good agreement. However, Seitner et al. report finding higher Sb abundances in four siderites than found by Smales et al. in the same meteorites. A non-homogeneous distribution of antimony in some siderites may explain the discordant results, although Fouché and Smales (1967) reported a coefficient of variation of only 5.3% in ten separate analyses of the Canyon Diablo siderite. Tanner and Ehmann report an abundance of 1.72 ppm Sb in the small spherules collected around the Canyon Diablo crater. As was the case for gold, this is considerably above the Sb abundance in the Canyon Diablo siderite itself.

Rankama and Sahama (1950) point out that antimony can combine with various heavy metals, such as Cu, Fe, Co and Ni to form compounds such as NiSb (nickel antimonide). This mechanism may, in part, account for the siderophilic properties of antimony. Smales et al (1967) notes a strong correlation between antimony abundances in siderites and abundances of copper, palladium, and arsenic. In terrestrial materials antimony is commonly found in the form of stibnite, Sb_2S_3, and mixed sulfides with copper, lead, and silver. There is also evidence that antimony may exhibit lithophilic tendencies, since it is found in the minerals of granitic pegmatites and oxgen compounds associated with minerals of niobium and tantalum. During the crystallization of bulk silicates, antimony apparently becomes concentrated in the terminal aqueous liquid phase.

References

Cameron, A.G.W. (1959) *Astrophys. J.* **129**, 676–699.

Clayton, D.D. and Fowler, W.A. (1961) *Ann. Phys. (N.Y.)* **16**, 51–68.

Fouché, K.F. and Smales, A.A. (1967) *Chem. Geol.* **2**, 105–134.

Hamaguchi, H., Nakai, T. and Endo, T. (1961) *Nippon Kagaku Zasshi* **82**, 1485–1489.

Kiesl, W., Seitner, H., Kluger, F. and Hecht, F. (1967) *Monatshefte f. Chemie* **98**, 972–992.

Kiesl, W. (1969) *Modern Trends in Activation Analysis*, U.S. Dept. Commerce, N.B.S. Spec. Publ. 312, vol. **1**, 302–307.

Larimer, J.W. and Anders, E. (1967) *Geochim. Cosmochim. Acta* **31**, 1239–1270.

Noddack, I. and Noddack, W. (1930) *Naturwiss.* **18**, 757–764.

Noddack, I. and Noddack, W. (1931) *Z. Physik. Chem.* **A154**, 207–244.

Noddack, I. and Noddack, W. (1934) *Svensk Kem. Tids.* **46**, 173–201.

Onishi, H. and Sandell, E.B. (1955) *Geochim. Cosmochim. Acta* **8**, 213–221.

Seitner, H., Kiesl, W., Kluger, F. and Hecht, F. (1968) 5th Communication (Preprint) from the Analytical Institute of the University of Vienna and the Reactor Center Seibersdorf of the "Österreichische Studiengesellschaft für Atomenergie G.m.b.H.".

Smales, A.A., Mapper, D., Morgan, J.W., Webster, R.K. and Wood, A.J. (1958) *Proc. Second U.N. Conf. Peaceful Uses Atom. En.*, Geneva **2**, 242–255.

Smales, A.A., Mapper, D. and Fouché, K.F. (1967) *Geochim. Cosmochim. Acta* **31**, 673–720.

Suess, H.E. and Urey, H.C. (1956) *Rev. Mod. Phys.* **28**, 53–74.

Tanner, J.T. and Ehmann, W.D. (1967) *Geochim. Cosmochim. Acta* **31**, 2007–2026.

Van Schmus, W.R. and Wood, J.A. (1967) *Geochim. Cosmochim. Acta* **31**, 747–765.

Vogt, J.R. and Ehmann, W.D. (1965) *Geochim. Cosmochim. Acta* **29**, 373–383.

Rankama, K. and Sahama, Th.G. (1950) *Geochemistry*, The University of Chicago Press, Chicago.

(Received 1 February 1969)

TELLURIUM (52)

Ithamar Z. Pelly* and Michael E. Lipschutz†

Department of Chemistry
Purdue University
Lafayette, Indiana 47907

TELLURIUM HAS BEEN DETERMINED in 77 meteorites (not including upper limits) and is present in these in trace quantities ranging from $0.05–8 \times 10^{-6}$ g/g. DuFresne (1960) determined both tellurium and Se spectrophotometrically; all other tellurium determinations have been made by techniques based upon neutron activation analysis. In all cases tellurium has been determined in conjunction with measurement of other trace elements: Se—Schindewolf (1960); Se, In—Akaiwa (1966); I—Kuroda et al. (1967); I, U—Goles and Anders (1960, 1961, 1962), Clark et al. (1967); U, halogens—Reed and Allen (1966), Reed and Jovanovic (1969); noble gases—Reynolds and Turner (1964); Kr, Xe—Merrihue (1966); Se, Zn, other elements—Greenland (1967); Case et al. (1971).

In meteorites, as on Earth, tellurium is chalcophile in geochemical behavior. Based solely on their own results, DuFresne (1960) and Greenland (1967) suggested that, in chondrites, tellurium is fractionated from its congeners, S and Se, indicating that other properties of tellurium affect its geochemistry. However, consideration of all of the pertinent data indicates that, with the possible exception of group E6, these chemically similar elements are not mutually fractionated (cf. Pelly and Lipschutz, Se chapter). Tellurium ranges from $0.017–0.09 \times 10^{-6}$ g/g in metal from 4 octahedrite samples and from $1.2–7.8 \times 10^{-6}$ g/g in sulfide inclusions from 4 octahedrites and Soroti (Goles and Anders, 1961, 1962) indicating its host mineral in iron meteorites to be troilite (FeS).

In chondrites, 25–65%, and in one achondrite, 95% of the tellurium can be removed by quite mild leaching (Goles and Anders, 1961, 1962; Reed

* Now at Dept. of Geology, Negev University, Beersheba, Israel.
† Also Department of Geosciences, Purdue University

TABLE 1 Tellurium in meteorites

Type	Number analyzed	No. of analyses	Te contents Range (10^{-6} g/g)	(Ref.)	Mean (10^{-6} g/g)	Mean (atoms/10^6 Si atoms)
chondrites						
C1	2	4	1.0–5.8	(1)	3.3	
	2	2	0.94–1.0	(2)	0.97	
	1	2	6.6–8.3[e]	(3)	*7.4*	
	1	2	3.3–3.4	(4)	3.4	
	2	8	0.94–5.8	All[h]	2.7	5.8
	2	10	(0.94–8.3)	All	(3.6)	(7.7)
C2	1	1	1.9	(1)	1.9	
	3	3	0.79–3.3	(2)	1.8	
	2	2	4.2–4.6[e]	(3)	*4.4*	
	1	2	1.8–2.0	(4)	1.9	
	2	5	1.2–2.6[a]	(5)	1.8	
	1	1	0.49[b, e]	(6)	0.49	
	3	11	0.79–3.3	All[h]	1.8	3.0
	4	14	(0.49–4.6)	All	(2.1)	(3.5)
C3	1	1	1.0	(1)	1.0	
	4	4	0.15–0.90	(2)	0.47	
	5	5	1.0–3.9[e]	(3)	*2.1*	
	1	2	1.2	(4)	1.2	
	4	7	0.15–1.2	All[h]	0.75	1.1
	5	12	(0.15–3.9)	All	(1.3)	(1.8)
C4	1	1	1.5	(1)	1.5	2.1
H3	3	3	3.2–7.4[e]	(3)	*5.2*	(6.7)
H4	4	4	0.81–2.0	(7)	1.3	1.7
H5	3	5	0.23–0.69	(1)	0.50	
	1	1	0.61	(2)	0.61	
	3	6	0.42–0.73[a]	(5)	0.52	
	1	1	0.74[e]	(6)	*0.74*	
	3	3	1.4–2.3	(7)	1.8	
	1	4	1.2–3.1	(8)	2.4	
	2	2	0.46–0.88	(9)	0.67	
	10	21	0.23–3.1	All[h]	1.1	1.4
	10	22	(0.23–3.1)	All	(1.0)	(1.3)
H6	1	1	0.87	(7)	0.9	1.2

TABLE 1 (cont.)

Type	Number analyzed	No. of analyses	Te contents Range (10^{-6} g/g)	(Ref.)	Mean (10^{-6} g/g)	Mean (atoms/ 10^6 Si atoms)
chondrites						
H (uncl.)	1	1	0.93	(2)	0.93	
	1	4	0.40–1.9	(8)	0.95	
	1	1	2.8	(10)	2.8	
	2	6	0.40–2.8	All	1.6	2.1
H	17	32	0.23–3.1	All[h]	1.1	1.4
	20	36	(0.23–7.4)	All	(1.5)	(1.9)
L3	1	1	2.8[e]	(3)	*2.8*	(3.3)
L4	1	1	0.64	(7)	0.6	0.7
L5	1	1	0.48	(5)	0.48	
	4	4	0.36–2.0	(7)	1.2	
	5	5	0.36–2.0	All	1.1	1.3
L6	3	4	0.26–0.87	(1)	0.51	
	4	4	0.60–0.89	(2)	0.77	
	2	4	0.20–0.50	(5)	0.41	
	1	2	0.27–0.67[e]	(6)	*0.47*	
	6	8	0.23–2.5	(7)	1.6	
	2	4	0.80–2.1	(8)	1.5	
	2	2	0.46–0.62	(9)	0.54	
	11	26	0.20–2.5	All[h]	1.0	1.2
	11	28	(0.20–2.5)	All	(0.97)	(1.1)
L	17	32	0.20–2.5	All[h]	1.0	1.2
	18	35	(0.20–2.8)	All	(1.0)	(1.2)
LL3	1	1	1.4	(2)	1.4	
	4[c]	4	1.4–4.3[e]	(3)	*2.6*	
	1	1	1.4	All[h]	1.4	1.6
	4	5	(1.4–4.3)	All	(2.4)	(2.8)
LL6	1	1	1.3	(7)	1.3	1.5
LL	2	2	1.3–1.4	All[h]	1.4	1.6
	5	6	(1.3–4.3)	All	2.2	(2.6)
E4	2	2	2.5–2.8	(1)	1.6	
	2	2	2.6–3.9	(2)	3.2	
	1	2	3.1–3.4	(4)	3.2	

TABLE 1 (cont.)

Type	Number analyzed	No. of analyses	Range (10^{-6} g/g)	(Ref.)	Te contents Mean (10^{-6} g/g)	Mean (atoms/ 10^6 Si atoms)
chondrites						
	2	5	1.8–3.4[a]	(5)	2.4	
	1	1	2.3[e]	(6)	*2.3*	
	1	1	3.2	(7)	3.2	
	2	12	1.8–3.9	All[h]	2.8	3.6
	2	12	(1.8–3.9)	All	(2.8)	(3.6)
E5	1	2	1.3–1.6[a]	(5)	1.4	1.6
E6	1	1	0.32	(1)	0.32	
	4	4	0.18–0.79	(2)	0.48	
	4	5	0.18–0.79	All	0.46	0.52
achondrites (calcium-rich)						
Eu	6	11	0.14–0.9[d]	(8)	0.31	
	1	1	0.32	(11)	0.32	
	6	12[f]	0.14–0.9	All	0.31	0.30
Ho	1	4[f]	0.05–0.32	(8)	0.16	0.15
Na	1	2	0.14–0.16	(8)	0.15	0.14
achondrites (calcium-poor)						
Au	3[g]	6[f]	0.22–0.8	(8)	0.39	0.31
Di	1	1	0.12	(8)	0.12	0.11
Mesosiderites						
M	3	7	0.36–1.4	(8)	0.76	

(1) Reed and Allen (1966)
(2) Greenland (1967)
(3) Case et al. (1971)
(4) Akaiwa (1966)
(5) Goles and Anders (1962)
(6) Merrihue (1966)
(7) DuFresne (1960)
(8) Clark et al. (1967)
(9) Schindewolf (1960)
(10) Kuroda et al. (1967)
(11) Reed and Jovanovic (1969)

and Allen 1966; Reed and Jovanovic, 1969) suggesting that this portion of the tellurium is located in water-soluble minerals. Although tellurium has not been determined in any separated mineral from stony meteorites, circumstantial evidence led Goles and Anders (1962) to suggest that the water-soluble host minerals might be epsomite ($MgSO_4$) in carbonaceous chondrites and oldhamite (CaS), lawrencite ($FeCl_2$) or MgS in chondrites and achondrites. The identification of MgS has been questioned by Keil and Snetsinger (1967) who have determined it to be a variety of alabandite [(Mn, Fe)S] and report yet another possible water-soluble mineral, niningerite [(Fe, Mg)S]. Interestingly the meteorite containing the highest proportion of leachable tellurium is a calcium-rich achondrite, Moore County (Reed and Jovonovic, 1969). The non-leachable tellurium may well be dispersed in such minerals as troilite. Whatever minerals may be involved, the correlation of I and Te over concentration ranges of 10^2 for each element suggests that both reside largely in an inhomogeneously-distributed minor phase or strongly associated phases in stony meteorites (Goles and Anders, 1960, 1962; Reed and Allen, 1966) and in the fine-grained matrix of C2 chondrites (Reynolds and Turner, 1964; Anders, 1964).

The data for 13 achondrites (including 2 finds), 3 mesosiderites (including 2 finds) and 12 carbonaceous, 43 ordinary (including 4 finds) and 7 enstatite chondrites are summarized in Table 1 and illustrated in Figure 1. Replicate analyses of the same meteorite, even those by the same investigator, often disagree by a factor of 2, further suggesting the inhomogeneous distribution of tellurium. Beyond this, certain of the results appear discrepant. The determinations by Reynolds and Turner (1964), Merrihue (1966) and, in some cases, DuFresne (1960) are accompanied by experimental uncertainties of up to a factor of 3. Other suspicious determinations include those by Case

[a] Some of these data were reported previously by Goles and Anders (1960, 1961).

[b] Recalculated from the results of Reynolds and Turner (1964) who list 0.73×10^{-6} g/g for this meteorite.

[c] Hamlet is assumed to be LL3.

[d] Not including a much higher replicate analysis for Juvinas, 5.2×10^{-6} g/g, which Clark et al. (1967) attribute to contamination.

[e] The quality of these data is uncertain (see text).

[f] Not including values of 0.030×10^{-6} g/g (Moore Co.), 0.023×10^{-6} g/g (Frankfort), and 0.025×10^{-6} g/g (Norton Co.) obtained by Reed and Jovanovic (1969) only from material remaining after leaching.

[g] Not including duplicate determinations of 2.2 and 2.3×10^{-6} g/g for Shallowater (the only aubrite find studied) which may be unusually high because of contamination (Clark et al., 1967).

[h] Doubtful data omitted.

et al. (1971), which systematically are very much higher than those of other investigators. For some ordinary chondrites, the results of DuFresne (1960) and Clark et al. (1967) are about a factor of 2 higher than other results for the same or similar chondrites; for other ordinary and enstatite chondrites they agree. Greenland's (1967) results for C1 and C3 chondrites are about a factor of 2 lower than others'; for C2, ordinary and enstatite chondrites they agree. For the remaining discussion, arithmetic group means will be used, although the variability of tellurium suggests that geometric means might be more appropriate. In either case the conclusions are the same.

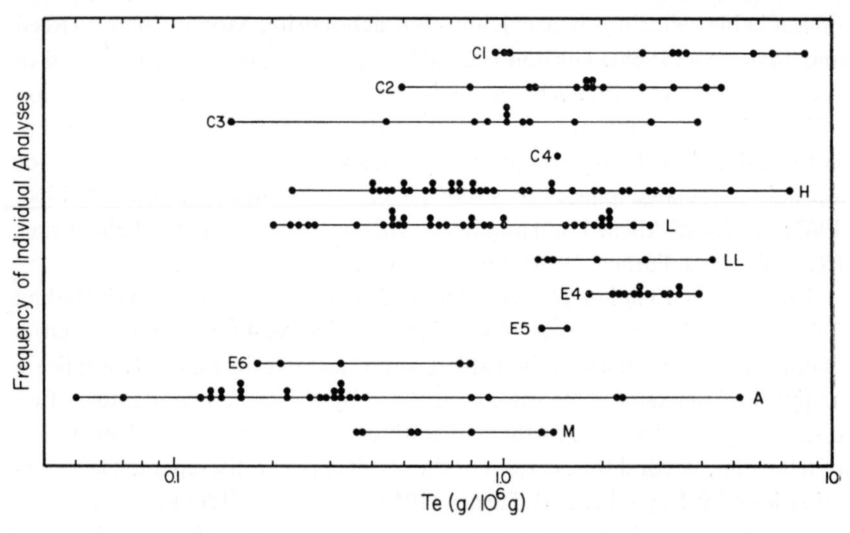

FIGURE 1 Tellurium concentrations in chondrites, achondrites (A) and mesosiderites (M).

Despite the ten-fold variation in the tellurium contents of ordinary chondrites, the ranges and mean values for the H-, L- and LL-group chondrites do not differ significantly (Table 1, Figure 1). Within each group, the unequilibrated ordinary chondrites (UOC) may be richer than equilibrated ones in tellurium. Unfortunately with but one exception all UOC were measured by Case et al. (1971). Assuming that these results are precise but systematically high, it appears that within each chondritic group tellurium is independent of chemical-petrologic type and of the degree of equilibration of the unequilibrated ordinary chondrites (Case et al., 1970). There appears to be no significant difference between the tellurium contents of the co-existing light and dark phases of Pantar (Reed and Allen, 1966), Fayetteville and

Leighton (Clark et al., 1967). The tellurium concentration in a single Bruderheim chondrule, 0.27×10^{-6} g/g, appears lower than the whole-rock datum, 0.67×10^{-6} g/g, (Merrihue, 1966) but the difference is well within the experimental uncertainties.

The mean abundances for C1 (Table 1) and all ordinary chondrites, 1.3 or 1.7 atoms/10^6 Si atoms (the latter value including all suspicious analyses) show tellurium to be a normally depleted element. The mean abundance in C1 is about twice that in C2 and about 4–5 times that in C3. The exact factors depend on inclusion or exclusion of suspicious data—their inclusion improving the already reasonable agreement with the predictions of Anders (1964). The C1 data suggest mean cosmic abundance values of 5.8 or 7.7 atoms/10^6 Si atoms. Neither Aller (1961) nor Müller (1968) list solar photospheric values for tellurium.

The ranges for the E-group chemical-petrologic types do not overlap and tellurium is progressively depleted in types E4–E6. In terms of Larimer and Anders' (1967) notation, relative to C1 chondrites the depletion factors of Types I and II are, respectively, 0.62 and 0.14 or 0.47 and 0.11 (if suspicious analysis are included).

The tellurium contents of the achondrites and mesosiderites are generally much lower than chondritic values (Table 1). For three of the four finds (excepting Lafayette) the higher tellurium contents suggest contamination, and if finds are disregarded the duplicate results for the mesosiderite Estherville yield a mean of 0.36×10^{-6} g/g (Clark et al., 1967). Clark et al. (1967) report no significant difference between the tellurium contents of the light and dark portions of the howardite Kapoeta, and further that the dark chondritic inclusion and white matrix of the unique achondrite Cumberland Falls have identical tellurium contents. There exist no whole rock analyses for iron meteorites nor has the isotopic composition of meteorite tellurium been determined.

Acknowledgement

This research was supported in part by the U.S. National Science Foundation, grant GA-1474.

References

Akaiwa, H. (1966) *J. Geophys. Res.* **71**, 1919.
Aller, L.H. (1961) *The Abundance of the Elements*. Interscience.
Anders, E. (1961) *Space Sci. Rev.* **3**, 583.
Case, D.R., Laul, J.C., Pelly, I.Z., Wechter, M.A., Schmidt-Bleek, F. and Lipschutz, M. E. (1971) in preparation.
Clark, R.S., Rowe, M.W., Ganapathy, R. and Kuroda, P.K. (1967) *Geochim. Cosmochim. Acta* **31**, 1605.
DuFresne, A. (1960) *Geochim. Cosmochim. Acta* **20**, 141.
Goles, G.G. and Anders, E. (1960) *J. Geophys. Res.* **65**, 4181.
Goles, G.G. and Anders, E. (1961) *J. Geophys. Res.* **66**, 3075.
Goles, G.G. and Anders, E. (1962) *Geochim. Cosmochim. Acta* **26**, 723.
Greenland, L. (1967) *Geochim. Cosmochim. Acta* **31**, 849.
Keil, K. and Snetsinger, K.G. (1967) *Science* **155**, 451.
Kuroda, P.K., Clark, R.S. and Ganapathy, R. (1967) *J. Geophys. Res.* **72**, 1407.
Larimer, J.W. and Anders, E. (1967) *Geochim. Cosmochim. Acta* **31**, 1239.
Merrihue, C. (1966) *J. Geophys. Res.* **71**, 263.
Müller, E. (1968) *Origin and Distribution of the Elements* (L.H. Ahrens, ed.) 155–176, Pergamon.
Reed, G.W. Jr. and Allen, R.O. Jr. (1966) *Geochim. Cosmochim. Acta* **30**, 779.
Reed, G.W. Jr. and Jovanovic, S. (1969) *Earth Planet. Sci. Letters* **6**, 316.
Reynolds, J.H. and Turner, G. (1964) *J. Geophys. Res.* **69**, 3263.
Schindewolf, U. (1960) *Geochim. Cosmochim. Acta* **19**, 134.

(Received 6 February; revised 1 June 1970)

IODINE (53)*

George W. Reed, Jr.

Argonne National Laboratory
Argonne, Illinois 60439

GEOCHEMICALLY IODINE, like bromine, is considered to be a dispersed element. It is found in all types of meteorites as a trace element. Iodine is a fractionated element in meteorites, with the enstatite and carbonaceous chondrites having the higher concentrations. However, the greatest concentrations of I in meteorites are found in the sulfide and phosphate mineral inclusions in iron meteorites. In 1950 Rankama and Sahama reviewed the geochemistry of I and emphasized the impossibility of diadochic replacement of the other halogens by iodine because of its large ionic radius (2.20 Å *vs.* 1.81 and 1.96 for Cl and Br, respectively). The tendency of the phosphate to concentrate I may be merely due to the fact that the metallic phase provides a less favorable site. The element's relative enrichment in troilite (FeS) is consistent with this. However, there are other pertinent experimental observations.

A significant amount of the I in chondritic meteorites is water leachable as shown by Goles and Anders (1962) and Reed and Allen (1966). As a consequence the former authors tentatively suggested sulfides such as oldhamite (CaS), and alabandite (MnS) or the chloride, lawrencite ($FeCl_2$), as possible iodine host minerals. An alternate explanation is that the I is present as a water-soluble salt on grain surfaces. Goles and Anders also found almost all the I in Bruderheim to be in −326 mesh fractions. This iodine may be associated with a fine-grained soluble mineral.

Chlorapatite is commonly found in hypersthene but rarely in bronzite chondrites. The I contents of these two classes are not significantly different. Thus in the ordinary chondrites, at least, the amount of I is independent of the presence of apatite and, in fact, the mineral is not even present in enstatite and carbonaceous chondrites, which have higher I concentrations.

* Based on work performed under the auspices of the U.S. Atomic Energy Commission.

TABLE 1 Iodine contents of chondrites

Meteorite classification*	No. of determi-nations	No. of meteorites analyzed	Range (ppm)	Mean (ppb)	At/10^6 Si§	Refer-ences
Hypersthene (L)	6	4	5–90	43†		A
	5	3	30–450	53‡	0.07	B
	6	5	6–17	12	0.014	C
	4	2	17–27	20		D
Bronzite (H)	5	2	21–65	45		A
	3	2	67–120	68	0.09	B
	1	1	–	9	0.011	C
Enstatite E6	2	1	17, 89	53	~0.06	C
	1	1	–	74	0.08	B
Enstatite E5	2	1	64, 100	82		A
Enstatite E4	4	2	140–300	208		A
	2	2	≤180, 470	–	0.40	B
	1	1	–	310	0.42	C
Carbonaceous-III	2	2	110, 220	170	0.23	B
	2	2	170, 260	215	0.30	C
Carbonaceous-II	1	1	–	550		B
	2	2	300, 480	390	0.66	C
Carbonaceous-I	4	2	230–1210	580	1.16	B
	1	1		400	0.83	C
Amphoterite (LL)	1	1		200	0.23	C

* Classical designations are used except for the enstatite chondrites which are labeled according to the Van Schmus–Wood classification (1967).
† Data excluded by authors, not averaged.
‡ 450 ppb value excluded.
§ Based on averages reported by the investigators.

A Goles and Anders (1962)
B Reed and Allen (1966)
C Goles et al. (1967)
D Clark et al. (1969).

Modern analyses of I in meteorites have all utilized the neutron activation technique already described. The analysis for iodine is complicated by the fact that, in addition to 25 min I^{128} formed from natural I^{127}, several fission-product iodine radioactivities are always produced and greatly complicate the extraction of the I^{128} data. The various investigators have discussed this problem. Goles and Anders (1962), Goles et al. (1968), and Reed and Allen (1966) have measured iodine in chondritic meteorites. The two former groups also measured iron meteorites. Clark et al (1969) measured I in achondrites and mesosiderites.

Another neutron activation approach has been explored by Reynolds and his collaborators (1968). Pile irradiation produces Xe^{128} through the reaction I^{127} (n, γ) $I^{128} \xrightarrow{\beta^-} Xe^{128}$. For quantitative determination of I this method requires that the Xe^{128} not escape from the sample. Hohenberg (1968) has shown that Xe concentrations measured mass spectrometrically in this procedure are low by an order of magnitude or more in many meteorites. He concludes that the I is in sites where Xe retention is low i.e. at grain boundaries, and this is consistent with the leachability of I in meteorites.

The iodine data for chondrites are summarized in Table 1. The spread in the measurements for a given class of chondrites may be an order of magnitude; however, duplicate determinations on a given meteorite usually agree within a factor of two. Thus I is fairly uniformly dispersed in a single meteorite, but the variation within a class suggests that the amount of iodine in the environment in which each meteorite of a given class equilibrated must have been quite different.

An interlaboratory comparison of data for a meteorite generally shows agreement within the amount of variation seen in a given laboratory. A striking exception is Reed and Allen's large values, 74 and 450 ppb for Bruderheim *vs.* 5–27 by Goles and Anders, 6 and 7 by Goles et al., 18 and 27 by Clark et al. and 7.8 by Merrihue. Goles et al. observed 11.2 ppm I in an Ivuna aliquot and discarded it as probably contamination. Such cases may indeed indicate contamination but they could also be due to an unusual concentration of I in that aliquot.

Iodine contents in the special group of light-dark gas-rich chondrites are summarized in Table 2. No clear case for enrichment in Pantar-dark can be made from Reed and Allen's data because the possible values for the light and dark fractions overlap. Goles et al.'s Pantar-light determination is lower than Reed and Allen's light or dark lower limits and infers an I enrichment in the dark phase. Clark et al.'s data indicate no I enrichment in Fayetteville-dark but a factor of 2.5 enrichment in Leighton-dark.

Iodine in iron meteorites and two mineral phases, troilite and chlor-apatite, from iron meteorites is given in Table 3. The amount of I in the metal phase is as great and greater than that found in ordinary chondrites. The above holds true for troilite which in some cases contains more I (1000 to 3600 ppb) than the carbonaceous chondrites (<600 ppb). As already men-

TABLE 2 Iodine in dark-light chondrites

Meteorite	No. of determinations	No. of meteorites	Concentration (ppb)	References
Pantar-Lt	1	1	$55 < x < 197$	Reed and Allen
Pantar-Lt	1	1	19	Goles et al.
Pantar-Dk	1	1	$128 < x < 551$	Reed and Allen
Fayetteville-Lt	2	1	77, 108	Clark et al.
Fayetteville-Dk	2	1	43, 162	Clark et al.
Leighton-Lt	2	1	292, 320	Clark et al.
Leighton-Dk	2	1	510, 990	Clark et al.

TABLE 3 Iodine contents of various phases in iron meteorites

Sample	No. of determinations	Concentration (ppb)	Reference
Metal			
Grant, Of	1	11	Goles and Anders
Toluca, Om	2	170, 320	Goles and Anders
Canyon Diablo, Og	1	28	Goles and Anders
Odessa, Og	1	130	Goles et al.
Sardis, Ogg	2	68, 130	Goles et al.
Troilite			
Grant	1	24	Goles and Anders
Soroti	1	50	Goles and Anders
Toluca	1	1030	Goles and Anders
Canyon Diablo	1	62	Goles and Anders
Sardis	1	3590	Goles and Anders
Apatite			
Mt. Stirling	1	1700	Reed and Allen

tioned, Goles and Anders early pointed out the chalcophilic behavior of I. The relatively large (1700 ppb) amount of I in chlorapatite from the Mt. Stirling iron meteorite has also been discussed.

All data on iodine in achondrites (Table 4) have been obtained by Clark et al. The range is so large in some classes that it is almost meaningless to average; for example, the enstatite achondrites Pena Blanca Spring and

TABLE 4 Iodine Contents of Achondrites and Mesosiderites (Clark et al.)

Meteorite classification	No. of Determinations	No. of Meteorites	Range (ppb)	Mean
Achondrites				
Enstatite				
(aubrites)	7	4	22–460*	101
Hypersthene				
(diogenites)	3	2	25–180	90
Basaltic				
(howardites)	4	1	34–46	37
(eucrites)	14	6	14–1000	200
				85†
Diopside-olivine				
(nakhlite)	3	2	86–180	130
Mesosiderites	7	3	28–170	89

* Range would be 22–140 if the light part of Cumberland Falls, a breccia containing achondrite and dark chondrite, is excluded. The chondrite portion gave 560 ppb iodine, thus incomplete separation of the phases may account for the 460 ppb result.

† Mean excluding the meteorites giving the extreme values. The range is reduced to 35–160 ppb.

Shallowwater gave 22.5 and 180 ppb I, respectively; two of the eucrites, Sioux County and Stannern, gave 14 and 830 ppb, respectively. The iodine in mesosiderites averages close to that found in the achondrites.

References

Clark, R.S., Rowe, M.W., Ganapathy, R. and Kuroda, P.K. (1967), "Iodine, uranium and tellurium contents in meteorites", *Geochim. Cosmochim. Acta* **31**, 1605–1613.

Goles, G.G. and Anders, E. (1962), "Abundances of iodine, tellurium and uranium in meteorites", *Geochim. Cosmochim. Acta* **26**, 723–737.

Goles, G.G., Greenland, L.P. and Jérome, D.Y. (1967) "Abundances of chlorine, bromine and iodine in meteorites", *Geochim. Cosmochim. Acta* **31**, 1771–1787.

Hohenberg, C.M. (1968) *Extinct radioactivities in meteorites*, Nininger Meteorites Competition Paper, University of California, Berkeley. See also C.M. Hohenberg and J.H. Reynolds, (1969) "Preservation of the iodine-xenon record in meteorites", *J. Geophys. Res.* **74**, 6679–6683.

Merrihue, C. M. (1966) "Xenon and krypton in the Bruderheim meteorite." *J. Geophys. Res.* **71**, 263–313.

Rankama, K. and Sahama, Th. G. (1950) *Geochemistry*, The University of Chicago Press, Chicago, Illinois, p. 759.

Reed, G. W. and Allen, R. O. (1966), "Halogens in chondrites", *Geochim. Cosmochim. Acta* **30**, 779–800.

Reynolds, J. H. (1967) "Isotopic abundance anomalies in the solar system", *Annual Review of Nuclear Science* Volume 17, Annual Review, Inc., Palo Alto, California, pgs. 253–316.

Van Schmus, W. R. and Wood, J. A. (1967) "A chemical-petrologic classification for the chondritic meteorites", *Geochim. Cosmochim. Acta* **31**, 747–765.

(Received 5 May 1970)

CAESIUM (55)

Gordon G. Goles

Center for Volcanology
and
Departments of Chemistry and Geology
University of Oregon
Eugene, Oregon 97403

THE EXCEPTIONAL VOLATILITY of Cs leads one to expect that this element would be sensitive to differences in accretion temperatures or subsequent thermometamorphism episodes in the histories of meteorites. Although there are regrettably few determinations of Cs contents, they confirm this supposition.

I have listed in Table 1 all published meteoritic Cs contents which I believe are likely to be reliable. They have been determined by isotope dilution or activation analysis techniques. Elemental ratios were calculated only in those cases where Rb or K had been determined in the same sample used for the Cs analysis, or in a split of the same sample. A critical review of older literature on Cs was given by Smales et al. (1964), who also studied experimentally the volatilization of this element from meteorites. Caesium contents in five chondritic finds as determined by Cabell and Smales (1957) and Smales et al. (1964) were excluded from the table. Chondrites are classified according to Van Schmus and Wood (1967) and achondrites and stony-irons are identified by the code introduced by Keil (1969).

One member of each of the C classes has been analysed by Smales et al. (1964). Their Cs contents decrease in a sequence which is consistent with the model of Larimer and Anders (1967). An unpublished value of $(0.06 \pm 0.01) \times 10^{-6}$ g Cs/g for the C3 chondrite *Pueblito de Allende* (Wakita and Schmitt, *Nature*, in press) lends support to this trend, although many more analyses are required before one can be confident of its reality.

Caesium contents and heavy alkali metal ratios in H and L chondrites exhibit some striking variations, which are correlated (but not in a simple

TABLE 1 Caesium contents and alkali metal ratios in meteorites

Meteorite (class)	Cs (10^{-6} g/g)	Reference	Rb/Cs (g/g)	K/Cs (10^{-3} g/g)
Ivuna (C1)	0.183 ± 0.005	(1)	12.4	
Mighei (C2)	0.125 ± 0.007	(1)	13.3	
Felix (C3)	0.045	(1)	30	
Bremervörde (H3)	0.192	(1)	18.9	
Beaver Creek (H4)	0.036 ± 0.002	(1)	37	
Ochansk (H4) "A"	0.100	(1)	16.4	
"B"	0.092		17.6	
	0.120 ± 0.008	(2)	18	
Allegan (H5)	0.064 ± 0.003	(1)	35	
Beardsley (H5)	0.193 ± 0.015	(3)	25.4	4.7
Forest City (H5)	0.106 ± 0.007	(3)	25.9	7.75
	0.098	(4)	31	
	0.0987	(5)	28.5	7.96
	0.107		26.1	7.36
Hessle (H5)	0.015	(1)	110	
Mount Browne (H6)	0.0046	(1)	340	
Limerick (H)	0.092	(1)	24	
	0.099	(4)	23	
Merua (H)	0.072	(1)	37	
Bjurböle (L4)	0.109 ± 0.009	(2)	27	
Chandakapur (L5)	0.204 ± 0.008	(2)	17	
Crumlin (L5)	0.006 ± 0.002	(1)	63	
Homestead (L5)	0.055 ± 0.010	(2)		
	0.07	(4)	45	
Knyahinya (L5)	0.0082 ± 0.0005	(1)	180	
Bruderheim (L6)	0.00296	(5)	885	298
	<0.01	(6)		
Château-Renard (L6)	0.0095 ± 0.0013	(2)	310	
Futtehpur (L6)	0.0111	(1)	180	
Holbrook (L6)	0.283 ± 0.009	(2)	8.1	
	0.146 ± 0.010	(3)	15.2	
Leedey (L6)	<0.00227 (minus some metal)	(5)	1250	399
Marion (L6)	0.1060 ± 0.0004	(1)	18	
Modoc (L6)	0.090 ± 0.007	(3)	38	
	0.08	(4)	37	
Atoka (L)	0.0045	(1)	670	
Krähenberg (LL5) "Light"	0.08	(7)	24	10.4
"Dark"	2.8		18	4.3

TABLE 1 (cont.)

Meteorité (class)	Cs (10^{-6} g/g)	Reference	Rb/Cs (g/g)	K/Cs 10^{-3} (g/g)
Olivenza (LL5)	0.074	(1)	30	
Dhurmsala (LL6)	0.023	(1)	100	
Mangwendi (LL6)	0.048 ± 0.002	(1)	49	
Abee (E4)	<0.237 (minus some metal)	(5)	14.6	3.66
Khor Temiki (Ae)	0.060	(1)	29	
Johnstown (Ab)	0.0076	(1)	18	
	0.007	(4)	~6	
Angra dos Reis (Aa)	0.000435 ± 0.000065	(5)	71.4	29.7
Bereba (Ap)	0.00577	(5)	30.8	44.7
Jonzac (Ap)	0.0144 ± 0.0004	(5)	28	22.8
Richardton (H5)	0.088 ± 0.006	(3)	34	9.3
Juvinas (Ap)	0.0057	(5)	29	56
Moore Co. (Ap)	0.0052 ± 0.0007	(3)	31	36
	0.00071 ± 0.00002	(5)	69	220
Nuevo Laredo (Ap)	0.020 ± 0.002	(3)	19	15
	0.0138	(5)	23.4	30
Pasamonte (Ap)	0.0112 ± 0.0010	(3)	19	38
"197g"	0.0114 ± 0.0003	(5)	23.5	28.8
"297y"	0.0076 ± 0.0002		25.5	36.5
Sioux Co. (Ap)	0.0118 ± 0.0010	(3)	15	27.3
	0.0092 ± 0.0002	(5)	22.1	33
Stannern (Ap)	0.0142	(5)	49.0	46.3
Estherville (M) (silicates)	0.032	(1)	58	
Imilac (P) (olivine?)	0.0027	(1)	12	

(1) Smales et al. (1964)
(2) Webster, Morgan and Smales (1958)
(3) Gast (1960)
(4) Cabell and Smales (1957)
(5) Tera et al. (1970)
(6) Shima and Honda (1967)
(7) Kempe and Müller (1969)

way) with petrographic grades as determined by Van Schmus and Wood (1967). See especially the Rb/Cs ratios; all samples with ratios greater than 100 are from chondrites of petrographic grade 5 or 6, while all samples from chondrites of petrographic grade 3 or 4 have Rb/Cs ratios less than 40. The K/Cs ratios are low when Rb/Cs are low and *vice versa.* However, not all chondrites of grades 5 or 6 have high Rb/Cs ratios; note especially *Holbrook.* Consequently, although it is likely that *Atoka* is an L6, no prediction of petrographic grade can be made for *Limerick* or *Merua.*

A wide range of Cs contents may be observed in samples with essentially the same Rb/Cs ratio. Compare especially the L chondrites *Chandakapur* and *Marion.* Probably at least two different kinds of mechanisms have operated to change contents of heavy alkali metals in these chondrites, one perhaps associated with mineral differentiation and the other with volatilization of Cs. The first might be principally responsible for changes in contents but only to a much lesser degree for changes in proportions of these elements, while the second might be principally responsible for changes in Rb/Cs and K/Cs ratios. In any case, it is clear that there are significant differences in the histories of the various chondrites of high petrographic grade listed in Table 1.

Clasts rich in K and Rb are probably present, in varied amounts, in all of the LL chondrites (see chapters on K and Rb). The limited data on Cs in LL stones suggest that volatilization, if it is evidenced by high Rb/Cs and K/Cs ratios, has not been the sole mechanism for generating these extreme differences in contents of heavy alkali metals.

Data on achondrites demonstrate that magmatic differentiation may be effective in changing Cs contents and K/Cs ratios, but has much less influence on Rb/Cs ratios. The range in Rb/Cs, excluding the doubtful value of Cabell and Smales (1957) for *Johnstown,* is from 15 to 71.4, while Cs contents range from 4×10^{-10} g/g to 2×10^{-8} g/g and K/Cs ratios range from 15×10^{3} to 22×10^{4}. Even within the class of eucrites, a similar contrast may be seen (compare data for *Moore Co.* and *Nuevo Laredo*). These observations reinforce the impression that Rb/Cs ratios will be very useful as guideposts to high-temperature episodes related to accretion or metamorphism, as distinct from magmatic differentiation, in the histories of meteorites.

References

Cabell, M.J. and Smales, A.A. (1957) "The determination of rubidium and caesium in rocks, minerals and meteorites by neutron-activation analysis." *Analyst* **82**, 390–406.

Gast, P.W. (1960) "Alkali metals in stone meteorites." *Geochim. Cosmochim. Acta* **19**, 1–4.

Keil, K. (1969) "Meteorite composition", chapter 4 in *Handbook of Geochemistry*, edited by K.H.Wedepohl (Springer-Verlag, Berlin) 78–115.

Kempe, W. and Müller, O. (1969) "The stony meteorite Krähenberg." In *Meteorite Research* (editor P.M.Millman, pub. by D.Reidel, Dordrecht, Holland), 418–428.

Larimer, J.W. and Anders, E. (1967) "Chemical fractionations in meteorites—II. Abundance patterns and their interpretation." *Geochim. Cosmochim. Acta* **31**, 1239–1270.

Shima, M. and Honda, M. (1967) "Distributions of alkali, alkaline earth and rare earth elements in component minerals of chondrites". *Geochim. Cosmochim. Acta* **31**, 1995–2006.

Smales, A.A., Hughes, T.C., Mapper, D., McInnes, C.A.J. and Webster, R.K. (1964) "The determination of rubidium and caesium in stony meteorites by neutron activation analysis and by mass spectrometry." *Geochim. Cosmochim. Acta* **28**, 209–233.

Tera, F., Eugster, O., Burnett, D.S. and Wasserburg, G.J. (1970) "Comparative study of Li, Na, K, Rb, Cs, Ca, Sr and Ba abundances in achondrites and in Apollo 11 lunar samples." *Geochim. Cosmochim. Acta Suppl. I*, 1637–1657.

Van Schmus, W.R. and Wood, J.A. (1967) "A chemical-petrologic classification for the chondritic meteorites." *Geochim. Cosmochim. Acta* **31**, 747.

Webster, R.K., Morgan, J.W. and Smales, A.A. (1958) "Caesium in chondrites." *Geochim. Cosmochim. Acta* **15**, 150–152.

(Received 4 September 1970)

BARIUM (56)

C. C. Schnetzler

Planetology Branch
Goddard Space Flight Center
Greenbelt, Maryland

THE FIRST MEASUREMENTS of the abundance of barium in meteorites were by von Engelhardt (1936), who used an emission spectrographic technique. Ranges, rather than absolute values, were reported in most cases with all eight analyzed chondrites falling between <1 ppm and 10 ppm Ba. Four of the eight chondrites were reported as having between 1 and 3 ppm. Two achondrites, Juvinas and Stannern, were reported as having 10 to 30 ppm and 48 ppm respectively. These general abundance levels have been subsequently confirmed by later, more accurate, analyses.

Pinson et al. (1953) reported emission spectrographic analyses of 21 chondrites as averaging 8 ppm barium, with a relative deviation of only 23%. They could not explain the large discrepancy (factor of 2 to 3) between their values and those of von Engelhardt. It now appears that the value they used for the Ba concentration in their standard, W-1, was about a factor of two higher than the most recent isotope dilution values (290 versus 158); this would seem to account for the discrepancy.

Since the late 1950's inherently more accurate methods of analysis, such as neutron activation and mass spectrometric isotope dilution, have been used to determine barium abundances in meteorites. All of the meteorite data known to us have been compiled in the accompanying table, with the exceptions of the early works of von Engelhardt (1936) and Pinson et al. (1953), mentioned above, and the emission spectrographic analyses of Moore and Brown (1963) and Greenland and Lovering (1965). The analyses reported in these latter two papers seem to be of variable quality. The data of Moore and Brown agree, qualitatively, with the data shown in the table as they found the median value of 43 chondrites "falls" to be 4.5 ppm. The higher

TABLE 1 Ba in meteorites, in ppm by weight. N.A. is neutron activation, E.S. is emission spectrograph, I.D. is mass spectrometric isotope dilution

	Hamaguchi *et al.* (1957)	Reed *et al.* (1960)	Gast (1965)	Duke and Silver (1967)	Schnetzler and Philpotts (1969)	Eugster *et al.* (1969)	Tara *et al.* (1969)
	N.A.	N.A.	I.D.	E.S.	I.D.	I.D.	I.D.
'Ordinary" Chondrites							
Forest City	3.7	3.5				3.37	3.65
Modoc	3.6	3.3					
Richardton	3.2						
Holbrook	4.0	3.5					
Beardsley		3.0					
Bruderheim						3.37	
Leedey						3.85	3.76
Enstatite Chondrites							
Abee		1.8					
Indarch		1.9				2.41*	2.41*
Carbonaceous Chondrites							
Mighei		2.5					
Orgueil		2.4					
Achondrites							
Binda				2			
Moore County			22	17	22.5		
Sioux County			25	20			
Juvinas				26	33.1		
Pasamonte			38	26		28.2	28.6
Stannern				62	57.8		
Nuevo Laredo	46	43	44	40		39.3	39.4
Serra de Mage					7.59		
Shergotty					32.0		
Angra dos Reis					26.4		
Jonzac					26.4		
Bununu					18.5		
Zmenj					11.4		
Mesosiderite							
Estherville				5			

* Magnetic iron removed.

TABLE 1 (cont.)

	Hamaguchi *et. al.* (1957)	Reed *et. al.* (1960)	Gast (1965)	Duke and Silver (1967)	Schnetzler and Philpotts (1969)	Eugster *et. al.* (1969)	Tara *et. al.* (1969)
	N.A.	N.A.	I.D.	E.S.	I.D.	I.D.	I.D.
Irons							
Canyon Diablo (troilite)		0.4– <0.006					
Toluca (troilite)		<0.1					
Campo del Cielo (El Taco) (silicate inclusion)						2.37	
Weekeroo Station (silicate inclusion)						8.70	

average for ordinary chondrite "falls" of about 11 ppm found by Greenland and Lovering is undoubtedly due, in part, to the high Ba value of 250 ppm used for their standard, W-1 (similar to the error in the work of Pinson et al., 1953). Both Moore and Brown and Greenland and Lovering found that the concentration of barium in chondrite "finds" was much more erratic, and generally higher, than in chondrite "falls".

The limited data in the table indicates that the barium abundance in "ordinary" chondrites is rather uniform at between 3 and 4 ppm, while enstatite chondrites and carbonaceous chondrites have slightly lower abundances of 2 to 3 ppm. Urey's (1952) choice of 3 ppm for the cosmic abundance of barium seems to be reasonable in light of the published chondrite data.

The basaltic achondrites have higher barium contents than the chondrites (with the exception of Binda) and, in contrast with the chondrites, they exhibit a good deal of variation in barium content. This might be due, in part, to variation in the proportion of plagioclase to pyroxene in the various achondrites. Analyses of plagioclase and pyroxene fractions of Juvinas and Moore County by Schnetzler and Philpotts (1969) have shown that the plagioclases contain about 15 times more barium (~70 ppm) than the pyroxene fractions (~4.5 ppm). However, the concentrations of barium in the liquids from which these achondrites crystallized probably were different also; for example, Serra de Mage (>90% plagioclase in the sample analyzed) has three times lower Ba content than Angra dos Reis (>90% pyroxene).

Barium analyses of troilite and silicate inclusions in several iron meteorites reflect the lithophilic character of this element.

The first measurement of the isotopic composition of Ba in meteorites was made, indirectly, by Reed et al. (1960). They reported that from their neutron activation analyses of barium they could determine no difference between the isotopic composition of barium in nine stone meteorites and terrestrial barium, within an accurarcy of 5%. Using a more accurate mass spectrometric technique, Krummenacher et al. (1962) showed that no anomalies existed in the five most abundant isotopes between terrestrial barium and barium from the Richardton chondrite within 30 per mil. However Umemoto (1962), using a similar solid-source mass spectrometric technique, reported an approximately linear deviation in meteoritic barium isotopic composition from terrestrial barium isotopic composition. The three meteorites measured—Bruderheim, Pasamonte, and Nuevo Laredo—all showed the same deviation, amounting to about 2.5 per mil/mass unit, or approximately a 2% higher Ba^{130}/Ba^{138} ratio than found in terrestrial barium analyzed under the same conditions. Recently Eugster et al. (1969) have made barium isotopic analyses using a double spiking technique which is designed to correct for any isotopic fractionation that might occur during the mass spectrometric analysis. Included in their samples were the three meteorites previously analyzed by Umemoto. They found that any differences between meteoritic and terrestrial barium isotopic abundances were less than 0.1% for all isotopes. Thus Eugster et al. (1969) concluded that the results of Umemoto were in error due to instrumental fractionation, and that the fractionation of barium in meteorites versus terrestrial samples in <0.2 per mil/mass unit.

In conclusion, it should be noted that Ba has not been accurately determined in a number of types of meteorites, including the olivine-pigeonite chondrites, the various classes of the Ca-poor achondrites, the nakhlites, and the pallasites.

References

Duke, M.B. and Silver, L.T. (1967) *Geochim. Cosmochim. Acta* **31**, 1637.

Eugster, O., Tera, F. and Wasserburg, G.J. (1969) *J. Geophys. Res.* **74**, 3897.

Gast, P.W. (1965) *Science* **147**, 858.

Greenland, L. and Lovering, J.F. (1965) *Geochim. Cosmochim. Acta* **29**, 821.

Hamaguchi, H., Reed, G.W. and Turkevich, K. (1957) *Geochim. Cosmochim. Acta* **12**, 337.

Krummenacher, D., Merrihue, C.M., Pepin, R.O. and Reynolds, J.H. (1962) *Geochim. Cosmochim. Acta* **26**, 231.

Moore, C.B. and Brown, H. (1963) *J. Geophys. Res.* **68**, 4293.
Pinson, W.H., Ahrens, J.H. and Franck, M.L. (1953) *Geochim. Cosmochim. Acta* **4**, 251.
Reed, G.W., Kigoshi, K. and Turkevich, A. (1960) *Geochim. Cosmochim. Acta* **20**, 122.
Schnetzler, C.C. and Philpotts, J.A. (1969) "Genesis of the Ca-rich achondrites in light of rare-earth and barium concentrations", in *Meteorite Research* (P. Millman, ed.), pp. 206–216. Reidel.
Tera, F., Burnett, D.S. and Wasserburg, G.J. (1969) to be published.
Umemoto, S. (1962) *J. Geophys. Res.* **67**, 375.
Urey, H.C. (1952) *Phys. Res.* **88**, 248.
Von Englehardt, W. (1936) *Chem. Erde* **10**, 187.

(Received 10 February 1970)

RARE EARTHS (51–71)

John A. Philpotts and **C. C. Schnetzler**

Planetology Branch
Goddard Space Flight Center
Greenbelt, Maryland

THE RARE-EARTH ELEMENTS, lanthanum through lutetium, constitute one of the most important groups of trace elements available as indicators of geochemical processes. This is because a) they are lithophilic in character, b) they are non-volatile, c) they have quite similar chemical properties but display small regular differences in mass and ionic radii, as a result of the lanthanide contraction, and d) the oxidation state is +3 under normal conditions (Eu can be reduced to +2 and Ce oxidized to +4 under conditions which occur in nature). In addition, this long series of elements is potentially quite useful in evaluating models of nucleosynthesis because some of the isotopes are produced by different processes, and some elements consist of isotopes which have vastly different thermal-neutron-capture cross-sections.

Several extensive reviews of rare-earth distributions in natural materials, including meteorites, have recently been published (Haskin et al., 1966; Haskin and Schmitt, 1967; and Haskin et al., 1968). It is not the intention of this review to repeat in detail, nor to elaborate on these excellent summaries. The interested reader should refer to these articles, particularly Haskin et al. (1966), for a more comprehensive review.

The first analyses of rare earths in meteorites were by Noddack (1935), who used an emission spectrographic technique. However, it was not until the early 1960's that data of satisfactory quality appeared, due largely to the development of neutron activation techniques (Schmitt et al., 1960; Bate et al., 1960; and Haskin and Gehl, 1962). The majority of meteorite rare-earth analyses have been by this method and appear in two papers by Schmitt et al. (1963, 1964). More recently, mass spectrometric isotope dilution techniques have been applied to meteorite rare-earth analyses (Philpotts et al., 1967;

Masuda, 1967; Masuda, 1968a, 1968b; and Schnetzler and Philpotts, 1969). The neutron activation and isotope dilution results agree, in most cases, to within ±10%, when comparable samples were reported, indicating neither method is subject to large systematic errors.

Schmitt and co-workers (1963, 1964) reported on rare earths in 19 chondrites—8 carbonaceous chondrites (types I, II, and III), 6 ordinary chondrites, 2 amphoterites, and 3 enstatite chondrites. In addition Haskin and Gehl (1962) analyzed one ordinary chondrite, Haskin et al. (1968) analyzed a chondrite composite and two individual chondrites, Schnetzler and Philpotts (1968) analyzed a chondrite composite, and Masuda (1968a) reported the analysis of the ordinary chondrite Modoc previously analyzed by Schmitt et al. (1964). The average rare-earth abundances for twenty chondrites, and for a composite of nine chondrites, are given in Table 1. As can be seen from

TABLE 1 Rare earth contents of chondritic meteorites, in ppm.

	(1)	(2)
La	0.30 ± 0.06	0.330 ± 0.013
Ce	0.84 ± 0.18	0.88 ± 0.01
Pr	0.12 ± 0.02	0.112 ± 0.005
Nd	0.58 ± 0.13	0.60 ± 0.01
Sm	0.21 ± 0.04	0.181 ± 0.006
Eu	0.074 ± 0.015	0.069 ± 0.001
Gd	0.32 ± 0.07	0.249 ± 0.011
Tb	0.049 ± 0.010	0.047 ± 0.001
Dy	0.31 ± 0.07	—
Ho	0.073 ± 0.014	0.070 ± 0.001
Er	0.21 ± 0.04	0.200 ± 0.005
Tm	0.033 ± 0.007	0.030 ± 0.002
Yb	0.17 ± 0.03	0.200 ± 0.007
Lu	0.031 ± 0.005	0.034 ± 0.002

(1) Average for 20 chondrites (Haskin et al., 1966).
(2) Composite of 9 chondrites (Haskin et al., 1968).
Mean deviations indicated by (±) values.

these data, the rare-earth elements are very low in the chondrites – none reach 1 ppm and the total of all 14 is only slightly over 3 ppm. The most important observation made by Schmitt et al. (1963, 1964) is that although the rare-earth abundances in these twenty chondrites vary by over a factor of three, the relative abundances in all the types of chondrites are the same, that is, there is very little *relative* fractionation of the rare-earths. Recently Schmitt et al. (1968) reported analyses of chondrules from four meteorites. Chondrules from the type III carbonaceous chondrite Mokoia and the unequilibrated LL-group chondrite Chainpur have rare earth abundances which are fractionated $\lesssim 20\%$ relative to average chondrite abundances. Forest City and Richardton (both H-group chondrites) chondrules have very significant fractionation, the lighter rare earths being depleted up to 50%. The total rare-earth concentrations in Forest City and Richardton chondrules are less than those in the whole meteorite. These data seemed to the authors to be consistent with diffusion of rare earths out of the chondrules during metamorphism.

The lack of relative fractionation exhibited by chondrites is also observed in the Ca-rich (basaltic) achondrites. Schmitt et al. (1963, 1964) determined rare-earth abundances in four eucrites, Philpotts et al. (1967) analyzed the howardite Bununu, and Schnetzler and Philpotts (1969) reported on three normal brecciated eucrites (two of which were previously analyzed by Schmitt and co-workers) and another howardite. These seven achondrites constitute a series with absolute rare-earth abundances from three to seventeen times the chondritic abundance, but in which the rare-earth elements show very little fractionation relative to chondrites. Schnetzler and Philpotts (1969) pointed out that the rare-earth data on the basaltic achondrites are consistent with their being part of an igneous differentiation series. Data on plagioclase and pyroxene fractions separated from Juvinas and Moore County were taken to indicate closed system competition; hypabyssal or extrusive crystallization was suggested. Rare earths from four rather unique Ca-rich achondrites—Serra de Mage, Moore County, Shergotty, and Angra dos Reis—were also reported by Schnetzler and Philpotts (1969). These are all unbrecciated achondrites and, in contrast with the brecciated achondrites mentioned above, they all exhibit fractionated rare-earth patterns. The authors concluded that these four unbrecciated achondrites appear to be cumulates and that the unfractionated brecciated achondrites appear to represent liquids. Furthermore, there is rare-earth evidence to suggest that Serra de Mage and Moore County are cumulates from a liquid of normal brecciated achondrite composition. Two other Ca-rich achondrites, the nakhlites Nakhla and Lafayette, have fractionated rare-earth patterns that are reminiscent of terrestrial basalts (Schmitt and Smith, 1963). The authors suggested that these two meteorites were the product of "terrestrial-like" vulcanism on the parent meteorite body.

Few samples of the remaining types of meteorites—the Ca-poor achondrites, mesosiderites, pallasites, and irons—have been analyzed for rare-earth distributions. Three Ca-poor achondrites (the hypersthene achondrites Johnstown and Shalka and the enstatite achondrite Norton County) were analyzed by Schmitt et al. (1963); all show fractionation of rare earths, with the lighter rare earths depleted relative to the heavy rare earths. In Norton County, the light rare earths are depleted monotonically from Gd to La, with relative fractionation of the heavy rare earths, while Johnstown and Shalka have monotonic decrease from Lu to about Dy or Gd with little fractionation of the light rare earths. Masuda (1967) reported on a detailed study of the rare earths in a number of fractions from Norton County and found that enstatite single crystals and polycrystalline material had notably different patterns. He interpreted this in terms of different crystallization mechanisms. The single crystal patterns were very similar to those of Johnstown and Shalka, and he concluded that these two meteorites were the result of precipitation of single crystals only. Two mesosiderites have been analyzed for rare-earth elements (Schmitt et al., 1963; Schmitt et al., 1964). The silicate portion of Estherville has rare-earth abundances approximately three times the average chondritic level, while Veramin has abundances approximately equal to chondrites. These two meteorites exhibit a small amount of rare-earth fractionation ($\sim 20\%$), but in opposite directions – in Estherville the lighter rare earths are enhanced relative to the heavy rare earths while in Veramin they are depleted. The olivine phase of two pallasites, Brenham and Thiel Mountains (both finds), were analyzed by Schmitt et al. (1964). Both are very fractionated, but are considerably different. Masuda (1968) re-examined Brenham and obtained a pattern very similar, in a relative sense, to Thiel Mountains. He concluded that the data of Schmitt and co-workers for Brenham were in error. Two partial rare-earth analyses of iron meteorites were reported by Schmitt et al. (1963). The rare-earth abundances were on the order of 10^{-4} to 10^{-5} ppm.

Comparison of the isotopic composition of a rare-earth element from meteoritic material with terrestrial composition was first attempted by Umemoto (1962). He found no difference between terrestrial cerium and that from the chondrite Bruderheim. Schmitt et al. (1963) noted that in their neutron activation procedure elemental abundances for several elements were determined from two isotopes. From the agreement of values they concluded that the isotopic ratios of Ce, Gd, Er, Yb and Lu agree within 5% with the ratios in terrestrial materials. The agreement in Yb values determined from Yb^{168} and Yb^{174} was most significant, as the thermal-neutron-capture cross section of Yb^{168} and Yb^{174} are 11,000 and 60 barns respectively, and the abundances of these isotopes would be particularily sensitive to different degrees of thermal-neutron irradiation. The most precise analyses of rare-earth isotopic composition were on Sm, Eu, and Gd in two ordinary chondrites, an achondrite, and a carbonaceous chondrite by Murthy and Schmitt (1963). Each of these elements contains isotopes having thermal neutron-capture cross sections which differ, (by factors of from 10^3 to 10^5), but the authors found no differences between terrestrial and meteoritic isotopic compositions, to within an experimental error of $\pm 0.5\%$.

In summary, the most remarkable result of rare-earth analyses in meteorites is that the bulk of meteorites (apparently all types of chondrites and most brecciated Ca-rich achondrites) have the same *relative* rare-earth abundances. If the various types of meteorites are related, the evolutionary process must have been rather simple.

References

Bate, G.L., Potratz, H.A. and Huizenga, J.R. (1960) *Geochim. Cosmochim. Acta* **18,** 101.

Haskin, L. and Gehl, M.A. (1962) *J. Geophys. Res.* **67**, 2537.

Haskin, L.A., Frey, F.A., Schmitt, R.A., Smith, R.H. (1966) "Meteoritic, Solar, and Terrestrial Rare Earth Distribution." In *Physics and Chemistry of the Earth,* VII (L.H. Ahrens, F. Press, S.K. Runcorn and H.C. Urey, eds.). Pergamon.

Haskin, L.A. and Schmitt, R.A. (1967) "Rare Earth Distributions." In *Researches in Geochemistry 2* (P.H. Abelson, ed.). Wiley.

Haskin, L.A., Haskin, M.A., Frey, F.A. and Wildeman, T.R. (1968) "Relative and absolute terrestrial abundances of the rare earths." In *Origin and Distribution of the Elements* (L.H. Ahrens, ed.) pp. 889–912. Pergamon.

Masuda, A. (1967) *Geochim. J.* **2**, 111.

Masuda, A. (1968a) *Earth Planet. Sci. Letters* **4**, 284.

Masuda, A. (1968b) *Earth Planet. Sci. Letters* **5**, 59.

Murthy, V.R. and Schmitt, R.A. (1963) *J. Geophys. Res.* **68**, 911.

Noddack, I. (1935) *Z. Anorg. Allgem. Chem.* **225**, 337.

Philpotts, J.A., Schnetzler, C.C. and Thomas, H.H. (1967) *Earth Planet. Sci. Letters* **2**, 19.

Schmitt, R.A., Mosen, A.W., Suffredini, C.S., Lasch, J.E., Sharp, R.A. and Olehy, D.A. (1960) *Nature*, **186**, 863.

Schmitt, R.A., Smith, R.H., Lasch, J.E., Mosen, A.W., Olehy, D.A. and Vasilevskis, J. (1963) *Geochim. Cosmochim. Acta* **27**, 577.

Schmitt, R.A. and Smith, R.H. (1963) *Nature* **199**, 550.

Schmitt, R.A. and Smith, R.H. and Olehy, D.A. (1964) *Geochim. Cosmochim. Acta* **28**, 67.

Schmitt, R. A., Smith, R. H. and Olehy, D. A. (1968) "Rare Earth Abundances in Meteoritic Chondrules." In *Origin and Distribution of the Elements* (L. H. Ahrens, ed.), pp. 273–292. Pergamon.

Schnetzler, C. C. and Philpotts, J. A. (1969) "Genesis of the Calcium-rich Achondrites in Light of Rare-earth and Barium Concentrations." In *Meteorite Research* (P. Millman, ed.) pp. 206–216. Reidel.

Umemoto, S. (1962) *J. Geophys. Res.* **67**, 375.

(Received 10 February 1970)

TANTALUM (73)

William D. Ehmann

Department of Chemistry,
University of Kentucky, Lexington, Kentucky
and
the Department of Chemistry,
Arizona State University, Tempe, Arizona

VERY LITTLE DATA exist in the literature for the abundance of tantalum in meteorites. RANKAMA (1944, 1948) reported a maximum abundance of tantalum in meteorites of 0.38 ppm, which is equivalent to an atomic abundance of 0.32 (Si = 10^6). Suess and Urey (1956) adopted a smaller value of 0.065 for the atomic abundance and mentioned that Rankama at that time supported this lower interpolated value. More recently tantalum has been determined in meteorites by Atkins and Smales (1960) and Ehmann (1965). While both of these groupt used neutron activation analysis, the determinations of the former were based mainly on beta counting and the latter on gamma-ray spectrometry.

Two stable isotopes of tantalum, ^{180}Ta (0.0123%) und ^{181}Ta (99.988%) exist in nature. The activation analysis determination of tantalum is based on thermal neutron irradiations in a nuclear reactor for periods of from 36 hours to 2 weeks to produce radioactive ^{182}Ta. This radionuclide has a half-life of 115 days and is a beta and gamma-ray emitter. The intense gamma-rays at 1.12 and 1.22 MeV are attractive for analytical work and the composite NaI (Tl) photopeak in this region was used in the determinations by Ehmann (1965). Interferences from iron, cobalt and scandium activities are serious in this energy region when using a NaI(Tl) detector and a specific radiochemical separation was necessary in both of the aforementioned studies.

Gordon et al. (1968) suggest use of the ^{182}Ta gamma-ray lines at 68 and 110 keV in the non-destructive determination of tantalum in geological

TABLE 1 Tantalum abundances in meteorites

Specimen	Classification	Ta, ppm	Reference
Murray	C2	0.017	(1)
Ochansk	H4	0.018, 0.018	(2)
Forest City	H5	0.022, 0.023, 0.027	(2) (1)
Plainview	H5	0.022, 0.022	(1)
Penokee	H	0.029	(1)
Bjurböle	L4	0.020, 0.021	(2)
Cynthiana	L4	0.021	(1)
Chandakapur	L5	0.018, 0.019, 0.020, 0.022	(2)
Bruderheim	L6	0.017	(1)
Chateau Renard	L6	0.024, 0.025, 0.027, 0.030	(2)
Holbrook	L6	0.024, 0.026	(2)
Long Island	L6	0.027, 0.027, 0.031	(2), (1)
Ness County	L6	0.022, 0.022	(2)
Elenovka	L	0.012, 0.011	(1)
Cumberland Falls	Ach.	<0.01	(1)
Johnstown	Ach.	0.007, 0.009	(2)
Norton County	Ach.	~ 0.004	(1)
Pasamonte	Ach.	0.12	(1)
Canyon Diablo	Sid.	0.0009, 0.001, 0.003, 0.004	(2)
Henbury	Sid.	0.004, 0.005	(2)
Odessa	Sid.	~ 0.0005	(1)
San Martin	Sid.	0.0006, 0.0009	(2)
Sikhote-Alin	Sid.	~ 0.001	(1)
Williamstown	Sid.	~ 0.003	(1)
Troilite from Canyon Diablo		~ 0.004	(1)
Olivine from Springwater		~ 0.004	(1)

(1) Ehmann (1965).
(2) Atkins and Smales (1960).

materials by means of activation analysis and Ge(Li) counting. In view of the difficult separation chemistry for tantalum it would probably be worthwhile to again attempt the determination of tantalum in meteorites via a simple group separation and Ge(Li) counting.

Due to the long half-life of ^{182}Ta, neither Atkins and Smales (1960) nor Ehmann (1965) were able to confirm the radiochemical purity of their separated tantalum by means of routine half-life determinations. Ehmann reported obtaining identical results on recounting several samples after a decay of 2 to 3 months, which suggests the samples were pure. Radiochemical purity checks by means of gamma-ray energy calibration and sample-standard spectral comparisons were done in both investigations. Atkins and Smales also found that sample and standard beta absorption curves compared well, again indicating radiochemically pure samples had been obtained. Ehmann points out that because of possible interferences, errors in the NaI(Tl) detector determinations would most likely be in the direction of the reported analyses being too high. He suggests confidence intervals of ± 10 to $\pm 20\%$ for his data.

Since the amount of tantalum data in meteorites is not large, all the data of Atkins and Smales (1960) and Ehmann (1965) are presented in Table 1. The classification given is that of Van Schmus and Wood (1967). Atomic abundances (Si $= 10^6$) derived from these data for major chondrite groups are given in Table 2, together with some interpolated or calculated data for solar

TABLE 2 Tantalum abundances in major chondrite classes

Class	Number of meteorites	Number of analyses	Mean abd. Ta, ppm	Atomic abundance Ta (Si $= 10^6$)*
Carbonaceous	1	1	0.017	0.019
H-group	4	8	0.023	0.021
L-group	9	21	0.021	0.018
Suess and Urey (1956)—Interpolated value				0.065
Cameron (1959)—Adjusted value				0.015
Clayton and Fowler (1961)—Calculated from theories of nucleosynthesis				0.030

* Based on silicon data of Vogt and Ehmann (1965).

system atomic abundances. In general, the data of Atkins and Smales and Ehmann are in excellent agreement. The few data available suggest only slight siderophilic and chalcophilic tendencies for tantalum in meteorites. It is interesting to note that tantalum has a very low abundance in the one pallasitic olivine sample analyzed, in spite of its apparent lithophilic tendencies. The high abundance in the calcium-rich achondrite Pasamonte suggests tantalum may be associated with the minerals pyroxene or plagioclase. Although no Type I carbonaceous chondrites have been analyzed for tantalum, the rather uniform atomic abundances among the three major classes of chondrites analyzed (Table 2) suggest a solar system atomic abundance of 0.02 to 0.03. This is consistent with the calculated value of Clayton and Fowler (1961).

References

Atkins, D.H.F. and Smales, A.A. (1960) *Anal. Chim. Acta* **22**, 462–478.

Cameron, A.G.W. (1959) *Astrophys. J.* **129**, 676–699.

Clayton, D.D. and Fowler, W.A. (1961) *Ann. Phys. (N.Y.)* **16**, 51–68.

Ehmann, W.D. (1965) *Geochim. Cosmochim. Acta* **29**, 43–48.

Gordon, G.E., Randle, K., Goles, G.G., Corliss, J.B., Beeson, M.H. and Oxley, S.S. (1968) *Geochim. Cosmochim. Acta* **32**, 369–396.

Rankama, K. (1944) *Bull. comm. geol. Finlande No. 133.*

Rankama, K. (1948) *Ann. Acad. Sci. Fennicae A*, **III**, *13.*

Suess, H.E. and Urey, H.C. (1956) *Rev. Mod. Phys.* **28**, 53–74.

Van Schmus, W.R. and Wood, J.A. (1967) *Geochim. Cosmochim. Acta* **31**, 747–765.

Vogt, J.R. and Ehmann, W.D. (1965) *Geochim. Cosmochim. Acta* **29**, 373–383.

(Received 1 February 1969)

TUNGSTEN (74)

Wiliiam D. Ehmann

Department of Chemistry,
University of Kentucky, Lexington, Kentucky
and
the Department of Chemistry,
Arizona State University, Tempe, Arizona

THE EARLIEST DATA for tungsten abundances in meteorites are those of Noddack and Noddack (1934). Since this paper also reported tungsten abundances in rocks that were more than a factor of ten higher than obtained by use of more modern analytical techniques, these early meteoritic determinations are certainly much too high. Suess and Urey (1956) discarded the data of the Noddacks and used an interpolated value of 0.49 (Si $= 10^6$) for tungsten in their compilation of atomic abundances.

More recently tungsten has been determined in meteorites by neutron activation analysis. Since the total number of determinations is small, all the data of Atkins and Smales (1960), Amiruddin and Ehmann (1962), and Rieder and Wänke (1968) are presented in Tables 1 and 2.

Stable tungsten consists of five isotopes-^{180}W (0.14%), ^{182}W (26.41%), ^{183}W (14.40%), ^{184}W (30.64%), and ^{186}W (28.41%). The determination by activation analysis involves irradiation of the specimens for several days in the thermal neutron flux from a nuclear reactor to produce radioactive ^{185}W and ^{187}W by thermal capture reactions. The determinations of Atkins and Smales (1960) were based principally on beta counting of 24-hour half-life ^{187}W. Amiruddin and Ehmann (1962) used scintillation spectrometry to measure the 0.686 MeV gamma-ray of ^{187}W and also used beta counting of 74-day ^{185}W. While the results of both counting techniques were comparable, Amiruddin and Ehmann noted a larger scatter among the beta-counting data and preferred to use the data obtained by gamma-ray spectrometry. Rieder and Wänke (1968) presumably also used gamma-ray spectrometry for ^{187}W,

TABLE 1 Tungsten abundances in stony meteorites

Specimen	Classification*	W, ppm	Reference
Murray	C2	0.13, 0.15	(2)
		0.22	(3)
Lancé	C3	0.27	(3)
Vigarano	C3	0.36	(3)
Dimmitt	H (3, 4)	0.26	(3)
Beaver Creek	H4	0.24	(3)
Ochansk	H4	0.14, 0.19	(1)
		0.13, 0.10	(2)
		0.22	(3)
Beardsley	H5	0.13, 0.14	(2)
		0.27	(3)
Forest City	H5	0.15, 0.15, 0.13, 0.14	(2)
Leighton (dark)	H5	0.46	(3)
Pantar (dark)	H5	0.34	(3)
Plainview	H5	0.17, 0.16	(2)
Pultusk	H5	0.14	(2)
Richardton	H5	0.13	(2)
Fayetteville (dark)	H	0.22	(3)
Ioka	L3	0.08, 0.12, 0.11	(2)
Bjurbole	L4	0.07, 0.09	(1)
Cynthiana	L4	0.16	(2)
Chandakapur	L5	0.11, 0.12, 0.13	(1)
La Lande	L5	0.13, 0.14	(2)
Melrose	L5	0.08, 0.09	(2)
Bruderheim	L6	0.13	(2)
Château Renard	L6	0.15, 0.15, 0.17	(1)
Harrisonville	L6	0.08, 0.08	(2)
Holbrook	L6	0.15, 0.15	(1)
		0.19	(2)
Long Island	L6	0.13, 0.16	(2)
Shaw	L6	0.07, 0.07	(2)
Elenovka	L	0.11	(2)
Jelica	LL6	0.09, 0.07	(2)
Cumberland Falls	Ach.	0.05, 0.12	(2)
Johnstown	Ach.	0.005, 0.008	(2)
Norton County	Ach.	0.09, 0.02, 0.06	(2)
Shallowater	Ach.	0.11, 0.12	(2)

* Classification according to Van Schmus and Wood (1967).

(1) Atkins and Smales (1960)

(2) Amiruddin and Ehmann (1962)

(3) Rieder and Wänke (1968)

TABLE 2 Tungsten abundances in siderites and separated meteoritic phases

Specimen	W, ppm	Reference
Canyon Diablo siderite	1.4, 1.5, 1.9	(1)
	1.5	(2)
Henbury siderite	0.75, 0.77	(1)
Odessa siderite	0.96	(2)
Sandia Mountains siderite	1.3	(2)
San Martin siderite	2.7, 2.6	(1)
Sikhote-Alin siderite	0.78	(2)
Williamstown siderite	1.4	(2)
Canyon Diablo troilite	0.020, 0.013*	(2)
Springwater olivine	1.9	(2)

* Regarded as upper limits due to the possibility of inclusion of small amount of metal phase.
(1) Atkins and Smales (1960)
(2) Amiruddin and Ehmann (1962)

although details on the indicator radionuclide used are not given in their paper. Specific radiochemical separations were used by all three groups.

The data of Atkins and Smales (1960) and Amiruddin and Ehmann (1962) are in good agreement, both groups obtaining a chondritic tungsten abundance of approximately 0.11 (Si = 10^6) for all chondrites analyzed. Amiruddin and Ehmann estimate an accountable error of ± 8 to 10% for their data. The data of Rieder and Wänke (1968) on nine carbonaceous and H-group chondrites are generally approximately 50% higher than reported by the other two groups. One explanation for the discordant results may be that the former two groups prepared their flux monitor standards from standard solutions containing tungsten. Atkins and Smales irradiated their standards in solution and Amiruddin and Ehmann dispersed their standard solution by evaporation on the surface of approximately 3 cm^2 of high purity aluminum foil. Rieder and Wänke do not describe the form in which their tungsten standards were irradiated. Atkins and Smales have discussed the unsuitability of pure compounds or metals as flux monitors for tungsten and tantalum, due to severe self-shielding effects. Flux monitor self-shielding would tend to yield higher abundance values.

In the absence of details concerning the manner of standard preparation in the work of Rieder and Wänke (1968) and in view of the close agreement among the data of Atkins and Smales (1960) and Amiruddin and Ehmann (1962), only the data of the latter two groups were used to compute the group average for chondrites given in Table 3. The tungsten abundance of 0.16 (Si = 10^6) in the carbonaceous chondrite Murray is in fair agreement with the value of 0.20 (Si = 10^6) calculated from theories of nucleosynthesis by Clayton and Fowler (1961).

TABLE 3 Tungsten abundances in major chondrite classes*

Class	Number of meteorites analyzed	Mean abundance W, ppm	Atomic abundance W (Si = 10^6)†
Carbonaceous	1	0.14	0.16
H-group	6	0.14	0.13
L and LL-group	15	0.12	0.20
	Suess and Urey (1956)		0.49
	Cameron (1959)		0.105
	Clayton and Fowler (1961)		0.20

* Based only on the data of Atkins and Smales (1960) and Amiruddin and Ehmann (1962). The data of Rieder and Wänke (1968) are approximately 50% higher (see Table 1).
† Based on silicon abundances of Vogt and Ehmann (1965).

Amiruddin and Ehmann (1962) noted a correlation of gross tungsten abundances to metal phase content in the stony meteorites. This fact, coupled to the observed high tungsten abundances in the siderites (0.8 to 2.7 ppm), suggests predominantly siderophilic tendencies for tungsten in meteoritic materials. However, a high value of 1.9 ppm was obtained in one sample of pallasitic olivine analyzed. Jeffery (1959) has noted that due to the difference in ionic radii of Si^{4+} in SiO_4^{4-} (r = 0.42 Å) and W^{6+} in WO_4^{2-} (r = 0.62 Å), tungsten is not readily accepted into a silicate structure, but rather would appear mainly as an accessory mineral in granitic rocks. Additional data would be required to determine the extent of lithophilic tendencies for tungsten in meteoritic material.

The low abundance of tungsten in Canyon Diablo troilite suggests that chalcophilic tendencies are small. In fact, Amiruddin and Ehmann (1962) note that the troilite abundance should be regarded only as an upper limit, due to the possibility of accidental inclusion of small flakes of metal phase. It would be desirable to have additional data for tungsten in separated meteoritic phases and also in the Type I carbonaceous chondrites to better establish its geochemical properties in meteoritic matter. It is possible that tungsten could be determined by activation analysis employing a simple group separation and high resolution Ge(Li) gamma-ray spectrometry, hence avoiding the somewhat tedious specific radiochemical separation of tungsten.

Addendum

Additional W abundances in three recent falls have recently been reported by our group. Averages of replicate analyses are: Allende – 0.15 ppm, Murchison – 0.14 ppm, and Lost City – 0.16 ppm.

Morgan, J.W., Rebagay, T.V., Showalter, D.L., Nadkarni, R.A., Gillum, D.E., McKown, D.M., and Ehmann, W.D. (1969) *Nature* **224**, 789–791.
Ehmann, W.D., Gillum, D.E., Morgan, J.W., Nadkarni, R.A., Rebagay, T.V., Santoliquido, P.M., and Showalter, D.L. (1970) *Meteoritics* **5**, 131–136.

References

Amiruddin, A. and Ehmann, W.D. (1962) *Geochim. Cosmochim. Acta* **26**, 1011–1022.
Atkins, D.H.F. and Smales, A.A. (1960) *Anal. Chim. Acta* **22**, 462–478.
Cameron, A.G.W. (1959) *Astrophys. J.* **129**, 676–699.
Clayton, D.D. and Fowler, W.A. (1961) *Ann. Phys. (N.Y.)* **16**, 51–68.
Jeffery, P.G. (1959) *Geochim. Cosmochim. Acta* **16**, 278–295.
Noddack, I. and Noddack, W. (1934) *Svensk Kem. Tidskrift* **XLVI,** 173–201.
Rieder, R. and Wänke, H. (1968) *Paper presented at the International Symposium on Meteorite Research IAEA*, Vienna, Austria, August 7–13, 1968. Published in *Meteorite Research*, D. Reidel Publ. Co., Dordrecht, Holland, 1969, pp. 75–86.
Suess, H.E. and Urey, H.C. (1956) *Rev. Mod. Phys.* **28**, 53–74.
Van Schmus, W.R. and Wood, J.A. (1967) *Geochim. Cosmochim. Acta* **31**, 747–765.
Vogt, J.R. and Ehmann, W.D. (1965) *Geochim. Cosmochim. Acta* **29**, 373–383.

(Received 1 February 1969; revised 22 August 1969)

RHENIUM (75)

John W. Morgan

Department of Chemistry,
University of Kentucky, Lexington, Kentucky

RHENIUM WAS DISCOVERED in 1925 by Noddack, Tacke and Berg. This element has two naturally occurring isotopes of mass number 185 and 187, whose atomic abundances are generally given as 37.07 and 62.93 per cent respectively (White and Cameron, 1948). More recent measurements of the 185/187 ratio have been reviewed by Riley (1967), and indicate a slightly higher abundance of the 185 isotope. Rhenium-187 is unstable to beta decay and has a very low transition energy. Measured values of the maximum beta energy vary considerably, the actual value probably lying between 1 and 3 keV (Watt and Glover 1962, Wolf and Johnston 1962). The low energy of the beta and the possibility of bound beta decay (Gilbert 1958) has made the direct determination of the half-life very difficult. The most reliable measurement has been made by geochemical methods by Hirt et al. (1963) who found a value of 4.3×10^{10} years. The decay of rhenium-187 to osmium-187 has been applied to the dating of iron meteorites (Herr et al. 1961) and to the examination of the chronology of nucleosynthesis (Clayton 1964).

The earliest study of the geochemistry of rhenium was made by Noddack and Noddack (1931), using an X-ray spectrographic method following a preliminary chemical concentration. As part of their survey they analysed the metal, troilite and silicate phases of meteorites and demonstrated the marked siderophile character of rhenium. The same workers reported an average value of 0.0023×10^{-6} g Re/g in chondritic meteorites (Noddack and Noddack 1934). For many years these were the only analyses available, until the early neutron activation study of iron meteorites by Brown and Goldberg (1949). These results indicated that the Noddack values were low by one or two orders of magnitude, probably due to loss of rhenium during the preliminary chemical concentration procedure. All subsequent studies

TABLE 1 Abundances of rhenium in chondrites

Group	Chondrite	Number of analyses	Rhenium, 10^{-9} g/g	References
C1	Ivuna	1	31	(3)
	Orgueil	2	37	(3), (4)
C2	Cold Bokkeveld	3	46	(3), (4)
	Mighei	4	56	(2), (3), (4)
	Murray	2	50	(4)
	Staroe Boriskino	2	50	(4)
C3	*Bencubbin* (inclusion)*	2	(63)†	(4)
	Lancé	3	60	(3), (4)
	Mokoia	2	58	(4)
	Ornans	1	58	(3)
	Warrenton	1	57	(4)
C4	Karoonda	2	59	(4)
E4	Abée	3	55	(3), (4)
	Indarch	3	51	(3), (4)
E5	Saint Marks	2	68	(4)
E6	Daniel's Kuil	1	53	(3)
	Hvittis	2	68	(4)
	Khairpur	3	61	(3), (4)
	Pillistfer	2	61	(4)
H3	Bremervörde	1	68	(3)
H4	Beaver Creek	1	86	(3)
	Ochansk	6	76	(2), (3), (4)
H5	Allegan	2	77	(4)
	Beardsley	2	66	(4)
	Forest City	1	99	(3)
	Pultusk	2	80	(4)
	Richardton	2	81	(4)
H6	Futtehpur	1	100	(3)
	Mount Browne	3	65	(3), (4)
	Zhovtnevyi	2	70	(4)
H	Limerick	1	82	(3)
L3	Khohar	2	43	(4)
L4	Saratov	2	41	(2)
L5	Barwell	1	63	(3)
	Bluff	1	(33)	(3)
	Chandakapur	1	44	(3)
	Crumlin	1	38	(3)
	Homestead	2	52	(4)
	Knyahinya	3	39	(3), (4)

TABLE 1 (cont.)

Group	Chondrite	Number of analyses	Rhenium 10^{-9} g/g	References
L6	Bruderheim	1	43	(3)
	Château Renard	1	67	(3)
	Grosslienthal	2	42	(2)
	Holbrook	1	50	(4)
	Long Island	1	(14)	(3)
	Modoc	1	23	(3)
	Mocs	3	54	(1), (4)
	Saint Michel	2	43	(4)
	Stavropol	1	47	(2)
L	Nikolskoe	2	41	(2)
LL3	Chainpur	2	31	(4)
	Ngawi	2	35	(4)
LL6	Benares	2	39	(4)

* Chondrite name *in italics* indicates a find.
† Value in parentheses not used in the calculation of averages in Table 2.
(1) Herr et al. (1961), whole meteorite abundance recalculated from metal and silicate abundances
(2) Perezhogin (1965)
(3) Fouché and Smales (1967)
(4) Morgan and Lovering (1964, 1967a)

of the rhenium abundance in meteorites have used the neutron activation method.

The abundance of rhenium in all the major chondrite groups has been studied by Morgan and Lovering (1964, 1967a). Perezhogin (1965) has communicated results for rhenium in one carbonaceous and five ordinary chondrites. The distribution of rhenium between the magnetic and non-magnetic fractions of a number of enstatite and ordinary chondrites was examined by Fouché and Smales (1967), who also reported whole rock analyses of six carbonaceous chondrites. Herr et al. (1961) analysed the metal phase of two chondrites and the silicate phase of one of them. An investigation of a large number of elements in chondrites was made by Kiesl et al. (1967) and Kiesl (1969), and analyses for rhenium were included. These results for rhenium tend to be low, especially in the carbonaceous chondrites, and have not been included in the present compilation of chondritic rhenium values. All the other values currently available are listed in Table 1, where the meteorites are grouped according to the classification of

Van Schmus and Wood (1967). An exception was made in the case of Mokoia, which was put in the C3 group because of the high silicon abundance.

Atomic abundances have been calculated relative to 10^6 atoms of silicon. Standard silicon values for each group were used, as recommended by

TABLE 2 Atomic abundances of rhenium in chondritic classes

Group	Number of chondrites analysed	Total number of analyses	Mean Re/10^6 Si
C1	2	4	0.052
C2	4	11	0.059
C3	4	7	0.057
C4	1	2	0.057
All C	11	24	0.057
E4	2	6	0.047
E5	1	2	0.057
E6	4	8	0.048
All E	7	16	0.049
H3	1	1	0.060
H4.	2	7	0.068
H5	5	9	0.069
H6	3	6	0.065
All H	12	24	0.067
L3	1	2	0.035
L4	1	2	0.033
L5	5	8	0.037
L6	8	12	0.038
All L	16	26	0.037
LL3	2	4	0.027
LL6	1	2	0.032
All LL	3	6	0.028
All H, L and LL	31	56	0.049
All chondrites	49	96	0.051

Mason (personal communication). The only exception was in the case of the E5 group enstatite chondrite St Marks, for which a silicon abundance of 18.0 per cent was employed. Mean atomic abundances for each chondrite group are shown in Table 2.

As might be expected for a siderophile element, the abundances in the ordinary chondrites decrease in the order H > L > LL. Despite the high iron content, the enstatite chondrites fall between the L and the H groups in rhenium content. The mean of all the ordinary chondrites analysed is identical to the mean enstatite chondrite abundance. The atomic abundance of the C1 chondrites is also very similar to the enstatite and ordinary chondrites, though the mean of all carbonaceous chondrites is some 10 per cent higher.

There is very little variation between the petrological classes within chemical groups, especially in the ordinary chondrites. The E5 class appears to be higher than the E4 and E6 classes by some 20 per cent; however, this may be a sampling problem. The C1 class seems lower than the other carbonaceous chondrites. This was observed by Morgan and Lovering (1967a), but was considered to be another sampling problem. However, the analyses of Fouché and Smales (1967) show the same tendency, so that the difference may well be real.

Observations of the rhenium abundance in the sun have not yet been made, so that it is not possible to compare solar and chondritic abundances directly (Aller, 1965). Calculations of solar abundances have been made by several groups, based on nucleosynthesis considerations. As some important parameters are not known, the calculations cannot be made *a priori*, but must be fitted to known solar abundances. Several studies have used chondritic values for normalization, or, for siderophilic elements, abundances in iron meteorites adjusted to chondritic nickel-iron contents (e.g. Cameron, 1959; Seeger et al. 1965). To use these for comparison with the chondrite values would be a rather circular procedure. Clayton and Fowler (1961) used elemental abundances derived from solar absorption spectra, and combined these with relative isotopic abundances from terrestrial and meteoritic material for normalization. These calculations provide a more independent comparison than those based upon meteoritic results. In Table 3, the abundances calculated by Cameron (1959), Clayton and Fowler (1961) and Seeger et al. (1965) are compared with the average chondrite values.

In comparisons of this type it is often a problem to decide which chondritic value is most representative of the abundance of the element under consideration in primitive solar dust. The C1 chondrites are often thought to be a reasonable approximation to the chemical composition of this material. Unfortunately this class of chondrites is rare and shows remarkable non-uniformity in the distribution of some elements.

For rhenium, it makes little difference whether one takes the C1 chondrites, the mean of all the ordinary chondrites, or the mean of all the chondrites

TABLE 3 Comparison of rhenium abundances in chondrites with calculated solar system abundances

Reference	Re atoms/ 10^6 Si atoms
Cameron (1959)	0.054
Clayton and Fowler (1961)	0.0p52
Seeger *et al.* (1965)	0.060
Corrected for decay of ^{187}Re	0.054
Chondritic abundances	
C1 chondrites	0.052
All C group	0.057
All E group	0.049
All ordinary chondrites (H, L and LL groups)	0.049
All chondrites	0.051

analysed, since about the same rhenium abundance is obtained. In this case, therefore, it is probably justifiable to take the mean of all the chondrites as a "recommended" value.

The good agreement between the chondritic value and the calculated values of Cameron (1959) and Seeger et al. (1965) might be expected since their estimates were based, at least in part, on meteoritic abundances. The abundance for rhenium given by Seeger et al. must be corrected for the decay of rhenium-187 since the isolation of solar material from the nucleosynthetic sources. The comparison of the average chondrite abundance with the estimate by Clayton and Fowler (1961) gives excellent agreement.

Fouché and Smales (1967) analysed the rhenium abundances in the separated magnetic and non-magnetic phases of the H, L and E group chondrites. These are summarized in Table 4. As might be expected from Prior's rule, the abundance of rhenium in the nickel-iron phase is approximately inversely proportional to the amount of metal present. The mean abundances of rhenium in the metal phase decrease in the order L > H > E.

Very few analyses have been published for rhenium in the achondritic meteorites. Morgan and Lovering (1964) reported duplicate analyses on the Bishopville aubrite, and Perzhogin (1965) analysed two samples of another enstatite achondrite, Norton County. Six achondrites were included in a survey of the distribution of rhenium by Morgan (1965). These results are summarized in Table 5. A single analysis of Stannern by Kiesl et al. (1967) has not been included, because of the analytical uncertainty of their method.

TABLE 4 Rhenium abundances in separated magnetic and non-magnetic phases of chondrites

Group	Chondrite	Per cent metal	Rhenium, 10^{-6} g/g	
			Metal	Silicate
E4	Abeé	29	0.21	0.003
	Indarch	25.5	0.20	0.007
E6	Daniel's Kuil	21	0.25	0.001
	Khairpur	19	0.23	0.002
Mean E group			0.22	0.003
H3	Bremervörde	21	0.35	0.010
H4	Beaver Creek	20.4	0.42	0.005
	Ochansk	20.2	0.32	0.021
H5	Forest City	20.1	0.47	0.010
	* *Gilgoin Station*	–	(0.18)†	–
H6	Futtehpur	20	0.51	0.004
	Mount Browne	21.1	0.30	0.006
H	Limerick	18.1	0.43	0.010
Mean H group			0.40	0.009
L5	Barwell	10.6	0.58	0.003
	Bluff	(5.8)	(0.50)	(0.011)
	Chandakapur	9.6	0.41	0.012
	Crumlin	5	0.74	0.003
	Knyahinya	5.0	0.55	0.005
L6	Bruderheim	8.4	0.48	0.004
	Château Renard	11.0	0.59	0.005
	Long Island	4.2	(0.73)	(0.026)
	Modoc	4.3	0.52	0.002
Mean L group			0.55	0.005

* Meteorite names *in italics* represent finds.
† Values in brackets have been omitted in the calculation of means.

As might be expected from the essentially siderophilic nature of rhenium, the abundances in the achondrites are generally considerably lower than those of the chondrites. There is one exception, Norton County, in which the rhenium abundance is as high as in an H group chondrite and higher than most enstatite *chondrites!* This result is very difficult to explain, and should

TABLE 5 Abundances of rhenium in achondrites

Group	Achondrite	Rhenium, 10^{-9} g/g	Reference
Angrite	Angra dos Reis	0.06; 0.08	(3)
Aubrites	Bishopville	0.22; 0.28	(1)
	Norton County	69; 72	(2)
Diogenites	Ellemeet	0.06; 0.13	(3)
	Johnstown	1.3; 0.3	(3)
Eucrite	Moore County	0.03; 0.09	(3)
Howardite	Binda	0.08; $\leq$0.08	(3)
Nakhlite	Nakhla	$\leq$0.06; 0.09	(3)

(1) Morgan and Lovering (1964)
(2) Perezhogin (1965)
(3) Morgan (1965)

be regarded with a certain reserve until it can be confirmed. Norton County does contain nickel-iron inclusions and it may be possible that this particular sample was rich in them. Other possible explanations are sample mislabelling, or perhaps contamination from the molybdenite samples analysed in the same work. Analytical error seems unlikely in view of the good agreement of the chondrite results with those of other laboratories.

It is interesting to compare the other achondrite results with those of separated non-magnetic fractions reported by Fouché and Smales (1967) and listed in Table 4. In all cases the achondrites are lower in abundance, indicating that there has been very efficient removal of the metal phase, and with it the rhenium, during the fractionation of the achondrites. Some of the achondrites still contain a small amount of metallic phase, and this may explain the wide variations in duplicate analyses observed for example in Johnstown. Binda contains some metal, and the very low abundances of rhenium found appears rather surprising at first sight. Lovering (1964) has shown, by electron microprobe analysis, that the metal phase is very poor in nickel, and that the metallic phase now present is due to a second stage reduction *after* the segregation of the original nickel-iron. As the results of Fouché and Smales (1967) show, rhenium, together with nickel and palladium, is extracted efficiently into the first metallic phase formed. Because of uncertainties introduced by the presence of a metallic phase of uncertain composition it is not at present possible to draw any rigorous parallels between the distribution of rhenium in the achondrites and in terrestrial rocks (Morris and Fifield 1961; Morgan and Lovering 1967b).

Herr et al. (1961) measured the age of the iron meteorites by the rhenium-osmium method. In order to construct their isochron they made a large number of rhenium analyses, generally taking between three and six measurements on each meteorite. More recently another extensive survey of the distribution of rhenium in iron meteorites has been made by Fouché and Smales (1966). These results are collected in Table 6 where meteorites are grouped by structural type, generally following the classification given by Hey (1966).

Several chemical classifications of the iron meteorites have been proposed. Goldberg et al. (1951) based their grouping on the quantization of gallium abundances; this was later extended by Lovering et al. (1957) who proposed the four well-known germanium-gallium groups. More recently the chemical classification has been intensively studied by Wasson (1967) and Wasson and Kimberlin (1967) and by Smales et al. (1967). In the present review the groupings proposed by Smales et al. will be followed because many of the rhenium analyses by Fouché and Smales (1966) were carried out on meteorites which were later classified.

There is no apparent grouping of the nickel-poor ataxites. Smales et al. (1967) grouped together seven hexahedrites, all of which come in the Lovering germanium-gallium class II. Negrillos also comes within this classification. Five of these meteorites, for which seven rhenium analyses are available, are closely grouped in rhenium abundance. Lombard has a rhenium abundance that lies within the range of this grouping and has been included. The mean of this group is $0.18 \pm 0.03 \times 10^{-6}$ g/g rhenium. Two of the remaining hexahedrites have abundances which are very similar to each other and which are the highest rhenium abundances so far found in meteorites. Possibly these represent a very fractionated subgrouping. Three coarsest octahedrites were found to be very similar chemically by Smales et al. (1967), and have very similar rhenium abundances. Of the other coarsest octahedrites listed in Table 6 only Linwood has a concordant rhenium abundance. The mean content of this group (5 analyses) is $0.26 \pm 0.09 \times 10^{-6}$ g/g. Two of the remaining coarsest octahedrites, Central Missouri and Sao Juliao de Moreira, have very low abundances and may be members of a rather rare fractionated group. The seven coarse octahedrites found to have chemical affinities previously have very similar rhenium concentrations. Odessa, though not classified by Smales et al., has a concordant rhenium abundance and has been added to this number. The mean rhenium abundance of this group is $0.19 \pm 0.04 < 10^{-6}$ g/g. The medium octahedrites do not exhibit any clusters of rhenium abundances. From this structural class Smales et al.

TABLE 6 Rhenium abundances in iron meteorites

Group	Meteorite	Classification Smales et al. (1967)	Number of analyses	Rhenium 10^{-9} g/g	References
Nickel-poor ataxites	Aswan		1	0.016	(2)
	La Primitiva		1	0.004	(3)
	Nedagolla		1	0.45	(3)
	Tombigbee River		1	0.0023	(3)
Hexahedrites	Bennett County	(a)	2	4.4	(3)
	Kopjes Vlei*	(a)	1	0.16	(3)
	Coahuila	(a)	2	1.3	(3)
	Lombard		3–6	0.16	(2)
	Mount Joy		3–6	0.026	(2)
	Negrillos		3–6	4.8	(2)
	San Martin	(a)	1	0.20	(2)
	Sikhote-Alin		1	$\ll$0.01	(2)
	Tocopilla	(a)	5	0.22	(2), (3)
	Uwet	(a)	1	0.15	(3)
	Walker County	(a)	1	0.15	(3)
Coarsest octahedrites	Arispe		1	0.85	(3)
	Central Missouri		1	0.0024	(2)
	Gladstone	(b)	1	0.27	(3)
	Linwood		3–6	0.27	(2)
	Otumpa*	(b)	2	0.34	(2), (3)
	Sao Juliao de Moreira		1	0.0023	(3)
	Seelasgen	(b)	2	0.11	(3)
Coarse octahedrites	Bendego		1	0.007	(3)
	Billings		1	0.43	(3)
	Bischtube	(c)	1	0.18	(3)
	Bohumilitz	(c)	1	0.17	(3)
	Canyon Diablo	(c)	6	0.24	(1), (2), (3)
	Cranbourne	(c)	6	0.14	(3)
	Odessa		5	0.25	(2)
	Pan de Azucar	(c)	1	0.25	(3)
	Wichita County	(c)	1	0.16	(3)
	Youndegin	(c)	1	0.17	(3)
Medium octahedrites	Breece	(d)	1	0.0041	(3)
	Campbellsville	(d)	1	0.0051	(3)
	Canton	(d)	1	1.1	(3)
	Caperr	(d)	2	0.020	(3)
	Cape York	(d)	1	0.43	(3)
	Carbo		3–6	1.3	(2)
	Casas Grandes		5	0.45	(2)

TABLE 6 (cont.)

Group	Meteorite	Classification Smales et al. (1967)	Number of analyses	Rhenium 10^{-9} g/g	References
	Chinautla	(e)	2	0.070	(3)
	Clark County	(e)	1	0.64	(3)
	Colfax	(e)	1	0.13	(3)
Medium octahedrites (continued)	Colomera		1	0.78	(2)
	Cumpas	(d)	1	0.21	(3)
	Descubridora	(d)	1	0.13	(3)
	Glorieta Mountain	(e)	1	0.0051	(3)
	Goose Lake		1	0.28	(1)
	Gun Creek	(e)	1	0.0023	(3)
	Henbury	(d)	6	1.3	(1), (2), (3)
	"Henbury"†		4	0.22	(2)
	Kingston	(e)	1	0.47	(3)
	La Caille	(e)	1	1.0	(3)
	Loreto		1	0.37	(2)
	Mount Edith	(d)	1	0.0050	(3)
	Puquios	(e)	1	1.6	(3)
	Rhine Villa	(d)	1	0.010	(3)
	Sam's Valley	(d)	1	0.0086	(3)
	Toluca	(d)	9	0.24	(1), (2)
			2	0.039	(3)
	Trenton		1	0.20	(2)
	Treysa‡		4	0.093	(2)
Fine octahedrites	Altonah		1	0.87	(1)
			2	0.14	(3)
	Boogaldi	(f)	1	0.062	(3)
	Bristol		1	0.18	(3)
	Cambria (Lockport)	(f)	1	0.085	(3)
	Carlton		1	0.016	(3)
	Gibeon	(f)	7	0.25	(2), (3)
	Grand Rapids	(f)	1	1.7	(3)
	Grant		1	0.0048	(2)
	Huizopa	(f)	1	0.26	(3)
	Moonbi	(f)	1	0.11	(3)
	Obernkirchen	(f)	1	0.31	(3)
	Otchinjau	(f)	1	0.23	(3)
Finest octahedrites	Bacubirito		1	0.46	(3)
	Ballinoo		1	0.71	(3)
	Mount Magnet		1	0.0054	(3)
	Salt River		1	0.72	(3)

TABLE 6 (cont.)

Group	Meteorite	Classification Smales et al. (1967)	Number of analyses	Rhenium 10^{-6} g/g	References
Brecciated octahedrites	Barranca Blanca		1	0.46	(3)
	Santa Rosa		1	0.0047	(3)
	Weekeroo Station		1	0.28	(3)
Granular Metabolite	Murnpeowie		1	0.16	(3)
Nickel rich ataxites	Babb's Mill (Troost)		1	2.8	(3)
	Cape of Good Hope		1	3.4	(3)
	Dayton		3–6	0.005	(2)
	Hoba		1	2.6	(3)
	Illinois Gulch		1	0.65	(3)
	Monahans		3–6	1.1	(2)
	Pinon		3–6	1.8	(2)
	San Cristobal		2	0.025	(3)
	Santa Catharina		1	0.03	(2)
			1	0.004	(3)
	Smithland		1	0.08	(3)
	Tlacotepec		3–6	2.9	(2)
Pallasite					
(iron)	Brenham		1	0.004	(2)
(troilite)	Odessa		1	0.031	(2)

* See reference (3) for a discussion of classification.
† Unidentified meteorite mis labelled "Henbury".
‡ There is some doubt about the authenticity of the various samples of this meteorite (Smales et al. (1967).

(a) Seven hexahedrites grouped by Smales et al. (1967).
(b) Three coarsest octahedrites grouped by Smales et al. (1967).
(c) Seven coarse octahedrites grouped by Smales et al. (1967).
(d) A group of twelve medium octahedrites selected by Smales et al. (1967).
(e) A group of eight medium octahedrites selected by Smales et al. (1967).
(f) Nine fine octahedrites grouped by Smales et al. (1967).

(1) Brown and Goldberg (1949); Goldberg and Brown (1950)
(2) Herr et al. (1961)
(3) Fouché and Smales (1966)

(1967) selected two groups of chemically similar meteorites. The first group, marked (d) in Table 6, consists of 12 specimens. These show no marked rhenium grouping, though five lie in the range 0.004 to 0.010 $\times 10^{-6}$ g/g with a mean of 0.007 $\pm$ 0.003 $\times 10^{-6}$ g/g. The second group selected consists of eight meteorites which show no similarity in rhenium content. Nine fine octahedrites were selected by Smales et al. (1967) as showing some chemical affinities, from which six were chosen as forming the Otchinjau subgroup. It was thought that Altonah and Bristol might also belong to this subgroup. All these meteorites (excluding the Herr et al. (1961) sample of Altonah) form a reasonably coherent cluster of rhenium abundances, the mean of which is $0.19 \pm 0.08 \times 10^{-6}$ g/g. Smales et al. (1967) found no apparent trace element grouping of four finest octahedrites analysed for eleven elements. Three of the four have rhenium contents which are reasonably close. The other specimen has a low abundance very close to the low abundance grouping of the medium octahedrites. The nickel-rich ataxites exhibit a very wide spread in rhenium abundance, with no really close groupings. The most noticeable feature of the distribution is the absence of samples with rhenium abundances between 0.1 and 0.6×10^{-6} g/g. This is just the region where most of the other iron meteorite groups show maxima in their frequency distributions. It is clear that the nickel-rich ataxites represent a very fractionated group of irons. The broad group of seven ataxites having high rhenium abundances between 0.6 and 4×10^{-6} g/g has a mean of $2.2 \pm 1.0 \times 10^{-6}$ g/g.

The remaining four specimens are brecciated octahedrites and a granular metabolite, which have no clearly defined structure on which to base a morphological classification. Murnpeowie is listed by Hey (1966) as showing signs of coarse octahedrite structure. The rhenium abundance fits in well with this group, although Smales et al. (1967) find little trace element similarity between Murnpeowie and Weekeroo Station on the one hand, and the coarsest and coarse octahedrites on the other. They do find some similarities between Barranca Blanca and the medium octahedrite Gun Creek, but there is no similarity in rhenium abundances. Chemical similarities have been noted between Santa Rosa and the Bendego medium octahedrite (Smales et al. 1967; Buchwald and Wasson, 1968); the rhenium abundances also correspond rather closely.

From the preceding discussion it can be seen that the rhenium abundances in chondrites and most groups of iron meteorites are now rather well known. Rhenium analyses are only lacking for the LL 4 and LL 5 groups. Further work could profitably be spent on the analysis of separated chondrite phases. It would be interesting to know the rhenium content of the magnetic phases of the LL and C groups, and particularly whether these groups follow the Prior's Law type of trend observed in the E, H and L groups. A detailed comparison of abundances in the iron meteorites and the chondrite metallic phases might be rewarding. In this context the similarity between many of the iron meteorite groupings and the enstatite chondrite metallic phase in their rhenium abundance may be significant.

There is still a great need for more analyses for rhenium on the achondrites, which have only been very sketchily examined. Particularly it is important to either confirm or disprove the very high rhenium abundance reported for the Norton County aubrite.

References

Aller, L.H. (1965) "The abundance of elements in the solar atmosphere." *Advances in Astronomy and Astrophysics* Vol. 3, pp. 1–25. Academic Press.

Buchwald, V.F. and Wasson, J.T. (1968) *Analecta Geologica* No. 3, 7.

Brown, H. and Goldberg, E. (1949) *Phys. Rev.* **76**, 1260.

Cameron, A.G.W. (1959) *Astrophys. J.* **129**, 676.

Clayton, D.D. (1964) *Astrophys. J.* **139**, 637.

Clayton, D.D. and Fowler, W.A. (1961) *Ann. Phys.* **16**, 51.

Fouché, K.F. and Smales, A.A. (1966) *Chem. Geol.* **1**, 329.

Fouché, K.F. and Smales, A.A. (1967) *Chem. Geol.* **2**, 105.

Gilbert, N. (1958) *Compt. Rend.* **247**, 868.

Goldberg, E. and Brown, H. (1950) *Anal. Chem.* **22**, 308.

Goldberg, E., Uchiyama, A. and Brown, H. (1951) *Geochim. Cosmochim. Acta* **2**, 1.

Herr, W., Hoffmeister, W., Hirt, B., Geiss, J. and Houtermans, F.G. (1961) *Z. Naturforsch.* **16a**, 1053.

Hey, M.H. (1966) *Catalogue of Meteorites* (3rd edition) British Museum.

Hirt, B., Herr, W. and Hoffmeister, W. (1963) "Age determinations by the rhenium-osmium method". *Radioactive Dating.* International Atomic Energy Agency pp. 35–43.

Kiesl, W., Sietner, H., Kluger, F. and Hecht, F. (1967) *Monatsh. Chem.* **98**, 972.

Kiesl, W. (1969) The 1968 International Conference; *Modern Trends in Activation Analysis.* Nat. Bur. Stand. (U.S.) Spec. Publ. 312, vol. 1, 302.

Lovering, J.F. (1964) *Nature* **203**, 70.

Morgan, J.W. (1965) *The application of activation analysis to some geochemical problems.* Unpublished thesis, Australian National University.

Morgan, J.W. and Lovering, J.F. (1967a) *Geochim. Cosmochim. Acta* **31**, 1893.

Morgan, J.W. and Lovering, J.F. (1967b) *Earth Planet. Sci. Letters* **3**, 219.

Morris, D.F.C. and Fifield, F.W. (1961) *Geochim. Cosmochim. Acta* **25**, 232.

Noddack, W., Tacke, I., Berg, O. (1925) *Naturwiss.* **13**, 567.

Noddack, I. and Noddack, W. (1931) *Z. Phys. Chem.* **154**, 207.

Noddack, I. and Noddack, W. (1934) *Svensk. Kem. Tidskr.* **46**, 173.

Perezhogin, G.A. (1965) *Zavodsk. Lab.* **31**, 402.

Riley, G.H. (1967) *J. Sci. Instrum.* **44**, 769.

Seeger, P.A., Fowler, W.A. and Clayton, D.D. (1965) *Astrophys. J. Supplement No. 97*, **XX**, 121.

Smales, A.A., Mapper, D. and Fouché, K.F. (1967) *Geochim. Cosmochim. Acta* **31**, 673.

Van Schmus, W.R. and Wood, J.A. (1967) *Geochim. Cosmochim. Acta* **31,** 747.
Wasson, J.T. (1967) *Geochim. Cosmochim. Acta* **31**, 161.
Wasson, J.T. and Kimberlin, J. (1967) *Geochim. Cosmochim. Acta* **31**, 2065.
Watt, D.E. and Glover, R.N. (1962) *Phil. Mag.* **7**, 105.
White, J.R. and Cameron, A.E. (1948) *Phys. Rev.* **74,** 991.
Wolf, G.J. and Johnston, W.H. (1962) *Phys. Rev.* **125**, 307.

(Received 10 July 1969)

OSMIUM (76)

John W. Morgan

Department of Chemistry,
University of Kentucky, Lexington, Kentucky

OSMIUM HAS seven naturally occurring isotopes, in the mass range between 184 and 192. These are listed in Table 1, with their percentage abundances (Nier, 1937). The isotopic abundance of osmium-187 can vary with origin from the decay of rhenium-187. Merz and Herr (1959) found that molybdenites rich in rhenium contained predominantly osmium-187, in one case with an isotopic abundance of 99.9 per cent. Herr et al. (1961) measured the 187/186 ratio in osmium extracted from osmiridium, and in a sample from the Bushveld found a ratio of 0.882 compared with the Nier value of 1.036. The decay of rhenium-187 to osmium-187 has important applications in the dating of iron meteorites (Herr et al. 1961) and to the chronology of nucleosynthesis (Clayton 1964).

The earliest determinations of osmium in meteorites were made by the Noddacks and Goldschmidt and Peters. These have been summarized by Goldschmidt (1958), who used the results to calculate a solar abundance of 1.7 atoms Os/10^6 atoms Si.

The first activation analyses for osmium in chondritic meteorites were made by Herr et al. (1961), who analysed the iron phase of the Mocs and Ramsdorf chondrites, and also the silicate phase of Mocs. Bate and Huizenga (1963) reported analyses for ten chondrites and separated chondritic magnetic and non-magnetic phases. Further neutron activation results in several important chondrites were reported by Morgan and Lovering (1964) and were followed by a more detailed study of 32 chondrites of all types (Morgan and Lovering 1967). Other neutron activation results on 9 carbonaceous and enstatite chondrites were given by Crocket et al. (1967). Sen Gupta (1968a) used a spectrophotometric technique to analyse several ordinary chondrites and the Abee enstatite chondrite.

TABLE 1 Abundance of isotopes in naturally occurring osmium

Mass Number	Per cent abundance (Nier 1937)
184	0.018
186	1.59
187	1.64*
188	13.3
189	16.1
190	26.4
192	41.0

* The abundance of osmium 187 is variable because of radiogenic enrichment by rhenium-187. See text for discussion.

All the results presently available for osmium in chondrites are collected in Table 2. The meteorites are arranged according to the classification of Van Schmus and Wood (1967), except Mokoia which has been put into the C3 group. Several of the ordinary chondrite results reported by Bate and Huizenga (1963) appear anomalously high, whereas their other results agree well with independent determinations. The reason lies in the sampling. All the discordant results are associated with samples having unusually high metal phase content. To take an extreme example, Holbrook as analysed by Bate and Huizenga has 36 per cent metal phase, but Keil and Fredriksson (1964) report only 8.3 per cent nickel-iron in this fall. The sample analysed by Bate and Huizenga contains more than twice as much osmium as most of the other L-group chondrites. On the other hand, specimens with a normal metal content show good agreement with the results of other groups. A sample of Allegan with 22 per cent metal (23.0 per cent given by Keil and Fredriksson 1964) agrees well with other osmium analyses for this chondrite. For these reasons, the results for samples with very high metal abundances have not been included in the calculation of mean abundances, and are shown in brackets in Table 2.

With the exception of the samples discussed above the agreement between laboratories is generally satisfactory. Particularly interesting is the concordance between the activation results and the spectrophotometric analyses

TABLE 2 Abundances of osmium in chondrites

Group	Chondrite	Number of analyses	Osmium, 10^{-6} g/g	Reference
C1	Alais	1	0.34	(3)
	Ivuna	2	0.61	(3)
	Orgueil	3	0.49	(3), (4)
C2	Cold Bokkeveld	4	0.63	(2), (3), (4)
	Mighei	3	0.72	(3), (4)
	Murray	2	0.69	(4)
	Nogoya	1	0.75	(3)
	Staroe Boriskino	1	0.54	(4)
C3	*Bencubbin (inclusion)**	2	(0.83)†	(4)
	Lancé	3	0.85	(3), (4)
	Mokoia	2	0.69	(4)
	Warrenton	1	0.80	(4)
C4	Karoonda	2	0.85	(4)
E4	Abée	3	0.64	(3), (4), (5)
	Indarch	3	0.63	(3), (4)
E5	Saint Marks	2	0.67	(4)
E6	Hvittis	2	0.73	(4)
	Khairpur	2	0.85	(4)
	Pillistfer	2	0.69	(4)
H4	Ochansk	2	0.88	(4)
		2	(1.05)	(2)
H5	Allegan	3	0.86	(2), (4)
		1	(1.21)	(2)
	Beardsley	2	0.79	(4)
		2	(1.02)	(2)
	Forest City	2	0.84	(2)
	Pultusk	2	0.92	(4)
	Richardton	2	0.81	(4)
			(1.05)	(2)
H6	Mount Browne	2	0.78	(4)
	Zhovtnevyi	1	0.66	(4)
H	*Belly River*	2	(0.72)	(5)
L3	Khohar	2	0.48	(4)
L4	Saratov	2	(0.68)	(2)
L5	Homestead	2	0.63	(4)
	Knyahinya	2	0.44	(4)

TABLE 2 (cont.)

Group	Chondrite	Number of analyses	Osmium, 10^{-6} g/g	Reference
L6	Bruderheim	1	0.53	(5)
		1	(0.66)	(2)
	Holbrook	2	0.50	(4)
		1	(1.24)	(2)
	Mocs	3	0.56	(1), (4)
		1	(0.89)	(2)
	Peace River	2	0.54	(5)
	Saint Michel	1	0.43	(4)
LL3	Chainpur	2	0.34	(4)
	Ngawi	2	0.32	(4)
LL6	Benares	2	0.50	(4)
	Benton	1	0.34	(5)

* Chondrite name *in italics* indicates a find.
† Value in parentheses not used in calculation of averages in Table 3.
(1) Herr et al. (1961), whole meteorite abundance recalculated from metal and silicate abundances.
(2) Bate and Huizenga (1963)
(3) Crocket et al. (1967)
(4) Morgan and Lovering (1967)
(5) Sen Gupta (1968)

This should go a long way to allay the fears expressed by Beamish et al. (1967) concerning the accuracy of the neutron activation method. The spectrophotometric method uses 20 g of sample which clearly limits its use; however, the agreement with activation methods indicates that the small samples used for the latter are generally reasonably representative.

In Table 3 atomic abundances related to 10^6 atoms of silicon are listed. Standard silicon values for each group were used (Mason, personal communication), except for Saint Marks, where a silicon abundance of 18.0 per cent was adopted.

In common with many strongly siderophilic elements, osmium in the ordinary chondrites decreases in the order H > L > LL. This trend cannot be extended to the enstatite chondrites. Many of these have a metal content as high as, or even higher than, the H group chondrites, and yet the osmium abundance in the E group lies between those of the L- and H-groups. The average abundance for the ordinary chondrites is very similar to that for the enstatite chondrites, however, the C group is considerably higher.

TABLE 3 Atomic abundances of osmium in chondritic classes

Group	Number of chondrites analysed	Total number of analyses	Mean Os/10^6 Si
C1	3	6	0.72
C2	5	11	0.75
C3	3	6	0.75
C4	1	2	0.80
All C	12	25	0.75
E4	2	7	0.55
E5	1	2	0.55
E6	3	6	0.58
All E	6	15	0.56
H4	1	2	0.77
H5	4	8	0.74
H6	2	3	0.64
All H	7	13	0.72
L3	1	2	0.38
L5	2	4	0.42
L6	5	9	0.42
All L	8	15	0.41
LL3	2	4	0.26
LL6	2	3	0.35
All LL	4	7	0.30
All ordinary chondrites	19	35	0.52
All chondrites	37	75	0.60

In most chemical groups there is little variation in abundance between the petrological classes. In the C group, however, there is a definite trend in osmium abundances such that C1 $<$ C2 $\approx$ C3 $<$ C4. Similar trends have been observed with other siderophilic elements, for example, rhenium, palladium, and nickel (Morgan and Lovering, 1967a; Fouché and Smales 1967; Wiik 1956).

Recent observations of solar spectra have provided an estimate of the solar abundance of osmium (Grevesse et al. 1968). Expressed to the base $\log H = 12.00$, $\log Os = 0.75$ and $\log Si = 7.74$, corresponding to 0.1 atoms osmium/10^6 atoms silicon. This is clearly much lower than the chondritic values. The errors of the spectral measurements are very large, and a rigorous comparison of meteorite and direct solar abundances must await an improvement in the precision of direct solar observation. Solar abundances have been calculated based on nuclear considerations, fitting the abundance curve to known abundances, either to well established spectral determinations (Clayton and Fowler, 1961) or to meteorite abundances (Cameron, 1959; Seeger et al. 1965). The chondritic values are compared to other estimates of the solar abundance of osmium in Table 4.

TABLE 4 Comparison of calculated, spectral and chondritic estimates for the solar abundance of osmium

Estimate	Os atoms/ 10^6 Si atoms
Nuclear calculation	
Cameron (1959)	0.64
Clayton and Fowler (1961)	0.51
Seeger et al. (1965)	0.61
Solar spectroscopy	
Grevesse et al. (1968)	$\sim$0.1
Chondritic abundances	
Cl chondrites	0.72
All C group	0.75
All E group	0.56
All ordinary chondrites (H, L and LL groups)	0.52
All chondrites	0.60

It is generally thought that the C1 chondrites are the best approximation available to the composition of primitive dust. Unfortunately, the empirical osmium values reported to date show a large scatter. Because of the sampling problem and the lack of very serious fractionation between the chondrite groups, the mean of all the chondrite values has been selected as a "recommended" chondritic abundance. This value agrees with the calculated values of Cameron (1959) and Seeger et al. (1965), although these do not really provide an independent test. The average chondrite value is about 15 per cent higher than the calculated solar abundance given by Clayton and Fowler

(1961). It seems possible that the Os-Ir-Pt peak may have been slightly under-estimated in the nucleosynthesis calculations. Sen Gupta (1968b) has discussed the abundance of platinum metals in chondrites, and considers that the chondrite osmium abundance is in good agreement with the abundance curves of Clayton and Fowler (1961). This misconception is probably because the chondritic abundance selected by Sen Gupta is not representative.

TABLE 5 Osmium abundances in separated magnetic and non-magnetic phases in chondrites

Group	Chondrite	Osmium, 10^{-6} g/g		References
		metal	silicate	
H4	Ochansk	1.3	0.12	(2)
H5	Forest City	1.7	0.30	(2)
L6	Mocs	3.7	0.35	(1)
L	Ramsdorf	2.6	–	(1)

(1) Herr et al. (1961)
(2) Bate and Huizenga (1963)

The distribution of osmium between the metal and silicate of the chondritic meteorites has not been studied in any detail. The few analyses that are available (Herr et al., 1961; Bate and Huizenga, 1963) are collected in Table 5. As might be expected osmium is concentrated in the metallic phase; however, there appears to be a surprisingly large amount of osmium in the non-magnetic fraction. Sen Gupta (1968b) considers that, because of the "considerable oxyphile character" of osmium, one might have expected an even *higher* content in the silicate phase. This seems a strange conclusion, in view of the very low abundances of osmium observed in most rocks (Morgan and Lovering 1967b) and in achondrites (Bate and Huizenga 1963; Morgan and Lovering 1964; Morgan 1965).

The osmium content of achondrites has not been studied in any great detail; Bate and Huizenga (1963) reported osmium analyses for a diogenite and a eucrite, Morgan and Lovering (1964) analysed the Bishopville aubrite and several other achondrite abundances were measured by Morgan (1965). A listing of these achondritic osmium abundances has been made in Table 6. The abundances are very much lower than in the chondrites. The levels in the achondrites are also considerably less than those found in the silicate phase of the chondrites (Herr et al. 1961; Bate and Huizenga, 1963), indicating that osmium is extracted with considerable efficiency into the metallic phase,

TABLE 6 Abundances of osmium in achondrites

Group	Achondrite	Number of analyses	Osmium, 10^{-9}g/g	References
Angrite	Angra dos Reis	2	0.78	(3)
Aubrite	Bishopville	2	5.0	(2)
Diogenites	Ellemeet	2	0.61	(3)
	Johnstown	4	7.9	(1), (3)
Eucrites	Moore County	2	0.44	(3)
	Nuevo Laredo	1	7.4	(1)
Howardite	Binda	2	$\leq$0.17	(3)
Nakhlite	Nakhla	2	0.70	(3)

(1) Bate and Huizenga (1963)
(2) Morgan and Lovering (1964)
(3) Morgan (1965)

and that the oxyphilic character inferred by Sen Gupta (1968b) may be greatly over-estimated.

The achondritic osmium abundances were very variable, not only within a single group but even between replicate analyses of the same meteorite. An extreme example is the case of Johnstown, where the abundances vary by a factor of almost 5. This wide variation may be due to the scattered distribution of rare metallic particles. The very low abundance of osmium in Binda may at first appear surprising, as this achondrite does contain a small but significant metal phase. It has been shown by microprobe analysis (Lovering 1964) that this metal is very low in nickel. This iron probably represents a second-stage reduction, after removal of the original nickel-iron and, with it, most of the osmium. The presence of a metal phase in many of the achondrites makes it difficult to discuss the fractionation of this element in terms of the abundances observed in terrestrial rocks (Bate and Huizenga 1963; Morgan and Lovering 1967b).

Most osmium analyses of iron meteorites which are available are the work of Herr et al. (1961); several other analyses have been reported by Sen Gupta (1968a). Earlier spectrophotometric determinations (Sandell, 1944; Sen Gupta and Beamish 1963) do not appear to be as accurate as the later work of Sen Gupta and have not been included in this compilation. The mean of two analyses of Henbury by Sen Gupta and Beamish is in good agreement with later results, even though the individual values differ by almost a factor of two. In Table 7 osmium abundances are arranged according to structural type, generally following the classification of Hey (1966).

TABLE 7 Osmium abundances in iron meteorites

Group	Meteorite	Number of analyses	Osmium, 10^{-6} g/g	References
Nickel-poor ataxite	Aswan	1	0.23	(1)
Hexahedrites	Lombard	3–6	1.1	(1)
	Mount Joy	3–6	0.51	(1)
	Negrillos	3–6	50	(1)
	San Martin	1	1.2	(1)
	Sikhote-Alin	1	≤0.025	(1)
	Tocopilla	4	1.3	(1)
Coarsest octahedrites	Central Missouri	1	0.25	(1)
	Linwood	3–6	4.5	(1)
	Otumpa*	1	3.7	(1)
Coarse octahedrites	Canyon Diablo	4	2.1	(1)
	Odessa	5	2.6	(1)
Medium octahedrites	Annaheim	2	3.7	(3)
	Carbo	3–6	16	
	Casas Grandes	5	3.2	(1)
	Colomera	1	12	(1)
	Henbury	1	13	(1)
	"Henbury"†	4	2.3	(1)
	Loreto	1	3.9	(1)
	Toluca	5	2.6	(1)
	Trenton	1	1.6	(1)
	Treysa	5	0.58	(1)
Fine octahedrites	Gibeon	6	2.9	(1)
	Grant	1	0.1	(1)
	Madoc	2	5.1	(3)
Nickel-rich ataxites	Dayton	3–6	0.47	(1)
	Klondike	2	17	(3)
	Monahans	3–6	13	(1)
	Pinon	3–6	28	(1)
	Santa Catharina	1	0.16	(1)
	Tlacotepec	3–6	40	(1)
Pallasite				
(iron phase)	Brenham	1	0.07	(1)
(troilite phase)	Odessa	2	0.005	(2)

* See Smales et al. (1967) for discuission of classification.

† Unidentified meteorite mislabelled "Henbury".

(1) Herr et al. (1961)

(2) Bate and Huizenga (1963)

(3) Sen Gupta (1968a)

Only one nickel-poor ataxite has been analysed for osmium, so that it is not possible to discern any chemical groupings for this type. Rather more results are available for the hexahedrites. Two hexahedrites, Tocopilla and San Martin, belong to the chemical grouping of seven members of this structural type (Smales et al. 1967). As with rhenium, Lombard has an abundance similar to members of this grouping. None of the other hexahedrites osmium analyses fit into this group. The mean of the three grouped abundances is $1.2 \pm 0.1 \times 10^{-6}$ g/g. The osmium content of the hexahedrite Negrillos is the highest yet found in a meteorite.

Three osmium abundances are listed for the coarsest octahedrites. Otumpa is listed under Campo del Cielo as a nickel-poor ataxite by Hey (1966). The classification of this iron has been discussed at length by Smales et al. (1967), who found it to be a coarsest octahedrite. This meteorite belongs to a chemical grouping proposed by these workers, and on the basis of rhenium abundances it appears that Linwood is another member. The osmium abundances of these two meteorites is very similar.

TABLE 8 Variation of the osmium 187/osmium 186 ratio in iron meteorites (Herr et al. (1961)

Group	Meteorite	Number of analyses	Osmium-187/ osmium-186 ratio
Hexahedrites	Negrillos	1	1.02
	Tocopilla	2	1.42
Coarsest octahedrite	Linwood	1	1.02
Coarse octahedrites	Canyon Diablo	2	1.12
	Odessa	2	1.09
Medium octahedrites	Carbo	1	1.06
	Casas Grandes	2	1.13
	Henbury	1	1.13
	"Henbury"*	1	1.07
	Toluca	2	1.06
	Treysa	1	1.20
Fine octahedrites	Gibeon	2	1.10
Nickel rich ataxite	Pinon	1	1.02
	Tlacotepec	1	1.00

* Unidentified meteorite mislabelled "Henbury".

Osmium abundances for only two coarse octahedrites have been listed, and both samples have similar abundances. Canyon Diablo belongs to the group of seven chemically similar coarse octahedrites selected by Smales et al. From rhenium and osmium abundances it seems that Odessa may also belong to this group.

Nine of the octahedrites in Table 7 form two rather well resolved clusters. The high osmium group consists of three meteorites with a mean of $14 \pm 2 \times 10^{-6}$ g Os/g. Six other medium octahedrites make up the low osmium group with a mean of $2.9 \pm 0.9 \times 10^{-6}$ g Os/g. Smales et al. (1967) also divided the medium octahedrites into two groups. Osmium analyses are available for only two of their samples, so comparisons between their classification and the osmium analyses are not possible. The work of Herr et al. (1961) shows that abundances of rhenium and osmium are very strongly correlated. It can be predicted with some confidence that the distribution of osmium will closely follow that of rhenium, which does not fall into the medium octahedrite groupings defined by the other elements studied by Smales et al. (1967).

Three of the four fine octahedrites have similar osmium abundances, with a mean of $3.6 \pm 1.3 \times 10^{-6}$ g Os/g. No finest octahedrites have been analysed for osmium.

Although there are six osmium analyses known for the nickel-rich ataxites there is no apparent grouping of the abundances. This is in general agreement with the discussion by Wasson (1967) and Smales et al. (1967). The only pallasite iron analysed fits none of the classifications mentioned above.

Measurements of the isotopic ratio of osmium 187/186 in a number of iron meteorites have been reported by Herr et al. (1961) and are listed in Table 8.

The variation in the osmium 187/186 ratio is correlated to the rhenium to osmium ratio. Herr et al. (1961) constructed an isochron of osmium 187/osmium 186 against rhenium 187/osmium 186. The slope of the line indicated an age of $4.0 \pm 0.8 \times 10^9$ years, and an initial osmium 187/osmium 186 ratio of 0.83. The physical significance of this apparent age has been discussed by Anders (1963).

In summary, it can be seen that the osmium abundances of the chondritic meteorites are quite well known, analyses being lacking only for the H3, L4, LL4 and LL5 groups. A careful study of the osmium distribution between the metal and silicate phases of the chondrites is needed, to investigate apparently high abundance of osmium in the silicate indicated by the few available analyses. The distribution of osmium in achondrites is poorly understood, many more analyses being required. The position is much better in the case of the iron meteorites, though certain groups such as the finest octahedrites, nickel-poor ataxites and pallasites need to be investigated in greater detail.

References

Anders, E. (1963) "Meteorite Ages." *The Moon, Meteorites and Comets* (The Solar System, Vol. IV) Chapter, 13, pp. 402–405. University of Chicago Press.

Bate, G.L. and Huizenga, J.R. (1963) *Geochim. Cosmochim. Acta* **27**, 345.

Beamish, F.E., Chung, K.S. and Chow, A. (1967) *Talanta* **14**, 1.

Cameron, A.G.W. (1959) *Astrophys. J.* **129**, 676.

Clayton, D.D. (1964) *Astrophys. J.* **139**, 637.

Clayton, D.D. and Fowler, W.A. (1961) *Ann. Phys.* **16**, 51.

Crocket, J.H., Keays, R.R. and Hsieh, S. (1967) *Geochim. Cosmochim. Acta* **31**, 1615.

Fouché, K.F. and Smales, A.A. (1967) *Chem. Geol.* **2**, 105.

Grevesse, N., Blanquet, G. and Boury, A. (1968) "Abondances solaires de quelques éléments représentatifs au point de vue de la nucléosynthèse". *Origin and Distribution of the Elements.* (L.H. Ahrens, ed.) pp. 117–182.

Goldschmidt, V.M. (1958) *Geochemistry* (edited by A. Muir). Clarendon Press.

Hey, M.H. (1966) *Catalogue of Meteorites.* British Museum (Natural History).

Herr, W., Hoffmeister, W., Hirt, B., Geiss, J. and Houtermans, F.G. (1961) *Z. Naturforschg.* **16a**, 1053.

Keil, K. (1962) *J. Geophys. Res.* **67**, 4055.

Keil, K. and Frederiksson, K. (1964) *J. Geophys. Res.* **69**, 3487.

Lovering, J.F. (1964) *Nature* **203**, 70.

Merz, E. and Herr, W. (1959) "Microdetermination of isotopic abundance by neutron activation." *Progress in Nuclear Energy Series IX.* Analytical Chemistry Vol. 1. pp. 137–144.

Morgan, J.W. (1965) *The application of activation analysis to some geochemical problems.* Unpublished thesis, Australian National University, Canberra.

Morgan, J.W. and Lovering, J.F. (1964) *Science* **144**, 835.

Morgan, J.W. and Lovering, J.F. (1967a) *Geochim Cosmochim. Acta* **31**, 1893.

Morgan, J.W. and Lovering, J.F. (1967b) *Earth Planet. Sci. Letters* **3**, 219.

Nier, A.O. (1937) *Phys. Rev.* **52**, 885.

Sandell, E.B. (1944) *Ind. Eng. Chem. Anal. Ed.*, **16**, 342.

Seeger, P.A., Fowler, W.A. and Clayton, D.D. (1965) *Astrophys. J. Supplement No. 97*, **XI**, 121.

Sen Gupta, J.G. (1968a) *Anal. Chim. Acta* **42**, 481.

Sen Gupta, H.G. (1968b) *Chem. Geol.* **3**, 293.

Sen Gupta, J.F. and Beamish, F.E. (1963) *Am. Mineralogist*, **48**, 379.

Smales, A.A., Mapper, D. and Fouché, K.F. (1957) *Geochim. Cosmochim. Acta* **31**, 673.

Van Schmus, W.R. and Wood, J.A. (1967) *Geochim. Cosmochim. Acta* **31**, 747.

Wasson, J.R. (1967) *Geochim. Cosmochim. Acta* **31**, 161.

Wiik, H.B. (1956) *Geochim. Cosmochim. Acta* **9**, 279.

(Received 12 July 1969)

IRIDIUM (77)

P. A. Baedecker

Institute of Geophysics and Planetary Physics
University of California, Los Angeles

In recent years a number of workers have carried out investigations concerning the distribution of iridium in meteorites. The iridium concentration in iron meteorites varies by approximately a factor of 6000, and Wasson (1968, 1969a) has found it valuable in his attempts to elucidate different groups of possibly genetically related iron meteorites. Studies concerning the distribution of iridium within the chondrite groups have provided information which is relevant to discussions concerning metal-silicate fractionation and oxidation-reduction processes which have been operative during the formation of the chondrites (Baedecker, 1967; Tandon and Wasson, 1968; Müller et al., 1971).

Goldschmidt (1954) has reviewed the early spectrographic work on meteorites, carried out by the Noddacks (1930, 1931ab, 1934) and Goldschmidt and Peters (1932). These authors determined iridium in the iron and troilite phases of a number of iron and chondritic meteorites. Goldschmidt assigned average values of 4 ppm and 0.4 ppm for the iridium content of meteoritic nickel-iron and troilite respectively.

Since the advent of activation analysis as a highly sensitive technique for the determination of trace elements, a great deal of data on the Ir content of meteorites has appeared in the literature. Hamaguchi et al. (1961) determined Ir in two bronzite chondrites, three irons, a pallasite and an achondrite by neutron-activation. A large number of analyses of different types of meteorites were published by Ehmann and coworkers (Rushbrook and Ehmann, 1962; Baedecker and Ehmann, 1965) who also employed neutron-activation techniques. Baedecker (1967) determined Ir in a large variety of meteoritic materials, but in the course of his investigations observed that chemical problems associated with the post-irradiation processing of the samples could lead to low results. He therefore suggested that his own results, and those of previous workers, represented a lower limit for the specimens analyzed. In the last three years, several workers have published activation analysis data for Ir, employing procedures which should overcome the difficulties encountered by earlier workers. In all cases the newer results are significantly higher than those of Hamaguchi et al., Baedecker and Ehmann, and Baedecker. The most recent analytical data for the chondrites are summarized in Table 1 and Figure 1.

TABLE 1 Ir abundances in chondrites

Meteorite type	No. of determinations	No. of meteorites analyzed	Concentration (10^{-6} g/g) Range	Mean	Abundance (Si = 10^6)	Reference
C1	6	3	0.30–0.48	0.40	0.57	Crocket
	3	2	0.45–0.63	0.49	0.70	Ehmann
C2	5	3	0.61–1.25	0.84	0.94	Crocket
	6	4	0.55–0.76	0.61	0.68	Ehmann
	1	1		0.79	0.88	Rieder
C3	1	1		0.73	0.69	Crocket
	4	3	0.56–0.78	0.65	0.61	Ehmann
	2	2	0.75–0.86	0.81	0.76	Rieder
H-group	15	9	0.64–0.84	0.77	0.66	Ehmann
	39	18	0.51–0.85	0.73	0.62	Müller
	4	4	0.86–1.00	0.92	0.79	Rieder
L-group	27	12	0.32–0.70	0.44	0.34	Ehmann
	44	17	0.31–0.59	0.46	0.36	Müller
	52	26	0.36–0.81	0.52	0.41	Tandon
LL-group	6	4	0.25–0.63	0.38	0.30	Ehmann
	23	10	0.18–0.44	0.33	0.26	Müller
E4	2	2	0.23–0.32	0.29	0.25	Crocket
	3	2	0.53–0.58	0.55	0.47	Ehmann
E5 and E6	8	5	0.57–0.70	0.64	0.48	Ehmann

The data of Crockett et al. (1967), Müller et al. (1971), Rieder and Wänke (1969) and Tandon and Wasson (1968) were obtained by neutron-activation analysis, involving chemical isolation of the ^{192}Ir activity induced in the sample. Ehmann and McKown (1969) have employed a "nondestructive" technique (involving coincidence counting of the ^{192}Ir activity) which

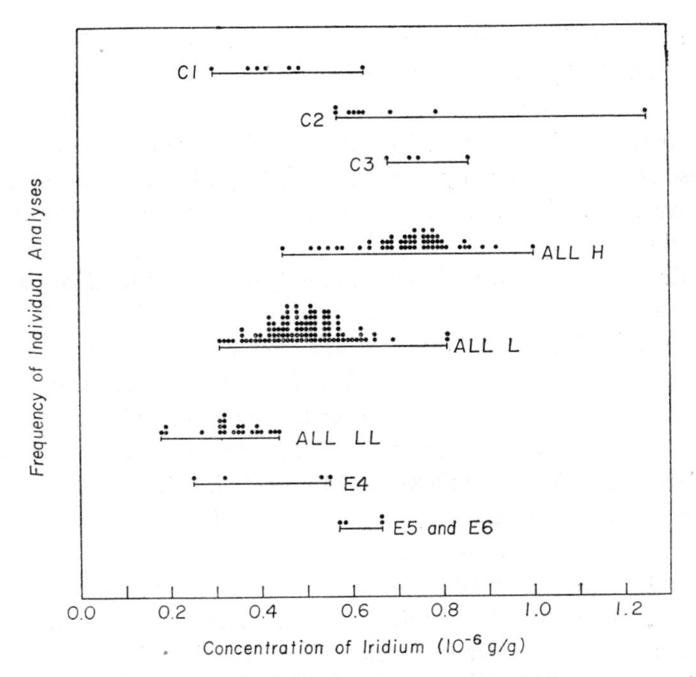

FIGURE 1 Frequency of individual analyses versus the iridium concentration for the various chondrite groups.

avoids those sources of error associated with the post-irradiation processing procedures. In addition to those data tabulated, Sen Gupta (1968), Kiesl et al. (1967) and Seitner et al. (1968) have also reported data for Ir in chondrites. Sen Gupta employed a spectrophotometric technique, and obtained results which were both higher and lower than the activation-analysis data. For example he obtained a value of 1.5 ppm for the Benton LL-group chondrite vs. a value of 0.25 ppm obtained by Müller for the same chondrite. He obtained a value of 0.10 ppm for Abee, vs. 0.54 ppm as obtained by Ehmann and McKown. Kiesl et al. (1967) and Seitner et al. (1968) determined Ir by neutron-activation, but did not treat the samples in a manner which would overcome the problems described by Baedecker. These workers obtained results which were in some cases in good agreement with those of other workers, but there were some large discrepancies. For example, they obtained a value of 0.27 for the Pultusk chondrite, vs. 0.74 as determined by Müller. On the other hand, their value of 1.58 ppm for the Pipe Creek H-group chondrite is much higher than has been obtained for the Ir content of other H-group chondrites. The agreement between the results summarized in

Table 1 is more satisfactory, but some discrepancies do exist. For example Crocket et al. (1967) report the Ir concentration of Cold Bokkeveld to be 1.25 ppm, while Ehmann and McKown obtained a value of 0.57 ppm for the same meteorite. The data of Rieder and Wänke are consistently higher than those of other workers by approximately 20%. Ehmann and McKown (1969) have summarized the possible sources of error which could account for these differences.

The Ir distribution in chondrites more closely follows the total iron content than the metal phase content of these meteorites (Baedecker and Ehmann, 1965). Therefore, if the chondrites formed from a common, homogeneous material, then the metal–silicate fractionation which occurred during the formation of these objects took place when the material was in a more highly reduced state than is presently observed for the chondrites (Baedecker, 1967). Tandon and Wasson (1968) and Müller et al. (1971) have used the Ir content as a measure of the degree of Fe–Si fractionation, and find some suggestion of a correlation between the Ir content and the oxidation state of the chondrites, as measured by the fayalite content of the olivine (Keil and Fredriksson, 1964). This is an indication that the chondrites formed under conditions such that the degree of metal–silicate fractionation and the redox state of the chondrites were not established independently.

The data of Baedecker (1967), Crockett et al. (1967) and Sen Gupta (1968) suggested that the enstatite chondrites were depleted in Ir relative to the other chondrite groups. The recent data of Ehmann and McKown (1969) show no such depletion. They find no significant correlation between the Ir content and the Fe/Si ratio, which varies from 1.13 to 1.87 within the enstatite chondrites analyzed.

Ir is strongly concentrated in the metal phases of most meteorites, as demonstrated in Table 2 by the data of Baedecker (1967) and Nichiporuk and Brown (1965) for the separated metal and silicate portions from chondrites and pallasites. However, the data for the non-magnetic fraction of the chondrites should not be accepted as an accurate indication of the Ir concentration in the silicate fraction of these meteorites, since it is very difficult to obtain silicate material which is completely free of metal. As stated above, the data of Baedecker (1967) are low, due to chemical problems associated with the technique. Nevertheless the data demonstrate the strong siderophilic character of this element. Ehmann et al. (1970) report average values of 0.065 and 0.22 for the Ir concentration in nodules of troilite from the Odessa and Canyon Diablo irons respectively. The same authors report the Ir concentration in both irons to be 2.16, suggesting a distribution ratio of

TABLE 2 Ir in separated fractions from meteorites

Material	No. of determin-ations	No. of meteorites analyzed	Concentration (10^{-6} g/g)		Phase/metal ratio		Reference
			Range	Mean	Range	Mean	
H-chondrites							
magnetic	2	2	0.9–1.4	1.2			Baedecker
	5	5	2.1–3.2	2.8			Nichiporuk
non-magnetic	2	2	0.04–0.2	0.1	0.047–0.12	0.088	Baedecker
L-chondrites							
magnetic	2	2	1.4–2.4	1.9			Baedecker
non-magnetic	2	2	0.04–0.07	0.06	0.028–0.031	0.030	Baedecker
E4 chondrites							
magnetic	1	1		0.9			Baedecker
non-magnetic	1	1		0.002		0.002	Baedecker
E5 and E6 chondrites							
magnetic	1	1		0.4			Baedecker
non-magnetic	1	1		0.04		0.01	Baedecker
Pallasite							
metal	2	2	0.02–2.8	1.4			Baedecker
	2	1		0.01			Hamaguchi
	34	16	0.006–10.0	0.74			Wasson
silicate	3	3	0.00008–0.001	0.0004	0.00005–0.054		Baedecker
Mesosiderite							
metal	2	1		3.4			Wasson
Troilite (Irons)	4	2	0.07–0.23		0.03–0.10		Ehmann
H-group							
dark	3	3	0.91–1.5	1.2			Rieder
Carbonaceous							
chondrules	2	2	0.1–0.3	0.2			Baedecker
matrix	1	1		0.6			Baedecker

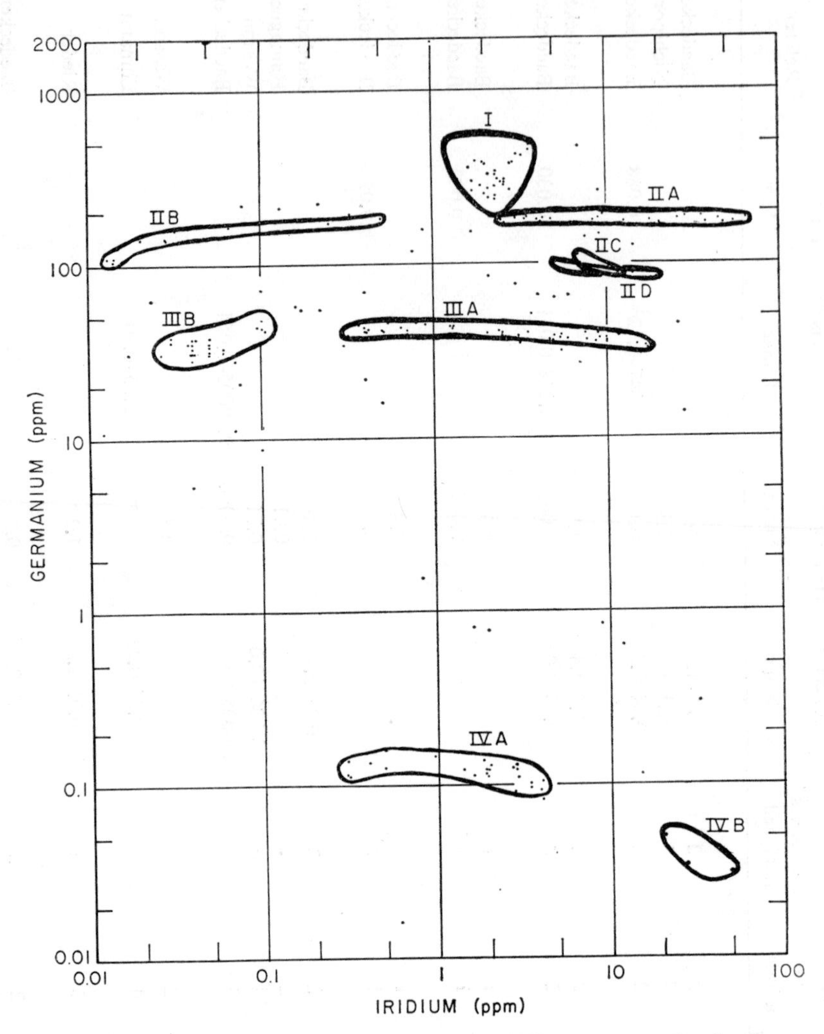

FIGURE 2 Germanium concentration versus iridium concentration for the iron meteorites. Data points for the members of the various chemical groups are enclosed in closed lines.

between 0.03–0.10 for Ir between troilite and metal, which demonstrates a distinct tendency of Ir to appear in the sulfide phase.

Rieder and Wänke (1969) have determined Ir in the dark portions of two H-group chondrites having a light-dark structure. There is no appreciable difference between their data for this material and the H-group chondrites as a whole.

Baedecker (1967) analyzed chondrules which were separated from a C2 and a C3 chondrite. The data for the chondrules was only slightly lower than the data for the meteorite samples as a whole.

Wasson (1968, 1969a) has found Ir to be a particularly useful element in helping to define groups of iron meteorites. Although there is considerable overlap between the Ir concentrations of different chemical groups, distinct groupings emerge when other trace elements are also considered. For example, those meteorites which fall into group III based on their Ga and Ge contents, can be subdivided into "chemical groups" IIIA (0.1–17 ppm Ir) and IIIB (0.014–0.1 ppm Ir) based on the content of Ir. Figure 2 is a plot of Ge vs. Ir for the nine chemical groups, based on the data of Wasson and co-workers, who determined Ir by neutron-activation.

Table 3 lists iron meteorite data by Wasson and coworkers (1967, 1968, 1969ab) and Cobb (1967) who employed a "non-destructive" neutron-activation technique, and Nichiporuk and Brown (1965) who employed spectrographic methods. Those irons which cannot be unambiguously assigned to one of the nine chemical groups resolved by Wasson and co-workers are listed as anomalous. In addition to the work cited in Table 3, Sen Gupta and Beamish (1963) and Sen Gupta (1968) have determined Ir in seven irons by a spectrophotometric technique. Their data appear to be consistently low, by approximately a factor of 2, when compared with the activation analysis results for the same meteorites.

The data which have been obtained for the achondrites by Baedecker (1967) are summarized in Table 4, along with the single analysis of the Stannern eucrite by Kiesl et al. (1967). The Ir content of most achondrites is of the order of a few ppb. The highest Ir concentrations observed for the achondrites were 0.14 ppm for the Shallowater aubrite, and 0.15 ppm for the Dyalpur ureilite, both of which have a relatively high metal phase content.

Grevesse et al. (1968) list a solar abundance for Ir of 3.0 relative to Si $= 10^6$ atoms, which is considerably higher than the abundance of 0.6–1.0 suggested by the carbonaceous chondrite data.

No determinations of the isotopic composition of meteoritic Ir have been reported.

TABLE 3 Ir in iron meteorites

Ga-Ge group	No. of determinations	No. of meteorites analyzed	Concentration (10^{-6} g/g) Range	Reference
I	7	5	1.2–5.0	Cobb
	6	3	1.6–2.2	Nichiporuk
	105	50	1.1–5.5	Wasson
II A	6	4	3.3–15	Cobb
	4	4	9.5–24	Nichiporuk
	61	29	2.3–59	Wasson
II B	2	1	<0.4–<0.5	Cobb
	2	2	<0.3–<0.4	Nichiporuk
	23	11	0.01–0.46	Wasson
II C	13	6	6.2–11	Wasson
II D	2	1	18–20	Cobb
	13	6	4.8–18	Wasson
III A	9	6	1.5–14	Cobb
	4	3	0.4–15	Nichiporuk
	116	55	0.30–17	Wasson
III B	1	1	20.7	Nichiporuk
	34	16	0.014–0.040	Wasson
IV A	3	2	1.6–2.5	Cobb
	3	3	1.0–2.4	Nichiporuk
	53	25	0.36–3.5	Wasson
IV B	1	1	8.5	Nichiporuk
	11	5	13–28	Wasson
Anom.	2	1		Cobb
	7	7		Nichiporuk
	187	89		Wasson

TABLE 4 Ir in achondrites

Meteorite type	No. of determinations	No. of meteorites analyzed	Concentration (10^{-6} g/g) Range	Mean	Abundance (Si = 10^6)	Reference
Ca-rich achondrites						
eucrites	1	1		0.02	0.01	Baedecker
	1	1		0.08	0.05	Kiesl
howardites	1	1		0.003	0.002	Baedecker
angrites	1	1		0.002	0.001	Baedecker
nakhlites	1	1		0.0008	0.0005	Baedecker
Ca-poor achondrites						
aubrites	3	3	0.002–0.14	0.05	0.03	Baedecker
diogenites	2	1	0.005–0.01	0.007	0.005	Baedecker
ureilite	1	1		0.15	0.11	Baedecker

References

Baedecker, P.A. (1967) Ph. D. Dissertation, University of Kentucky, 110 pp. USAEC Technical Report ORO-2670-17.
Baedecker, P.A. and Ehmann, W.D. (1965) *Geochim. Cosmochim. Acta* **29**, 329–342.
Cobb, J.C. (1967) *J. Geophys. Res.* **72**, 1329–1341.
Crocket, J.H., Keays, R.R. and Hsieh, S. (1967) *Geochim. Cosmochim. Acta* **31**, 1615–1623.
Ehmann, W.D. and McKown, D.M. (1969) *Anal. Lett.* **2**, 49–60.
Ehmann, W.D., Baedecker, P.A. and McKown, D.M. (1970) *Geochim. Cosmochim. Acta* **37**, 493–507.
Goldschmidt, V.M. (1954) *Geochemistry*, Oxford University Press, Oxford, England.
Goldschmidt, V.M. and Peters, C. (1932) *Nachr. Ges. Wiss. Göttingen, Math.-Phys. Kl.* 377.
Grevesse, N., Blanquet, G. and Boury, A. (1968) *Origin and Distribution of the Elements.* L.H. Ahrens editor, Pergamon Press, Oxford, England, pp. 177–182.
Hamaguchi, H., Nakai, T. and Kamenoto, Y. (1961) *Nippon Kagaku Zasshi* **82**, 1489–1493.
Keil, K. and Fredriksson, K. (1967) *J. Geophys. Res.* **69**, 3781–3815.
Kiesl, W., Seitner, H., Kluger, F. and Hecht, F. (1967) *Monatsh. Chem.* **98**, 972–992.
Müller, O., Baedecker, P.A., and Wasson J.T. (1971) *Geochim. Cosmochim. Acta*, in press.
Nichiporuk, W. and Brown, H. (1965) *J. Geophys. Res.* **70**, 459–470.
Noddack, I. and Noddack, W. (1930) *Naturwiss.* **18**, 757.
Noddack, I. and Noddack, W. (1931a) *Z. Phys. Chem.* **154**, 207.
Noddack, I. and Noddack, W. (1931b) *Z. Phys. Chem. Bodenstein Festbd.* 890.
Noddack, I. and Noddack, W. (1934) *Svensk Kem. Tidskr.* **46**, 173.
Rieder, R. and Wänke, H. (1969) *Meteorite Research* (ed. P.M. Millman), 75–86.
Rushbrook, P.R. and Ehmann, W.D. (1962) *Geochim. Cosmochim. Acta* **26**, 649–657.
Seitner, H., Kiesl, W., Kluger, F. and Hecht, F. (1968) preprint.
Sen Gupta, J.G. (1968) *Anal. Chim. Acta* **42**, 481–488.
Sen Gupta, J.G. and Beamish, F.E. (1963) *Am. Mineralogist* **48**, 379–389.
Tandon, S.N. and Wasson, J.T. (1968) *Geochim. Cosmochim. Acta* **32**, 1087–1109.
Wasson, J.T. (1968) *J. Geophys. Res.* **73**, 3207–3211.
Wasson, J.T. (1969a) *Geochim. Cosmochim. Acta* **33**, 859–876.
Wasson, J.T. (1969b) unpublished data.
Wasson, J.T. and Goldstein, J.I. (1968) *Geochim. Cosmochim. Acta* **32**, 329–339.
Wasson, J.T. and Kimberlin, J. (1967) *Geochim. Cosmochim. Acta* **31**, 2065–2093.
Wasson, J.T. and Sedwick, S.P. (1969) *Nature* **222**, 22–24.

References for Tables

Baedecker and Ehmann (1965), Baedecker (1967).
Cobb (1967).
Crocket et al. (1967).
Ehmann and McKown (1969), Ehmann et al. (1969).
Hamaguchi et al. (1961).
Kiesl et al. (1967).

Müller and Wasson (1969).
Nichiporuk and Brown (1965).
Rieder and Wänke (1969).
Tandon and Wasson (1968).
Wasson (1968, 1969ab), Wasson and Goldstein (1968), Wasson and Kimberlin (1967), Wasson and Sedwick (1969).

(Received 28 July 1969)

PLATINUM (78)

William D. Ehmann

Department of Chemistry, University of Kentucky, Lexington, Kentucky and the Department of Chemistry, Arizona State University, Tempe, Arizona

PLATINUM, BEING a precious metal, was sought for in many early analyses of the meteorites. Some of the early data on platinum and platinum group metals in meteorites is summarized by Hawley (1939). Goldschmidt (1958) estimated abundances of 20 ppm in meteoritic nickel-iron, 2 ppm in troilite, and 3.24 ppm in average meteoritic material made up of 10 parts silicate, 2 parts nickel-iron, and 1 part troilite. Goldschmidt also pointed out the serious problem of potential laboratory contamination in the early platinum analyses, due to the presence of platinum laboratory-ware in most chemical laboratories. He regarded platinum to be principally siderophilic in meteoritic materials, and felt that early reports of platinum in meteoritic silicates were due to accidental inclusion of small amounts of metal or troilite phases.

More recently platinum data in meteorites have been published by Mason and Wiik (1961) and Nichiporuk and Brown (1965) using emission spectrography, Sen Gupta and Beamish (1963) and Sen Gupta (1968) using spectrophotometric methods, and Hamaguchi, Nakai and Kamemoto (1961), Baedecker and Ehmann (1965), and Crocket, Keays and Hsieh (1967) using neutron activation analysis. Since the number of analyses of chondrites is small, most of the available data are presented in Table 1. Two chondrite analyses of Hamaguchi et al. (1.0 ppm and 1.0 ppm) were not included, since this writer did not have access to the original publication which identified the chondrites analyzed.

Platinum in nature consists of six isotopes–^{190}Pt (0.0127 %), ^{192}Pt (0.78 %), ^{194}Pt (32.9 %), ^{195}Pt (33.8 %), ^{196}Pt (25.3 %) and ^{198}Pt (7.21 %). In the activation analysis procedures for platinum, thermal neutron irradiations in a nuclear reactor for periods of several days are used to produce radio-

TABLE 1 Platinum abundances in chondrites

Specimen	Classification*	Pt, ppm	Reference
Alais	C1	0.6	(3)
Ivuna	C1	1.7, 0.9, 1.1	(3)
Orgueil	C1	1.0	(3)
Cold Bokkeveld	C2	1.4	(3)
Mighei	C2	1.3	(3)
Murray	C2	1.5	(2)
Nogoya	C2	1.0	(3)
Lancé	C3	2.0, 1.5, 1.7	(3)
Abee	E4	1.3	(2)
		1.3, 1.3	(3)
		0.70, 0.75	(4)
Indarch	E4	1.1, 1.4	(3)
Atlanta	E5	~2.2	(2)
Forest City	H5	~1.5, ~2.0	(2)
Plainview	H5	1.6	(2)
Richardton	H5	~1.7	(2)
Belly River	H	0.80, 0.75	(4)
Bruderheim	L6	0.60	(4)
		~0.9	(2)
Colby	L6	1.4	(2)
Harleton	L6	~0.7, ~1.1	(2)
Holbrook	L6	0.44	(1)
Peace River	L6	0.52, 0.55	(4)
Benton	LL6	0.80	(4)

* Classification according to Van Schmus and Wood (1967).

(1) Mason and Wiik (1961)
(2) Baedecker and Ehmann (1965)
(3) Crocket, Keays and Hsieh (1967)
(4) Sen Gupta (1968)

active ^{199}Pt and its radioactive daughter ^{199}Au. Ordinarily the 3.15 day half-life ^{199}Au is counted by means of its 158 KeV gamma-ray, using scintillation spectrometry. However, ^{199}Au is also produced by double neutron capture reactions on gold. Therefore, corrections must be made on the observed ^{199}Au counting rate in the radiochemically separated gold sample derived from the irradiated specimens. This correction may be derived from analysis of the spectrum from a simultaneously irradiated gold standard. Both Crocket et al. (1967) and Baedecker and Ehmann (1965) point out that their chondrite data should be regarded as only approximate, largely based on the uncertainties in this correction. Crocket et al. regard their data as good to a factor of two. Baedecker and Ehmann state error limits of ±20% for four chondrites and regard the remainder of their chondrite analyses as only order of magnitude determinations. Work on an improved method for platinum determinations in meteorites is currently in progress (Gillum and Ehmann, unpublished).

TABLE 2 Platinum abundances in chondrite classes*

Class	Number of meteorites analyzed	Mean Pt abundance, ppm	Atomic abundance Pt (Si = 10^6)†
C1	3	0.9	1.2
C2	4	1.3	1.4
C3	1	1.7	1.6
Enstatite	3	1.5	1.2
H-group	4	1.5	1.3
L- and LL-group	6	0.8	0.6
	Suess and Urey (1956), based on early chondrite analyses and interpolation		1.625
	Cameron (1959), adjusted chondritic values		1.28
	Clayton and Fowler (1961), calculations based on theories of nucleosynthesis		0.80

* Based on all data of Table 1, although it may be noted that the abundances reported by Sen Gupta (1968) and Mason and Wiik (1961) are somewhat lower than the data obtained by activation analysis.

† Using silicon data of Vogt and Ehmann (1965).

Platinum abundances in the principal chondrite classes derived from the data of Table 1 are given Table 2. Based on the few analyses available, it is interesting to note that the platinum atomic abundances (Si = 10^6) in the carbonaceous chondrites appear to increase in the order C1, C2, C3. This is the reverse of the trend noted for a number of other siderophilic metals. Larimer (1967) does not list platinum among the "depleted elements" in meteorites. Crocket et al. (1967) compute a condensation temperature of 1650 °K (1 atm.) for platinum, suggesting it would be among the first few metals to condense in the cooling solar nebula. The platinum atomic abundance in the Type I carbonaceous chondrites based on this compilation is in reasonably good agreement with the adjusted abundance value of Cameron (1959).

Nichiporuk and Brown (1965) report a range of platinum abundances of 7.9 to 9.1 ppm in the metal phases of five chondrites. Baedecker and Ehmann (1965) report an abundance of 0.71 ppm in separated Canyon Diablo troilite. Platinum abundances in siderites (Table 3) range up to 29.3 ppm. Therefore, based on the available data, platinum would appear to be principally siderophilic in meteoritic matter. Additional data on platinum abundances in separated meteoritic phases would, however, be desirable.

TABLE 3 Platinum abundances in siderites*

Classification	Meteorites analyzed	Range of abundances Pt, ppm	Mean Pt abundance, ppm
Fine octahedrites	6	0.5– 6.1	3.6
Medium octahedrites	10	2.4–17.7	7.9
Coarse octahedrites	5	5.0–23.5	12.8
Nickel-rich ataxites	2	11.8–23.5	17.7
Hexahedrites	6	4.0–29.3	19.2

* Based on data of Sen Gupta and Beamish (1963) Nichiporuk and Brown (1965), Baedecker and Ehmann (1965), and Sen Gupta (1968). Values based on mean abundances for each meteorite. Individual determinations exhibited a slightly larger range of values.

Addendum

Recently, Gillum and Ehmann (1971) have obtained replicate Pt analyses on 39 chondrites. The average values in ppm for the various chondrite classes are: C1–1.01, C2–1.00, C3–1.55, C4–1.78, LL–0.80, L–1.10, H–1.49 and E–1.66.

Gillum, D.E. and Ehmann, W.D. (1971). Unpublished data.

References

Baedecker, P.A. and Ehmann, W.D. (1965) *Geochim. Cosmochim. Acta* **29**, 329–342.
Cameron, A.G.W. (1959) *Astrophys. J.* **129**, 676–699.
Clayton, D.D. and Fowler, W.A. (1961) *Ann. Phys. (N.Y.)* **16**, 51–68.
Crocket, J.H., Keays, R.R. and Hsieh, S. (1967) *Geochim. Cosmochim. Acta* **31**, 1615–1623.
Goldschmidt, V.M. (1958) *Geochemistry*, Oxford University Press, London.
Hamaguchi, H., Nakai, T. and Kamemota, Y. (1961) *Niippon Kagaku Zasshi* **82**, 1489–1493; as reported in *Chem. Abstr.* **56**, 6650a.
Hawley, F.G. (1939) *Contrib. Soc. Res. Meteorites* **2**, 132–137.
Larimer, J.W. (1967) *Geochim. Cosmochim. Acta* **31**, 1215–1238.
Mason, B. and Wiik, H.B. (1961) *Geochim. Cosmochim. Acta* **21**, 276–283.
Nichiporuk, W. and Brown, H. (1965) *J. Geophys. Res.* **70**, 459–470.
Sen Gupta, J.G. (1968) *Anal. Chim. Acta* **42**, 481–488.
Sen Gupta, J.G. and Beamish, F.E. (1963) *Am. Mineralogist* **48**, 379–389.
Suess, H.E. and Urey, H.C. (1956) *Rev. Mod. Phys.* **28**, 53–74.
Van Schmus, W.R. and Wood, J.A. (1967) *Geochim. Cosmochim. Acta* **31**, 747–765.
Vogt, J.R. and Ehmann, W.D. (1965) *Geochim. Cosmochim. Acta* **29**, 373–383.

(Received 1 February 1969)

GOLD (79)

William D. Ehmann

Department of Chemistry, University of Kentucky,
Lexington, Kentucky and the Department of Chemistry,
Arizona State University, Tempe, Arizona

GOLD IN NATURE is monoisotopic at mass number 197 and exhibits strongly siderophilic properties. Early data on the abundance of gold in meteoritic materials were summarized by Goldschmidt (1958). He reported an average abundance of 4 ppm in meteoritic nickel-iron and 0.7 ppm in meteoritic troilite. Goldberg, Uchiyama and Brown (1951) first used the sensitive technique of neutron activation analysis to determine gold in a wide range of iron meteorites. They reported a range of 0.094 to 8.7 ppm and a mean abundance of 1.4 ppm for siderites. Goldschmidt estimated a mean meteoritic abundance of 0.7 ppm based on average meteoritic matter being made up of 10 parts silicate (in which the gold abundance was assumed to be negligable), 2 parts nickel-iron, and 1 part troilite. Using the lower mean siderite abundance of Goldberg et al., this average abundance would be approximately 0.3 ppm.

In the last ten years numerous investigators have published data on gold abundances in meteoritic materials. Neutron activation analysis has been the analytical method of choice for the majority of the recent work. This is only natural, since the technique couples high sensitivity for gold with inherent freedom from reagent and laboratory contamination. This method involves the irradiation of the meteorite specimen in the thermal flux of neutrons from a nuclear reactor for a period of 3 to 5 days to produce radioactive ^{198}Au, which has an easily handled half-life of 64.8 hours. Ordinarily, radiochemical separations employing a carrier are required to eliminate interfering activities before counting the 0.412 MeV gamma-rays resulting from the decay of ^{198}Au.

The principal problem in the determination of gold by activation analysis is one of representative sampling. Ordinarily, small powdered samples of

TABLE 1 Gold abundances in various classes of stony meteorites

Classification*	Number of meteorites analyzed	Total number of analyses	Mean Au abundance in ppm†	Atomic Au abundance (Si = 10^6)‡
C1	3	10	0.14	0.19
C2	6	14	0.18	0.19
C3	4	9	0.18	0.17
C4	1	1	0.15	0.13
All C	14	34	*0.17*	0.18
E4	2	6	0.40	
E5	2	3	0.35	
E6	4	6	0.27	
All E	8	15	*0.32*	0.26
H3	2	3	0.23	
H4	4	13	0.22	
H5	9	17	0.23	
H6	2	2	0.21	
All H	20	41	*0.23*	0.19
L4	2	6	0.13	
L5	5	7	0.14	
L6	16	24	0.17	
All L	26	42	*0.16*	0.13
LL3	1	1	0.15	
LL4	1	1	0.23	
LL6	1	1	0.12	
All LL	3	3	*0.17*	0.12
Achondrites–Ca poor	2	4	0.006–0.030	–
Achondrites–Ca rich	5	5	0.002–0.060	–

* Classification of Van Schmus and Wood (1967), where available. Mean values for an entire group often contain analyses on meteorites not classified in this reference.

† Based on data of Vincent and Crocket (1960), Shcherbakov and Perezhogin (1964), Baedecker and Ehmann (1965), Crocket, Keays and Hsieh (1967), Kiesl et al. (1967), Kiesl (1968), Rieder and Wanke (1968), Seitner et al., (1968), and Ehmann, Baedecker and McKown (1970). Several discordant abundances were omitted from the means where other replicate analyses indicated the possibility of analytical error. Schaudy et al. (1967 and 1968) have recently reported a number of gold analyses for many of the same meteorites as used by Kiesl et al. (1967) and Seitner et al. (1968). In many cases the gold data of Schaudy et al. are higher than those reported by Kiesl et al. and Seitner et al. although all analyses were performed at the same institute and apparently with the same meteorite powders. Certain L-group analyses by Schaudy et al. are higher than any others previously reported for this group. For this present compilation the data of Schaudy et al. (1967 and 1968) have not been included.

‡ Using silicon data of Ehmann and Durbin (1968).

0.5 gram or less are used in order to minimize the effect of sample self-shielding. Since gold is strongly siderophilic, it concentrates in the metallic nickel-iron phase in the chondritic meteorites. When only small samples of chondrites are used and the metal phase aggregates are large, the chance addition or exclusion of even one metal particle can lead to divergent results in replicate analyses. Means obtained by numerous replicate analyses are therefore to be desired. Problems associated with the determination of noble metals by activation analysis have also been recently discussed by Beamish et al. (1967). Fire-assay preconcentration followed by spectrophotometric measurements (Sen Gupta, 1968) has been used for a number of other noble elements. While this technique would minimize the sampling problem, if used for gold it would require extravagant (15 to 20 grams) amounts of valuable meteoritic material. Numerous replicate activation analyses for gold in chondrites may be done with several grams of material. This material may be selected via microscopic examination to be reasonably representative of the meteorite as a whole.

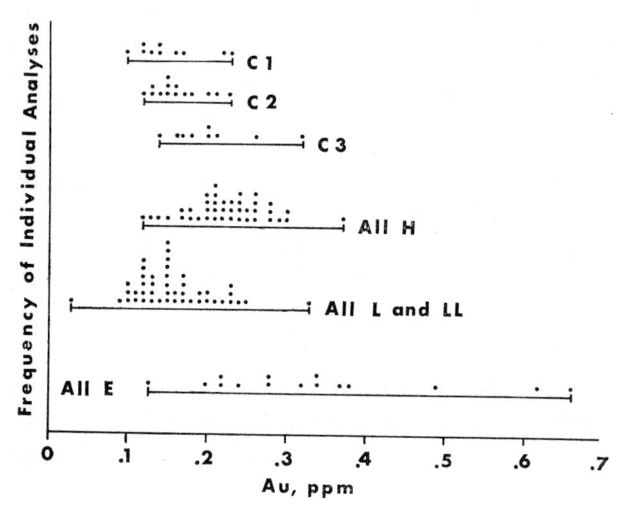

FIGURE 1 Distribution of individual gold determinations for the principal classes of chondrites. Discordant determinations have been deleted where other replicate data are normal.

Gold abundances in various classes of stony meteorites are summarized in Table 1. Atomic abundances relative to Si = 10^6 are also included in this table. The distribution of individual analyses is indicated in Figure 1. The atomic abundance in the Type I carbonaceous chondrites (Class C1 in the classification of Van Schmus and Wood, 1967) is most often taken as representative of average solar system primordial material. The relative atomic abundance of 0.19 of this compilation is somewhat higher than the value of 0.145 selected by Suess and Urey (1956) and Cameron (1959). It is 50% higher than the value of Clayton and Fowler (1961), based on calculation from theories of nucleosynthesis.

Gold is listed as one of nine elements exhibiting a "normal" depletion pattern of 1/0.6/0.3 in the groups C1/C2/C3 (Larimer and Anders, 1967). Using the data of Table 1, the atomic abundance depletion ratios relative to C1 are 1/1/0.9. While the experimental data are not too numerous for these meteorites and do exhibit some scatter, it is questionable to group gold among the "normally depleted elements" based on the data available. Larimer and Anders found these same nine elements were depleted in the ordinary chondrites with respect to C1 chondrites by a factor between 0.15 and 0.25. On the basis of relative atomic abundances in Table 1, this factor for gold appears to be closer to 0.8 or 0.9, again making the inclusion of gold among the depleted elements somewhat questionable.

Fouché and Smales (1967) analyzed a number of separated magnetic (metal) and non-magnetic phases of H-group, L-group, and enstatite chondrites. The average metal phase gold abundance found for the H-group was 1.37 ppm, for the L-group was 2.32 ppm, and for the enstatite chondrites was 2.21 ppm. A similar enrichment in the L-group metal phase is reported by Ehmann, Baedecker and McKown (1970). Both of the afore mentioned groups report low and widely varying (0.005 to 0.053 ppm) gold abundances in the non-magnetic phases of these three groups of chondrites. Additional chondritic metal phase gold abundances are given by Vincent and Crocket (1960), and Rieder and Wänke (1968). Gold abundances in three pallasite olivine phases ranged from 0.001 to 0.09 ppm (Ehmann, Baedecker and McKown, 1970). Abundances in the pallasite metal phases ranged from 1.5 to 2.3 ppm. All these data confirm the strongly siderophilic character of gold in meteoritic matter.

Gold abundances in the troilite phase of siderites have recently been reported by Shcherbakov and Perezhogin (1964), Baedecker and Ehmann (1964), Linn and Moore (1967), and Ehmann, Baedecker and McKown (1970). The troilite gold abundances reported are generally less than 0.07 ppm, although one determination of 0.23 ppm was reported by Ehmann, Baedecker and McKown for Sardis troilite. Linn and Moore also report a gold abundance of less than 0.07 ppm in a Brenham schreibersite inclusion.

Rieder and Wänke (1968) have found that essentially equal concentrations of gold in separated light and dark portions of gas-rich bronzite chondrites, in contrast to indium which is enriched in the dark portions. Ehmann, Baedecker and McKown (1969) found evidence of a depletion of gold in chondrules extracted from Chainpur and Al Rais as compared to concentrations in the matrix or the meteorite as a whole.

Gold abundances in large groups of siderites have been reported by Goldberg et al. (1951), Fouché and Smales (1966), Cobb (1967), and Linn and Moore (1967). Additional data on various siderites is also to be found in many of the papers dealing mainly with abundances in stony meteorites (for example, Hofler and Sorantin, 1967). Data of different groups for the same meteorite often show a considerable scatter, which Linn and Moore attribute in part to an inhomogeneous distribution of gold in the siderites. The range of gold abundances in various classes of siderites based on analyses by Fouché and Smales (1966) appear to be representative of most of the available data and are presented in Table 2. Only the data for meteorites falling

TABLE 2 Gold abundances in some distinct structural groups of siderites (As given in Fouché and Smales, 1966)

Class	Number of meteorites	Range of Au abundances ppm
Hexahedrites	7	0.43–0.75
Coarsest octahedrites	3	1.42–1.77
Coarse octahedrites	7	0.97–1.61
Medium octahedrites	12	0.50–2.20
Fine octahedrites	6	0.55–2.12

within the five structural and trace element groups defined by Smales, Mapper and Fouché (1967) are given here. Many other siderites, while apparently similar structurally to other members of these groups, were dissimilar with respect to their abundances of a wide variety of trace elements and were not included in the above groupings. Fouché and Smales feel their gold data are consistent with these five basic groupings. They also note that there is apparently a strong correlation of gold abundances with arsenic and palladium abundances in siderites. There is also a rather weak correlation of gold with antimony.

Ehmann, Baedecker and McKown (1970) have found that gold is enriched in the small metallic spherules associated with the Canyon Diablo crater with respect to the Canyon Diablo meteorite itself. This is probably due to oxidation and subsequent loss of iron in the oxidation of the spherules, with the more noble element, gold, concentrating in the residual metal. They have also reported that the gold content of a metallic spherule extracted from Philippine tektite falls within the range observed for the siderites. Choy (1966) has observed gold enrichment in the magnetic portion of dust collected on greased nylon mesh on an offshore island. It is suggested that this enrichment may be indicative of an extraterrestrial origin for the magnetic portion of the dust.

Addendum

Recently, replicate determinations of Au in 39 chondrites have been obtained in our laboratory. The group means are in excellent agreement with the data summarized in Table 1.

Gillum, D.E. and Ehmann, W.D. (1971). Unpublished data.

References

Baedecker, P.A. and Ehmann, W.D. (1965) *Geochim. Cosmochim. Acta* **29**, 329–342.
Beamish, F.E., Chung, K.S. and Chow, A. (1967) *Talanta* **14**, 1–32.
Cameron, A.G.W. (1959) *Astrophys. J.* **129**, 676–699.
Choy, T.K. (1966) *Anal. Chim. Acta* **34**, 372–374.
Clayton, D.D. and Fowler, W.A. (1961) *Ann. Phys. (N.Y.)* **16**, 51–68.
Cobb, J.C. (1967) *J. Geophys. Res.* **72**, 1329–1341.
Crocket, J.H., Keays, R.R. and Hsieh, S. (1967) *Geochim. Cosmochim. Acta* **31**, 1615–1623.
Ehmann, W.D., Baedecker, P.A. and McKown, D.M. (1970) *Geochim. Cosmochim. Acta* **34**, 493–507.
Ehmann, W.D. and Durbin, D.R. (1968) *Geochim. Cosmochim. Acta* **32**, 461–464.
Fouché, K.F. and Smales, A.A. (1966) *Chem. Geol.* **1**, 329–339.
Fouché, K.F. and Smales, A.A. (1967) *Chem. Geol.* **2**, 105–134.
Goldberg, E., Uchiyama, A. and Brown, H. (1951) *Geochim. Cosmochim. Acta* **2**, 1–25.
Goldschmidt, V.M. (1958) *Geochemistry*, Oxford University Press, London.
Hofler, H. and Sorantin, H. (1967) *Chem. Geol.* **2**, 273–278.
Kiesl, W., Seitner, H., Kluger, F. and Hecht, F. (1967) *Monatshefte f. Chemie* **98**, 972–992.
Kiesl, W. (1968) Paper presented at the *1968 International Conference Modern Trends in Activation Analysis*, Gaithersburg, Maryland, October 7–11, 1968. Published in *Modern Trends in Activation Analysis*, **N. B. 5.** Special Publ. **312**, pp. 302–307 (1969).

Larimer, J.W. and Anders, E. (1967) *Geochim. Cosmochim. Acta* **31**, 1239–1270.

Linn, T.A. and Moore, C.B. (1967) *Earth Plan. Sci. Letters* **3**, 453–456.

Rieder, R. and Wänke, H. (1968) *Paper presented at the International Symposium on Meteorite Research IAEA*, Vienna, Austria, August 7–13, 1968. Published in *Meteorite Research*, D.Reidel Publ. Co., Dortrecht, Holland 1969, pp. 75–86.

Schaudy, T., Kiesl, W. and Hecht (1967) *Chem. Geol.* **2**, 279–287.

Schaudy, T., Kiesl, W. and Hecht (1968) *Chem. Geol.* **3**, 307–312.

Seitner, H., Kiesl, W., Kluger, F. and Hecht, F. (1968) 5th Communication (Preprint) from the Analytical Institute of the University of Vienna and the Reactor Center Seibersdorf of the "Österreichische Studiengesellschaft für Atomenergie G.m.b.H.".

Sen Gupta, J.G. (1968) *Anal. Chim. Acta* **42**, 481–488.

Shcherbakov, Yu.G. and Perezhogin, G.A. (1964) *Geochemistry USSR (English Translation)* **6**, 489–496.

Smales, A.A., Mapper, D. and Fouché, K.F. (1967) *Geochim. Cosmochim. Acta* **31**, 673–720.

Suess, H.E. and Urey, H.C. (1956) *Rev. Mod. Phys.* **28**, 53–74.

Van Schmus, W.R. and Wood, J.A. (1967) *Geochim. Cosmochim. Acta* **31**, 747–765.

Vincent, E.A. and Crocket, J.H. (1960) *Geochim. Cosmochim. Acta* **18**, 143–148.

(Received 22 August 1969)

MERCURY (80)*

George W. Reed, Jr.

Argonne National Laboratory
Argonne, Ill. 60439

THE GEOCHEMISTRY of Hg is to a large extent governed by its association with late-stage mineral deposits. It is associated with sulfide-containing phases and the most abundant Hg-containing mineral is the sulfide, cinnabar (HgS). Redox conditions in the earth may also lead to the reduction of Hg ions, which accounts for its natural occurrence as the metal.

The concentrations of Hg in meteorites range from $\sim 10^{-8}$ gm/gm to $> 10^{-6}$ gm/gm. Its content does not appear to be correlated with the amount of sulfide, specifically troilite (FeS), nor necessarily with the redox conditions during formation of the various types and classes of meteorites. Even its volatility does not appear to be a unique determinant of its distribution in meteoritic matter. Except for the carbonaceous Type I and II chondrites, meteorites appear to be products of igneous processes; the achondrites may be directly derived from such. The latter contain amounts of Hg that span the range found in the chondrites, including the carbonaceous chondrites.

Modern analyses of Hg in meteorites have relied on the highly sensitive and contamination-free technique of neutron activation analysis. The standard chemical separation involves HF dissolution of the sample in the presence of Hg carrier solution. Total Hg measurements based on this technique have been reported by Reed, Kigoshi, and Turkevich (1960), Ehmann and Huizenga (1959), and Ehmann and Lovering (1967). The former investigators attempted to remove and determine Hg that would volatilize at about 100 °C for two samples. The volatilization of Hg from irradiated samples in stepwise heating from 110° to 1200 °C in a closed system has been used by Reed and Jovanovic (1967, 1968). That negligible Hg remained in the samples was established by dissolution of the sample at 300 °C in a sealed fused-silica

* Work performed under the auspices of the U.S. Atomic Energy Commission.

bomb containing Hg carrier, HCl and $HClO_4$. The volatilization approach was attractive as a possible way of differentiating between contamination and indigenous Hg. The problem of contamination has been considered in detail by Reed and Jovanovic (1967) and does not seem to be serious.

These investigators have attributed the release patterns observed in the volatilization experiments to the presence of Hg in at least two types of sites. The release from one type is triggered by a threshold process such as the chemical decomposition of a compound or a solid state transformation; the release from another type site is diffusion-regulated.

Mercury contents of chondrites are listed in Table 1. The extremely large range observed within a given class of meteorite makes averaging meaningless. Even for a given meteorite large variations are obtained in the same laboratory. Reed and Jovanovic (1967) tabulated the results of replicate analyses on a number of meteorites. For Orgueil they obtained 2.40, 15.0, 213 ppm; for Bjurbole 0.77 → 5.99 ppm; four other meteorites gave concentrations varying from less than 10% to a factor of 5. Ehmann reports values of Orgueil of 17.3 (Ehmann and Huizenga, 1959) and 114 ppm (Ehmann and Lovering, 1967). Ehmann and Lovering (1967) also report 0.69 and 7.3 ppm for Mokoia. These variations have nothing to do with the laboratory or method used for analysis, for example, Reed and Jovanovic (1967) and Ehmann and Lovering (1967) report, respectively, 213 and 114 ppm for Orgueil, 4.9 and 6.8 ppm for Mighei, 0.80 and 1.28 ppm for Karoonda, 0.044 and 0.064 ppm for Bruderheim, and 0.015 and 0.028 ppm for Peace River. These all agree within a factor of two. There are also cases of as poor agreement between laboratories as there are within a given sample measured in the same laboratory.

Reed and Jovanovic (1967) observed that the ratio of the amount of Hg released above 450 °C to the total amount tended to be relatively constant, and surmised that since contamination Hg would enter as a volatile it should be lost at lower temperatures. This trend towards a constant ratio was not dependent on the type of meteorite or, therefore, on whether it contained high, low, or no metallic iron, whether its sulfur was free or combined, or what its mineral composition was.

The fractionation of Hg between chondrite classes does not follow the patterns observed for other elements. Even the ordinary chondrite classes exhibit differences, with the bronzites tending to be higher in Hg on the average than the hypersthenes. In fact, a number of bronzite chondrites contain as much Hg as the various types of carbonaceous chondrites including one member, Ivuna, of Type I. Only Orgueil averages higher. The

TABLE 1 Mercury in chondrites

Meteorite classification	Number of determinations	Number of meteorites	Range (ppm)	Mean (ppm)	At/10^6 Si†	Ref.‡
Hypersthene (L)	2	2	0.054, 0.174	0.114		A
	22	16	0.015–5.99			B
	4	3	0.028–0.16	0.084		C
Bronzite (H)	2	2	0.074, 0.081	0.078		A
	20	13	0.26–13.90			B
	3	2	0.076–0.36	0.262		C
Amphoterite (LL)	3	2	0.24–0.84	0.55		B
Enstatite (E5, 6)	1	1		0.80		B
	3	3	0.16–0.72	0.44	0.47	C
(E4)	3	2	0.16–1.4	0.48		B
	1	1		1.52	1.3	C
	1	1		0.004		D
Carbonaceous-I	4	2	2.4–213.0	4.96, 111.2*	29	B
	1	1		114		C
	1	1		17.3		D
Carbonaceous-II	2	2	1.57, 6.82	4.4		C
	1	1		4.9	4.4	B
	1	1		3.77		E
Carbonaceous-III	2	2	0.80, 1.6	1.2		C
	3	3	0.69–7.3	2.6	2.8	B
	1	2	2.43, 5.00	3.72		E

* The 4.96 ppm value is for Ivuna, the 111.2 ppm is the weighted average for three determinations on Orgueil.

† Abundances given by authors, data selected by authors.

‡ A = Ehmann and Huizenga, B = Reed and Jovanovic, C = Ehmann and Lovering, D = Reed, Kigoshi, and Turkevich, E = Kiesl and Hecht.

enstatite chondrites fall within the range of the hypersthene chondrites. Further details of this type are pointed out by Reed and Jovanovic (1967).

It is clear, therefore, that the concentrations of Hg observed do not reflect an initial distribution but are the result of modifications of this distribution. Reed and Jovanovic (1967, 1968) have explored the significance of this.

Considering the volatility of Hg, it is of interest that chondrules which are most probably high-temperature condensates and the total meteorite including the fine-grained crystalline matrix have very similar total Hg concentrations (Table 2).

TABLE 2 Mercury in chondrules (Reed and Jovanovic, 1967)

Meteorite	No. of deter-minations	Concentration Chondrules (ppm)	Concentration Whole Meteorite (ppm)
Bjurbole	1	1.29	2.87 (0.77–5.99)
Allegan	1	1.52	5.70 (3.05–7.90)
Tieschitz	1	2.01	7.22
Soko-Banja	1	0.91	0.84

The most extensive study of Hg in achondrites has been reported by Ehmann and Lovering (1967). Their data, along with the other two analyses reported, are listed in Table 3. The large variation found among the eucrites parallels that in the chondrites and spans about the same concentration range. The range of values reported, even the high value of 9 ppm for the eucrite, Stannern, have been confirmed in an unpublished study by Jovanovic and Reed of the volatilization of Hg from achondrites. Ehmann and Lovering (1967) discuss the Hg contents of achondrites on the basis of their possible genesis, but in view of the large range of concentrations observed in the eucrites more extensive sampling of all classes would seem to be a prerequisite to such an analysis.

The low contents of Hg found in troilite from iron meteorites (Table 4) appear to eliminate this mineral as the site of most of the Hg found in

TABLE 3 Mercury in achondrites

Meteorite classification	No. of deter-minations	No. of meteorites	Range (ppm)	Mean (ppm)	Ref.*
Hypersthene (diogenite)	1	1		0.121	A
	1	1		0.43	C
Basaltic (eucrites)	1	1		0.078	B
	4	4	0.21–9.12	4.27	C
Basaltic (howardites)	1	1		0.66	C
Enstatite (aubrite)	1	1		0.054	C
Nakhlite	1	1		0.23	C
Angrite	1	1		2.51	C

* A = Ehmann and Huizenga (1959), B = Reed, Kigoshi, and Turkevich (1960), C = Ehmann and Lovering (1967).

TABLE 4 Mercury in troilite from iron meteorites

Meteorite	Number of determinations	Concentration (ppb)	Ref.
Canyon Diablo	2	7.5, 5.9	A
Toluca	1	<34	A
Odessa	1	700	B
Odessa (troilite + graphite)		320	B
Mt. Joy	1	500	C
Magura	1	2210	C
Brownfield	1	90	C
Mocs (chondrite) troilite—metal grain	1	1100	C

A = Reed, Kigoshi and Turkevich (1960), B = Reed and Jovanovic (1967), C = Kiesl and Hecht (1969).

meteorites, unless the troilite grains in the chondrites have much higher Hg contents on the average than that reported for iron meteorites. The 1.1 ppm Hg in a troilite-metal grain from the chondrite Mocs reported by Kiesl and Hecht is consistent with this possibility.

References

Ehmann, W.D. and Huizenga, J.R. (1959) "Bismuth, thallium, and mercury in stone meteorites by activation analysis", *Geochim. Cosmochim. Acta* **17**, 125–135.

Ehmann,W.D. and Lovering, J.F. (1967) "The abundance of mercury in meteorites and rocks by neutron activation analysis", *Geochim. Cosmochim. Acta* **31**, 357–376.

Kiesl, W. and Hecht, F. (1969) "Meteorites and the high temperature origin of terrestrial planets", in *Meteorite Research* (P.M. Milliman, editor), Reidel Pub. Co., Holland.

Noddack, I. and Noddack, W. (1931) "Die Geochemie des Rheniums", *Z. Phys. Chim.* **154**, 223–244.

Reed, G.W., Kigoshi, K. and Turkevich, A.L. (1960) "Determinations of concentrations of heavy elements in meteorites by activation analysis", *Geochim. Cosmochim. Acta* **20**, 122–140.

Reed, G.W. and Jovanovic, S. (1967) "Mercury in chondrites", *J. Geophys. Res.* **72**, 2219–2228.

Reed, G.W. and Jovanovic, S. (1968) "Thermal history of meteorites" in *Origin and Distribution of the Elements* (L.H. Ahrens, editor) Pergamon Press, Oxford, pp. 321–328.

(Received 25 September 1970)

THALLIUM (81)

Michael E. Lipschutz

Departments of Chemistry and Geosciences
Purdue University
Lafayette, Indiana 47907

THALLIUM IS PRESENT in meteorites in quite variable, ultratrace concentrations ranging from 10^{-11}–10^{-7} g/g. Relatively few determinations exist, most of which were made by neutron activation and the remainder by isotope dilution. In almost all cases where the same meteorite has been studied by both techniques, isotope dilution yields somewhat higher results. Thallium determinations or upper limits have been reported for 28 different chondrites and 2 achondrites; and, in some cases, replicate samples were analyzed (Table 1). No "whole-rock" analyses of iron or stony-iron meteorites have been published.

In meteorites, as on the Earth, Tl appears to be primarily chalcophilic, residing in troilite (FeS) as a dispersed element. The ionic radius of Tl^+ is quite similar to that of K^+ and therefore Tl is found in terrestrial K-containing minerals such as feldspars (Rankama and Sahama, 1950; Goldschmidt, 1954; Mason, 1958). Thus, in meteorites, Tl may also have some lithophile character but this cannot yet be established since the only separated mineral phases in which it has been determined are troilite and iron. Reed et al. (1960) reported 7.8 and 13×10^{-9} g/g in troilite inclusions from Canyon Diablo and 200×10^{-9} g/g FeS in Toluca while Ostic et al. (1969) determined 1.5 and 0.25×10^{-9} g/g in troilite and iron, respectively, from Canyon Diablo. Anders and Stevens (1960) have reported that in troilite from Canyon Diablo and Toluca and in samples of Mighei (C2), Abee (E4) and Beardsley and Richardton (H5) the Tl^{205}/Tl^{203} ratios are the same as that of terrestrial Tl to within 1%. Ostic et al. (1969) determined the Tl^{205}/Tl^{203} ratios in troilite and iron from Canyon Diablo and in a sample of Plainview to be the same as terrestrial Tl to within 0.6%.

TABLE 1 Thallium determinations in meteorites

Chemical-Petrologic Type	Meteorite*	Tl Measurements† (10^{-9} g/g)	Ref.	Mean Tl Contents‡ (10^{-9} g/g)	Mean Tl Contents‡ (atoms/10^9 Si atoms)
C1	Ivuna	63	(1)		
	Orgueil	67, 93; 140	(1), (2)	91	121
C2	Mighei	45; 97; 140	(1), (2), (3)		
	Murray	42, 44	(1)	74	77
C3	Felix	28	(1)		
	Grosnaja	52	(1)	35	31
	Vigarano	24	(1)		
E4	Abee	71, 96; 150	(2), (3)		
	Indarch	120	(2)	110	89
H3	Bremervörde	6.7	(1)		
	Clovis # 1	0.69	(1)	*7.2*	*5.8*
	Sharps	80	(1)		
H5	Beardsley	1.9, 4.4, 4.7	(4)		
		5.8	(3)		
	Fayetteville [a]	4.7; 0.92	(1)		
	Forest City	0.14, 0.39, 0.51	(2)		
		0.30, 0.41, 0.51	(4)	*1.1*	*0.91*
	Plainview	6.0	(1), (5)		
	Richardton	0.94	(3)		
L3	Khohar	2.0	(1)		
	Krymka	51	(1)	*7.2*	*5.3*
	Mezö-Madaras	3.6	(1)		
L4	*Barratta*	0.25	(1)	0.2	0.2
L6	Bruderheim	(< 0.06)	(1)		
	Holbrook	0.42, 0.45, 0.56, 3.8	(2)		
		(0.91), 1.2	(4)	*0.40*	*0.29*
	Modoc	0,24, 0.41, (< 4.2)	(4)		
		0.03, 0.14, 0.23, 0.83	(2)		
LL3	Chainpur	15	(1)		
	Hamlet [b]	1.1	(1)	*3.9*	*2.9*
	Ngawi	32	(1)		
	Parnallee	0.45	(1)		
Ah	Johnstown	0.40, 0,81, 0.98	(4)	0.7	0.4
Ap	*Nuevo Laredo*	0.57, 0.75	(2)	0.7	0.4

In chondrites, the Tl contents clearly vary with chondritic type (Table 1, Figure 1), the carbonaceous and enstatite chondrites having the highest concentrations. In contrast to the large variations in the ordinary chondrites, the Tl contents in the C1, C2, C3 and E4 types are rather constant, varying

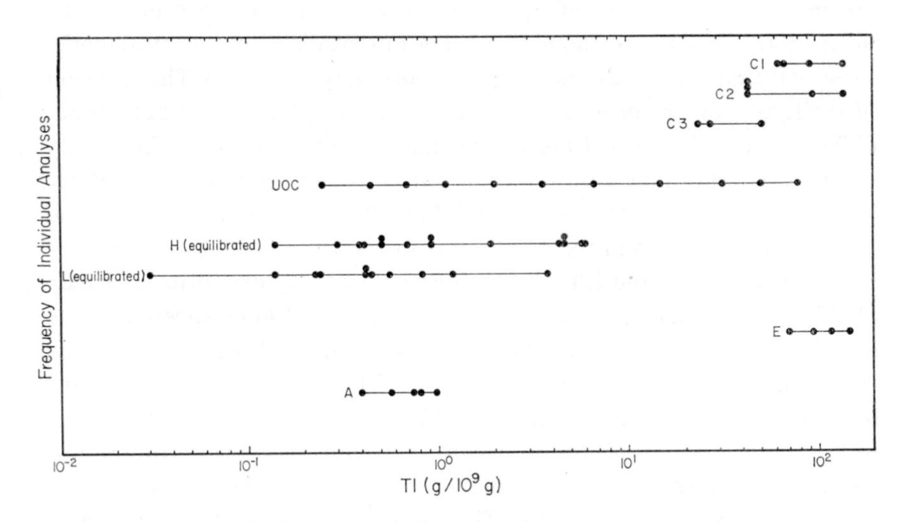

in the extremes by about a factor of 2–3. Relative to the ordinary chondrites, both the enstatite and carbonaceous chondrites are rich in FeS (Mason, 1962). Laul et al. (1970) have noted that in C1 and C2 chondrites those few Tl determinations reported prior to their study seem somewhat too high. If those values for the carbonaceous chondrites are disregarded the mean Tl abundances in C1 and C2 are reduced to 74 and 44×10^{-9} g/g or 96 and 46 atoms/10^9 Si atoms respectively and the mean abundance in C1 is about

* Meteorites listed in italics are finds, all others are observed falls.

† Values in parentheses are upper limits or suspicious (see original references) and have been omitted in computing mean values for the group.

‡ Values in italics are geometric means.

[a] Petrologic type unknown. The first value listed is for the dark phase; the second, for the co-existing light phase.

[b] Classified as LL [3, 4).

(1) Laul et al. (1970).

(2) Reed et al (1960). Some of the data reported were previously listed in a preliminary fashion in Reed et al. (1958).

(3) Anders and Stevens (1960).

(4) Ehmann and Huizenga (1959).

(5) Ostic et al. (1969).

twice that in C2 and about four times that in C3 (Laul et al., 1970) as predicted by Anders (1964). Neither Aller (1961) nor Müller (1968) list solar photospheric values for Tl.

The Tl concentrations in the H-, L- and LL- group chondrites vary in the extreme cases by a factor of 10^3 (Table 1), with the range in values of the unequilibrated ordinary chondrites being intermediate to and overlapping those of the carbonaceous and equilibrated ordinary chondrites. The precision of replicate analyses of individual ordinary chondrites has not been good, differences of a factor of 10 being encountered often. Thus part, if not all, of this variability may be ascribed to sample inhomogeneity, although at levels of 10^{-11}–10^{-9} g/g experimental difficulties may play a role. Because of the wide variations in the Tl contents of the H-, L- and LL-groups unequilibrated and H- and L-group equilibrated ordinary chondrites (a factor of about 10^2 in each case) the means listed in Table 1 are geometric means rather than arithmetic ones. If Barratta (L4) is included with the other unequilibrated L-group chondrites, the mean Tl contents of this group are reduced to 3.1×10^{-9} g/g or 2.3 atoms/10^9 Si atoms. Thus there appears to be no significant dependence of mean Tl content with chondritic group, either in the unequilibrated or equilibrated ordinary chondrites. It does seem clear however that the mean Tl contents in the unequilibrated H- and L-groups are significantly higher than those in the corresponding equilibrated chondrite groups, despite the similarity in the K and FeS contents of these chondrites (Mason, 1965; Dodd et al., 1967).

It is quite interesting to note that the Tl contents in the unequilibrated ordinary chondrites decrease exponentially with increasing equilibration of the silicates (Laul et al., 1970). Based on the available data it is impossible to state whether in the equilibrated ordinary chondrites the Tl contents also vary with increasing equilibration (i.e. chemical-petrologic type). If such a dependence is present it is probably not as pronounced as that evident in the unequilibrated ordinary chondrites. In the unequilibrated ordinary chondrites, the Tl concentrations generally can be correlated with those of bismuth, carbon and primordial argon, krypton and xenon except in the most unequilibrated chondrite studied of each group, where both bismuth and, to a lesser extent, carbon appear underabundant (Laul et al., 1970).

The Tl concentrations have been determined in co-existing light and dark portions of a single H-group chondrite, Fayetteville, which is rich in primordial noble gases (Zähringer, 1969). In this meteorite the concentrations of primordial noble gases are 3–300 times higher in the dark portions than in the light. Thallium, too, appears to be richer in the dark portion, 4.7×10^{-9} g/g compared with 0.92×10^9 g/g in the light (Laul et al., 1970). The Tl contents in the hypersthene achondrite Johnstown and in the eucrite Nuevo Laredo appear to be similar to those in equilibrated ordinary chondrites.

Acknowledgements

This research was supported in part by the U.S. National Science Foundation, grant GA-1474.

References

Aller, L.H. (1961) *The Abundance of the Elements*. Interscience.
Anders, E. (1964) *Space Sci. Rev.* **3**, 583.
Anders, E. and Stevens, C.M. (1960) *J. Geophys. Res.* **65**, 3043.
Dodd, R.T., Van Schmus, W.R. and Koffman, D.M. (1967) *Geochim. Cosmochim. Acta* **31**, 921.
Ehmann, W.D. and Huizenga, J.R. (1959) *Geochim. Cosmochim. Acta* **17**, 125.
Goldschmidt, V.M. (1954) *Geochemistry*, Oxford Univ. Press.
Laul, J.C., Pelly, I. and Lipschutz, M.E. (1970) *Geochim. Cosmochim. Acta* **34**, 909.
Mason, B. (1958) "*Principles of Geochemistry*", Wiley.
Mason, B. (1962) "*Meteorites*", Wiley.
Mason, B. (1965) *Am. Museum Novitates* 2223.
Müller, E. (1968) *Origin and Distribution of the Elements* (L.H. Ahrens, ed.) 155–176, Pergamon.
Rankama, K. and Sahara, Th.G. (1950) *Geochemistry*, University of Chicago Press.
Reed, G.W., Kigoshi, K. and Turkevich, A. (1958) Paper 953 in *Proceedings of the Second International Conference on the Peaceful Uses of Atomic Energy, Geneva* **28**, 486, United Nations (New York).
Reed, G.W., Kigoshi, K. and Turkevich, A. (1960) *Geochim. Cosmochim. Acta* **20**, 122.
Zähringer, J. (1968) *Geochim. Cosmochim. Acta* **32**, 209.

(Received 25 September 1969; revised 1 June 1970)

LEAD (82)

Virginia M. Oversby

Lamont-Doherty Geological Observatory
*of Columbia University**
Palisades, New York

LEAD OCCURS as a dispersed element in all classes of meteorites. The concentration is greatest in the troilite phase of iron meteorites and least in the metallic phases of these meteorites. Interpretation of lead concentration data is complicated by the possibility of terrestrial contamination. However, even in cases where contamination is known to be small, adjacent troilite nodules from a single slice of meteorite have shown concentrations which differ by a factor of three (Oversby, 1970). Spacially distant samples of troilite from the same meteorite, with primitive isotopic compositions indicating lack of terrestrial contamination, show even larger differences in concentration of lead (cf. Table 1).

The isotopic composition of lead extracted from iron meteorites may be used as a guide to interpretation of concentration data. Table 1 summarizes all published data on lead concentrations from iron meteorites in which the isotopic composition of lead was determined. Cobb (1964) used alpha activation analysis; all other studies employed isotope dilution mass spectrometry. The isotopic data, shown in Figure 1, has been recently discussed in detail (Oversby, 1970).

The most primitive isotopic composition found in the Canyon Diablo meteorite is interpreted to be the isotopic composition of lead at the time the iron meteorites formed. The concentration of uranium in iron meteorites, both in metallic and troilite phases, is so low relative to the lead concentration that in-situ decay of uranium and thorium could not have significantly altered the isotopic composition of these leads in times on the order of 5×10^9 yrs. Samples with primitive isotopic compositions will thus give lead

* Lamont-Doherty Contribution No. 1358.

TABLE 1

Meteorite	Phase*	206/204	207/204	208/204	Conc. 10^{-6} g/g	Source
Canyon Diablo	tr.	9.46	10.34	29.44	18	(1)
	met.	9.43	10.58	29.80	0.14	(2)
	tr.	9.61	10.39	29.87	4.6	(4)
	met. + tr.	10.0	10.4	29.5	0.22 (m) 3.7 (tr)	(9)
	tr.	17	–	38	–	(7)
	tr.	9.34^{6}	10.21^{8}	28.96	5.76	(11)
	tr.	9.70	10.47	29.44	1.88	(11)
	tr.	10.10	10.82	30.62	4.76	(11)
	tr.	18.60	15.55	38.16	515	(11)
Odessa	met. + tr.	10.6	10.5	29.2	0.1 (m) 5 (tr)	(9)
Toluca	†tr.	17.15	15.33	37.42	59	(5)
	†tr.	17.22	15.64	38.51	59	(5)
	met.	9.87	10.70	30.36	0.16	(3)
	‡tr.	16.87	15.03	37.11	–	(8)
	tr.	9.31	10.30	29.60	4.4	(10)
	met.	9.6	10.3	29.4	0.1	(9)
	tr.	10.11	10.64	29.59	2.17	(11)
	tr.	12.80	12.85	35.01	1.66	(11)
Augustinovka	met.	16.80	15.20	37.30	0.19	(3)
Baquedano	tr.	17.84	15.38	37.47	1.40	(11)
Bishtyube	met.	9.80	10.74	30.08	0.18	(3)
	tr.	17.72	15.47	38.40	7.5	(3)
Bogou	tr.	9.87	10.77	30.78	3.68	(11)
Boguslavka	met.	17.39	16.11	37.33	0.02	(3)
Burgavli	tr.	9.79	10.68	30.27	7.7	(3)
	met.	9.34	10.58	30.28	0.24	(3)
Chebankol	met.	17.68	15.76	38.49	0.03	(3)
Chinga	met.	16.89	15.27	35.38	0.03	(3)
Gibeon	tr.	17.52	15.30	36.78	1.57	(11)
Grant	tr.	17.55	15.23	37.08	0.69	(11)
Gressk	met.	17.99	15.84	38.23	0.04	(3)
	tr.	18.07	15.87	38.23	2.00	(3)
Henbury	tr.	9.55	10.38	29.54	5	(1)
	met.	18.13	15.96	38.70	0.20	(2)
	tr.	18.41	15.78	39.00	4.9	(2)

TABLE 1 (cont.)

Meteorites	Phase	206/204	207/204	208/204	Conc. 10^{-6} g/g	Source
Santa Catharina	met.	17.99	15.67	38.51	0.40	(3)
Sardis	tr.	9.37	10.22	29.19	–	(6)
Sikhote-Alin	met.	17.89	15.84	38.19	0.03	(3)
	met.	17.55	15.60	37.97	0.03	(3)
Staunton	tr.	18.47	15.63	38.64	13.9	(11)
Toubil	met.	17.49	15.56	37.62	0.20	(3)
Yardymly (Aroos)	met.	10.14	10.97	30.18	0.17	(3)
	tr.	10.01	10.85	30.78	2.3	(3)

* tr. = troilite; met. = metal
† same nodule
‡ Average of three closely similar analyses.

(1) Patterson, et al., 1953; 1955.
(2) Starik, et al., 1959.
(3) Starik, et al., 1960.
(4) Chow and Patterson, 1961.
(5) Marshall and Hess, 1961.
(6) Murthy and Patterson, 1962.
(7) Cobb, 1964.
(8) Murthy, 1964.
(9) Marshall and Feitknecht, 1964.
(10) Ostic, 1966.
(11) Oversby, 1970.

concentration data truly representative of the meteorite. Samples with radiogenetic lead, unsupported by uranium in the meteorite, represent disturbed meteorites. The disturbance was most probably due to terrestrial contamination (Oversby, 1970).

The concentration of lead in the metallic phase of iron meteorites ranges from 0.02 to 0.40 × 10^{-6} g/g. When only samples with primitive isotopic composition are included the range is 0.1 to 0.24 × 10^{-6} g/g. The variation within Canyon Diablo covers most of the concentration range, making correlation of lead concentration with meteorite class a pursuit of dubious merit. Several meteorites have lead with radiogenic isotopic composition in the metallic phase averaging 0.03 × 10^{-6} g/g. The true lead content in the metal must have been less than 0.01 × 10^{-6} g/g. One of these (Chebankol) is a coarse octahedrite, as is Canyon Diablo. These two meteorites differ in

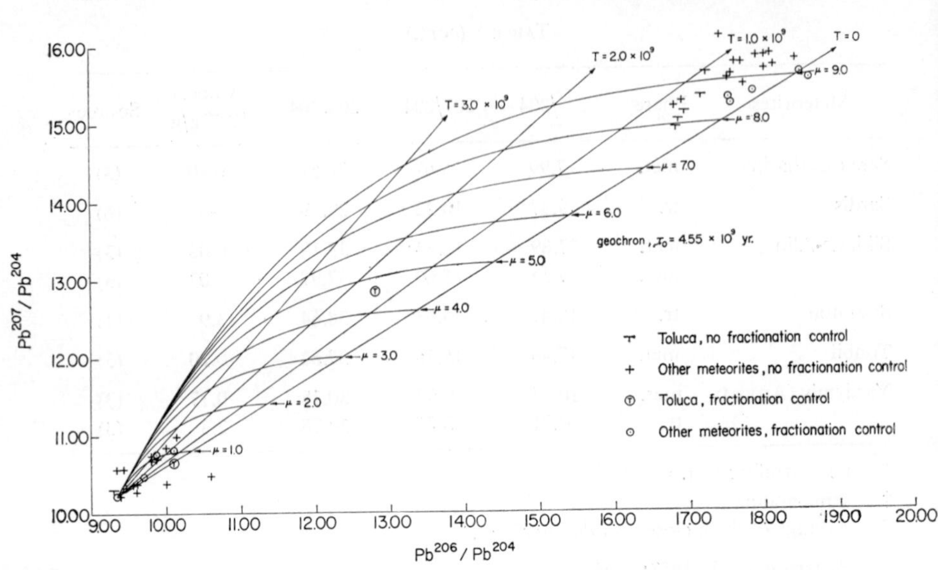

FIGURE 1 Pb^{207}/Pb^{204} vs. Pb^{206}/Pb^{204} for iron meteorites. Single stage growth lines for $T_0 = 4.55 \times 10^9$ yr and values of $U^{238}/Pb^{204} = \mu$ from 1 to 9 are shown for reference. Lines labelled $T = 1.0 \times 10^9$ yr, etc. are isochrons.

metallic lead content by a factor of at least 20 indicating extremely heterogeneous distribution of lead among iron meteorites with closely similar Fe-Ni ratios.

The concentration of lead in troilite modules from iron meteorites shows an even wider range. Including all measurements, the range is 0.69 to 515 $\times 10^{-6}$ g/g. If only samples with primitive isotopic compositions are considered, the range is 1.9 to 18 $\times 10^{-6}$ g/g. The extremes both occur in nodules from Canyon Diablo. Again, no correlation with meteorite class can be made.

The extreme heterogeneity of lead distribution in iron meteorites coupled with the possibility of contamination makes assignment of a meaningful "average" concentration impossible. One way to partially overcome the difficulties would be to use the most frequently observed concentration as representing the average. For metallic phases, most observations lie between 0.1 and 0.2 $\times 10^{-6}$ g/g. For troilite, most observations fall between 1.5 and 5.0 $\times 10^{-6}$ g/g. Reasonable average values would then be 0.15 $\times 10^{-6}$ g/g for metal and 3.5 $\times 10^{-6}$ g/g for troilite. Both of these values are probably too high, since there are a number of samples with very low lead contents coupled with radiogenic isotopic compositions.

TABLE 2

Meteorite	Class*	Conc. Pb 10^{-6} g/g	206/204	207/204	208/204	% Blank in analysed sample	Source
Mighei	C2	–	(14.42)	(13.15)	(34.58)	(?)†	6
		1.5	–	–	31.1	–	10
Mokoia	C2	0.93	10.75	11.13	30.90	~10	6
Murray	C2	–	10.2	–	27.6	(?)†	7
		4.6	10.38	10.94	30.25	~10	6
Orgueil	C1	–	(11.78)	(11.00)	(30.45)	(?)†	6
		3	–	–	30.3	–	10
Abée	E4	3	–	–	–	–	10
Indarch	E4	2.61	10.09	10.87	30.18	~10	6
		2.2	–	–	26.8	–	10
Elenovka	L	0.48	21.54	16.94	39.86	10–20	4
		–	19.8	–	37.3	?	7
Holbrook	L6	0.32	17.52	15.52	38.93	?	3
		0.036; 0.32; }	{ 21.2	15.2	35.5	20	5
		0.21; 0.51; }	{ 16.7	14.6	35.7	15	5
		0.49; 0.12 }	{ 16.0	14.0	33.8	19	5
		0.4	–	–	–	–	10
Kunashak	L6	0.53	19.64	16.24	40.04	10–20	4
		0.49	18.58	15.97	39.06	15	8
Mocs	L6	0.3	20.0	15.6	38.0	15	2

TABLE 2 (cont.)

Meteorite	Class*	Conc. Pb 10^{-6} g/g	206/204	207/204	208/204	% Blank in analysed sample	Source
Modoc	L6	0.9	19.48	15.76	38.21	?	1
		0.3	20.4	16.5	38.8	15	2
		0.06		–	–	–	10
Pervomaisky	L	0.06	19.19	15.76	38.67	60	8
Saratov	L4	0.40	19.53	16.70	40.25	10–20	4
Beardsley	H5	0.13	13.7	12.4	31.8	32	5
		0.14	–	–	–	–	10
Forest City	H5	0.4	19.27	15.95	39.05	?	1
		0.095	–	–	–	?	3
		0.093; 0.086	–	–	–	?	5
		0.12	–	–	–	?	9
		0.1	–	–	–	–	10
Orlovka	H	0.38	18.86	16.38	40.09	20	8
Plainview	H5	0.46	14.9	13.6	34.0	13	5
Richardton	H5	0.13; 0.02; 0.20; 0.2	27.6‡	22.1	48.5	?	3
		0.019; 0.091	(38.16)†	(27.70)	(56.27)	50	5
		0.1	–	–	–	?	9
Zhovtnevyi	H	0.20	22.45	17.95	41.95	38	8
Nuevo Laredo	eucrite	0.7	50.28	34.86	67.97	?	1
		–	34.4	25.8	52.0	15	2
		0.5	–	–	–	–	10

TABLE 2 (cont.)

Meteorite	Class*	Conc. Pb 10^{-6} g/g	206/204	207/204	208/204	% Blank in analysed sample	Source
Pasamonte	eucrite	0.4	18.2	15.5	36.8	15	2
Norton County	aubrite	0.57	22.75	15.87	37.70	10–20	4
		–	24	–	45	?	7
		–	31	–	48	?	7
Pesyanoe	aubrite	0.36	–	–	–	?	4
Bondoc	mesosiderite	0.07	20.11‡	18.53	39.12	20	9
Brahin	pallasite metal	0.07	18.19	15.61	38.72	30	8
	silicate	0.33	17.94	15.49	38.02	20	8

* Van Schmus and Wood chemical-petrologic classification, 1967.
‡ Standard error on 206/204 = ±12%.
‡ Based on 4.3 γ Pb; 13 γ Pb leached from surface had 206/204 = 18.4; 207/204 = 15.7.
† Composition not corrected for blank.

(1) Patterson, 1955
(2) Edwards and Hess, 1956
(4) Marshall and Hess, 1958
(3) Starik et al., 1958
(5) Hess and Marshall, 1960
(6) Marshall, 1962
(7) Cobb, 1964
(8) Sobotovich et al., 1964
(9) Marshall, 1968
(10) Reed et. al., 1960

Table 2 is a summary of published data on lead in chondrites, achondrites and stony-iron meteorites. Carbonaceous and enstatite chondrites show the highest concentrations of lead, ranging from 0.9 to 4.6×10^{-6} g/g. Uranium concentrations for these six meteorites range from 9×10^{-9} g/g for Abee to 17×10^{-9} g/g for Mighei (Morgan and Lovering, 1968), giving a range of Pb/U atomic ratios from 77 for Mokoia to 480 for Murray. Using the measured Pb and U concentrations and assuming the initial lead isotopic composition 4.5×10^9 years ago was identical to that for iron meteorites we can calculate an expected isotopic composition for Mokoia lead. The result is $Pb^{206}/Pb^{204} \cong 10.0$ and $Pb^{207}/Pb^{204} \cong 10.5$. Since the other carbonaceous and enstatite chondrites have higher Pb/U ratios, their isotopic compositions should be correspondingly less radiogenic. All of the measured isotopic compositions show lead more radiogenic than expected on the basis of the measured lead and uranium concentrations. The extent of laboratory contamination during the lead analysis was estimated to be approximately 10% for Mokoia, Murray and Indarch, and the measured isotopic composition was corrected accordingly (Marshall, 1962). High lead blanks are usually quite variable, and it is possible that the extent of contamination was underestimated for these samples. Corrections of 15 to 20% would produce the expected isotopic compositions. It is also possible that some lead contamination was introduced into the meteorite by the soil in which it landed or during storage and handling since its discovery. The lead on the surface of these meteorites was found to be very radiogenic compared to that in their interiors (Marshall, 1962).

Bronzite and hypersthene chondrites have lead concentrations ranging from 0.1 to 0.5×10^{-6} g/g. Data for individual analyses, including the original investigator's estimate of laboratory contamination, are given in Table 2. Uranium concentrations for H and L group chondrites range from 10 to 15×10^{-9} g/g (Morgan and Lovering, 1968). The range of Pb/U atomic ratios is 11.5 to 38, indicating that the most radiogenic isotopic composition to be expected from these meteorites is $Pb^{206}/Pb^{204} \cong 15.0$, $Pb^{207}/Pb^{204} \cong 13.6$. All of the meteorites except Beardsley show lead compositions more radiogenic than can be explained by the measured uranium and lead concentrations. Many of these meteorites have Pb^{207}/Pb^{204} ratios higher than observed in modern terrestrial rocks. This results in some cases where up to 90% of the lead in the meteorites cannot be accounted for by simple closed system growth of Pb from U and Th. Since many of these meteorites are finds, it is possible that terrestrial lead from ground water and soils has been added to the meteoritic lead. The large blanks for the lead analyses, ranging from 10 to 60% of the total lead, also represent a major source of uncertainty in both the isotopic composition and lead concentration data. The actual meteoritic abundance of lead may be a factor of two or more lower than that calculated from the present measurements of lead concentration in ordinary chondrites.

The amount of data on lead in achondrites and stony-irons is very limited. As can be seen in Table 2, the achondrites have lead concentrations slightly higher than those found in ordinary chondrites. Reed et al. (1960) measured the lead and uranium concentrations in a single sample of Nuevo Laredo and found Pb $= 0.5 \times 10^{-6}$ g/g and U $= 1.5 \times 10^{-7}$ g/g. This gives a Pb/U atomic ratio of 3.8, indicating that the lead in Nuevo Laredo can probably be explained by simple closed-system growth of lead from uranium and thorium plus a primordial component similar to that in iron meteorites. The abundance data for achondrites may thus be more accurate than that for chondrites even though a much smaller amount of data is available.

TABLE 3 Abundance of Pb in stony meteorites relative to 10^6 Si atoms calculated from data in Table 2

Chondrites	No. of determ.	Range 10^{-6} g/g	Mean 10^{-6} g/g	Atoms/10^6 Si
Carbonaceous I	1		3	4
Carbonaceous II	3	0.9–4.6	2.3	2.4
Carbonaceous III	0	–	–	?
Bronzite	6	0.08–0.46	0.24	0.19
Hypersthene	7	0.06–0.51	0.37	0.27
Enstatite, Type I	2	2.2–3	2.6	2.1
Type II	0	–	–	?
Achondrites				
Calcium-poor, enstatite	2	0.36–0.57	0.46	0.22
Calcium-rich, eucrites	2	0.4–0.7	0.5	0.30

Table 3 gives a summary of the abundance data for lead in stone meteorites, calculated relative to 10^6 Si atoms. Carbonaceous chondrites show the highest abundance of lead relative to silicon, followed closely by Type I enstatite chondrites. Hypersthene chondrites and bronzite chondrites have a factor of ten lower lead abundance than carbonaceous and enstatite chondrites. The hypersthenes appear to have a slightly greater abundance than

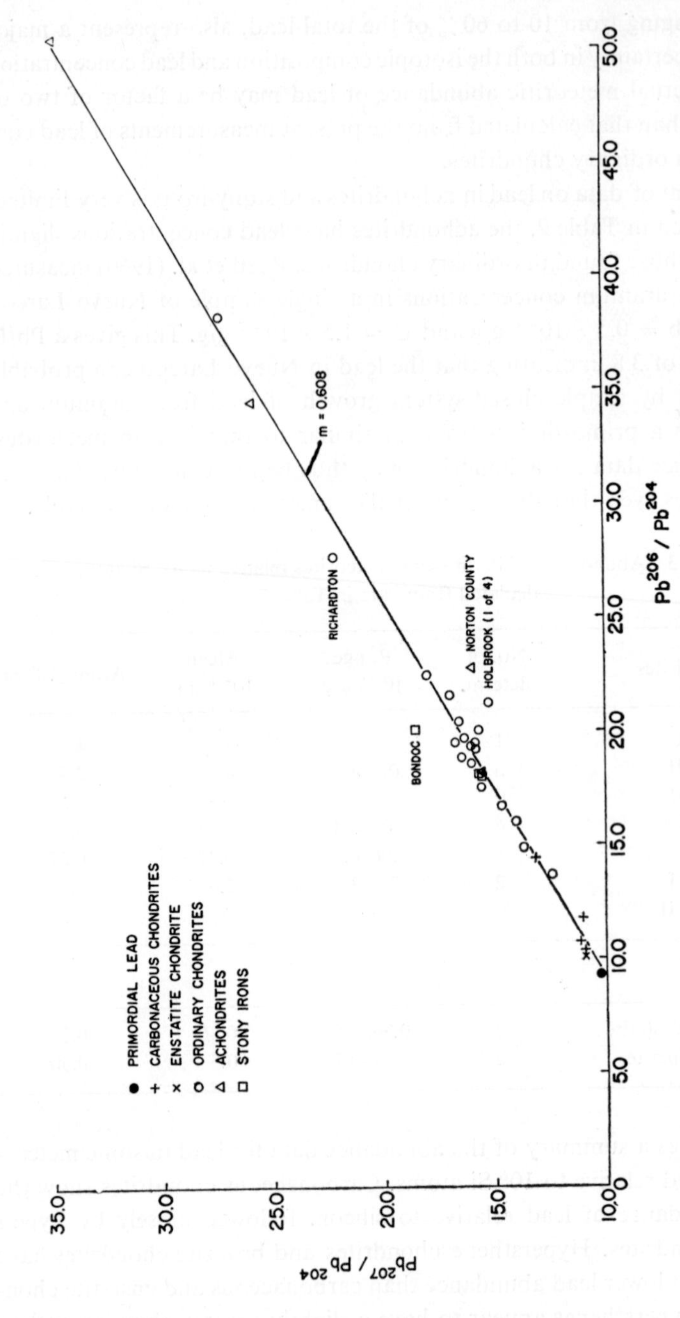

FIGURE 2 Pb^{207}/Pb^{204} vs. Pb^{206}/Pb^{204} for stone and stony iron meteorites. Primordial lead composition is taken as the most primitive Canyon Diablo lead. The least squares isochron (excluding the four labelled points) corresponds to average age of 4.6×10^9 yr.

the bronzites; however, this difference may not be significant, considering the uncertainties in the measurements. Achondrites have a higher lead concentration than ordinary chondrites, and also a higher silicon content. As a result, the lead abundance relative to silicon overlaps that of ordinary chondrites.

The isotopic composition data for stone meteorites are plotted in Figure 2. The bulk of the data can be fitted to a single straight line within the experimental limits of the data. The slope of this line is 0.608, corresponding to an age of 4.6×10^9 yrs. for stone meteorites. Four samples, including 1 out of 4 analyses of the Holbrook meteorite fall clearly off the line. The age of stone meteorites obtained from lead isotopic data, despite the large uncertainties in the measurements, agrees remarkably well with recent, more precise, work on the Rb/Sr ages of chondritic meteorites which gave an age of 4.7×10^9 yrs. for bronzite (H group) chondrites (Kaushal and Wetherill, 1969) and 4.5×10^9 yrs for hypersthene (L group) chondrites (Gopalan and Wetherill, 1968).

References

Chow, T.J. and Patterson, C.C. (1961) *Geokhimiya* **12**, 1124.
Cobb, J.C. (1964) *J. Geophys. Res.* **69**, 1895.
Edwards, G. and Hess, D.C. (1956) *Nuclear Processes in Geologic Settings*, Proceedings of the Second Conference, Nat. Acad. Sci., Nat. Res. Coun., 100.
Gopalan, K. and Wetherill, G.W. (1968) *J. Geophys. Res.* **73**, 7133.
Hess, D.C. and Marshall, R.R. (1960) *Geochim. Cosmochim. Acta* **20**, 284.
Kaushal, S.K. and Wetherill, G.W. (1969) *J. Geophys. Res.* **74**, 2717.
Marshall, R.R. (1962) *J. Geophys. Res.* **67**, 2005.
Marshall, R.R. (1968) *Geochim. Cosmochim. Acta* **32**, 1013.
Marshall, R.R. and Feitknecht, J. (1964) *Geochim. Cosmochim. Acta* **28**, 365.
Marshall, R.R. and Hess, D.C. (1958) *J. Chem. Phys.* **28**, 1258.
Marshall, R.R. and Hess, D.C. (1961) *Geochim. Cosmochim. Acta* **21**, 161.
Morgan, J.W. and Lovering, J.F. (1968) *Talanta* **15**, 1079.
Murthy, V.R. (1964) in *Isotopic and Cosmic Chemistry*. North-Holland Publ. Co.
Murthy, V.R. and Patterson, C.C. (1962) *J. Geophys. Res.* **67**, 1161.
Ostic, R.G. (1966) *J. Geophys. Res.* **71**, 4060.
Oversby, V.M. (1970) *Geochim. Cosmochim. Acta* **34**, 65.
Patterson, C.C. (1955) *Geochim. Cosmochim. Acta* **7**, 151.
Patterson, C.C., Brown, H., Tilton, G. and Inghram, M. (1953) *Phys. Rev.* **92**, 1234.
Patterson, C., Tilton, G. and Inghram, M. (1955) *Science* **121**, 69.
Reed, G.W., Kigoshi, K. and Turkevich, A. (1960) *Geochim. Cosmochim. Acta* **20**, 122.
Sobotovich, E.V., Lovtsyus, G.P. and Lovtsyus, A.V. (1964) *Meteoritika* **24**, 29.

Starik, I.E., Shats, M.M. and Sobotovich, E.V. (1958) *Doklady Akad. Nauk. SSSR* **123**, 424.

Starik, I.E., Sobotovich, E.V., Lovtsyus, G.P., Shats, M.M. and Lovtsyus, A.V. (1959) *Doklady Akad. Nauk. SSSR* **128**, 688.

Starik, I.E., Sobotovich, E.V., Lovtsyus, G.P., Shats, M.M. and Lovtsyus, A.V. (1960) *Doklady Akad. Nauk. SSSR* **134**, 555.

Van Schmus, W.R. and Wood, J.A. (1967) *Geochim. Cosmochim. Acta* **31**, 747.

(Received 13 July 1969)

BISMUTH (83)

Michael E. Lipschutz

Departments of Chemistry and Geosciences
Purdue University
Lafayette, Indiana 47907

BISMUTH IS PRESENT in meteorites in quite variable, ultratrace concentrations ranging from 10^{-10}–10^{-7} g/g. Thus, neutron activation has been the only technique used for recent analyses of stony meteorites. Bismuth determinations have been reported for 33 different chondrites and an achondrite; and, in some cases, replicate samples were analyzed (Table 1). In addition, Reed et al. (1960) list upper limits of about 2×10^{-9} g/g in three samples of the eucrite Nuevo Laredo and "high" concentrations in two samples of Indarch (E4). No "whole-rock" analyses of iron or stony-iron meteorites have been published.

In meteorites, as in the Earth, Bi is primarily chalcophile and probably resides in troilite (FeS) as a dispersed element. Reed et al. (1960) report 40 and 180×10^{-9} g/g in troilite inclusions from Canyon Diablo and Toluca, respectively. Bismuth has not been determined in any other separated mineral phase. Bismuth probably has some lithophile character in meteorites but its host mineral is unknown. Since stable Bi is monoisotopic, there have been no studies of its isotopic composition.

In chondrites, the Bi contents clearly vary with chondritic group (Table 1, Figure 1). As a group, carbonaceous chondrites are richest in Bi and the mean abundance in C1 is about twice that in C2 and about four times that in C3 (Laul et al., 1970a) as predicted by Anders (1964) for such a strongly-depleted element. Considering its compositional variability in samples of ordinary chondrites, the Bi contents of members of the carbonaceous chondrite types are strikingly constant (Table 1). The greatest variation observed is in type C3 chondrites. It may be, however, that type C3 is complex, consisting of Ornans and Vigarano sub-types (Van Schmus, 1969). Laul et al.

TABLE 1 Bismuth determinations in meteorites

Chemical-petrologic type	Meteorite*	Bi measurements† (10^{-9} g/g)	Ref.	Mean Bi contents‡ (10^{-9} g/g)	Mean Bi contents‡ (atoms/10^9 Si atoms)
C1	Ivuna	110, 130	(1)		
	Orgueil	110; 130	(1), (2)	120	160
C2	Cold Bokkeveld	65	(1)		
	Mighei	79; (180)	(1), (2)	69	71
	Murray	62, 71	(1)		
C3	Felix	39, 43	(1)		
	Grosnaja	68	(1)		
	Lancé	49	(1)	43	37
	Ornans	24	(1)		
	Vigarano	51	(1)		
	Warrenton	31	(1)		
E4	Abee	68, 93	(2)	80	60
H3	Bremervörde	25	(1)		
	Clovis # 1	4.6	(1)		
	Prairie Dog Creek	7.2	(1)	*16*	*13*
	Sharps	89	(1)		
H5	Beardsley	3.0, 4.1	(2)		
		2.7, 3.3, 3.8	(3)		
	Fayetteville [a, b]	11; 8	(1)	*2.5*	*2.0*
	Forest City	0.7, 1.1	(2)		
		0.15, 0.18, 0.50	(3)		
	Pantar [b]	8; 3; 8.8; 1.6	(4)		
	Plainview	3.0, 3.8, 4.1	(3)		
		15	(1)		
L3	Khohar	31	(1)		
	Krymka	17	(1)		
	Mezö-Madaras	120	(1)	*40*	*29*
L4	*Barratta*	1.6	(1)	2	1
L6	Bruderheim	1.4	(1)		
	Holbrook	1.7, 2.6, 7.6	(2)		
		(0.49), 2.1	(3)	*1.3*	*1.0*
		2.3	(1)		
	Modoc	0.4, 1.0, 1.2, 6.2	(2)		
		0.07, 0.33, 3.0	(3)		

TABLE 1 (cont.)

Chemical-petrologic type	Meteorite*	Bi measurements† (10^{-9} g/g)	Ref.	Mean Bi contents‡ (10^{-9} g/g)	Mean Bi contents‡ (atoms/10^9 Si atoms)
LL3	Chainpur	65	(1)		
	Hamlet [c]	6.5	(1)		
	Ngawi	24	(1)	*19*	*14*
	Parnallee	12	(1)		
LL6	Ensisheim [b]	0.55; 1.0	(4)	*0.7*	*0.5*
Ah	Johnstown	1.2, 1.5, 5.6	(3)	3	2

* Meteorites listed in italics are finds, all others are observed falls.
† Values in italics are geometric means.
‡ Values in parentheses are suspicious (see original reference for Holbrook) and have been omitted in computing mean values for the group.

[a] Petrologic type unknown.
[b] In each pair, the first value listed is for the dark phase; the second, for the co-existing light phase.
[c] Classified as LL (3, 4).

(1) Laul et al. (1970a)
(2) Reed et al. (1960). Some of the data reported were previously listed in a preliminary fashion in Reed et al. (1958)
3) Ehmann and Huizenga (1959)
(4) Reed (1963)

(1970a) have tentatively noted that members of the Vigarano sub-type (Grosnaja and Vigarano) seem richer in Bi than members of the Ornans sub-type (the remaining C3 chondrites listed in Table 1). Abee, the only E4 chondrite measured (Reed et al., 1960), contains Bi in amounts comparable to those in C2 chondrites. Relative to the ordinary chondrites, both the enstatite and carbonaceous chondrites are rich in troilite (Mason, 1962). Neither Aller (1961) nor Müller (1968) list solar photospheric values for Bi although Grevesse et al. (1968) indicate an upper limit for Bi in the solar photosphere, <57 atoms/10^9 Si atoms, which is lower than the Bi abundance derived from analysis of C1 chondrites (Table 1.)

The Bi concentrations in the H-, L-, LL-group chondrites vary in the extreme cases by about a factor of 10^3 (Table 1). For these groups, the Bi contents of the unequilibrated ordinary chondrites are, in general, markedly higher than those in the equilibrated ordinary chondrites, despite the similarities in their FeS contents (Mason, 1965; Dodd et al., 1967). Because of the wide variation in the Bi contents of the unequilibrated ordinary chondrites (a factor of 10^2), the means calculated in Table 1 are geometric means

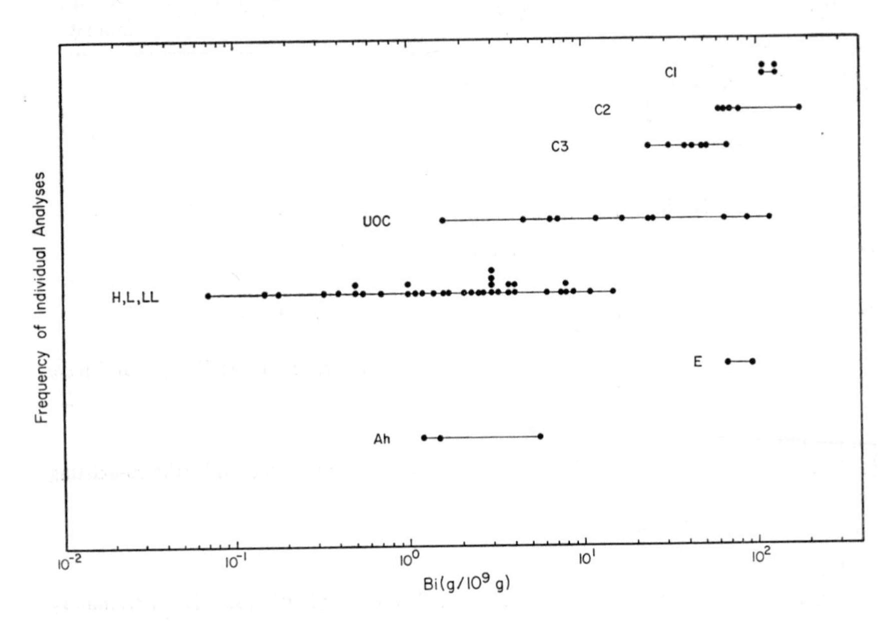

rather than arithmetic ones. If Barratta (L4) is included with the other unequilibrated L-group chondrites, the geometric mean Bi contents of the unequilibrated L-group are reduced to 18×10^{-9} g/g and 13 atoms/10^9 Si atoms. Thus, there appears to be no significant dependence of mean Bi concentration with chondritic group, either in the unequilibrated or equilibrated ordinary chondrites. In general, the concentrations in the unequilibrated ordinary chondrites decrease about exponentially with increasing equilibration of the silicates (Laul et al., 1970a). However, assuming this exponential decrease, the most unequilibrated members of each group (H, L, and LL) studied seem to have relatively low Bi contents (Laul et al., 1970a). It is interesting to note that, in the unequilibrated ordinary chondrites, the Bi concentrations generally seem well correlated with those of thallium, carbon and primordial argon, krypton and xenon (Laul et al., 1970a, b).

The Bi concentrations in the equilibrated ordinary chondrites are, in general, quite low and vary in the extreme cases by a factor of about 10^2 (Figure 1). For at least one meteorite alone, Modoc (L6), the extreme analyses of replicate samples vary by the same factor (Table 1). Part, if not all, of this variation must be due to sample inhomogeneity although, at levels of $\sim 10^{-9}$ g/g, experimental difficulties may play a role. Based on the data available (Table 1) it is impossible to state whether the Bi concentrations in the equilibrated ordinary chondrites vary with chemical-petrologic type (i.e. equilibration). If such a dependence is present, it is certainly not as pronounced as that evident in the unequilibrated ordinary chondrites. Two equilibrated H-group chondrites, Fayetteville and Pantar, are known to be quite rich in primordial noble gases (Zähringer, 1969). These contain a pronounced light-dark structure in which the dark regions contain $\sim$300 times more primordial noble gases than the light regions. The Bi concentrations of co-existing dark and light regions have been determined and, in both chondrites, the Bi concentration (in units of 10^{-9} g/g) in the dark phase are higher than those in the light (Pantar; 8 ± 3 vs. 3 ± 1 and 8.8 ± 0.3 vs. 1.6 ± 0.1 [Reed, 1963]: Fayetteville; 10.9 ± 0.2 vs. 7.8 ± 0.2 [Laul et al., 1970a]). The LL6 chondrite Ensisheim also contains a light-dark structure but this chondrite is not rich in primordial noble gases. Reed (1963) has found that the dark region in this chondrite contains less Bi than the light (0.55 ± 0.05 vs. $1.0 \pm 0.1 \times 10^{-9}$ g/g). The reasons for the apparent Bi enrichment in the dark, gas-rich regions of the H-group chondrites are, as yet, unknown. The Bi concentrations in the hypersthene achondrite Johnstown (Table 1) and the limits in the eucrite Nuevo Laredo (Reed et al., 1960) are similar to the Bi concentrations in equilibrated ordinary chondrites.

Acknowledgments

This research was supported, in part, by the U.S. National Science Foundation, grant GA-1474.

References

Anders, E. (1964) *Space Sci. Rev.* **3**, 583.

Dodd, R.T., Van Schmus, W.R. and Koffman, D.M. (1967) *Geochim. Cosmochim. Acta* **31**, 921.

Ehmann, W.D. and Huizenga, J.R. (1959) *Geochim. Cosmochim. Acta* **17**, 125.

Grevesse, N., Blanquet, G. and Boury, A. (1968) *Origin and Distribution of the Elements* (L.H. Ahrens, ed.) 177–182, Pergamon.

Laul, J.C., Case, D.R., Schmidt-Bleek, F. and Lipschutz, M.E. (1970a) *Geochim. Cosmochim. Acta* **34**, 89.

Laul, J.C., Pelly, I.Z. and Lipschutz, M.E. (1970b) *Geochim. Cosmochim. Acta.* **34,** 909.

Mason, B. (1962) *Meteorites*, Wiley.

Mason, B. (1965) *Am. Museum Novitates* 2223.

Reed, G.W. (1963) *J. Geophys. Res.* **68**, 3531.

Reed, G.W., Kigoshi, K. and Turkevich, A. (1958) Paper 953 in *Proceedings of the Second International Conference on the Peaceful Uses of Atomic Energy, Geneva,* **28**, 486, United Nations (New York).

Reed, G.W., Kigoshi, K. and Turkevich, A. (1960) *Geochim. Cosmochim. Acta* **20**, 122.

Van Schmus, W.R. (1969) *Meteorite Research* (P.M. Millman, ed.) 480–491. Reidel.

Zähringer, J. (1968) *Geochim. Cosmochim. Acta* **32**, 209.

(Received 25 July 1969; revised 1 June 1970)

THORIUM (90)

John W. Morgan

Department of Chemistry
University of Kentucky,
Lexington, Kentucky

THORIUM IN NATURE is essentially mono-isotopic; the long-lived radionuclide thorium-232 has a half life of 1.41×10^{10} years (Farley, 1960). Other thorium isotopes exist only as part of the thorium-232, uranium-235 and uranium-238 decay series, and have not been studied in meteorites. In falls one would expect the short-lived thorium radionuclides to be in equilibrium with their long-lived ancestors, and their abundances are therefore accessible by calculation from a knowledge of the abundances of the parental elements. Disequilibrium probably exists in weathered finds, but it is not immediately apparent what purpose a study of this would serve.

The abundance of thorium in the early solar system material, together with that of uranium, plays a very special part in discussions of *r* process nucleosynthesis. The progenitors of this pair of elements lying in a region around atomic weight 250 are expected to have increased stability because of nuclear deformation (Fowler 1969), analogous to that found in the rare earth region. Clearly a comparison of calculated thorium abundances with those found empirically is a severe test of the tenet of a deformation hump between the closed shells at $N = 126$ and $N = 184$. The decay of thorium-232 since the formation of the elements requires that a chronological model be constructed before such a comparison can be made. Fowler and Hoyle (1960) used the observed uranium-238/uranium-235 ratio and an estimated thorium/uranium ratio together with their calculated ratios to construct a self-consistent time scale, based on a continuous synthesis model.

In view of the importance of the meteoritic thorium abundance, it is surprising that so little information has been available until very recently, especially for the rare and perhaps more significant carbonaceous and enstatite chondrites. Part of the reason for this may lie in the analytical problems involved. Van Dijk et al. (1953) used nuclear emulsions to measure the alpha activity of a stone meteorite (Monze), and calculated a thorium abundance of 0.1×10^{-6} g/g. Other studies by Dalton et al. (1953) and Dalton and Thomson (1954) used chemical pre-concentration followed by radiochemical measurements to investigate abundance in a number of iron meteorites. The results of these determinations all were systematically high. Reasbeck and Mayne (1955) reported thorium abundances for two chondrites and the Brenham pallasite. These results also tended to be high, though the value for Akaba (credited to Dalton) agrees with more recent determinations when the rather large experimental error is taken into account.

Patterson (1955) studied the lead abundance and isotopic composition in two chondrites and the Nuevo Laredo achondrite, and was able to estimate uranium and thorium contents from this measurements: these results are systematically high.

Bate et al. (1957, 1958, 1959) applied neutron activation to the problem of the analysis of meteorites for thorium, and established the abundances in ordinary H and L group chondrites, two iron meteorites, and in two achondrites.

Other applications of this method have been reported. In a series of investigations reported by Lovering and Morgan (1964, 1969) and Morgan and Lovering (1964, 1965, 1967, 1968) the distribution of thorium in all major groups of the chondritic and achondritic meteorites was studied. Wakita et al. (1967) reported three analyses of two ordinary chondrites.

The radioactivity of meteorites was measured by Rowe et al. (1963) by gamma ray spectrometry. In the spectra of some achondrites the 2.62 MeV photopeak of thallium-208 was observed, from which thorium abundances were estimated. The thorium content of chondrites is too low to be detected by the equipment used by Rowe et al. The gamma spectrometric method for the determination of potassium, thorium and uranium has been widely applied to the study of terrestrial rocks in which these elements are fairly abundant, and for which the large samples required are readily obtainable. In the study of meteorites, gamma spectrometry does not give thorium values with the precision and accuracy of the activation method. The sophisticated gamma spectrometer at the Lunar Receiving Laboratory has apparently been used to determine thorium in the Allende chondrite (King et al. 1969); however, one can scarcely imagine such an expensive instrument being used for a general study of thorium in meteorites.

Recently (Sen Gupta 1967) determined thorium in several chondritic and iron meteorites by a spectrophotometric method using Arsenazo III. This technique requires about 5 g of iron meteorite or 15 to 30 g of chondrite per determination. The results for thorium are high by a factor of two for the chondrites and by several orders of magnitude for the siderites.

The most reliable values for thorium in the chondritic meteorites appear to be the neutron activation results reported by Bate et al. (1957, 1959) and by Morgan and Lovering (1967, 1968). These abundances are assembled in Table 1, where the specimens are arranged according generally to the classification of Van Schmus and Wood (1967). Comparisons of the sets of results from the two groups of investigators are possible only for the ordinary chondrites, where agreement is very good. One result reported in the first study by Bate et al. (1967) for the Holbrook chondrite has been rejected as being probably due to contamination. This is a serious problem in the study of meteoritic thorium abundances, as this element is relatively concentrated in many common rocks and minerals. Morgan and Lovering (1968) critically evaluated their own analyses against objective criteria and were obliged to reject several of their results as erroneous because of probable contamination.

In most chondritic groups the analyses show good internal consistency, both for replicate analyses of the same meteorite and for different specimens within the same group. This is not the case for either the C1 or C3 groups. As discussed in detail by Morgan and Lovering (1968), the analyses for Bencubbin, Lancé and Warrenton are not above suspicion. Although there was no objective reason to reject these rather high results it was pointed out that the specimens were not prepared especially for the thorium study. There is clearly a need for a carefully conducted re-examination of the thorium abundance in the C3 groups chondrites.

In the C1 group most of the analyses agree well except for those of Orgueil, which show considerable scatter. The significance of this was discussed in detail by Morgan and Lovering (1967), who concluded that the inhomogeneity was probably real, and weighted the means accordingly.

Mean atomic abundances, relative to 10^6 atoms of silicon, have been calculated for the chondrite groups, and are summarized in Table 2. Standard silicon abundances (Mason, personal communication) have generally been used. In the C1 and C2 groups the weighted mean values given by Morgan and Lovering (1967) have been retained for the sake of consistency. Non-standard silicon values were also used for Renazzo and Saint Marks, where silicon abundances of 15.8 and 18.0 per cent, respectively, were employed.

The mean abundances for each petrological group within a particular chemical class are very similar, except for the C3 group. In view of the doubts

TABLE 1 Thorium abundances in chondritic meteorites

Group	Chondrite	Number of analyses	Thorium, 10^{-9} g/g	References
C1	Alais	2	37	(3)
	Ivuna	6	29	(3)
	Orgueil	8	53	(3)
	Tonk	2	31	(3)
C2	Cold Bokkeveld	2	40	(3)
	Mighei	2	46	(3)
	Murray	2	45	(3)
	Nawapali	2	38	(3)
	Renazzo	4	41	(3)
	Staroe Boriskino	2	39	(3)
C3	*Bencubbin (inclusion)**	2	(76)†	(3)
	Lancé	2	120	(3)
	Mokoia	2	61	(3)
	Warrenton	1	79	(3)
C4	Karoonda	2	57	(3)
E4	Abée	2	30	(3)
	Indarch	2	29	(3)
E5	Saint Marks	4	30	(3)
E6	Hvittis	2	31	(3)
	Khairpur	2	41	(3)
	Pillistfer	2	42	(3)
H4	Forest Vale	2	42	(3)
	Ochansk	2	41	(3)
H5	Allegan	2	39	(3)
	Beardsley	4	41	(1), (2), (3)
	Pultusk	2	41	(3)
	Richardton	4	34	(1), (2), (3)
H6	Mount Browne	2	41	(3)
	Zhovtnevyi	2	42	(3)
L3	Khohar	2	44	(3)
L5	Farmington	2	38	(3)
	Homestead	2	40	(3)
	Knyahinya	2	49	(3)
L6	Holbrook	1	(90)	(1)
		4	41	(2), (3)

TABLE 1 (cont.)

Group	Chondrite	Number of analyses	Thorium, 10^{-9} g/g	References
L6	Modoc	3	41	(1), (2)
	Perpeti	2	40	(3)
	Saint Michel	2	44	(3)
LL3	Chainpur	2	43	(3)
	Ngawi	2	43	(3)
LL6	Bandong	2	50	(3)
	Benares	2	44	(3)
	"Bialystok"‡	2	50	(3)

* Meteorite names in italics represents find.

† Values in brackets have been omitted in the calculation of mean atomic abundances in Table 3.

‡ Bialystok is generally considered to be a howardite, but the sample so labelled in reference 3 is an amphoterite and was assigned to the LL6 class.

(1) Bate et al (1957).
(2) Bate et al. (1969)
(3) Lovering and Morgan (1964); Morgan and Lovering (1967, 1968)

expressed earlier concerning the control of several C3 group specimens, this apparent anomaly may be an artifact.

The mean abundances of the three chemical classes if the ordinary chondrites (the H, L and LL groups) are identical. Between the chemical groups, thorium abundances decrease in the order C > (H, L, LL) > E, a relationship which holds even if the doubtful C3 specimens are excluded.

Calculations of the solar system abundance of thorium based on nuclear systematics applied to the *r* process have been made by Hoyle and Fowler (1963) and Seeger et al. (1965). To relate thorium to silicon, which is not formed by the *r* process, a normalization factor r' was estimated. Clayton (1963) derived an expression for r' based on the observed solar abundance of lead. Clayton adopted the Helliwell (1961) lead abundance of 2.5 atoms/10^6 silicon atoms for his calculations. Müller (1968) reviewed solar abundances and concluded that the value of 1.4 reported by Mutschlecner (1962) was a superior estimate of the solar lead abundance. The thorium value has been recalculated, following Clayton, by Morgan and Lovering (1967) using this lower lead value.

TABLE 2 Atomic abundances of thorium in chondritic classes

Group	Number of chondrites analysed	Total number of analyses	Mean Th/ 10^6 Si
C1	4	18	0.043*
C2	6	14	0.038*
C3	3	5	0.068
C4	1	2	0.044
All C	14	39	0.046
E4	2	4	0.022
E5	1	4	0.021
E6	3	6	0.024
All E	6	14	0.022
H4	2	4	0.029
H5	5	15	0.028
H6	2	4	0.029
All H	9	23	0.029
L3	1	2	0.028
L5	3	6	0.027
L6	4	11	0.027
All L	8	19	0.027
LL3	2	4	0.028
LL6	3	6	0.030
All LL	5	10	0.030
All ordinary chondrites	22	49	0.028
All chondrites	42	105	0.034

* Weighted average from Morgan and Lovering (1967).

Recently Grevesse (1969) measured several heavy elements in the solar spectrum. A value of $r' = 0.325$ estimated from the lead abundance is in good agreement with the number adopted by Fowler and Hoyle (1963). A direct measurement of the thorium abundance in the sun was also made by Grevesse, who feared that the observed spectral abundance may be under-estimated, due to thermal and gravitational diffusion at the bottom of the convective zone, and applied an arbitrary correction factor of ~ 1.5.

In Table 3, the calculated and spectroscopic results are compared with chondritic values. There is clearly a wide descrepancy between the meteoritic abundances and other estimates. Even the lowest calculated abundance is about twice as high as the highest chondritic group average. The activation method for thorium has been widely applied to rocks and agrees well with other methods (for example Morgan and Heier 1966). It is very improbable that the activation results are low by a factor of three. The resolution of this wide disagreement is a pressing problem for meteoriticists and nuclear astrophysicists alike.

The activation method has been applied to most achondrite classes (Bate et al. (1957, 1959; Morgan and Lovering 1964; Lovering and Morgan 1969). These results are compiled in Table 4. Some results by gamma spectrometry (Rowe et al. 1963) have been reported. These have been listed separately in Table 4. These values are not of the same precision and accuracy as the

TABLE 3 Comparison of calculated, spectral and chondritic estimates for the solar abundance of thorium

Estimate		Thorium/ 10^6 silicon
r process calculation		
Hoyle and Fowler (1963)		0.13
Seeger et al. (1965)*		0.20
Clayton (1963) (Lead = 2.5)†		0.18
Morgan and Lovering (1967) (Lead = 1.4)‡		0.099
Solar spectroscopy		
Grevesse (1969)		0.19 [0.28]#
Chondrites		
Morgan and Lovering (1967)	C1 and C2	0.040
This work	All C	0.046
	All E	0.022
	All ordinary chondrites	0.028
	All chondrites	0.034

* Value given by Fowler (1969), apparently recalculated from the original work.

† Lead value from Helliwell (1969).

‡ Lead value from Mutschlecner (1962).

\# Corrected for thermal and gravitational diffusion at the bottom of the convective zone.

TABLE 4 Abundance of thorium in achondrites

Group	Achondrite	Activation: Number of analyses	Activation: Th, 10^{-9} g/g	Activation: Ref.	Spectrometry: Th, 10^{-9} g/g	Spectrometry: Ref.
Angrite	Angra dos Reis	2	970	(4)		
Aubrites	Bishopville	5	47	(4), (5)		
	Norton County	4	4	(5)		
	Mean	9	28			
Diogenites	Ellemeet	2	4	(4)		
	Johnstown	2	6	(1), (2)		
		2	30	(4)		
	Shalka	2	20	(5)		
	Mean	8	15			
Eucrites	Bereba	4	340	(5)		
	Emmaville	2	370	(5)		
	Juvinas	2	370	(5)	600	(3)
	Luotolax*	2	310	(5)		
	Nuevo Laredo	6	490	(1), (2), (4)	470	(3)
	Pasamonte				520	(3)
	Sioux County				350	(3)
	Stannern	2	680	(5)	500	(3)
	Mean	18	430		490	
	Mean (without Luotolax)	16	450			
Eucrites (unbrecciated)	Moore County	2	62	(4)		
	Serra de Mage	2	53	(5)		
	Mean	4	58			
Howardites	Binda†	2	63	(4)		
	Frankfort	2	120	(5)		
	Mean	4	94			
	Mean (Frankfort and Luotolax)	4	220			
Nakhlite	Nakhla	2	190	(4)		
Sherghottite	Shergotty	2	470	(5)		
Ureilite	*Goalpara*‡	2	3	(5)		

activation results and have not been included in the general averages. The classification followed in Table 4 is that of Hey (1966).

In many achondrite groups the variation in abundance is large and the number of analyses small, so that averages are not especially meaningful. In one case (Johnstown) analyses on different pieces of the same fall vary by a factor of five. This is unlikely to be analytical error, as duplicates are internally consistent.

In the eucrites a reasonable number of analyses are available. There is a bimodal distribution in thorium abundances in this group. The unbrecciated eucrites are characterized by low abundances. The remaining eucrites are brecciated and have high thorium contents. There is some uncertainty about the classification of some calcium-rich achondrites, which may influence the discussion of thorium distribution. This is particularly important in the case of the howardites. If the Hey (1966) classification is followed, there are thorium analyses available for two howardites, Binda and Frankfort. If these are taken to be typical of all howardites the mean abundance in this group lies between the two classes of eucrite.

In a detailed study of the eucrites Duke and Silver (1967) found that Binda was a monomict breccia, and classified it as a eucrite. However, the mineralogy is distinct from that of other brecciated eucrites, and it must be regarded as a unique type. The low thorium abundance is in agreement with other trace element abundances reported by Duke and Silver. The chemical composition of Frankfort is not typical of howardites, as it contains an unusually low amount of calcium. It is possible that the thorium abundance is also atypically low. The achondrite Luotolax has been reclassified as a howardite by Duke and Silver (1967) and may provide the best estimate presently available for the thorium abundance in that group.

There are a few reliable analyses available for the thorium content of the olivine phase of the pallasites (Bate et al. 1957; Lovering and Morgan 1969). These are summarized in Table 5. The abundances are extremely low and

* Classified as a howardite by Duke and Silver (1967).

† Classified as a monomict brecciated eucrite of unusual composition by Duke and Silver (1967).

‡ Name in italics indicates a find.

(1) Bate et al. (1957)
(2) Bate et al. (1959)
(3) Rowe et al. (1963)
(4) Morgan and Lovering (1964)
(5) Lovering and Morgan (1969)

there is an apparent discrepancy between the two sets of analyses of Brenham. This meteorite is a find, and terrestrial contamination of some samples by at least the rare earth elements has been demonstrated (Masuda 1968). The difference in thorium content may be a reflection of real inhomogeneity, though the potassium content between different samples of Brenham olivine has been shown to be rather uniform (Morgan and Goode 1966).

TABLE 5 Thorium abundances in pallasite olivines

Pallasite olivine	Number of analyses	Th, 10^{-9} g/g	References
Brenham	1	11	(1)
	2	3	(2)
Huckitta	2	2	(2)

(1) Bate et al. (1957)
(2) Lovering and Morgen (1969)

Only one neutron activation study of the distribution of thorium in iron meteorites has been made; Bate et al. (1958) examined a coarsest octahedrite and a brecciated hexahedrite. To ensure that protactinium-233 was actually being detected they counted the uranium X-ray region, the 310 keV gamma region and measured the gross beta activity. The results are shown in Table 6.

The preceding summary indicates that the thorium abundance in chondrites is well established. A re-examination of the abundances of the C3 chondrites is needed to verify the reportedly high values. Other chondrite groups lacking thorium analyses are the H3, L4, and LL4 and 5. It is not likely that these will differ significantly from the averages in Table 2. In the achondrites there

TABLE 6 Thorium abundances in iron meteorites (Bate et al. 1958)

Group	Meteorite	Th, 10^{-12} g/g		
		100 keV U X-ray	310 keV Pa γ-ray	Gross beta
Coarsest octahedrite	Arispé	18	16	11
		6	6	5
Brecciated hexahedrite	Sandia Mountains	10	11	9
	Mean	11	11	8

is a reasonable coverage of the eucrites, but other groups are only sparsely represented. In particular more analyses of typical howardites are required.

Although the iron meteorites have not been widely investigated, the thorium abundances appear extremely low and there would have to be strong motivation to justify the undertaking of any large number of difficult analyses involved. A survey of the thorium abundances in troilites would be interesting; information on this phase appears to be completely lacking at present.

References

Bate, G.L., Huizenga, J.R. and Potratz, H.A. (1957) *Science* **126**, 612.

Bate, G.L., Potratz, H.A. and Huizenga, J.R. (1958) *Geochim. Cosmochim. Acta* **14**, 118.

Bate, G.L., Huizenga, J.R. and Potratz, H.A. (1959) *Geochim. Cosmochim. Acta* **16**, 88.

Clayton, D.D. (1963) *J. Geophys. Res.* **68**, 3715.

Dalton, J.C., Golden, J., Martin, G.R., Mercer, E.R. and Thomson, S.J. (1953) *Geochim. Cosmochim. Acta* **3**, 272.

Dalton, J.C. and Thomson, S.J. (1954) *Geochim. Cosmochim. Acta* **5**, 74.

Duke, M.B. and Silver, L.T. (1967) *Geochim. Cosmochim. Acta* **31**, 1637.

Farley, T.A. (1960) *Can. J. Phys.* **38**, 1059.

Fowler, W.A. (1969) *The role of neutrons in astrophysical phenomena.* Preprint.

Fowler, W.A. and Hoyle, F. (1960) *Ann. Phys.* **10**, 280.

Grevesse, N. (1969) *Solar Physics* **6**, 381.

Helliwell, T.M. (1961) *Astrophys. J.* **133**, 566.

Hey, M.H. (1966) *Catalogue of Meteorites* (3rd edition) British Museum.

Hoyle, F. and Fowler, W.A. (1963) "On the abundance of uranium and thorium in Solar System material". *Isotopic and Cosmic Chemistry*. (Craig, H., Miller, S., Wasserburg, G.J., eds.) North-Holland.

King, E.A., Jr., Schonfeld, E., Richardson, K.A. and Eldridge, J.S. (1969) *Science*, **163**, 928.

Lovering, J.F. and Morgan, J.W. (1964) *J. Geophys. Res.* **69**, 1979.

Lovering, J.F. and Morgan, J.W. (1969) *Uranium and thorium in achondrites.* In press.

Masuda, A. (1968) *Earth Planet. Sci. Letters* **5**, 59.

Morgan, J.W. and Goode, A.D.T. (1966) *Earth Planet. Sci. Letters* **1**, 110.

Morgan, J.W. and Heier, K.S. (1966) *Earth Planet. Sci. Letters* **1**, 158.

Morgan, J.W. and Lovering, J.F. (1964) *J. Geophys. Res.* **69**, 1989.

Morgan, J.W. and Lovering, J.F. (1965) *J. Geophys. Res.* **70**, 2002.

Morgan, J.W. and Lovering, J.F. (1967) *Nature* **213**, 873.

Morgan, J.W. and Lovering, J.F. (1968) *Talanta* **15**, 1079.

Müller, E.A. (1968) "The solar abundances." *Origin and Distribution of the Elements.* (L.H. Ahrens, ed.) Pergamon Press. p. 156.

Mutschlecner, P. (1962) Thesis, University of Michigan.

Patterson, C.C. (1955) *Geochim. Cosmochim. Acta* **7**, 151.

Reasbeck, P., and Mayne, K.I. (1955) *Nature* **176**, 186.
Rowe, M.W., van Dilla, M.A. and Anderson, E.C. (1963) *Geochim. Cosmochim. Acta* **27**, 983.
Seeger, P.A., Fowler, W.A. and Clayton, D.D. (1965) *Astrophys. J. Supplement No. 97*, XI, 121.
Sen Gupta, G.A. (1967) *Anal. Chem.* **39**, 18.
Van Dijk, T., de Jager, C., and de Metter, J. (1953) *Mem. 8⁰ de la soc. roy. sci. Liège 4th series* **13** Fasc. III 495.
Van Schmus, W.R. and Wood, J.A. (1967) *Geochim. Cosmochim. Acta* **31**, 747.
Wakita, H., Nagasawa, H., Uyeda, S. and Kuno, H. (1967) *Earth Planet. Sci. Letters* **2**, 377.

(Received 12 July 1969)

URANIUM (92)

John W. Morgan

Department of Chemistry,
University of Kentucky,
Lexington, Kentucky

URANIUM HAS two long-lived isotopes which occur in nature, uranium-235 and uranium-238, of which the latter is by far the most abundant at the present time. Uranium-234 is almost invariably present, as a decay product of uranium-238. The isotopic abundances and half-lives of these three nuclides are listed in Table 1. In meteorite falls it is safe to assume that

TABLE 1 Half lives and atomic abundances of the naturally occurring uranium isotopes (Lederer et al. 1968)

Isotope	Atomic abundance per cent	Half life years
^{234}U	0.006*	2.47×10^5
^{235}U	0.7	7.1×10^8
^{238}U	99.3	4.51×10^9

* Equilibrium value

uranium-234 is in secular equilibrium with its parent radionuclides; disequilibrium may exist in some finds as a result of terrestrial weathering.

The abundances of the two long-lived uranium isotopes in early solar system material have a relevance to *r* process nucleosynthesis similar to that of thorium (Fowler 1969; Fowler and Hoyle 1960). This has been discussed in the chapter on thorium and will not be repeated here.

The abundance of uranium in meteorites has been of interest for many years because of the cosmochronological implications. Davis (1950) used a vacuum fusion technique to extract radon from a number of samples of meteoritic material, and from these measurements derived radium and uranium abundances. A nuclear emulsion technique for the measurement of alpha particle emission was employed by Van Dijk et al. (1953) to estimate the uranium and thorium content in a chondrite. The results of these two radiometric techniques have since been shown to be rather high.

A considerable effort was expended by Paneth (1953) and his coworkers on the study of uranium, thorium and helium abundances in iron meteorites. Uranium abundances were determined by fluorimetry or radon counting (Dalton et al. 1953; Dalton and Thomson 1954). The results for the iron meteorites have since been shown to be too high, though the uranium abundance in the Akaba chondrite (Reasbeck and Mayne 1955) which was also analysed fluorimetrically by Dalton agrees very well with later determinations on the same meteorite.

Several analyses for uranium in meteoritic material were made by stable isotope dilution by Tilton (Patterson et al. 1953); this method is capable of very good precision and accuracy, and these results agree well with later work. Neutron activation analyses on a number of stony and iron meteorites were reported by Starik and Shats (1956). Their technique was rather an unusual one in that the uranium was separated *before* neutron irradiation, the actual assay being performed by counting fission fragments during neutron bombardment. In principle this method should be a very sensitive one; however, the preliminary separation introduces the possibility of contamination, and loses one of major advantages of conventional neutron activation analyses. This is very probably the reason that these results are much higher than those found in more recent studies. In a later investigation of the lead-uranium ages of meteorites Starik et al. (1958) reverted to a fluorimetric technique to determine the uranium abundance in several meteoritic types. These results are also high by an order of magnitude or more compared to values now accepted generally.

Hamaguchi et al. (1957) used two different methods of neutron activation to determine uranium in the ordinary chondrites. One was based on the fission of uranium-235, and used barium-140 for the indicator radionuclide. The other used the reaction

$$^{238}\text{U}\,(\text{n}, \gamma) \quad ^{239}\text{U} \xrightarrow{\beta-} \quad ^{239}\text{Np}$$

which is inherently more sensitive than the fission product approach, although the carriere-free radiochemical separations need considerably more skill to perform. The fission product method was employed by Hernegger and Wänke (1957) and Ebert et al. (1957), the former group using barium-140 and the latter a novel method based on xenon-133. The results of all these analyses established a uranium abundance in the ordinary chondrites of about 10×10^{-9} g/g. This value was given further confirmation by König and Wänke (1959), who also used the xenon-133 method.

Reed et al. (1960) analysed two carbonaceous and two enstatite chondrites by the barium-140 method. Analyses for lead made on the same meteorites required irradiation in a fast neutron flux. Reed et al. consider the accuracy of the new results to be inferior to their earlier activation work (Hamaguchi et al. 1957), due to interference by the fission of thorium-232. Reed (1963), Wakita et al. (1967), and Lavrukhina et al. (1968) have made further analyses by the same method. A modification of the fission product type of analysis has been applied to the simultaneous determination of iodine, tellurium, and uranium, all of which give rise to iodine radionuclides as a result of neutron irradiation (Goles and Anders 1962; Reed and Allen 1966; Clark et al. 1967). This method has the advantage that the three elements are obtained by the radiochemical separation of only one element.

On the other hand, the activity due to uranium must be unscrambled from a multicomponent decay curve, which impairs the sensitivity of the determination. As a result, large samples must be used and the reported precision is about 10 per cent at best and frequently 20 per cent or more. The value of this method as far as uranium is concerned is considerably less than the barium and xenon methods, and is far inferior to the neptunium-based technique. Nevertheless, when carefully applied, the accuracy of the radio-iodine determinations is comparable with other analyses by more precise methods (Goles and Anders 1962).

Another facet of the fission process has been applied to the analysis of uranium in chondrites by Amiel et al. (1967). Some short-lived fission product nuclides transmute by beta decay to highly excited states which in turn rapidly decay by neutron emission. If samples are counted by BF_3 detectors a short time (typically 25 to 30 sec) after neutron irradiation, the integral count is proportional to the amount of fissile material in the sample. This method has been shown (Gale, 1967) to be very accurate for uranium at the levels found in common terrestrial rocks; however, for chondritic abundances the precision is poor (10–15 per cent). The results reported by Amiel et al. have a bias towards high values which may partly be due to the large number of finds analysed. High results are also obtained if the neutron detectors have insufficient shielding against gamma rays (Gale, 1967). A major advantage cited for the delayed neutron method is that it is "non-destructive", in that

TABLE 2 Uranium abundances in chondritic meteorites

Group	Chondrite	Number of analyses	U, 10^{-9} g/g	References
C1	Alais	2	10	(14)
	Ivuna	8	9	(10), (14)
	Orgueil	10	17	(7), (14)
	Tonk	2	11	(14)
C2	Cold Bokkeveld	2	11	(14)
	Mighei	6	15	(7), (8), (10), (14)
	Murray	5	14	(7), (14)
	Nawapali	2	11	(14)
	Renazzo	4	12	(14)
	Staroe Boriskino	2	11	(14)
C3	*Bencubbin** (inclusion)	2	(19)†	(14)
	Lancé	4	19	(9),, (10), (14)
	Mokoia	2	14	(14)
	Warrenton	1	18	(14)
C4	Karoonda	2	14	(14)
E4	Abée	7	11	(7), (8), (14)
	Indarch	7	11	(7), (8), (10), (14)
E5	Saint Marks	5	8	(8), (14)
E6	Hvittis	2	6	(14)
	Khairpur	2	16	(14)
	Pillistfer	2	6	(14)
H4	Forest Vale	2	11	(14)
	Ochansk	2	11	(14)
H5	Allegan	4	11	(10), (14)
	Beardsley	6	11	(6), (7), (14)
	Beddgelert	2	25	(4), (6)
	Plainview	2	(13)	(8)
	Pultusk	4	12	(5), (6), (14)
	Richardton	8	12	(3), (8), (11), (14)
	Stalldalen	1	11	(8)
H6	Mount Browne	2	12	(14)
	Zhovtnevyi	2	15	(14)
H	Breitscheid	4	13	(4), (5), (6)
L3	Khohar	2	13	(14)
L4‡	*Barratta*	2	(25)	(11)
	Goodland	1	(18)	(11)
	McKinney	1	(16)	(11)

TABLE 2 (cont.)

Group	Chondrite	Number of analyses	U, 10^{-9} g/g	References
L5	*Arapahoe*	2	(18)	(11)
	Ergheo	1	21	(8)
	Farmington	3	13	(11), (14)
	Homestead	2	12	(14)
	Knyahinya	2	19	(14)
	Wickenburg	1	(25)	(11)
L6#	Akaba	3	8	(2), (4), (5)
	Bruderheim	9	15	(8), (10), (11), (12), (13), (15)
	Holbrook	11	15	(3), (7), (14), (15)
	Mocs	1	11	(8)
	Modoc	4	11	(1), (3), (6)
	Peace River	2	24	(13)
	Perpeti	2	11	(14)
	Saint Michel	2	16	(14)
LL6	Chainpur	2	14	(14)
	Ngawi	2	11	(14)
	Bandong	2	13	(14)
	Benares	2	12	(14)
	"Bialystok"+	2	13	(14)

* Meteorite names in italics represent finds.

† Values in brackets have been omitted in the calcultation of means.

‡ The silicate phase of the L4 chondrite Saratov has been analysed by Vinogradov (1961) who finds 8.1 and 6.7×10^{-6} g/g respectively in duplicate samples. These values have not been included in the whole rock chondrite averages.

Lavrukhina et al. (1968) report "0.0011 ± 0.004 ppm" for the L6 chondrite Kunushak. As the standard deviation is much larger than the actual value it is suspected that there is a typographical error and the true value is probably 11×10^{-9} g/g.

\+ Bialystok is generally considered to be a howardite, but the sample analysed in reference (14) is undoubtedly an amphoterite and has been assigned to the LL6 group.

damage is limited to that caused by short neutron irradiation. This is true, but several g of sample are needed, whereas using the neptunium method a precise and accurate analysis can be made on 100 mg or less of material.

A protracted study of uranium and thorium abundances in chondrites has been made by Lovering and Morgan (1964) and Morgan and Lovering (1967, 1968). The neptunium-239 method used for the uranium analysis is generally conceded to be the most sensitive activation method available at present.

This work, summarized in the 1968 paper, reported over 100 analyses of 44 different chondritic specimens. A critical evaluation of the results led to the rejection of 16 of the analyses as probably being in error due to terrestrial contamination. The results of chondritic uranium analyses thought to be reasonably reliable are listed in Table 2.

The abundances for the ordinary chondrites generally agree with the value of 10×10^{-9} g/g established by Hamaguchi et al. (1957), Hernegger and Wänke (1957), and Ebert et al. (1957). Many of the results for the finds are high, and should be rejected. The problem of terrestrial contamination for uranium is a serious one. The chondrite Beddgelert yielded a high uranium analysis, but similar results were found in two different studies, so that it is likely that this is a true determination of the abundance. The specimen of Peace River analysed by Clark et al. (1967) also gives a high abundance. The duplicates agree reasonably well, so this may be real. A single analysis of Ergheo by Goles and Anders (1962) is high and needs to be confirmed by further analysis. Three analyses, by separate groups, of the chondrite Akaba give very reproducible results averaging 8×10^{-9} g/g. This is significantly lower than the average abundance than any other ordinary chondrite, and is more the sort of level found in the enstatite chondrites.

If the abundances in different chondrite groups are to be compared, it is necessary to reduce all the weight analyses to atomic abundances. Following the usual convention, these have been calculated relative to 10^6 atoms of silicon.

The work of Morgan and Lovering (1968) covered more groups than have been investigated in all other studies combined. Averages calculated from these results are more directly comparable than those resulting from analytical data from a variety of different methods. In addition standard deviations for the same chondritic groups are generally lower for this one study that those derived from all analytical data. On the other hand it might be felt that the use of all analytical results gives a wider sampling coverage and is therefore preferable. In Table 3 average atomic abundances for the chemical-petrologic classes of chondrite are shown, for the results of Morgan and Lovering (1967) alone, and also for all the analyses listed in Table 2.

The agreement for the two sets of averages is very good. This is not entirely unexpected as many of the fission product results have been shown to be accurate, even where the precision has been only fair. In addition over half of all the analyses listed were made by Morgan and Lovering. If averages are being compared it can be seen that it makes little difference which set of numbers is used.

TABLE 3 Atomic abundances of uranium in chondritic meteorites

Group	(Morgan and Lovering, 1967, 1968)			All analyses in Table 2		
	Number of chondrites	Number of analyses	U/10^6 Si	Number of chondrites	Number of analyses	U/10^6 Si
C1	4	18	0.013	3	22	0.013
C2	6	14	0.011	6	21	0.011
C3	3	5	0.014	3	7	0.013
C4	1	2	0.010	1	2	0.010
All C	14	39	0.012	14	52	0.012
E4	2	4	0.007	2	14	0.008
E5	1	4	0.005	1	5	0.005
E6	3	6	0.006	3	6	0.006
All E	6	14	0.006	6	24	0.007
H4	2	4	0.008	2	4	0.008
H5	4	8	0.008	6	25	0.010
H6	2	4	0.009	2	4	0.009
All H	8	16	0.008	11	36	0.009
L3	1	2	0.008	1	2	0.008
L5	3	6	0.009	4	8	0.010
L6	3	6	0.008	8	34	0.009
All L	7	14	0.009	13	44	0.009
LL3	2	4	0.008	2	4	0.008
LL6	3	6	0.008	3	6	0.008
All LL	5	10	0.008	5	10	0.008
All ordinary chondrites	20	40	0.009	29	90	0.009
All chondrites	40	93	0.009	49	166	0.010

A comparison of the mean values of the ordinary chondrites shows clearly that the abundances in the H, L, and LL groups respectively are very uniform. Between the other major chemical groupings the abundances decrease in the order C > (H, L, LL) > E. Within the E, H, L and LL groups there is little or no variation with petrological type. In the C group the situation is more complicated. The difference in average abundance between the C1 and C2 groups has been shown to be insignificant statistically (Morgan and Lovering, 1967). As far as can be gathered from the duplicate analysis of a single sample, the abundance in the C4 group is also very similar. Uranium is apparently enriched in the C3 group. This mean abundance is largely based on the results of Morgan and Lovering (1968), who expressed some reservation about the control of the samples used. Their high value was supported by a single result for Lancé reported by Reed (1963). Later work on the same meteorite by Reed and Allen (1967) yielded a very much lower answer, more in keeping with the ordinary or enstatite chondritic abundances. It is clear that a systematic investigation of uranium in the C3 group using carefully selected specimens and employing a sensitive and accurate analytical method is urgently required.

Uranium analyses have been carried out on some separated phases of chondritic meteorites. The discovery of large enrichments of rare gases in the dark parts of meteorites showing "light-dark" structure (König et al., 1961) has led to investigations of other trace element distributions between the two phases. The distribution of uranium has been studied by Reed (1963), Nix and Kuroda (1967), and Clark et al. (1967) in several chondrites showing this type of structure. The results of these studies are shown in Table 4.

TABLE 4 The distribution of uranium between the light and dark parts of chondrites

Group	Chondrite	Uranium, 10^{-9} g/g		Reference
		light	dark	
H5	Leighton	51; 18	47; 22	(3)
	Pantar	13	14	(1)
H	Fayetteville	16; 188	29; 69	(3)
		29; 33; 32	23; 22; 18	(2)
LL6	Ensisheim	17	16	(1)

(1) Reed (1963)
(2) Clark et al. (1967)
(3) Nix and Kuroda (1967)

Clark et al. also list an analysis for the dark part of Cumberland Falls. This meteorite is a whitleyite, a breccia containing fragments of white aubrite and black chondrite, and does not show light-dark structure in the sense usually accepted. The abundance of 2×10^{-9} g/g found in the chondrite part of Cumberland Falls is very much lower than that of other chondrites, if one doubtful result for Hvittis (Reed and Allen 1967) is excluded.

The picture presented by the analyses of the chondritic light and dark parts is not a clear one. Reed's results for Pantar and Ensisheim indicate that the uranium abundances in the light and dark parts are very similar, and if a possible positive analytical bias of about 30 per cent is considered, in good agreement with analyses of other chondrites which do not show a light-dark structure. The analyses by Clark et al. (1967) show such wide variation even between duplicates of the same phase that valid comparisons are not possible. The results are generally higher than those found by Reed (1963) and one is strongly inclined to suspect a contamination problem. The values reported by Nix and Kuroda (1967) for the Fayetteville chondrite are much more precise than those of Clark et al. The dark parts analyses by Nix and Kuroda are just within the range of the abundances found by other workers in the ordinary chondrites, but the uranium content of the light parts appears rather high.

Studies of the distribution of uranium in chondritic phases have been made possible at a uniquely high resolution by a neutron activation technique based on the formation of fission tracks in mica or plastic film in contact with a surface of the meteorite under investigation. In an early study Hamilton (1966) showed that much of the uranium in the Saint Marks enstatite chondrite is isolated in small segregations of very high uranium content. A more quantitative examination made by Fleischer (1968) provided a wealth of information on the dispersion of uranium throughout chondritic minerals and phases. Examination of another enstatite chondrite, Abbe, confirmed the presence of segregations of areas of high uranium abundance. In three L3 chondrites investigated a correlation was found between the dispersion of the iron content of the olivines (Dodd et al. 1967) and the dispersion of uranium in chondrules. Abundances found in chondrites by Fleischer are summarized in Table 5.

Large variations in abundance are apparent; Fleischer (1968) points out that none of the single regions studied (equivalent to 0.2 μg) has the same abundance for the average of the whole chondrite. It is not surprising, therefore, that some of the mean chondrite values differ from the bulk analyses mady by other neutron activation methods. Generally the fission track values tend to be high. This is possible due in part to sampling, and perhaps to a subjective inclination to study the more interesting areas.

Like thorium, the question of the solar system abundance of uranium is one that is at present unresolved. A consistent chondritic value for uranium has now been obtained, but this has proved incompatible with calculations based on *r* process nucleosynthesis (Hoyle and Fowler 1963; Clayton 1963;

Seeger et al. 1965). Direct spectral observation has provided only an upper limit for the solar abundance of uranium (Grevesse 1969). The abundance calculated by Clayton (1963) is based upon the solar abundance of lead. The value of 2.5 lead atoms/10^6 silicon atoms reported by Helliwell (1961) was

TABLE 5 Uranium distribution in chondrite phases by fission tracks (Fleischer 1968)

Group	Chondrite	Uranium, 10^{-9} g/g			
		Mean	Chondrules	Matrix	Other features
C2	Murray	195		400 (opaque)	19 (crystals)
E4	Abée	17			
L3	Bishunpur	19	98; 14	<5	
	*Ioka**	14	20 (olivine) 61 (crystal) boundaries) 15; 25	23 (opaque)	
	Khohar	31	95†; 33; 98	26 (opaque)	
	Mezo-Madaras	31	36 (olivine) 230 (opaque areas)		
L4	*Barratta*			6 (opaque) 75 (transparent)	
L6	Bruderheim	20		20 (general fine grained areas)	400 (opaque crystal) 4,000 (unidentified) region) 600,000 (crystal)

* Chondrite name in italics indicates a find.

† This chondrule was examined in considerable detail by Fleischer (1968); the original work gives the full analysis.

used for the original calculation. Recently Müller (1968) reviewed solar abundances and proposed that the value of 1.4 for the solar abundance of lead (Mutschlecner 1962) was the best presently available. Following Clayton, Morgan and Lovering (1967) recalculated the solar uranium abundance using the lower lead value. The solar lead abundance was also measured by Grevesse (1969), who calculated a normalization factor, r', in essential agreement with the value adopted by Hoyle and Fowler (1963). The calculated, spectroscopic, and chondritic abundances are compared in Table 6.

TABLE 6 Comparison of calculated, spectroscopic and chondritic estimates for the solar abundance of uranium

Estimate	Uranium/ 10^6 silicon
r process calculation	
Hoyle and Fowler (1963)	0.034
Seeger et al. (1965)*	0.049
Clayton (1963) (lead = 2.5)†	0.048
Morgan and Lovering (lead = 1.4)‡	0.026
Solar spectroscopy	
Grevesse (1969)	≤0.11
Chondrites	
C1 and C2 (Morgan and Lovering 1967)	0.012
All C	0.012
All E	0.007
All ordinary chondrites	0.009
All chondrites	0.010

* Value given by Fowler (1969) apparently recalculated from the original work.
† Lead value from Helliwell (1961).
‡ Lead value from Mutschlecner (1962).

It can be seen that all the chondritic group averages are in serious disagreement with the calculated values. A large number of chondrite uranium analyses have been made, and it seems unlikely that any significant change will come about as a result of further analyses.

Many of the studies of the abundance of uranium in chondrites have included the analysis of one or two achondrites (Patterson et al., 1955; Hamaguchi et al., 1957; König and Wänke, 1969; Reed et al., 1960; Nix and Kuroda, 1967). In addition, Bate and Huizenga (1963) studied ruthenium in a number of meteorites by neutron activation, and, from the fission-product ruthenium activity, were able to derive the uranium abundance in the Johnstown diogenite. The only large scale investigations of uranium in achondritic meteorites which have been reported to date have been by Lovering and Morgan (1969; Morgan and Lovering 1964) and by Clark et al. (1967). One of the authors of the latter work has questioned the quality of the analyses, (Kuroda 1969). The method used by Clark et al. requires resolution of a complex decay curve, and, according to Kuroda insufficient time was available to follow the decay, so that interference from activities due to tellurium could have caused serious errors. Kuroda believes that this may be the reason for the large scatter in the results by Clark et al. (1967). In view of Kuroda's criticism Fisher (1969) has used a fission track technique to check the abundances in certain meteorites analysed by Clark et al. Though the use of fission track counting introduces a sampling problem, the results indicate that some of the results reported by Clark et al. may be in error. With this in mind, these values have been critically evaluated, and accepted only if they satisfy two of the three following criteria;

a) replicate determinations agree within 20 per cent

b) the determinations agree well with those by other workers on the same or similar meteorites

c) the tellurium abundance is low (about 0.1×10^{-6} g/g or less).

All the available analyses for uranium in the achondritic meteorites are collected in Table 7, where they are classified according to Hey (1966).

There is a reasonably representative coverage of most classes of achondrites, although the howardite analyses are few and possibly not of very typical specimens.

The aubrites form a consistent grouping in uranium abundance, which is in marked contrast to the large scatter observed in the case of thorium. The Norton County value by Tilton (1951) has been recalculated from the original analytical data and is not the result given by Patterson et al. (1955). This has been discussed more fully elsewhere (Morgan 1965).

The abundances in the diogenites are not clearly defined. Three of the values reported for Johnstown agree well, and a duplicate analysis of Shalka yielded a very similar result. One value each of Ellemeet, Johnstown, and Shalka also are in accord, yet the two groups of analyses differ by a factor of five. It is clear that there is either a sampling problem, or possibly some contamination, yet it is hard to understand why such a bimodal distribution of abundances should result.

The eucrites habe been divided into brecciated and unbrecciated types (Duke and Silver 1967). It can be seen from Table 7 that the uranium abundances also fall into two distinct groupings. The brecciated eucrites fall mainly within the range 100 to 200×10^{-9} g/g. There are two noticeable exceptions, thes reults for Pasamonte and Sioux County obtained by a xenon fission product method (König and Wänke, 1959). The value of Sioux County has been confirmed by Fisher (1969), but the Pasamonte result does not agree

TABLE 7 Abundance of uranium in achondrites

Group	Achondrite	Number of analyses	Uranium, 10^{-9} g/g	References
Angrite	Angra dos Reis	2	200	(6)
Aubrites	Cumberland Falls*	2	(15)†	(8)
	Bishopville	5	5	(6), (10)
	Norton County	3	4	(1), (10)
	Pena Blanca Spring	2	(15)	(8)
		1	8	(9)
	Shallowater‡	2	(55)	(8)
	Mean	9	5	
Diogenites	Ellemeet	2	2	(6)
	Johnstown	2	2	(3)
		4	11	(5), (6), (8)
	Shalka	2	3	(8)
		2	11	(10)
	Mean	12	7	
Eucrites (brecciated)	Bereba	2	94	(10)
	Emmaville	2	120	(10)
	Juvinas	2	(200)	(8)
		2	99	(10)
	Luotolax	2	90	(10)
	Nuevo Laredo	12	130	(2), (4), (6)
	Pasamonte	1	54	(3)
		4	130	(7), (9)
		3	(80)	(8)
	Petersburg	2	100	(8)
	Sioux County	2	52	(3), (9)
		5	(120)	(8)
	Stannern	5	190	(8), (9), (10)
	Mean	34	130	
Eucrites (unbrecciated)	Moore County	3	22	(6), (9)
		2	(45)	(8)
	Serra de Mage	2	16	(10)
	Mean	5	19	
Howardites	Binda	2	23	(6)
	Frankfort	2	46	(10)
	Kapoeta (light)	2	(64)	(9)
	(dark)	2	(77)	(9)
	Mean	4	35	
Nakhlites	*Lafayette*	2	(45)	(9)
	Nakhla	3	46	(6), (9)

TABLE 7 (cont.)

Group	Achondrite	Number of analyses	Uranium, 10^{-9} g/g	References
Sherghottite	Shergotty	2	110	(10)
Ureilite	*Goalpara*	2	<0.5	(10)

* Cumberland Falls is a breccia of aubrite and chondrite, and is classed as a whitleyite. The aubrite phase trace element composition may not be typical and has not been included in the mean.

† Values in brackets have been omitted in the calculation of mean abundances. Name in italics indicates a find.

(1) Tilton (1951) Norton County value recalculated from samples I and II of the original work
(2) Hamaguchi et al. (1957)
(3) König and Wänke (1959)
(4) Reed et al. (1960)
(5) Bate and Huizenga (1963)
(6) Morgan and Lovering (1964; 1965)
(7) Nix and Kuroda (1967)
(8) Clark et al. (1967)
(9) Fisher (1969)
(10) Lovering and Morgan (1969)

well with later analyses by Nix and Kuroda (1967), although it lies just within the rather wide range of results reported by Clark et al. (1967).

The abundances of uranium in the two specimens of unbrecciated eucrites analysed are in good agreement, and are very much lower than those found in the brecciated type. The analyses of Moore County by Clark et al. (1967) support the depletion found in these meteorites, at least in a qualitative way.

The low abundance in Moore County has caused some difficulties in the plutonium-xenon method for dating meteorites, where the abundance of uranium-238 is taken as a measure of the initial plutonium abundance. (Plutonium-244 is actually a precursor of thorium-232, and in principle it would appear that the thorium abundance in a meteorite might provide a superior estimate of this initial abundance). Kuroda (1969) finds that, using the *measured* uranium abundance for Moore County, the plutonium-xenon chronology for this meteorite is not concordant with the plutonium-xenon age estimate. To achieve the required concordance, the uranium content of Moore County must be assumed, in the face of analytical and petrological evidence to the contrary, to be equal to the average abundance in the *brecciated* eucrites.

In a recent investigation of the petrology of eucrites and howardites, Duke and Silver (1967) reclassified Luotolax and Petersburg as howardites. The mean abundance of the brecciated eucrites is not significantly altered by the removal of these two specimens. The mean abundance of the howardites depends considerably on which classification is used. Following Hey (1966), the mean lies between the brecciated and unbrecciated eucrites. Binda has been described by Duke and Silver as a monomict breccia, and classed as a unique type of eucrite. They also point out that Frankfort is not a typical howardite, as it has a very low calcium content. It is possible, therefore, that the typical abundance of uranium in howardites is more closely represented by Luotolax and Petersburg, which have a mean content of 96×10^{-9} g/g.

The three analyses for uranium in Nakhla agree well and yield an average abundance similar to that of the Frankfort howardite. Lafayette is a find, although apparently a very freshly fallen specimen, and its uranium abundance agrees well with that of Nakhla.

Only one sherghottite has been analysed for uranium, the type specimen Shergotty. In spite of mineralogical and petrological differences, the uranium content is quite similar to that of the brecciated eucrites, and also the putative howardites, Luotolax and Petersburg.

The only analyses for uranium in a ureilite were carried out on a find, in default of the availability of a better specimen. The abundance found is very low, indicating that the terrestrial contamination of the specimen by uranium was not likely to have been very serious.

Fleischer (1968) has examined the uranium content of two eucrites by the fission track technique. The results are shown in Table 8. It will be noted that in neither case is the average abundance the same as that of bulk meteor-

TABLE 8 Uranium in achondritic phases by the fission track method (Fleischer 1968)

Meteorite	Phase	Uranium, 10^{-9} g/g
Juvinas	Average	65 ($\pm$40%)
Stannern	Average	820
	"General"	300
	Region A	5
	Region B	18
	Region C	1900
	inclusion	700,000

ite analyses. Juvinas is lower than the activation results of Lovering and Morgan (1969); however, if the rather large error in the fission track measurement is taken into consideration, there is no essential disagreement. The fission track average for Stannern is some four times higher than the highest activation value. An inspection of Table 8 reveals a variation between different phases of more than 10^5, so that the calculation of a meaningful average from the small area investigated by the fission track study involves a formidable sampling problem.

Although many investigations have been made by a variety of methods of the uranium content of iron meteorites, it was the work of Reed and Turkevich (1955) that first indicated that levels of around 10^{-10} g/g or less were present. A more exhaustive study later confirmed the extreme depletion of this element (Reed et al., 1958). Using a less sensitive method, Goles and Anders (1962) investigated the uranium abundance in both the metal and troilite phases of several iron meteorites. Other analyses for uranium in troilite have been made by Patterson et al. (1953) and Reed et al. (1960). The results of these analyses are collected in Table 9.

The abundance in the troilite phases are surprisingly high, indicating that uranium has distinct chalcophilic tendencies under reducing conditions. It would be interesting to know how much uranium is in the troilite phase of the enstatite chondrites. The enrichment in the troilite of the iron meteorites is greatest in Soroti, an unusual specimen which contains about 50% troilite. In this troilite sample the uranium abundance is higher than in many chondrites.

Only one analysis of a pallasite iron has been reported (Reed et al. (1957); two pallasite olivines have also been analysed for uranium (Lovering and Morgan 1969). These results are shown in Table 10.

Clark et al. (1967) reported analyses of several mesosiderites; however, they give no details of sampling and it is not clear what phases were analysed. Because of this, and the doubts expressed by Kuroda (1969) concerning the validity of Clark's method, these results have not been included in this summary. The sample of Bencubbin listed as a C3 chondrite in Table 2 is actually a carbonaceous inclusion from a mesosiderite.

The distribution of uranium in certain phases of the Vaca Muerta mesosiderite has been studied using the fission track method by Fleischer et al. (1965); the results are given in Table 11.

To sum up: most chondrite groups are well represented by uranium analyses, though there is still a need for a close examination of the high uranium abundances reported for the C3 chondrites. In addition, analyses

TABLE 9 Uranium abundances in metal and troilite phases of iron meteorites

Group	Meteorite	Metal phase		Troilite phase	
		Uranium, 10^{-9} g/g	Ref.	Uranium, 10^{-9} g/g	Ref.
Hexahedrite	Coahuila	<0.16	(3)	–	–
Coarsest octahedrites	Arispe	0.04	(3)	–	–
	Sardis	–	–	7	(4)
Coarse octahedrite	Canyon Diablo	0.005–0.014	(3)	9	(1)
		<0.06	(4)	4	(4)
Medium octahedrites	Carbo	0.003–0.070	(3)	–	–
	Henbury	<0.007	(3)	–	–
	Tamarugal	<0.2	(2)	–	–
		0.003–0.035	(3)	–	–
	Thunda	<0.1	(2)	–	–
		0.005–0.32	(3)	–	–
	Toluca	<0.15	(4)	10	(4)
Fine octahedrite	Grant	<0.5	(4)	7	(4)
Iron plus troilite	Soroti	–	–	17	(4)

(1) Patterson et al. (1955)
(2) Reed and Turkevich (1955)
(3) Reed et al. (1958)
(4) Goles and Anders (1962)

TABLE 10 Uranium abundances in metal and olivine phases of pallasites

Pallasite	Metal phase		Olivine phase	
	Uranium, 10^{-9} g/g	Reference	Uranium, 10^{-9} g/g	Reference
Brenham	0.005–0.12	(1)	4.0	(2)
Huckitta	–	–	0.7	(2)

(1) Reed et al. (1958)
(2) Lovering and Morgan (1969)

TABLE 11 Uranium abundances in some phases of Vaca Muerta mesosiderite (Fleischer et al. 1965)

Mineral	Uranium, 10^{-6} g/g
Zircon A	800–4000
B	1–10
Whitlockite	≤ 90
Chromite	0.5–1
	Uranium, 10^{-9} g/g
Rutile	30
Tridymite	3–50
Hypersthene	0.7–3
Anorthite	0.3–0.9

are needed for the H3, L4, and LL4 and 5 groups. A reasonable number of analyses are available for the achondrites, but typical howardites are under-represented. The abundance of uranium in the iron meteorites is well established, both in metal and troilite phases. The problem of the solar abundance of uranium and the wide disagreement between empirical and calculated abundances is still a vexed question.

References

Amiel, S., Gilat, J. and Heymann, D. (1967) *Geochim. Cosmochim. Acta* **31**, 1499.

Bate, G.L. and Huizenga, J.R. (1963) *Geochim. Cosmochim. Acta* **26**, 345.

Clark, R.S., Rowe, M.W., Ganapathy, R. and Kuroda, P.K. (1967) *Geochim. Cosmochim. Acta* **31**, 1605.

Clayton, D.D. (1963) *J. Geophys. Res.* **68**, 3715.

Dalton, J.C., Golden, J., Martin, G.R., Mercer, E.R. and Thomson, S.J. (1953) *Geochim. Cosmochim. Acta* **3**, 272.

Dalton, J.C. and Thomson, S.J. (1954) *Geochim. Cosmochim. Acta* **5**, 74.

Davis, G.L. (1950) *Am. J. Sci.* **245**, 693.

Dodd, R.T., Van Schmus, W.R. and Koffman, D.M. (1967) *Geochim. Cosmochim. Acta* **31**, 921.

Duke, M.B. and Silver, L.T. (1967) *Geochim. Cosmochim. Acta* **31**, 1637.

Ebert, K.H., König, H. and Wanke, H. (1957) *Z. Naturforsch.* **12a**, 763.

Fisher, D.E. (1969) *Nature* **222**, 1156.

Fleischer, R.L. (1968) *Geochim. Cosmochim. Acta* **32**, 989.

Fleischer, R.L., Naeser, C.W., Price, P.B., Walker, R.M. and Marvin, U.B. (1965) *Science* **148**, 629.

Fowler, W.A. (1969) *The role of neutrons in astrophysical phenomena.* Preprint.

Fowler, W.A. and Hoyle, F. (1960) *Ann. Phys.* **10**, 280.

Gale, N.H. (1967) "Development of delayed neutron technique as rapid and precise method for determination uf uranium and thorium at trace levels in rocks and minerals, with applications to isotope geochronology." *Radioactive Dating and Methods of Low-level Counting*, IAEA Vienna, paper SM 87/38, pp. 431.
Goles, G.G. and Anders, E. (1962) *Geochim. Cosmochim. Acta* **26**, 723.
Grevesse, N., (1969) *Solar Phys.* **6**, 381.
Hamaguchi, H., Reed, G.W. and Turkevich, A. (1957) *Geochim. Cosmochim. Acta* **12**, 337.
Hamilton, E.I. (1966) *Science* **151**, 570.
Helliwell, T.M. (1961) *Astrophys. J.* **133**, 566.
Hernegger, F., and Wänke H. (1957) *Z. Naturforsch.* **12a**, 759.
Hey, M.H. (1966) *Catalogue of Meteorites* (3rd edition) British Museum, London.
Hoyle, F. and Fowler, W.A. (1963) "On the abundance of uranium and thorium in Solar System material." *Isotopic and Cosmic Chemistry* (Craig, H., Miller, S. and Wasserburg, G.J., eds.) North-Holland, Amsterdam.
König, H. and Wänke, H. (1959) *Z. Naturforsch.* **14a**, 866.
König, H., Keil, K., Hintenberger, H., Wlotska, F. and Begeman, F. (1961) *Z. Naturforsch.* **16a**, 1124.
Kuroda, P.K. (1969) *Nature* **221**, 726.
Lavrukhina, A.K., Kashkarov, L.L., Kolesov, G.M. and Genaeva, L.I. (1968) *Acta Chim. Acad. Sci. Hung.* **57**, 353.
Lederer, C.M., Hollander, J.M. and Perlman, I. (1968) *Table of Isotopes* (6th edition) Wiley, New York.
Lovering, J.F. and Morgan, J.W. (1964) *J. Geophys. Res.* **69**, 1979.
Lovering, J.F. and Morgan, J.W. (1969) *Uranium and thorium in achondrites.* In press.
Morgan, J.W. and Lovering, J.F. (1964) *J. Geophys. Res.* **69**, 1989.
Morgan, J.W. and Lovering, J.F. (1965) *J. Geophys. Res.* **70**, 2002.
Morgan, J.W. and Lovering, J.F. (1967) *Nature* **213**, 873.
Morgan, J.W. and Lovering, J.F. (1968) *Talanta* **15**, 1079.
Müller, E.A. (1968) "The solar abundances." *Origin and Distribution of the Elements.* (L.H.Ahrens, ed.) p. 156.
Mutschlecner, P. (1962) Thesis, University of Michigan.
Nix, J.F. and Kuroda, P.K. (1967) unpublished work, see *Nucl. Sci. Abstracts* **21**, 8230.
Paneth, F.A. (1953) *Geochim. Cosmochim. Acta* **3**, 257.
Patterson, C., Brown, H., Tilton, G., and Inghram M. (1953) *Phys. Rev.* **82**, 1234.
Reasbeck, P. and Mayne, K.J. (1955) *Nature* **176**, 186.
Reed, G.W. and Turkevich, A. (1955) *Nature* **176**, 794.
Reed, G.W., Hamgauchi, H. and Turkevich, A. (1958) *Geochim. Cosmochim. Acta* **13**, 248.
Reed, G.W., Kigoshi, K. and Turkevich, A. (1960) *Geochim. Cosmochim. Acta* **20**, 122.
Reed, G.W. (1963) *J. Geophys. Res.* **68**, 3531.
Reed, G.W. and Allen, R.O. (1966) *Geochim. Cosmochim. Acta* **30**, 779.
Starik, I.E. and Shats, M.M. (1956) *Geochemistry (Geokhimiya)* No. **2**, 140.
Starik, I.E., Shats, M.M. and Sobovich, E.V. (1958) *Soviet Phys.* "*Doklady*" **3**, 1086.
Seeger, P.A., Fowler, W.A. and Clayton, D.D. (1965) *Astrophys. J. Suppl.No. 97*, **11**, 121.
Tilton, G.R. (1951) *U.S. Atomic Energy Commission Report AECD* **3182.**
Van Dijk, T., de Jager, C. and de Metter, J. (1953) *Mem. 8^0 de la soc. roy. sci. Liège, 4th series.* **13**, Fasc. lII, 495.

Van Schmus, W.R. and Wood, J.A. (1967) *Geochim. Cosmochim. Acta* **31,** 747.
Vinogradov, A.P. (1961) *Geochemistry (Geokhimiya)* No. **1**, 1.
Wakita, H., Nagasawa, H., Uyeda, S. and Kuno, H. (1967) *Earth Planet. Sci. Letters*, **2,** 377.

References for Table 2

(1) Tilton (1951); Patterson et al. (1953)
(2) Dalton quoted by Reasbeck and Mayne (1955)
(3) Hamaguchi et al. (1957)
(4) Hernegger and Wänke (1957)
(5) Ebert et al. (1957)
(6) König and Wänke (1959)
(7) Reed et al. (1960)
(8) Goles and Anders (1962)
(9) Reed (1963)
(10) Reed and Allen (1966)
(11) Amiel et al. (1967)
(12) Nix and Kuroda (1967)
(13) Clark et al (1967).
(14) Lovering and Morgan (1964); Morgan and Lovering (1967, 1968)
(15) Wakita et al. (1967)

(Received 14 July 1969)

INDEX OF METEORITES